Introduction to
Modern Photogrammetry

Introduction to Modern Photogrammetry

Edward M. Mikhail and James S. Bethel
Purdue University

J. Chris McGlone
Carnegie Mellon University

John Wiley & Sons, Inc.
New York / Chichester / Weinheim / Brisbane / Singapore / Toronto

Acquisitions Editor   *Wayne Anderson*
Marketing Manager   *Katherine Hepburn*
Senior Production Editor   *Michael Farley*
Senior Designer   *Maddy Lesure*
Production Management Services   *Publication Services*

This book was set in *Times Roman* by *Publication Services* and printed and bound by *Hamilton Printing*. The cover was printed by *Phoenix Color.*

The book is printed on acid-free paper.

Library of Congress Cataloging in Publication Data:
Edward M. Mikhail, James S. Bethel, and J. Chris McGlone
    Introduction to Modern Photogrammetry
    Mikhail.—
        p.   cm.
    Includes bibliographic references.

L.C. Call no.   Dewey Classification No.   L.C. Card No.

ISBN 0-471-30924-9

Printed in the United States of America

10 9 8 7 6 5 4 3 2 1

Preface

The practice of photogrammetry today bears little resemblance to that during its formative years in the last century. While the basic mathematical principles of photogrammetry remain unchanged, their implementation and application for production purposes have drastically changed. The photogrammetric equations, formerly embodied in precise analog solutions, now exist as programs in general-purpose digital computers; the implications of this advance are still being explored.

Image acquisition, formerly limited to film cameras in aerial or terrestrial applications, now includes digital imagery obtained from a variety of platforms, including terrestrial, aerial, and satellite, covering portions of the electromagnetic spectrum far outside the visible region. Imagery is now widely available through Internet distribution, making its application practical within a number of areas and facilitating its use by non-photogrammetrists. Image exploitation, once entirely dependent on skilled analysts, is increasingly becoming a joint effort of man and computer. The changes in photogrammetry parallel those in related geospatial fields, such as remote sensing, geographic information systems, and computer vision. Developments in photogrammetry and these related fields have been intertwined.

This book is intended to serve as a rigorous introduction to the basic principles and practice of photogrammetry and as a guide to the related areas in which photogrammetry plays a role or which affect the practice of photogrammetry. As such, it is suitable as an introductory text for undergraduate students with no prior exposure to photogrammetry, as well as an advanced text for graduate-level study. It will also serve as a valuable reference for scientists, engineers, and practitioners who use imagery to solve geospatial problems. Our hope is that this book will provide a useful understanding of current photogrammetric technology that will serve as a solid foundation for integrating the inevitable advances in the years ahead.

Chapters 1 and 2 give an overview of photogrammetry and present some techniques to derive useful though approximate quantitative data from imagery. Chapter 3 describes the physics of image acquisition and the design and calibration of cameras and sensing systems for photogrammetric applications. Chapter 4 develops the mathematical relationships that govern the geometric exploitation of imagery. The application of these mathematical models to solve practical problems in point determination, including multi-image triangulation, is presented in Chapter 5. Chapter 6 covers relevant concepts from the fields of digital image processing and computer vision in the context of photogrammetry's application to digital imagery. Chapter 7 addresses the topic of photogrammetric instruments and systems , including coverage of softcopy-based systems. Photogrammetric products, both hardcopy and softcopy, are reviewed in Chapter 8. Close-range and non-topographic photogrammetric applications are presented in Chapter 9. A review of statistical pattern recognition for remote sensing applications is presented in Chapter 10. The technologies behind the growing use of such active sensors as SAR, IFSAR, and LIDAR are described in Chapter 11. Mathematical details that, although considered important, would interrupt the flow of ideas in the main text are provided in the appendices for easy reference. Finally, Appendix G gives a collection of C-language software, MATLAB® code, and sample imagery implementing some of the more common photogrammetric operations, from the elementary to the more advanced. Also included in this appendix is a listing of World Wide Web resources.

The authors wish to acknowledge with thanks the many colleagues, both in photogrammetry and in related fields, who pointed out the need for, and continuously provided encouragement to write, this book. Special thanks are due to Dr. David Landgrebe, Professor of Electrical and Computer Engineering, Purdue University, for contributing the outstanding Chapter 10 on Analysis of Multispectral and Hyperspectral Image Data; and to Mr. Thomas Ager of the National Imagery and Mapping Agency (NIMA), Dr. Neil Carender of Veridian ERIM International, Mr. Richard D'Alessandro of Lockheed Martin, and Mr. Gregg Kunkel of BAE Systems for the excellent Chapter 11 on Active Sensing Systems. These two chapters add significantly to the value of this book, both as a text and as a reference book. Several excellent suggestions were made by the manuscript reviewers, Dr. Peggy Agouris, Dr. Peter Boniface, Dr. Clive Fraser, and Dr. Ayman Habib, for which the authors are grateful. The authors express their deep appreciation to the instrument and system manufacturers and other organizations who generously provided imagery, illustrations, and other support. Among these are: Cyberware, Inc; General Atomics Aeronautical Systems, Inc.; Geodetic Services, Inc.; LH Systems, LLC; Lockheed Martin Fairchild Systems, Inc.; Oy Mapvision Ltd.; Recon Optical, Inc.; Space Imaging, Inc.; SPOT Image Corp.; TerraPoint, LLC; Terrasim, Inc.; The MathWorks, Inc.; the United States Geological Survey; and Z/I Imaging Corp. Thanks are also due to Dr. Steven Cochran, Dipl.-Ing. Matthias Hemmleb, and to Dr. Jeffrey Shufelt for their helpful contributions.

During the checking of the page proofs, substantial editing assistance was provided by Dr. Henry Theiss of Purdue University. The authors are grateful to him for his contributions and expertise. Dr. Changno Lee, Mr. Ahmed Elaksher, and Mr. Ade Mulyana, also of Purdue University, provided valuable help in reviewing of proofs and problem solutions. Thanks to Cheryl Kerker, the Geomatics Area Secretary at Purdue University, who tirelessly worked on typing a significant part of the manuscript. The authors would like to acknowledge with gratitude the outstanding editorial and production support provided by Publication Services, Inc.

Finally, a very special note of thanks is expressed to our families: Laverne Mikhail and the Mikhail children, Dorothy and Graham Bethel, and Ellen and Leah McGlone for their support and patience during the writing of this book. Chris also thanks his colleagues in the Digital Mapping Laboratory, especially David McKeown, for allowing him to be a part of a stimulating and challenging environment.

Edward M. Mikhail
James S. Bethel
J. Chris McGlone

Contents

Chapter 1

Introductory Concepts

1.1   DEFINITIONS

Traditionally, *photogrammetry* has been defined as the process of deriving (usually) metric information about an object through measurements made on photographs of the object. The closely related area *photo interpretation* is defined as the extraction of qualitative information about the photographed objects by human visual analysis and evaluation of photographs. As more *sensing* techniques were developed to provide imagery in a wider region of the electromagnetic spectrum than that which interacts with the photographic emulsion, the term *remote sensing* was introduced. It expands upon classical photo interpretation by employing computer analysis techniques in addition to human visual interpretation, and by applying such techniques to types of imagery other than photography.

Modern photogrammetry covers a considerably wider domain. Imagery of all types, both *passive*, such as photography, and *active* (i.e., providing its own energy source), such as radar imaging, is used. The imagery may be collected either in *hardcopy form* (e.g., on film) or in *digital form* by electro-optical sensors. The analysis may be performed on single images or on overlapping (*stereo*) imagery. The photogrammetric processing and analysis systems may operate on the hardcopy form (either originals, or digital images written on film) or directly on digital images (again, either collected directly in digital form or converted from hardcopy to digital form by scanning).

The fundamental task of photogrammetry is to rigorously establish the geometric relationship between the image and the object as it existed at the time of the imaging event. Once this relationship is correctly recovered, one can then derive information about the object strictly from its imagery. This relationship can be established by various means, which can be broadly classified into two categories: *analog,* using optical, mechanical, and electronic components; or *analytical,* where the modeling is mathematical and the processing is digital. Analog solutions are increasingly being replaced by analytical/digital solutions.

The availability of imagery in digital form, along with the processing increasingly done on digital computers, opens up many possibilities for adaptations of techniques from other disciplines. The first of these is *image processing,* where the processing is performed on the digital image by a computer. It addresses several tasks, such as *compression,* whereby the original image is converted to another form needing less storage; *enhancement* and *restoration* to improve the quality of a noisy image; and *segmentation* and *description,* which entail conversion to *parts* or *primitives*, measurement of their properties, and description of the image in terms of primitives and their properties. A second related discipline is *pattern recognition,* computer analysis of images that results in automatic extraction of salient patterns from the images and the recognition of the patterns by comparison to an externally available typical pattern set in a dictionary. *Artificial intelligence* involves the study of techniques to allow computers to do things at which, at the present time, people are better. When these techniques are applied to imagery, we often speak of *image understanding* (IU) and *computer vision* (CV). The major objective of these is to develop meaningful descriptions of physical objects from their images. IU/CV methods are increasingly being incorporated into photogrammetric systems in the form of automated tools that lessen the burden on the human operator, improve robustness, and increase efficiency.

1.2   PHOTOGRAMMETRIC SYSTEMS

A photogrammetric project involves two general functions: (1) acquisition and preparation of imagery and support data, and (2) processing of the imagery to derive the required products. The first function encompasses several operations, such as project definition, specifications, and planning; acquisition of suitable imagery; preprocessing of the imagery in preparation for use in the photogrammetric reduction or processing system; and collection of other, supporting data such as survey-derived ground control. The second function includes deciding which photogrammetric system to use based on the specified products expected from the project.

In general, photogrammetric projects can be broadly classified as those requiring satellite imagery, those requiring airborne imagery, and close-range or industrial applications. The majority of photogrammetric operations have historically involved the use of aerial photographs, and to a significant extent, current photogrammetric operations still do. For metrical work, the photographs must be taken with a precision *cartographic aerial camera* whose geometric characteristics have been determined by a calibration process. A modern cartographic camera is shown in Fig. 1-1. An aerial photograph taken with such a camera is shown in Fig. 1-2. This type of camera is referred to as a *frame* camera since it exposes an entire frame in essentially one instant of time. The following subsection is devoted to photography by aerial frame cameras. Electronic imaging systems are more practical for satellite imagery than photographic systems because of the difficulty of retrieving the exposed film from space, compared with transmitting the digital images to ground stations. One of the most successful to date is the SPOT system, which uses a linear array in a so-called pushbroom imaging mode, as shown in Fig. 1-3. Close-range and industrial applications are characterized by having the cameras or sensors on or very near the surface of the Earth, usually stationary, although some applications rely on motion in the form of image sequences. Film-based, video, and digital cameras are all being used for a variety of applications. These are discussed in Chapter 9, after the various sensing systems are introduced in Chapter 3.

Figure 1-1 Cartographic aerial camera. Courtesy of Z/I Imaging.

Figure 1-2 Aerial photograph made with cartographic camera.

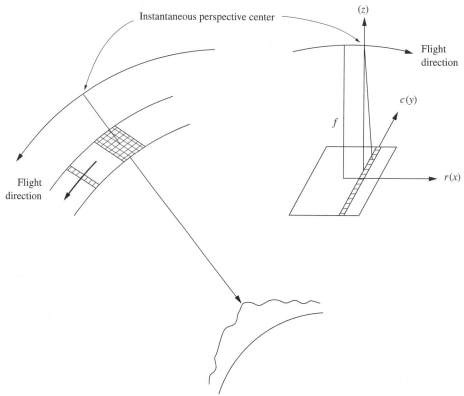

Figure 1-3 SPOT pushbroom imaging system.

The contribution of photogrammetry and its sister activity, remote sensing, is depicted in the overview diagram in Fig. 1-4. Image acquisition is one of the earlier steps, which provides an important source of data. The photogrammetric system is at the heart of information extraction, which provides several of the components or *layers* in the information system, usually a Geographic Information System, or GIS, covered in Chapter 8.

1.2.1 Aerial Frame Photography

Although a cartographic aerial frame camera is, as shown in Fig. 1-1, a complex and expensive equipment system, it is treated from a geometric viewpoint in a relatively simple manner. Figure 1-5 shows a three-dimensional schematic of exposing a square-format aerial photograph. For purposes of geometry and mathematical modeling, the camera lens is represented by a single point, called the *perspective center,* even though the lens assembly is composed of many optical elements. The *optical axis* of the camera lens is most often vertical in aerial photography. Consequently, an essentially square area of the terrain is photographed by the square format of the aerial photograph.

When aerial photography is used for mapping, flight lines are laid out on a flight map with a spacing that will cause the photographs to cover a common strip of ground. This overlap between flight strips, called *sidelap* and shown in Fig. 1-6, amounts to about 25% of the width of the area covered by the photograph.

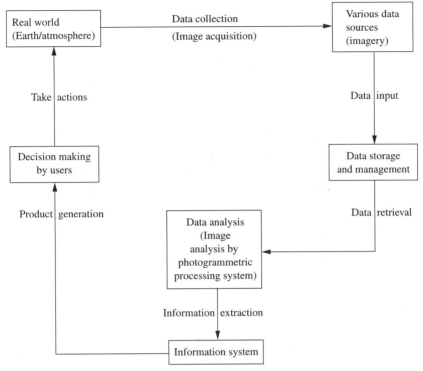

Figure 1-4 Information cycle.

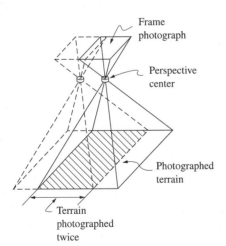

Figure 1-5 Frame camera imaging.

Each photograph in the line of flight covers an area that overlaps the area covered by the previous photograph by about 60%. This overlap along the line of flight, called forward overlap or simply *overlap*, is shown in Figs. 1-6 and 1-7. The large overlap between successive photographs serves three primary purposes. First, it provides coverage of the entire ground area from two viewpoints, which is necessary for stereoscopic viewing, as explained in Chapter 2. Two photographs successively exposed along a flight line are called a *stereo pair* of photographs. Second, it allows all but the central

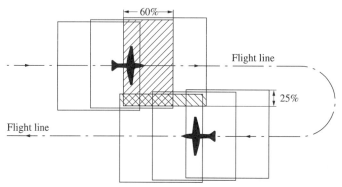

Figure 1-6 Photographic overlap.

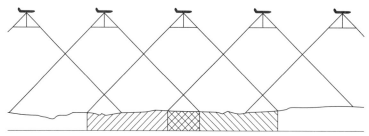

Figure 1-7 Overlap along flight line.

portion of each photograph to be discarded in the construction of *mosaics*. Third, the small overlap area between *alternate* photographs is necessary for establishing supplemental ground control by photogrammetric triangulation, as discussed in Chapter 6.

1.2.2 Photogrammetric Processing

A photograph or an image, although useful as a record, is most often used as an input to a processing system, the output of which determines the type of product needed. During the past several decades, many systems have been developed to provide users with a wide range of products. One class of photogrammetric processing systems is suitable for processing a single image at a time, and another class of systems includes those that process two or more images simultaneously, usually allowing for stereo viewing. The early versions of the latter class were mainly analog in nature, called *stereoplotters,* and used hardcopy photographs, usually in pairs, as input. Their successors, called *photogrammetric workstations,* operate on digital images and can therefore process multiple images. These are discussed in Chapter 7. All systems essentially implement the fundamental photogrammetric (projective) relationship between the object and its images. Once this relationship is established, various types of data and information may then be extracted as output of the system.

1.3 PHOTOGRAMMETRIC APPLICATIONS AND PRODUCTS

The potential applications of the art and science of photogrammetry are almost limitless. Astronomers expose and measure star plates in order to classify star magnitudes and determine star parallaxes. Engineers rely on large-scale topographic maps for all types of planning and construction. Modern topographic maps are invariably produced by photogrammetry. Almost all phases of modern highway design, location, construction, and maintenance are conducted wholly or in part by photogrammetry. Large-scale, small-contour-interval maps compiled from aerial photographs are used as the basis for the geometric design of highways. Location surveys are made with reference to points whose ground positions have been determined photogrammetrically. Earthwork quantities are calculated from measurements taken either from the design map or directly from stereo models formed in the photogrammetric processing system. The need for corrective action (to improve the condition of the pavement or to stop erosion of the banks) in a highway maintenance program can be determined from interpretation of frame or continuous-strip photography. Engineers also depend on photogrammetric solutions to such problems as the measurement of structural deformations, wave shapes, sedimentation in channels, and vehicular movements.

Photogrammetry is applied to industrial inventory of coal piles, wood pulp, and mineral deposits and to agricultural inventory of crops. Hydrologists depend on the photogrammetric analysis of slopes, ground coverage, watershed areas, and snow depths in order to determine runoff quantities for water supply studies.

Archaeologists depend on analysis of aerial and terrestrial photographs and other imagery for the measurement and interpretation of ruins dating to antiquity. Image interpretation of geological forms and features is infinitely more efficient than field methods. Urban planners depend on aerial and space images for land use studies. Nuclear physicists must track energy particles in cloud chambers by measuring their traces on photographs.

Mapping of landing sites on lunar or planetary surfaces as well as the geological exploration of the moon and planets are photogrammetric functions.

Land surveyors employ aerial photographs and photogrammetric methods in varying degrees of complexity. Planning for surveys, the identification and location of boundary lines and corners, and identification of vegetation and soil types can be performed by examination and interpretation of imagery. If land boundary corners have been marked on the ground by some kind of distinctive targets prior to the aerial photography, the positions of these corners can be located with a high degree of accuracy using analytical photogrammetric methods. Arbitrarily targeted points can be located photogrammetrically and then used as control points for further surveying. Complete land subdivisions can be staked out with very little field surveying using these targeted points as reference points.

The principles of very-close-range photogrammetry are applied to diverse problems in the field of medicine. This branch of close-range photogrammetry is called *biostereometrics*. Problems such as measuring body motion, body surfaces, contours, change in body shape, posture, and movement of teeth are but a few that can be solved by this method.

The collection of diverse types of information into a database that is used in a GIS is performed by photogrammetric systems. Products derived from various photogrammetric processing systems fall into two broad categories: *image products* and *point and vector products*. Photogrammetric products are described further in Chapter 8.

1.3.1   Image Products

These products are in the form of images that depict the original object, which may well be three-dimensional, by a two-dimensional representation. The following are several image products:

    a. *Aerial photographs,* which have uses in all varieties of interpretation and general planning.

b. Any directly acquired panchromatic or color images, in which the color is a code for a particular band of the electromagnetic spectrum. These can be used for general interpretation and overall planning of the imaged area.

c. *Mosaics,* continuous pictures of the terrain constructed by assembling individual photographs together in a composite. A mosaic is thus a series of contiguous perspective views of the ground.

d. *Rectified imagery,* in which tilt effects have been removed. Effects of relief are present, and, like the original image or photograph and the mosaic, the scale varies within the pictorial product.

e. *Orthophotos,* pictures of the ground prepared from a pair of overlapping images in such a manner that the perspective aspect of the picture has been removed. The orthophoto can be used as a planimetric map because it has a constant scale. If contour lines are superimposed over the imagery, then the resulting orthophoto map is used as a topographic map.

f. *Orthophotomosaics,* products where a series of contiguous individual orthophotos are assembled together into one continuous image. It has uniform scale throughout and is used as a map.

g. *Radar mosaics,* composites of the output of a series of radar images taken along consecutive flight lines into continuous radar images of the ground.

Modern photogrammetric workstations also produce additional useful products such as auxiliary perspective and oblique views of the terrain, and color-coded (thus pictorial) elevation maps.

1.3.2 Point and Vector Products

Point products are those in which each individual point is given by three coordinates, either in an object space reference coordinate system or, in some cases, in an arbitrary coordinate system. They include

1. Supplementary control, usually derived by photogrammetric triangulation.

2. Targeting, where the locations of prespecified targets are determined.

3. Digital elevation model (DEM), which is a digital representation of the terrain surface, given by
  a. Random data points, at locations where significant elevation changes occur, or
  b. Gridded data points, where a uniform grid is established in the horizontal (XY) plane, and the elevation at each grid intersection is given.

Vector or line products represent features of either the natural terrain or human-made infrastructure objects. Examples of these are all types of line maps, such as

a. *Planimetric maps,* which contain only the horizontal positions of ground features. They are prepared to varying degrees of accuracy, depending on the photogrammetric technique employed.

b. *Contour maps,* in which each line represents the intersection of a level surface and the terrain. This is the earlier method of depicting the shape of terrain, preceding the introduction of the DEM concept. These may be compiled directly using a photogrammetric instrument or derived from a DEM.

c. *Topographic maps,* which show both the planimetric features of the terrain and the shape and elevation of the ground by means of contour lines. The

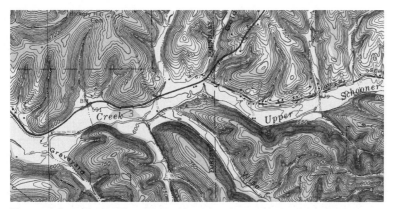

Figure 1-8 Topographic map produced by photogrammetric methods.

topographic map shown in Fig. 1-8 was compiled using photogrammetric methods.

d. *Thematic maps,* which delineate a particular *theme.* Any number of thematic maps can be made of one given area. For example, delineations of the transportation network, drainage patterns, vegetative types, and slopes can be made, resulting in four separate thematic maps. These may now constitute layers in a Geographic Information System, or GIS.

e. *Profiles,* which represent the intersection line of a plane, usually vertical, and the terrain surface. The profile could be in any orientation within the reference horizontal (*XY*) plane.

f. *Three-dimensional object models,* or *wire frames.* The increased application of photogrammetric technology in combination with the tools of computer vision to various industrial problems resulted in this type of product. The salient features of the imaged object are represented by nodes and lines intersecting in these nodes.

1.4   SOURCES OF PHOTOGRAMMETRIC INFORMATION

Photogrammetric activities are conducted by or for federal, state, and county agencies and by private mapping firms. The major federal agencies in the United States engaged in photogrammetry are

> U.S. Geological Survey (USGS)
>
> National Geodetic Survey (NGS), formerly U.S. Coast and Geodetic Survey
>
> U.S. Forest Service (USFS)
>
> Bureau of Land Management (BLM)
>
> U.S. Army Corps of Engineers
>
> National Imagery and Mapping Agency (NIMA), formerly Defense Mapping Agency (DMA)
>
> National Aeronautics and Space Administration (NASA)

The primary federal agencies in Canada are

> Geomatics Canada, Earth Sciences Sector, Department of Natural Resources
>
> Canada Center for Remote Sensing, Department of Natural Resources

Most of the aerial photography used by federal and other governmental agencies is obtained under contract with private aerial photography firms. The U.S. Geological Survey has developed the Aerial Photographic Summary Records System (APSRS), which catalogs the aerial photographic coverage of the various federal agencies by quadrangles. This is the most complete source of information regarding existing photographic coverage. The offices of the National Cartographic Information Center, which maintains this system, are located in the four regions of the USGS: at Reston, Virginia; Rolla, Missouri; Denver, Colorado; and Menlo Park, California. The USGS encourages agencies on the state and local levels to contribute information to the APSRS. Information pertaining to coverage of the Landsat and other orbiting satellite systems can be obtained from the EROS Data Center in Sioux Falls, South Dakota.

The Internet and, especially, the World Wide Web have revolutionized the availability and distribution of aerial imagery and digital cartographic data. For instance, the USGS now has its holdings catalogued online in the Global Land Information System (GLIS). With GLIS, World Wide Web users can search for photography or map data by place name, map name, or geographic coordinates. Once the data are located, delivery by downloading or shipment may be requested.

Private companies are also active in the online distribution of cartographic imagery and data. Although the marketplace is changing rapidly, two notable sources as of this writing are Microsoft's TerraServer and Space Imaging's Carterra. In both cases, imagery archives can be searched online. With Space Imaging's system, the acquisition of new satellite data can actually be ordered online.

Reflecting the importance of Web based information, a selection of relevant and current links is given in Appendix G. A browser compatible version of this list is included on the companion CD, and updated versions can be obtained through `www.wiley.com/college/mikhail`.

Information of a technical nature can be obtained from the American Society for Photogrammetry and Remote Sensing (ASPRS) through its technical committee structure. The ASPRS is a technical-professional society whose aim is to advance the science and professional practice of photogrammetry. It has representation in all governmental and private organizations performing photogrammetric work, and can provide a wealth of information pertaining to all phases of photogrammetry on request. The ASPRS has published a series of manuals for the profession. These are *The Manual of Photogrammetry,* 4th edition (1980); *The Manual of Photogrammetry,* 5th edition (due 2001); *The Manual of Photographic Interpretation* (1960); *The Manual of Color Aerial Photography* (1968); *The Manual of Remote Sensing* (1999); *The Manual of Non-Topographic Photogrammetry* (1988); and *Digital Photogrammetry: An Addendum to the Manual of Photogrammetry* (1996). Additionally, the ASPRS publishes a monthly journal, *Photogrammetric Engineering and Remote Sensing,* which contains technical articles of current interest to ASPRS members.

The Canadian counterpart to the ASPRS is the Canadian Institute of Geomatics (CIG). Information can be obtained through its Committee on Photogrammetry. Technical articles pertaining to photogrammetry are published quarterly in *Geomatica,* the journal of the CIG.

Because of the extensive use of photogrammetry in highway engineering, state highway agencies are fruitful sources of information regarding aerial photographic coverage, especially at large scales. Many county agencies have acquired countywide photographic and orthophotographic coverage. They usually have control of the aerial negatives, and the engineer can arrange to acquire photography through the appropriate county agency.

Although the ASPRS can furnish information regarding commercial aerial photographers and photogrammetric firms in a given area, it is usually more direct to consult the classified section of the local telephone book for this information.

1.5 HISTORY

One can trace the earliest roots of photogrammetry to the Renaissance painters, particularly Leonardo da Vinci, who studied the principles involved in the geometric analysis of pictures in the late 1400s. The next significant development was projective geometry, which forms the mathematical basis of photogrammetry from passive imaging systems; notables include Desargues, Pascal, and Lambert, from the mid-1600s to the mid-1700s. This was followed by the production of usable photographs by Niepce and Daguerre in the early 1800s. In 1858, Nadar in France captured photographs of the countryside from a balloon. In 1895, Laussedat created the first suitable camera and procedure, which he called *metrophotographic,* for making photogrammetric measurements, and is thus regarded as the *father of photogrammetry.* The word *photogrammetry* first appeared in a paper on photographic surveying by Meydenbauer in 1893. The first photographs from an aircraft were taken by Wilbur Wright in 1909 over Centoceli, Italy. The first third of the twentieth century saw the development of early analog plotters and the first highly corrected wide-angle lens for use in aerial cameras. In 1911, von Orel and Zeiss produced the stereoautograph for plotting from terrestrial photographs. Heinrich Wild produced the Autograph A1 in 1923. Bauersfield and Zeiss produced the Stereoplanigraph, also in 1923. The Multiplex was introduced by Zeiss in 1933. The current American Society for Photogrammetry and Remote Sensing was originally founded as the American Society of Photogrammetry in 1934. The next third of the twentieth century saw accelerated development of various first-order analog plotters, with parallel development of the mathematical basis for photogrammetric triangulation and its error characteristics. The Wild A7 was first produced in 1949. Bean at Bausch & Lomb developed the Balplex in 1951. The PG2 from Yzerman and Kern came in 1960.

The idea for radial line triangulation was worked out by Adams in the 1890s. Many implementations followed. The USGS developed the *lazy daisy* in 1950. Hellmut Schmid and Duane Brown developed the principles of multistation analytical photogrammetry in the late 1950s. The analytical plotter was developed by Uki Helava in 1961 and continued to improve for two decades as electronic computational means became progressively more powerful and more economical to use.

During the last two decades of the twentieth century, photogrammetry has experienced very significant changes caused by advances in optics, electronics, imaging, and computer technologies, which made possible the next logical development, the digital photogrammetric workstation. This represents a marked jump in processing, where the emulsion-based photograph, which has always been the basis of all photogrammetric activity, is now being replaced by a purely digital record. The fundamentals of photogrammetry remain unchanged, but the operational environment has changed significantly. With the imagery in digital form there now exists an almost limitless opportunity for photogrammetry to continue to flourish. Semiautomated and automated tools continue to be developed to lighten the burden of feature extraction. Classification of remote sensing image data is progressively being combined with photogrammetry, which will ultimately lead to real-time image analysis. High-resolution imagery of various types, from satellites, aircraft, and unmanned aerial vehicles, will become abundant, and the photogrammetrist will be busy providing a myriad of services and products from the currently regular to many new ones for use in a variety of applications and visualization environments.

REFERENCES

AMERICAN SOCIETY FOR PHOTOGRAMMETRY AND REMOTE SENSING. 1968. *The Manual of Color Aerial Photography.* Menasha, WI: George Banta Co.

ANSON, A. 1975. Photogrammetry as a science and as a tool (an index). *Photogrammetric Engineering and Remote Sensing* 41:225–236.

GREVE, C., ed. 1996. *Digital Photogrammetry: An Addendum to the Manual of Photogrammetry.* Bethesda, MD: American Society for Photogrammetry and Remote Sensing.

RYERSON, R., ed. 1999. *The Manual of Remote Sensing,* 3rd edition. Bethesda, MD: American Society for Photogrammetry and Remote Sensing.

SLAMA, C., ed. 1960. *The Manual of Photographic Interpretation.* Bethesda, MD: American Society for Photogrammetry and Remote Sensing.

SLAMA, C., ed. 1980. *The Manual of Photogrammetry.* 4th edition. Bethesda, MD: American Society for Photogrammetry and Remote Sensing.

Chapter 2

Elementary
Photogrammetry

2.1 PERSPECTIVE PROJECTION

The basic model used to describe the geometry of a frame photograph is that of a *pinhole camera*. Before the days of mass-produced, inexpensive cameras, children would often construct such a device from a shoe box, with a piece of aluminum foil for the pinhole aperture and a sheet of cut film taped to the opposite end of the box. As indicated in Fig. 2-1, each visible point in the object space reflects a ray of light that passes through the "ideal" pinhole aperture and forms a unique image point on the film. Such a camera has unlimited depth of field but requires very long exposure times. In practical cameras, the aperture must be enlarged to admit light sufficient for a short exposure time, and this requires the introduction of refracting lens elements to maintain a sharp image. The pinhole concept illustrates the basic *frame* imaging principle of the *collinearity* of the three points that define each ray: the object point, the pinhole aperture, and the image point. Stated another way, the three-dimensional points in the object space are transformed via a perspective projection into the two-dimensional image plane, with the pinhole acting as the perspective center. The *principal distance* (or focal length for fixed-focus lenses) is the distance along the optical axis from the perspective center to the image plane, as presented in detail in Section 2.4.

The vertical frame photograph, as employed in aerial photogrammetry and remote sensing, is shown viewed from the side in Fig. 2-2. In a perspective projection,

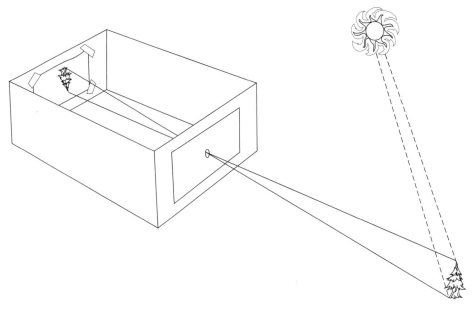

Figure 2-1 Pinhole camera.

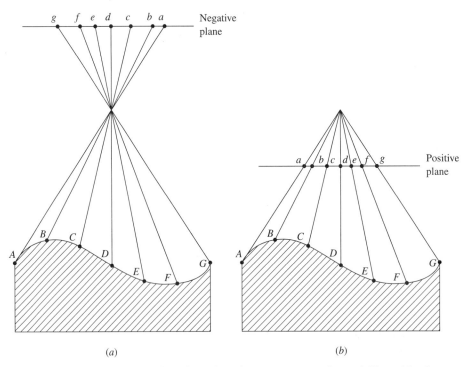

Figure 2-2 Perspective projection of terrain points onto (*a*) negative and (*b*) positive image planes.

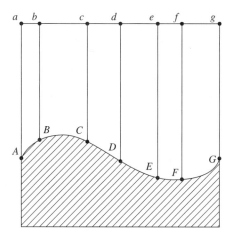

Figure 2-3 Orthographic projection of terrain points onto a map plane.

it is completely equivalent to refer to the *positive* image plane in Fig. 2-2b or to the *negative* image plane in Fig. 2-2a. The term *projection* implies a transformation of a higher-dimensional space (3-D in the object) into a lower-dimensional space (2-D in the photograph). A map represents another way of projecting 3-D terrain features into a 2-D plane. A map must present all points in their planimetrically correct positions, as in the *orthographic projection* shown in Fig. 2-3, which projects points orthogonal to a reference surface. For vertical photographs, terrain relief causes an image point to be displaced from its planimetrically correct position. The building corners in Fig. 2-4a have a single planimetrically correct representation on a map or orthographic projection of given scale, whereas the same corners have a multitude of representations, depending on the elevation that is considered, when depicted in a perspective projection, as in Fig. 2-4b and 2-4c.

In addition to this displacement due to height difference, the other principal source of discrepancy between the map and the photograph is displacement due to tilt, as shown in the simulated photo image of a rectangular grid pattern in Fig. 2-5. As mentioned in Chapter 1, map products produced by photogrammetry can be in the form of vector maps with features depicted using point and line symbols, in the form of images with displacement errors corrected, or possibly in a combined format with vectors overlaying the raster image.

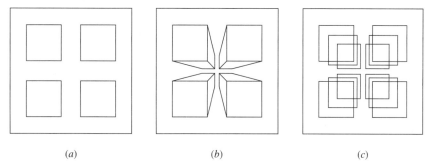

(a)   (b)   (c)

Figure 2-4 Orthographic projection (*a*) compared with perspective projections (*b*) and (*c*).

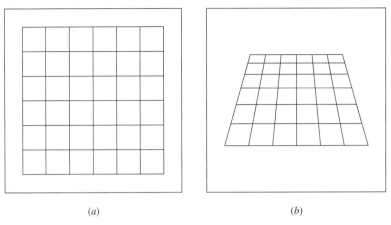

(a)   (b)

Figure 2-5 Tilt effects and displacement due to tilt.

2.2 SCALE AND COVERAGE

Everyone is familiar with the concept of map *scale*. A map has a scale of 1:24,000 if 1 unit on the map corresponds to 24,000 units on the ground (or, more accurately, on the projection surface). The scale is best and most unambiguously expressed as a unitless ratio such as 1:24,000 or 1/24,000, also known as a *representative fraction*. Engineers in the United States often express this scale as 1 inch = 2000 feet, however this form should be discouraged since it depends on particular units that are not widely used elsewhere. Vertical photographs over flat terrain have a simple characterization of scale, as shown in Fig. 2-6. The ratio of image distance $\overline{ab}$ to object distance $\overline{AB}$ gives the scale of the photograph.

$$\text{scale} = \frac{\text{image distance}}{\text{object distance}} \tag{2-1}$$

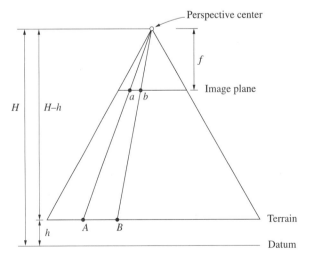

Figure 2-6 Elements related to scale in a vertical photograph.

From the geometry of the figure, the scale can also be expressed as the ratio of the focal length to the flying height above the terrain.

$$scale = \frac{f}{H - h} \qquad (2\text{-}2)$$

For this ideal case, scale is independent of the direction in which it is measured. There is a simple relationship between scale and coverage. If the scale is reduced by a factor of n, then the area covered is increased by a factor of n^2, so for example, going from a photo scale of 1:5000 to 1:10,000 yields four times the area of coverage. For actual photographs rather than ideal ones, the concept of scale becomes more complicated. Nominally vertical photographs (i.e., within a few degrees of vertical) are considered to have a nominal scale, often obtained by the simple methods of Eqs. 2-1 and 2-2. For tilted or oblique photos, scale varies with direction. For example, in the background of an oblique photograph the scale in the direction of view is smaller than the scale in the transverse direction. Artists refer to this effect as *foreshortening*. A concept from map projections is useful here. Tissot's Indicatrix (Maling, 1973; Snyder, 1989) is defined as the projection of an infinitely small circle on the Earth's surface into a map projection plane. In a conformal projection the projected figure remains a circle, but in general it becomes an ellipse. Its shape conveys, in graphical form, the distortion present in the projection. Borrowing this concept, one can project a matrix of such circles from a horizontal object plane into a tilted photograph. The relative magnitudes of projected radii in different directions give a graphical indication of the scale variation. These ellipses are shown for two simulated photographs of a grid pattern in Fig. 2-7. Thus, on a tilted photograph, not only is the scale variable from point to point, it is also variable at a single point, depending on direction. In general, the scale at a point for any direction can be expressed as

$$scale = \frac{ds_i}{ds_o} \qquad (2\text{-}3)$$

where ds_o is the magnitude of the differential displacement vector on the object plane and ds_i is the magnitude of the corresponding differential displacement vector in the image plane, as shown in Fig. 2-8. Further refinement of the concept of scale due to terrain slope and aspect could be developed, but in practice this refinement is not particularly useful.

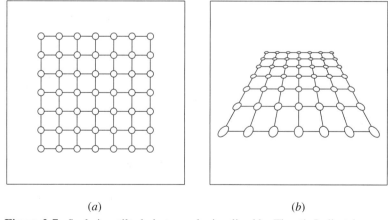

(a)   (b)

Figure 2-7 Scale in a tilted photograph visualized by Tissot's Indicatrix.

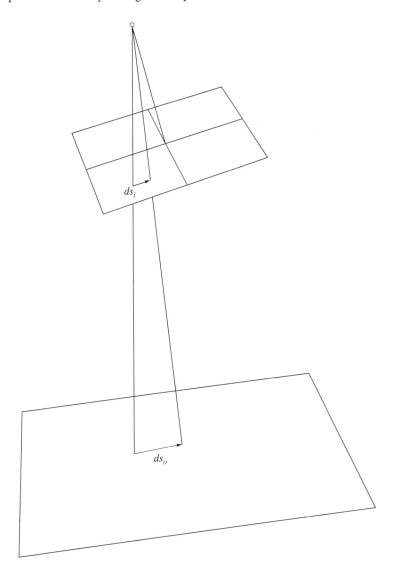

Figure 2-8 Scale at a point in an arbitrary direction.

EXAMPLE 2-1

A vertical photograph is taken at an altitude of 800 m above sea level. The terrain is flat and has an elevation of 237 m. The camera focal length is 50 mm. What is the scale?

SOLUTION   From Eq. 2-2,

$$\text{scale} = \frac{f}{H - h} = \frac{0.050 \text{ m}}{800 \text{ m} - 237 \text{ m}} = 8.8810 \times 10^{-5} = \frac{1}{11{,}260}$$

This can also be expressed as 1:11,260. Photo scales are often expressed as nominal values, rounded to two or three significant figures: 1:11,260 ≈ 1:11,300.

EXAMPLE 2-2

An oblique photograph is taken which includes the image of a road intersection, as shown in Fig. E2-2. The road widths on the ground are both 10 m. Image measurements are taken in two directions as shown, $a = 5$ mm and $b = 2$ mm. What are the photo scales along direction a and direction b?

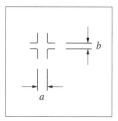

Figure E2-2

SOLUTION   A finite approximation to Eq. 2-3 gives

$$\text{scale}_a = \frac{\Delta S_{\text{image}}}{\Delta S_{\text{object}}} = \frac{0.005}{10} = \frac{1}{2000}$$

and

$$\text{scale}_b = \frac{0.002}{10} = \frac{1}{5000}$$

2.3   VANISHING POINTS

In a perspective image, lines that are parallel in the object space appear in the image to converge toward a single point. Lines in the object space which are parallel to the image plane "converge" to a point located at infinity. These points of convergence are referred to as *vanishing points*. There will be as many vanishing points as there are distinct and visible sets of parallel lines in the object space (see Fig. 2-9). Vanishing

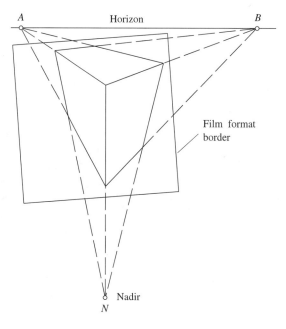

Figure 2-9 Vanishing points in a perspective projection.

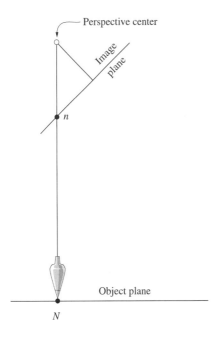

Figure 2-10 Object and image nadir points.

points *A* and *B* in Fig. 2-9, produced by horizontal parallel lines in the object space, define the *horizon*. Vanishing point *N*, from vertical lines in object space, defines a special point known as the *nadir point*. If we hung a plumb line from the perspective center, it would intersect the image plane at the *image nadir point*, and the object plane at the *ground nadir point*, as shown in Fig. 2-10.

2.4   IMAGE COORDINATE SYSTEM

The optical axis of a camera passes through the perspective center (actually the rear nodal point of the lens system) and intersects the image plane at a point called the *principal point* or the *principal point of autocollimation,* PPA. The distance from the perspective center to the image plane measured along the optical axis is the *principal distance,* PD. For aerial cameras and other fixed-focus cameras, the principal distance is equal to the *focal length f.* For close-range cameras, the principal distance is greater than the focal length and changes with focus setting.

Fiducial marks in the image plane provide fixed reference positions visible in the image. The intersection of lines joining opposite fiducial marks (or similar specification) defines a point known as the *fiducial center,* FC. This point is close to, but rarely coincides with, the principal point. The image coordinate system is defined as having its origin at FC, with the *x*-axis direction explicitly defined as either through a side fiducial mark, or bisecting the lines joining the corner fiducial marks, etc. The *y*-axis then lies at right angles to the *x*-axis in the image plane so as to form a right-handed coordinate system. The small offsets from FC to PPA are determined during camera calibration and are referred to as (x_0, y_0). Figure 2-11 shows the relationships between these elements. When we want to extend the coordinates of a point from a 2-D image coordinate to a 3-D sensor coordinate, it becomes $(x - x_0, y - y_0, -f)$, with implicit origin at the perspective center. As mentioned in Chapter 3, after calibration a principal point of best symmetry, PPS, may be used in a similar manner.

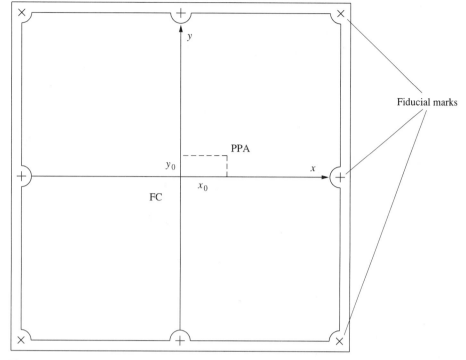

Figure 2-11 Elements defining the image coordinate system.

EXAMPLE 2-3

A nominal coordinate system is defined on a photograph with respect to the fiducial marks, with origin at FC, as shown in Fig. E2-3. By camera calibration, either in a laboratory or by self-calibration in a block adjustment, the principal point is determined to be at location PPA, with coordinates $(x_0, y_0) = (0.015, -0.005)$. If a point P has coordinates $(x, y) = (75.542, 26.381)$ with respect to FC, what are its coordinates with respect to PPA?

SOLUTION

$$x' = x - x_0 = 75.542 - 0.015 = 75.527$$
$$y' = y - y_0 = 26.381 + 0.005 = 26.386$$

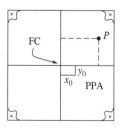

Figure E2-3

2.5   RELIEF DISPLACEMENT

As described in Section 2.3, a vertical object (for example, a flagpole) will appear in an aerial photograph to be lying along a line radial to the image nadir point (or principal point in a truly vertical photograph). The magnitude of the displacement in the image between the top and the bottom of the feature is its *relief displacement* and is related to the height of the feature and to its distance from the nadir point. A vertical feature at the nadir point will not be displaced at all, while a vertical feature at the edge of a vertical photograph would appear to lean away from the nadir point. We can derive an expression for the relationship between object height and relief displacement using the geometry depicted in Fig. 2-12. We may write two expressions for distance D in Fig. 2-12, in terms of radial image distances r_B and r_T,

$$\frac{r_B}{D} = \frac{f}{H}, \quad D = \frac{Hr_B}{f}$$

and

$$\frac{r_T}{D} = \frac{f}{H-h}, \quad D = \frac{r_T(H-h)}{f}$$

and set the two expressions for D equal to each other,

$$D = \frac{Hr_B}{f} = \frac{r_T(H-h)}{f}$$

$$Hr_T - hr_T - Hr_B = 0$$

$$H(r_T - r_B) = hr_T$$

$$\frac{H\Delta r}{r_T} = h \tag{2-4}$$

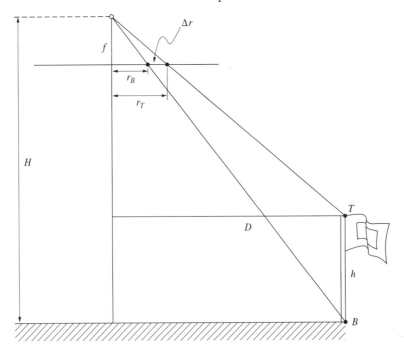

Figure 2-12 Relief displacement.

The interpretation of Eq. 2-4 is that the flying height above the base of the object times the relief displacement in the photograph divided by the radial distance from the principal point to the top of the object is equal to the height of the object. The units will be consistent if all object units are the same (e.g., meters), and all image units are the same (e.g., millimeters). Thus we have a way to estimate the height of an object from a single photograph with one auxiliary piece of information, the flying height. The flying height can often be estimated by independent knowledge of scale and the focal length, by the altimeter in the data strip, or possibly by GPS position data recorded at the time of exposure.

EXAMPLE 2-4

The flying height above the base of the building shown in Fig. E2-4 is 500 m for a vertical photograph. If the image measurements of the building as shown are $\Delta r = 4$ mm and $r_T = 75$ mm, what is the height of the building?

SOLUTION   From Eq. 2-4,

$$h = H\frac{\Delta r}{r_T} = 500 \text{ m } \frac{4 \text{ mm}}{75 \text{ mm}} = 26.7 \text{ m}$$

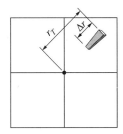

Figure E2-4

2.6 PARALLAX AND STEREO

2.6.1 Parallax

The concept of *parallax* is well known in an intuitive way to every passenger of a moving vehicle. Nearby objects (e.g., fenceposts, utility poles) move by quickly while distant objects (e.g., hills and mountains) seem to move very slowly if at all. Parallax is the apparent shift in the position of an object due to a shift in the position of the observer. We can quantify the notion of parallax by imagining the scene through a car window with a coordinate system fixed to the window frame with the positive x-axis in the direction of travel. Between time T_1 and time T_2, the apparent change in the position of an object along the x-direction would be a measurement of the parallax, as shown in Fig. 2-13. If the views presented in Fig. 2-13 were photographs taken at times T_1 and T_2, with the previously described coordinate system fixed on the camera frame, then the parallax of a particular object could be expressed as

$$p = x_{\text{left}} - x_{\text{right}} \tag{2-5}$$

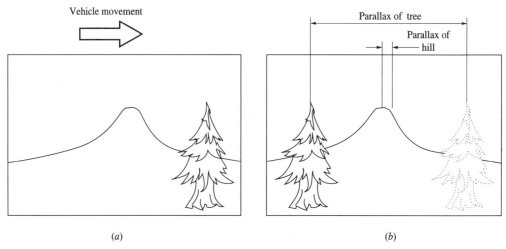

(a)   (b)

Figure 2-13 Intuitive concept of parallax from sequential images (*a*) at time T_1 and (*b*) at time T_2.

This situation is identical to the case of two vertical aerial photographs taken consecutively along a flight line, with some overlap of coverage between the two. In Fig. 2-14, point *A* is imaged onto the two frame photographs at a_l and a_r, with the *x* coordinate origins at the principal points, O_1 and O_2. The parallax at point *A* is expressed by Eq. 2-5 and can be shown graphically by transferring the right image point to the left image as shown in Fig. 2-15. If we assume the ideal geometry of Fig. 2-15, then we can use the parallax to obtain information about the height of point *A*. Observing the similarity between triangles (L_1, L_2, A) and (L_1, a_l, a'_r) we may infer that

$$\frac{B}{H-h} = \frac{p}{f} \tag{2-6}$$

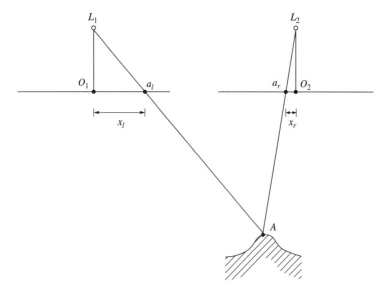

Figure 2-14 Parallax from image measurements.

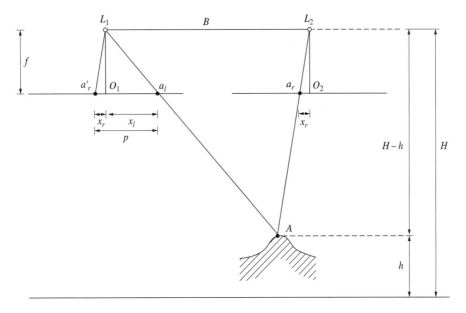

Figure 2-15 Parallax at a point.

or

$$H - h = \frac{Bf}{p} \tag{2-7}$$

which confirms what we have already stated intuitively, that parallax and proximity are inversely related. Large parallax implies close proximity (high elevation) while small parallax implies a distant object (low elevation). Equation 2-7 is rarely used in exactly the form given here. More often, parallax is measured at two nearby points and the difference in parallax is used to compute the difference in elevation. Since we are making numerous assumptions in this derivation, including vertical photographs, equal exposure station heights, etc., such computations would only be used for approximations or for planning purposes. For point 1 with elevation h_1 and parallax p_1 and point 2 with elevation h_2 and parallax p_2,

$$\Delta h = h_2 - h_1 = (H - h_1) - (H - h_2) = \frac{Bf}{p_1} - \frac{Bf}{p_2}$$

$$\Delta h = Bf\left(\frac{1}{p_1} - \frac{1}{p_2}\right) = Bf\left(\frac{p_2 - p_1}{p_1 p_2}\right)$$

$$\Delta h = Bf\left(\frac{\Delta p}{p_1 p_2}\right) \tag{2-8}$$

The product $p_1 p_2$ in Eq. 2-8 can be approximated by choosing a nominal flying height $(H - h)$ above the region of interest and rearranging Eq. 2-6,

$$p_1 p_2 \approx p^2 = \left[\frac{Bf}{H - h}\right]^2$$

Substituting this into Eq. 2-8,

$$\Delta h = Bf \left[\frac{\Delta p}{\left[\dfrac{Bf}{H-h} \right]^2} \right] = \frac{(H-h)\Delta p}{\left[\dfrac{f}{(H-h)} \right] B}$$

But $f/(H-h)$ is equal to the nominal photo scale, and $[f/(H-h)]B$ is the base distance at photo scale, b. Distance b can be determined on the photo by stereo transfer of the principal point of one photo onto the other photo. Distance b can then be measured between the principal point and the (transferred) *conjugate principal point*. Making this substitution yields

$$\Delta h = \frac{(H-h)\Delta p}{b} \tag{2-9}$$

The value of $H-h$ in Eq. 2-9 can be estimated from independent sources as stated earlier. Thus Eq. 2-9 can be used to obtain estimates of height differences between ground points from parallax measurements on the two photographs. In the past, a special device called a *parallax bar* was used to easily measure parallax while viewing the two photos stereoscopically.

EXAMPLE 2-5

The flying height above the base of the building shown in Fig. E2-5 is 500 m. The transferred principal point permits the measurement of the base, b. From the given measurements, determine the height of the building by parallax.

SOLUTION   The parallax at point A, from Eq. 2-5, is

$$P_A = 40 \text{ mm} - (-30 \text{ mm}) = 70 \text{ mm}$$

The parallax at point B is

$$P_B = 42 \text{ mm} - (-32 \text{ mm}) = 74 \text{ mm}$$

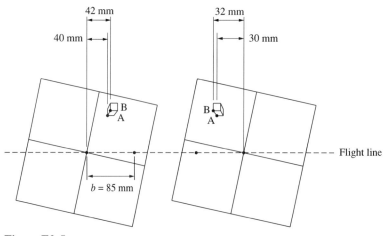

Figure E2-5

Using Eq. 2-9, the height of the building is

$$\Delta h = (H - h) \frac{P_B - P_A}{b} = 500 \text{ m} \frac{(74 \text{ mm} - 70 \text{ mm})}{85 \text{ mm}} = 23.5 \text{ m}$$

2.6.2   Stereo

2.6.2.1   Stereo Vision

Stereo vision is the capability of the human binocular visual system to perceive depth when presented with the images from the left and right eyes. The brain interprets the parallax or *retinal disparity* in a manner conceptually similar to the principles of height inference presented in the last section. In Fig. 2-16 a nearby object generates a *parallactic angle* ϕ_1 and an image displacement d_1, while an object further away generates a parallactic angle ϕ_2 and an image displacement d_2. The difference between d_1 and d_2 is very similar to our definition of parallax, and this is closely related to the magnitude of the parallactic angle. The visual system translates this parallax or parallactic angle into an impression of depth. There are numerous other depth cues which contribute to our ability to recognize proximity, but this one is of primary interest to us. In stereo photogrammetry we present the observer with a pair of images through a viewing system that guarantees that each eye sees only the image meant for it, as in Fig. 2-17. The larger parallax corresponding to point a will make it seem nearer to the viewer than point b. The viewer will thus perceive a three-dimensional model of the

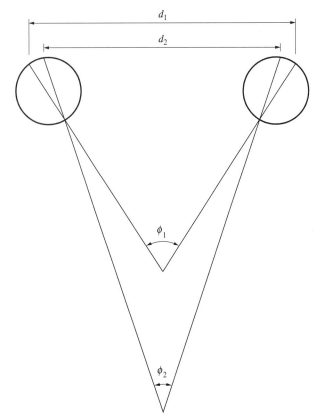

Figure 2-16 Parallax or retinal disparity in the visual system.

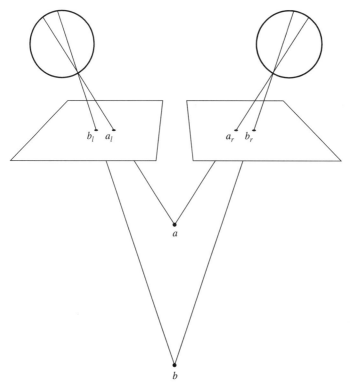

Figure 2-17 Height perception from stereo viewing of overlapping images.

features in the photographs as if it were a miniature scale model of the scene. Figure 2-17 also suggests the principle of the *floating mark*. If we superimpose a small measuring mark in both the left and right image planes and allow the operator to adjust their spacing, then the visual system will fuse them into a single object in the stereo model which appears to move up and down as the spacing is decreased or increased respectively. This effectively puts a *measuring mark* into the stereo view that the operator can translate in all three dimensions in order to point to features of interest in the model.

2.6.2.2   Vertical Exaggeration

If we rearrange Eq. 2-7 and assume that the object is at the datum plane, we obtain

$$p = f \frac{B}{H} \tag{2-10}$$

and see that the parallax is proportional to the quantity B/H or the *base-height ratio*. The simulated photographs in Fig. 2-18 have the same scale at ground level. The pair of simulated photographs in Fig. 2-18*b* were taken at twice the height above the ground with a lens of twice the focal length as the pair in Fig. 2-18*a*, but with the same base distance. Notice that the parallaxes of the building appear to be one-half in Fig. 2-18*b*. This demonstrates that the base-height ratio of the photographs will directly influence the perceived depth or height of objects in the scene. Conventional mapping photography has a base-height ratio of about 0.6. The human visual system, on the other hand, has an eye base of around 70 mm and a perceived distance to the

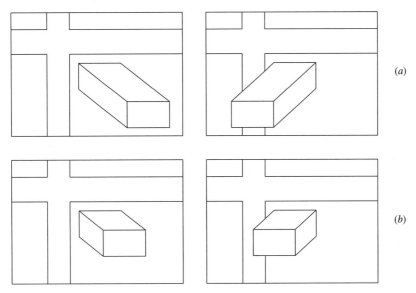

Figure 2-18 Influence of base-height ratio on parallax and depth visualization.

stereo model of 250 to 500 mm (it cannot be measured because it is a virtual image). Thus it is expecting a base-height ratio much less than that presented in conventional mapping photography. This difference causes the z-dimension in the stereo model to appear exaggerated or stretched. Note that this visual impression has no adverse impact on measuring accuracy.

2.6.2.3 Techniques for Effecting Image Separation

One of the functions of the optical system in a stereo viewing instrument or system is to present the left image to the left eye and the right image to the right eye with a minimum amount of "crosstalk." There is currently a renewed interest in this matter due to the emergence of softcopy or digital photogrammetric systems, virtual reality displays, and stereo viewing of CAD/CAM models. Table 2-1 gives a list of techniques which have been used over the years of development of photogrammetric instrumentation.

2.7 IMAGE OVERLAP

Aerial photography for conventional topographic mapping is often specified as having 60% *forward overlap* and 30% *side overlap*. The overlap percentage conveys the fraction of an image that is common with the adjacent image. The forward overlap describes this relationship between adjacent photographs along the flight line. The side overlap describes this relationship between a pair of photographs in two adjacent flight lines. In cases where one wishes to have imagery with reduced relief displacements (for mosaicking or orthophoto production, for example), a specification might call for 80% forward overlap and 60% side overlap. The imagery for the mosaic could then be taken from a small region around the center of each image. For a given camera, a given image scale, and given overlap requirements, the distances between exposures and the distances between flight lines will

Table 2-1   Stereo Image Separation Techniques

Technique	Description
Anaglyph	Red and blue spectacle lenses used to view corresponding images either from projection or from printed hardcopy
Mechanical shutter	Rotating shutter alternates viewing between the left and right eye and simultaneously alternates the projection of the left and right image
Polarized light	Spectacle lenses polarized in orthogonal planes with the left and right images presented via illumination through corresponding filters
Separate optical trains	Binocular viewing system with a separate viewing microscope for each eye
Electronic shutter	Conceptually similar to the mechanical shutter technique, but the shuttering is done via a liquid crystal display (LCD), which alternately blocks and passes light through the left and right spectacle lenses in coordination with an alternate left/right video display
Polarized video display	Conceptually similar to the polarized light method, except the images are presented alternately on a video monitor with a synchronized LCD filter which changes the polarization between left and right, to match the spectacle lenses of the observer

be fixed. This is a common calculation in project planning. Figure 2-19 shows the relevant geometry, which will serve for either forward or side overlap.

In Fig. 2-19, w is the image format size, W is this dimension scaled to the ground, B is the base distance, H is the flying height above the terrain, and f is the focal length. Given W and B the overlap fraction R can be obtained from the geometry of Fig. 2-19,

$$R = 1 - \frac{B}{W} \tag{2-11}$$

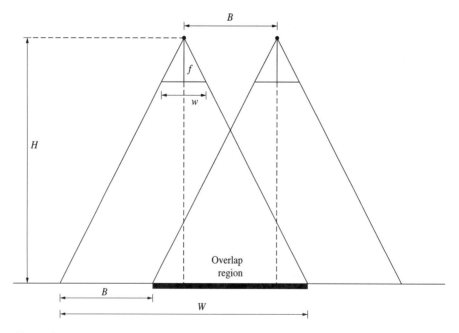

Figure 2-19   Image overlap.

and given W and R, one can compute the required base distance, B.

$$B = W(1 - R) \tag{2-12}$$

2.8 EPIPOLAR PLANES AND LINES

A pair of overlapping photographs are depicted in Fig. 2-20 with object point A and corresponding image points a_1 and a_2. In general, the photographs will not be exactly vertical and their exposure positions or exposure stations will have different heights. The plane defined by the three points L_1, L_2, and A is known as an *epipolar plane*. The two lines where this plane intersects the two photographs are referred to as *epipolar lines*, and moreover they are *conjugate* or corresponding epipolar lines. There are, in fact, an infinite number of epipolar lines corresponding to the family of planes containing line $\overline{L_1 L_2}$. A subset of these will intersect the two photo planes. The epipolar lines can be determined after the photographs have been *relatively oriented*. If a ground point is imaged onto a given epipolar line in one photograph, then we are guaranteed that it will lie in the conjugate epipolar line in the other photograph. There are two significant applications of these concepts in photogrammetry. First, digital images which have been sampled or resampled along epipolar lines can be presented in a video or softcopy stereo system with the assurance that there will be no *y-parallax* or misregistration in the *y*-dimension. Note that this is a somewhat inexact but nevertheless entrenched usage of the term *parallax*. See the related development of image normalization in Chapter 7. The second application is also in digital photogrammetry. In attempting to automate the stereo perception process, image matching techniques are

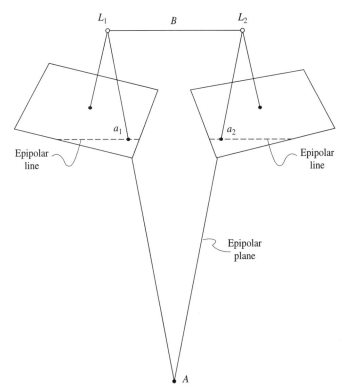

Figure 2-20 Epipolar plane and corresponding epipolar lines.

used to locate conjugate image points and features in overlapping images. If we use the point on one photo as the reference, and search for its conjugate point on the other photo, then we can restrict the effort to a one-dimensional search along the epipolar line, rather than a much more time-consuming two-dimensional search throughout the entire image. These applications will become increasingly important as digital photogrammetry and automated techniques become more prevalent.

PROBLEMS

2.1 A camera on an orbiting space platform has an altitude above the ground of 400 km. The lens has a focal length of 2 m. After return to earth, the film is scanned such that the square image pixel is 10 μm by 10 μm. What is the scale of the photograph? What is the size of the pixel projected to the ground at nadir?

2.2 For a camera with focal length 152.4 mm and the standard 23 cm format size, what height above ground is necessary for a vertical photograph to cover an area of 9 km^2 (i.e., 3 km by 3 km)?

2.3 You wish to determine the approximate height of a building visible in a photograph. The flying height above the the base of the building is 500 m. In the nominally vertical photograph the radial distance from the principal point to the top of the building is 80 mm. The relief displacement of the building is 4 mm. What is the approximate height of the building?

2.4 Two overlapping vertical photographs contain the image of a church steeple. The parallax at the top of the steeple is 100 mm. The parallax at the bottom of the steeple is 95 mm. The flying height above the ground is 450 m. The image base (the distance between the principal point and the transferred conjugate principal point) is 80 mm. What is the approximate height of the steeple?

2.5 Aerial photography with 60% forward overlap and 30% side overlap is needed for a mapping project. The focal length is 152.4 mm. The desired image scale is 1:4000. What is the spacing between exposures along the flight line? What is the spacing between flight lines? If the aircraft has a ground speed of 190 km/hr, what is the time between exposures along the flight line?

REFERENCES

KRAUS, K. 1993. *Photogrammetry.* Bonn: Dümmler.

MALING, D. 1973. *Coordinate Systems and Map Projections.* London: George Philip & Son, Ltd.

MOFFITT, F., and MIKHAIL, E. 1980. *Photogrammetry.* New York: Harper & Row.

SLAMA, C. 1980. *The Manual of Photogrammetry.* Bethesda, MD: American Society for Photogrammetry and Remote Sensing.

SNYDER, J. 1989. *Map Projections: A Working Manual.* Professional Paper 1395. Washington, DC: US Geological Survey.

WOLF, P., and DEWITT, B. 2000. *Elements of Photogrammetry.* New York: McGraw-Hill.

Chapter 3

Photogrammetric Sensing Systems

The term *photogrammetric sensor* has traditionally referred to a calibrated aerial frame camera. However, advances in camera technology, especially the introduction of electronic imaging sensors, have led to the widespread use of sensors with a variety of imaging geometries and spectral properties. The related development of analytical triangulation methods and digital photogrammetric workstations has increased the number of types of sensors that can be exploited with photogrammetric methods. Advanced mathematical techniques now allow the use of uncalibrated cameras in a number of situations. This chapter provides an overview of sensor technology, including the physical principles of optics and sensing, a discussion of image acquisition geometries, and a detailed look at the construction of the standard aerial mapping camera.

3.1 THE PHYSICS OF REMOTE SENSING—ELECTROMAGNETIC ENERGY

Photogrammetry is a form of remote sensing, since it uses electromagnetic radiation captured by a sensor in the form of an image. To understand photogrammetric sensors we must therefore first understand the geometric and radiometric properties of electromagnetic radiation. Physical optics deals with the radiometry and also the wave nature of light, making use of Fourier transform methods (Goodman, 1996) to analyze optical systems, and is beyond the scope of this book. Geometric optics (Hecht et al., 1997), which is more relevant to photogrammetric issues, deals with the geometric aspects of electromagnetic radiation and will be discussed in the next section.

The measurement of the fundamental quantitative properties of transmitted electromagnetic energy is called *radiometry;* when we deal with those wavelengths visible to the eye, we term this *photometry* and refer to the energy as light. Two properties are of particular importance in discussing photogrammetric sensors: the amount of energy and the wavelength of the energy.

3.1.1   Radiometry—Quantifying Electromagnetic Energy

To design a sensor or to relate a sensor's readings to scene characteristics, we must be able to describe the amount of radiation reflected or emitted by the scene and transmitted to the sensor. The *radiant energy, Q,* is measured in *joules* and represents the ability to do work, such as heating an object or exposing a photographic film. The *radiant flux, ϕ,* is the time rate of flow of the energy, *dQ/dt,* measured in watts, or joules per second. When the radiant flux impinges upon a surface, the incident flux density or *irradiance, E,* is expressed as flux per unit area, in watts per meter squared. The radiant flux leaving a surface is the *exitance, M,* also measured in watts per meter squared.

For point sources, the radiant intensity, *I,* is the radiant flux per unit solid angle emitted in a particular direction. This is measured in units of watts per steradian. (A *steradian* is a measure of solid angle analogous to the radian as a measure of two-dimensional angle. While a radian is the angle subtended by an arc equal to the radius, a steradian is the solid angle subtended by an area equal to the radius squared. There are 2π radians in a circle and 4π steradians in a sphere.)

For an extended source we define the *radiance, L,* as the radiant flux per unit solid angle in a given direction, analogous to the radiant intensity of a point source. However, since the source has finite extent, we must normalize by the projected area in the direction of interest, making the units watts per steradian per meter squared. If the radiance does not depend on the viewing angle, we say that the surface is *Lambertian.*

3.1.2   The Electromagnetic Spectrum

Figure 3-1 shows an overview of the electromagnetic spectrum with the wavelength ranges of the various regions. The properties of electromagnetic waves vary with their wavelength (or frequency, which is equal to the speed of light divided by the wavelength). For remote sensing, we are interested in how particular wavelengths indicate the object properties under investigation by their emittance or absorption characteristics; in photogrammetry, our main concern is the resolution and interpretability of the imagery. In both cases, we must have sensors that can accurately record the energy in an interpretable form.

Most photogrammetric sensors work within the visible or near-visible portions of the spectrum (Fig. 3-1). While film can record radiation up to about 0.9 micrometers, which is in the near-infrared region, solid-state sensors can detect radiation into the far-infrared, up to about 12 micrometers. The near-infrared corresponds mostly to reflected radiation, while the middle (from 3 to 5.5 micrometers) and far-infrared (from 8 to 14 micrometers) regions correspond mostly to emitted energy due to the heat energy of the object. Atmospheric transmission varies with wavelength and only certain

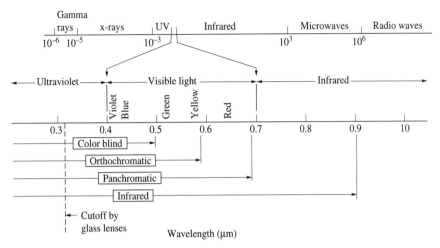

Figure 3-1 Portion of electromagnetic spectrum useful for imaging.

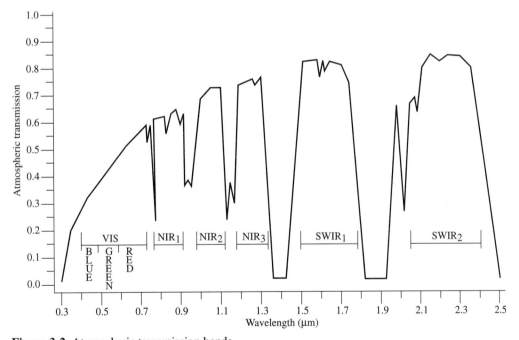

Figure 3-2 Atmospheric transmission bands.

atmospheric windows exist within the infrared, as shown in Fig. 3-2. The ultraviolet region, at the other end of the visible spectrum, is seldom used for photogrammetry or remote sensing due to its absorption by the atmosphere.

3.2 OPTICS

To gain a basic appreciation of photogrammetric lens design, we will work with the branch of optics known as *geometric optics* (Hecht et al., 1997). Geometric optics defines light in terms of *rays*, lines in space corresponding to the flow direction of the radiant energy. Rays can be related to the wave nature of light if we think of them as perpendiculars to the wavefront at a particular point. Light waves coming from a point at infinity are essentially planar, and the rays are therefore parallel. This is known as *collimated* light. *Physical optics* is concerned with the wave nature of light. Analysis of light waves allows us to model phenomena such as diffraction effects and interference fringes, but in other contexts we will need to think of light as particles. For instance, when studying solid-state imaging devices we will treat light as photons striking the sensor.

3.2.1 Refraction

When a light ray passes through the interface between two media with different optical properties, the light ray is bent or *refracted*. The amount of refraction is determined by the *index of refraction* of each medium.

The index of refraction of a substance is the ratio of the speed of light in a vacuum (2.99792458×10^8 m/s) to the speed of light in the medium. For air at standard temperature and pressure, the index of refraction is approximately 1.00029. The types of glass used in lenses typically have refractive indices between 1.4 and 1.7. The index of refraction is a function of wavelength. This explains why a prism separates white light into its different constituent colors, since each wavelength is refracted by a different amount.

To quantify the amount of refraction, we define two angles with respect to the surface normal at the point of incidence (Fig. 3-3): the *angle of incidence, i,* which is the angle between the incident light ray and the normal, and the *angle of refraction, r,* which is the angle between the normal and the refracted light ray. *Snell's law* gives the relationship between the incident and refracted angles as

$$n_i \sin i = n_r \sin r \qquad (3\text{-}1)$$

where n_i is the index of refraction on the incident side of the interface and n_r is the index of refraction on the other side of the interface. The incident ray, the refracted ray, and the surface normal are coplanar, and the ray in the medium with the higher refractive index is closer to the normal. If the ray is traveling from a medium with a higher

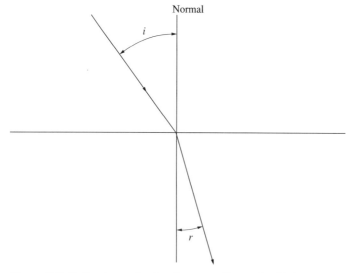

Figure 3-3 Reflection and refraction according to Snell's law.

refractive index to one with a lower refractive index, the refracted ray will be bent away from the normal until, at some value of the incident angle, the angle of refraction will be 90 degrees. From that point on, the ray will be totally reflected. This angle is known as the *critical angle*.

EXAMPLE 3-1 *Snell's Law*

A ray of light travels from glass with a refractive index of 1.4 into the air, with an assumed refractive index of 1.0. If the angle of incidence is 30 degrees, what is the angle of refraction? At what angle of incidence will total reflection occur?

SOLUTION Rearranging Eq. 3-1,

$$\sin r = \frac{1.4 \sin i}{n_r} = \frac{1.4 \sin (30°)}{1.0}$$

$$r = \sin^{-1}(0.7) = 44.4°$$

At the critical angle, the angle of refraction is 90 degrees, so sin r = 1 and the critical angle, i_c, is

$$\sin i_c = \frac{n_r}{n_i} = \frac{1.0}{1.4}$$

$$i_c = \sin^{-1}(0.71429) = 45.6°$$

Therefore, once the angle of incidence reaches 45.6 degrees, the ray is totally reflected.

3.2.2 Imaging with a Thin Lens

Snell's law deals with a single ray crossing an interface. We can think of a lens as an object that refracts a bundle of light rays from a point, according to Snell's law, to converge at another point. For each ray to be refracted to the same point, the lens surfaces must be curved. Spherical surfaces are most commonly used, for ease of design and fabrication.

The simplest form of lens is the *thin lens*, where we neglect the physical thickness of the lens and are concerned only with the radii of its surfaces. As shown in Fig. 3-4, if collimated light from a point at infinity passes through the lens parallel to the optical

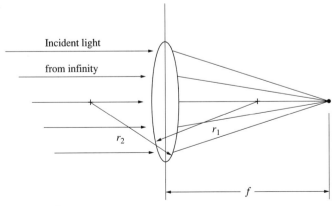

Figure 3-4 Thin lens.

axis, it is focused to a point at a distance f from the lens. The distance f is called the *focal length* of the lens, and can be determined from the radii of the two lens surfaces and its index of refraction, n, using the *lens maker's equation*

$$\frac{1}{f} = (n - 1)\left(\frac{1}{r_1} + \frac{1}{r_2}\right) \tag{3-2}$$

where, as shown in Fig. 3-4, r_1 is the radius of the front surface, coming from the left, and is positive if the center of curvature is to its right as shown, while r_2 is the radius of the back surface and is positive if the center of curvature is to the left as shown. If the focal length is positive, the lens is referred to as a *converging lens,* since the rays will converge to a point and form a real image which can be viewed on a screen or produce an image on film. If the focal length is negative, as with a concave lens, the rays will diverge and form a *virtual image.* A virtual image is formed on the same side of the lens as the object, and can only be viewed by looking through the lens.

If the object point is not located at infinity, the rays from the point will not be parallel entering the lens. By tracing the path of the rays from the object through the lens we can determine where they intersect to form an image. The distance from the lens to the image, measured parallel to the optical axis, is given by the *thin lens equation*

$$\frac{1}{p} + \frac{1}{q} = \frac{1}{f} \tag{3-3}$$

where p is the object distance and q is the image distance. Notice that when the object distance is equal to infinity, the image distance is equal to the focal length as discussed above. Conversely, when the object distance is equal to the focal length, the image distance is equal to infinity. This is the principle of the collimator, which is used in camera calibration (Section 3.8) to produce a target that appears to be at infinity.

3.2.3 Imaging with a Thick Lens

While the thin lens equation is simple, real lenses have finite thickness (Fig. 3-5). Modifying Eq. 3-2 to include the lens thickness t yields

$$\frac{1}{f} = (n - 1)\left(\frac{1}{r_1} + \frac{1}{r_2} - \frac{(n - 1)t}{(nr_1r_2)}\right) \tag{3-4}$$

For thick lenses we define two special planes, the *principal planes,* perpendicular to the optical axes. If a ray is directed toward a point m in one principal plane, it will

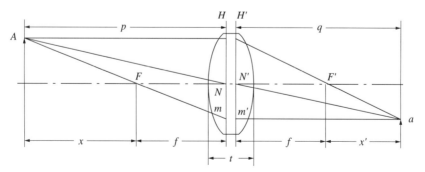

Figure 3-5 Thick lens.

appear to leave the lens from a point m' in the other principal plane. The points m and m' are the same distance from the optical axis. If the refractive index is the same on both sides of the lens, the principal planes intersect the optical axis at the *nodal points*. If a ray intersects one of the nodal points, it will appear to leave the other nodal point in the same direction.

We can use the thin lens equation to model a thick lens by measuring the focal length and the object and image distances from the principal planes. In effect, we put the two principal planes together and define an equivalent thin lens, which we model with the thin lens equation above (Eq. 3-3).

EXAMPLE 3-2   *Thin and Thick Lens Calculations*

The front surface of a simple lens, which is made of glass with an index of refraction of 1.4, has a radius of +150 mm and the back surface has a radius of +125 mm. The focal length of the lens is therefore

$$\frac{1}{f} = (1.4 - 1)\left(\frac{1}{150} + \frac{1}{125}\right)$$

$$f = 170.45 \text{ mm}$$

If an object is located at a distance of 1.0 m from the lens, an image will be formed at a distance of

$$\frac{1}{1.0} + \frac{1}{q} = \frac{1}{0.17045}$$

$$\frac{1}{q} = \frac{1}{0.17045} - \frac{1}{1.0}$$

$$q = 0.205 \text{ m}$$

Notice that the image distance is greater than the focal length. As the object distance increases, the image distance decreases until, when the object is at infinity, the image distance is equal to the focal length.

Suppose this same lens is to be treated as a thick lens, and the distance between the two surfaces along the optical axis is 15 mm. The focal length is now calculated using the thick lens formula (Eq. 3.4):

$$\frac{1}{f} = (1.4 - 1)\left(\frac{1}{150} + \frac{1}{125} - \frac{(1.4 - 1)15}{1.4 \times 150 \times 125}\right)$$

$$f = 173.153 \text{ mm}$$

3.2.4   Aperture

The *aperture stop* of a lens defines the "diameter" of the lens, which determines the amount of light the lens can collect. Imaging lenses are equipped with an adjustable aperture or diaphragm that controls the amount of light admitted and therefore the exposure. In lens design, the absolute diameter of the aperture is less important than the relative sizes of the aperture diameter and the focal length. We therefore speak of the *f number,* the ratio of the focal length to the aperture diameter, written f/# or 1:#.

The standard f stops marked on the aperture rings of most lenses (1, 1.4, 2, 2.8, 4, 5.6, etc.) increase by the square root of two. Increasing the f number by the square root of two reduces the diameter by the square root of two and halves the area of the aperture. Since the amount of light admitted depends on the area of the aperture, increasing the aperture by one f stop (from 5.6 to 4, for instance) means that the shutter speed should be halved (from 1/500 to 1/1000 second, for instance) to maintain the same exposure.

3.2.5   Field of View

Photogrammetric sensors are often described in terms of their *field of view,* the angular size of the cone in space that the sensor can image. A lens with a *normal* field of view has roughly the same cone of vision that the human eye does, about 50 degrees. Most photogrammetric sensors use *wide-angle* lenses, with fields of view of about 90 degrees. The geometry of intersecting image rays from multiple images is much stronger (the position of the intersection point is better defined) when the images represent a very wide cone of rays, since the intersection angles of the rays in object space are nearer a right angle. Super-wide-angle lenses, with fields of view from 110 to 130 degrees, are also in use.

3.2.6   Illumination

For physical reasons, the intensity of illumination at the edges of the image is less than that in the center of the image by a factor of the fourth power of the cosine of the off-axis angle ϕ. Three factors contribute to this:

1. The inverse-square law of illumination. Light spreading from a point source uniformly fills a sphere centered at the source. The area of a sphere increases as the radius squared. Since the same amount of light is covering a larger area, the illuminance decreases with distance from the source. In a camera, the distance from the lens to the edge of the format is greater than that from the lens to the center of the image by a factor of $\cos^2\phi$.

2. The circular lens aperture appears as an ellipse from the off-axis point. The apparent area of the aperture is therefore reduced by a factor of $\cos\phi$.

3. Lambert's law states that the illumination is reduced by a factor of $\cos\phi$ due to the obliquity of the rays striking the focal plane.

These factors combine to reduce the illumination by a total factor of $\cos^4\phi$, which can be significant for a wide-angle lens. Modern lens design techniques reduce this effect greatly. In some cases an anti-vignetting filter, which is shaded in the center and clear at the edges, is used in front of the lens to even out the exposure across the image.

3.2.7   Lens Aberrations

Any system of lenses consisting only of spherical surfaces will have *aberrations,* deviations from the theoretically perfect imaging geometry. The five primary aberrations are

1. *Spherical aberration,* the dependence of focal length on aperture, which results in a point being imaged as a blurred circle. This can be reduced by closing (stopping down) the aperture.

2. *Coma,* the degradation due to the difference in focus for axial and off-axis rays, which results in a point source being imaged as a comet shape.

3. *Astigmatism*, the variation in focus with orientation. Objects oriented radially with respect to the optical axis are focused differently than objects oriented tangentially.

4. *Field curvature*, which occurs if the region of best focus is not a plane.

5. *Distortion*, the change in magnification with off-axis distance.

The above aberrations are monochromatic; lenses may also exhibit *chromatic aberration*, due to the change in the index of refraction with wavelength. The current state-of-the-art in lens design reduces these aberrations to negligible levels and, except for distortion, lens aberrations are not a consideration.

3.2.8 Lens Distortion

Lens distortion is the aberration most relevant to photogrammetric practice. While other aberrations mainly affect image quality and must be removed by the lens designer, lens distortion directly affects the metric accuracy of the image and must be corrected by the photogrammetrist.

Radial distortion is the radial displacement of an imaged point from its theoretically correct position or, equivalently, a change in the angle between a ray and the optical axis. Radial distortion is determined by a calibration procedure (see Section 3.8). A typical radial distortion curve is shown in Fig. 3-6, with the distortion in millimeters shown as a function of the radial distance from the principal point. The distortion can also be plotted as a function of the angle from the optical axis, since the angle α between the optical axis and the ray is a function of the focal length f and the distance d of the point from the optical axis.

$$\alpha = \tan^{-1} \frac{d}{f} \tag{3-5}$$

Radial distortion can have both positive (outward, away from the principal point) and negative (inward) values. Positive radial distortion is often referred to as *pincushion* distortion, since an imaged square will appear to have its sides curve toward the middle, and negative distortion is termed *barrel* distortion, since the sides of the square will be bowed outward relative to the corners, as shown in Fig. 3-7.

A radial distortion curve may be unbalanced, with most of the curve either positive or negative. It may be more desirable to have the maximum positive and negative values in the distortion curve nearly equal, as when using analog plotters with a

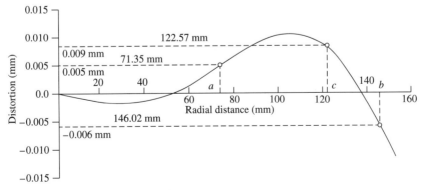

Figure 3-6 Typical radial distortion curve.

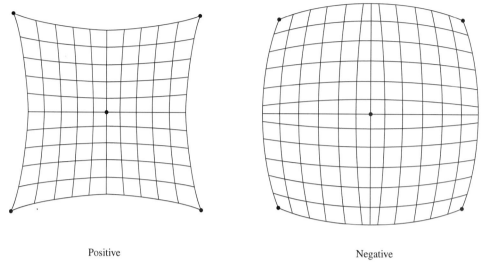

Positive   Negative

Figure 3-7 Positive and negative radial distortion.

limited ability to correct for lens distortion. From Eq. 3-5 we see that by properly specifying the focal length we can make the distortion at a point take any value. Therefore, we can calculate a *calibrated focal length* based on equalizing the maximum positive and negative distortions or minimizing the distortion values by least-squares or some other criterion. This calculated focal length is sometimes referred to as the *camera constant*.

When correcting image coordinates for radial distortion, the distortion curve can be stored as a table of radial distances and corresponding distortion values. The correction is then determined by table lookup and interpolation. Another method is to express the distortion as a polynomial function of odd powers of the radial distance.

$$d_r = k_1 r^3 + k_2 r^5 + k_3 r^7 + \cdots \tag{3-6}$$

The distortion can then be evaluated directly from the calculated radial distance. Note that a linear term is usually not included in this expression, since inclusion of the linear term is equivalent to adjusting the equivalent focal length to balance the distortion curve.

Tangential distortion is not a lens aberration as such, but is a consequence of errors in assembly of the lens components affecting the rotational symmetry of the lens. For this reason, tangential distortion is often referred to as *decentering* distortion. While not generally significant in current mapping lenses, it is common in commercial lenses with variable focus or zoom.

Decentering distortion has radial and tangential components, which vary as the vector from the principal point to the point of interest varies with respect to the *axis of maximum tangential distortion* (Fig. 3-8). When the vector to the point of interest is parallel to the axis of maximum tangential distortion, the tangential component, δ_t, is maximum and the radial component, δ_r, is zero. When the vector to the point is perpendicular to the axis, the radial component is maximized and the tangential component is zero.

$$\begin{aligned}\delta_r &= 3(J_1 r^2 + J_2 r^4 + \cdots)\sin(\phi - \phi_0)\\\delta_t &= (J_1 r^2 + J_2 r^4 + \cdots)\cos(\phi - \phi_0)\end{aligned} \tag{3-7}$$

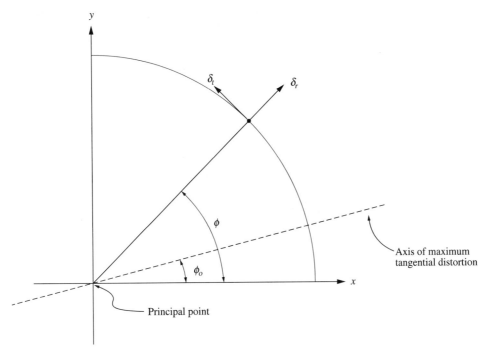

Figure 3-8 Tangential distortion.

where ϕ is the angle between the positive x-axis of the image and the vector to the point of interest and ϕ_0 is the angle between the position x-axis and the axis of maximum tangential distortion.

EXAMPLE 3-3 *Lens Distortion*

We are given a photograph taken with a lens whose radial and decentering distortion coefficients are $k_1 = 3.47222 \times 10^{-9}$, $k_2 = 1.20563 \times 10^{-13}$, $k_3 = 0.0$, $J_1 = 6.94444 \times 10^{-8}$, $J_2 = 4.82253 \times 10^{-12}$, and $\phi_0 = 45°$. Calculate the distortion at image coordinates $x = 57.239$, $y = -77.918$ (relative to the principal point).

SOLUTION The radial distance r is

$$\sqrt{57.239^2 + 77.918^2} = 96.6826 \text{ mm}$$

The radial distortion is

$$d_r = k_1 r^3 + k_2 r^5$$
$$= (3.47222 \times 10^{-9})(96.6826)^3 + (1.20563 \times 10^{-13})(96.6826)^5$$
$$= 0.0042 \text{ mm}$$

To calculate the decentering distortion, we first calculate the angle between the positive x-axis and the radial line to the point of interest:

$$\phi = \tan^{-1}\left(\frac{y}{x}\right)$$

$$= \tan^{-1}\left(\frac{-77.918}{57.239}\right)$$

$$= 306.301°$$

We can now calculate the radial and tangential components of the decentering distortion:

$$\delta_r = 3(J_1 r^2 + J_2 r^4)\sin(\phi - \phi_0)$$
$$= 3[(6.94444 \times 10^{-8})\,96.6826^2 + (4.82253 \times 10^{-12})\,96.6826^2]\sin(306.301 - 45)$$
$$= -0.0032$$

$$\delta_t = (J_1 r^2 + J_2 r^4)\cos(\phi - \phi_0)$$
$$= [(6.94444 \times 10^{-8})\,96.6826^2 + (4.82253 \times 10^{-12})\,96.6826^4]\cos(306.301 - 45)$$
$$= -0.0002$$

The total radial correction is $0.0042 - 0.0032 = 0.001$ mm. The x and y coordinate corrections for the radial distortion are

$$\delta_x = \frac{x}{r}\,\delta_r = \frac{57.239}{96.6826}0.001 = 0.0006$$

$$\delta_y = \frac{y}{r}\,\delta_r = \frac{77.918}{96.6826}0.001 = 0.0008$$

The corrections will be added to the x and y coordinates, although this small correction may not be considered significant in many applications. The tangential correction is insignificant. If the coordinates were to be corrected, the x and y corrections would be

$$\delta_x = -\delta_t \sin\phi = -(-0.0002)\sin 45° = 0.0001$$

$$\delta_y = \delta_t \cos\phi = -0.0002\cos 45° = -0.0001$$

3.2.9   Depth of Field and Depth of Focus

Although we speak of a lens being "in focus" at a specific distance, objects not located precisely at that distance can also be imaged with acceptable quality. For any given focus setting, there exist near and far points where the amount of blur reaches unacceptable amounts. The distance between the near and far points is referred to as the *depth of field* for that focus setting, and is a function of the lens focus setting, the lens aperture, and the amount of blur acceptable. We quantify the blur in terms of the *circle of confusion,* as shown in Fig. 3-9. As the aperture is decreased, the depth of field is increased. The near and far points of acceptable focus, p_N and p_F, can be calculated for a given focus distance p, circle of confusion c, aperture D, and focal length f using:

$$p_N = \frac{p}{1 + (p - f)c\dfrac{D}{f^2}}$$

$$p_F = \frac{p}{1 - (p - f)c\dfrac{D}{f^2}}$$

$$(3\text{-}8)$$

At the *hyperfocal distance,* the depth of field extends to infinity. By setting the denominator of the equation for p_F to zero, so that p_F goes to infinity, and then solving for p, we obtain the hyperfocal distance p_H.

$$p_H = \frac{f^2}{cD} + f$$

$$(3\text{-}9)$$

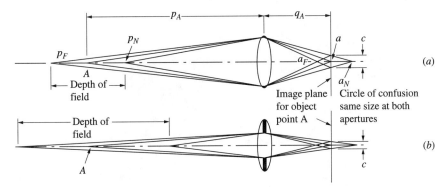

Figure 3-9 Depth of field as a function of diameter of aperture stop.

When the camera is focused at the hyperfocal distance, the near point of acceptable focus is one-half the object distance or 0.5 p_H in Eq. 3-9. Depth of field is not generally a problem for aerial photography due to the large object distances involved, but it must be considered for close-range photogrammetry.

   Depth of focus is a concept similar to depth of field, but instead refers to how far the focal plane can be moved in a particular situation without unacceptable blurring.

EXAMPLE 3-4   *Calculation of Depth of Field and Hyperfocal Distance*

What is the depth of field of a lens if the object distance is 5 m, the focal length is 75 mm, the diameter of the circle of confusion is 0.050 mm, and the f number is 5.6? What is the hyperfocal distance?

SOLUTION   By Eqs. 3-8,

$$p_N = \frac{5000}{1 + \dfrac{(4925 \times 0.050 \times 5.6)}{75^2}} = 4016\,\text{mm} = 4.016\,\text{m}$$

$$p_F = \frac{5000}{1 - \dfrac{(4925 \times 0.050 \times 5.6)}{75^2}} = 6624\,\text{mm} = 6.624\,\text{m}$$

The depth of field is therefore $6.624 - 4.016 = 2.608$ m.
   From Eq. 3-9, the hyperfocal distance is

$$p_H = \frac{75^2}{0.050 \times 5.6} + 75 = 20164\,\text{mm} = 20.164\,\text{m}$$

If the lens is focused at H, the near limit of the depth of field, $p_N,$ is

$$p_N = \frac{20164}{1 + \dfrac{(4925 \times 0.050 \times 5.6)}{75^2}} = 16194\,\text{mm} = 16.194\,\text{m}$$

3.2.10   Lens Design

Lens design, like many other fields, has been revolutionized by the use of computers. For years, new lens designs were based on variations of existing designs or were

Figure 3-10 Mapping camera lens design. The space in the middle is for the shutter. Courtesy of LH Systems, LLC.

dependent upon the insights and instincts of experienced designers. Design optimization was limited by the amount of calculation required for *ray tracing,* computing the paths of rays at different angles through the optical system.

With the advent of computer-aided design techniques, lens performance and complexity has improved dramatically. Figure 3-10 shows a schematic of a current mapping camera lens design. Consider the differences between modern lenses and the Metrogon lens, designed and built during World War II. The Metrogon had a maximum aperture of f/6.3 and was suitable only for panchromatic film, while current lenses can work at f/4.0, allowing photography at higher shutter speeds or at lower light levels, and are well-corrected for use with color film. The maximum radial distortion of the Metrogon lens was 110 micrometers, with resolution of from 20 to 24 line pairs per millimeter (lp/mm). Modern lenses have nearly zero distortion and resolution of 100 to 120 lp/mm.

3.2.11 Reflective Optical Systems

Refractive optics (lenses) are not suitable for all imaging systems, particularly those designed for a wide range of wavelengths. Since the amount of refraction depends on the wavelength, an uncorrected refractive optical system will focus light of different colors on different planes. This chromatic aberration is typically well-corrected in today's cameras, but becomes more difficult to remove as the range of wavelengths increases. The absorption of radiation by refractive optical elements is also an issue at longer wavelengths. The properties of reflective optics do not change as a function of wavelength, so a wide spectrum can be imaged with a reflective optical train. For this reason, multispectral imaging systems typically use reflective optical trains.

Reflective elements may also be used for mechanical reasons. Scanning systems often use a mirror as the scanning element, or a mirror may be used to "fold" an optical train with a long effective focal length so that it fits within a confined space.

Reflective elements come in several shapes other than the familiar planar mirrors. A parabolic mirror has the property that collimated light (coming from infinity) entering the mirror on axis converges upon a single point focus. Conversely, light emitted from the focal point will be reflected into a plane wave. Projectors often use a parabolic reflector with the light source located at the focus. Hyperboloid and ellipsoid mirrors may also be used in sensor optics.

3.2.12 Dispersive Optical Elements

For multispectral and hyperspectral sensors the incoming light is split into some number of bands. One approach is to use a different camera for each band, with filters on each which pass only one band. However, the more common approach is to split the light into bands after it has entered the optical system by using dispersive optical elements.

The simplest example of a dispersive optical element is a prism, which breaks white light into its constituent wavelengths due to the variation of the refractive index with wavelength. Another type of dispersive element is the *diffraction grating*, which consists of a surface with fine parallel grooves ruled across it. The rulings diffract incident light of different wavelengths by different amounts. When either type of dispersive element is used in the optical train in a multispectral sensor, the light is spread across the sensing array. Each element of the array records the intensity of only the wavelength band falling on it.

3.3 SENSING

After the energy reflected or emitted from the scene has passed through the optics and has been focused, it must be recorded. The most common means of recording imagery is photographic film. Film continues to improve in resolution, speed, and color fidelity, but even the best of modern films are not suitable for all purposes. Imaging tube systems, seldom used today, provided the capability of transmitting pictures in real time to obtain pictures without retrieving the camera—an important consideration for space applications. Some photogrammetric applications using imaging tube video cameras have been studied, but these were limited by the low resolution and geometric instability of imaging tube systems.

With the development of solid-state electronics, solid-state imaging array chips became a reality. These offer high geometric stability and spectral characteristics that can be tailored to the bands of interest. While limited resolution, in terms of number of pixels, has been a problem, the continuing refinement of semiconductor manufacturing techniques makes possible larger and larger arrays.

3.3.1 Photographic Film

In its simplest form, photographic film consists of a light sensitive coating, the *emulsion,* applied to a plastic or glass *base.* There may also be an *anti-halation layer* behind the base to prevent reflections from the base back into the emulsion.

The base must not deform during exposure and processing, or the geometry of the image will be compromised. Current aerial films use bases of stable plastics that are not affected by changes in temperature or humidity. Glass plates are sometimes used in terrestrial photography to obtain the ultimate in dimensional stability and flatness. Aerial cameras using glass plates have been used in the past, but the difficulties in handling glass plates, along with the improvements in plastic film bases, have made their use extremely rare.

Great care must be taken in handling and developing photographic film so that it is dried uniformly and not subjected to uneven stresses. Uniform film deformation can be corrected to some extent by using the calibrated fiducial marks or reseaux, if they are present, but nonuniform film deformations will cause distortions in the final mapping results.

The emulsion contains minuscule grains of silver halides, light-sensitive salts. When a silver halide crystal is exposed to light, a chemical change occurs and a *latent image* is formed; when the film is developed, the crystal is reduced to elemental silver. The proportion of silver halide crystals changed to silver depends upon the intensity of the light striking the film, yielding an image with varying shades of density.

Silver halide grains range in size from a few tenths of a micrometer up to a few micrometers. A "faster" film, which requires less exposure time for a given amount of illumination, has larger, but fewer, grains. Images taken with such film have lower resolution and will therefore appear to be "grainy" under magnification. Commercial film speed is measured in terms of ASA or DIN ratings. Aerial film speed is measured in accordance with the Aerial Film Speed standard (AFS).

An unmodified emulsion is sensitive mostly to blue and ultraviolet light. By incorporating dyes into the emulsion, the spectral sensitivity of the emulsion can be extended into the green portion of the visible spectrum (an *orthochromatic* emulsion), across the entire visible spectrum (a *panchromatic* emulsion), or into the near-infrared region (up to 900 nanometers) of the spectrum (Fig. 3-1).

To understand how color films work, we must understand the concepts of additive and subtractive color. White light can be produced by adding together red, green, and blue *primary colors.* Adding pairs of the primary colors produces new colors: blue and green make cyan, blue and red make magenta, and red and green make yellow. These new colors can be used to subtract their *complementary color,* the primary color not used in the mixture. So, for example, since green and blue make cyan, the complementary color of cyan is red. Cyan subtracts red from white light, yellow subtracts blue, and magenta subtracts green. A stack of cyan, yellow, and magenta filters would appear black, since all three primary colors would be subtracted.

Color film is based on a subtractive process. Three different emulsion layers are used, each of which is sensitive to a different portion of the spectrum. The emulsion layers are typically arranged as shown in Fig. 3-11, although other configurations are possible. Since all of the emulsion layers are sensitive to blue light, the blue-sensitive layer is on top and a yellow filter is added below it to subtract the blue light. The next

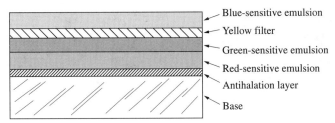

Blue-sensitive emulsion
Yellow filter
Green-sensitive emulsion
Red-sensitive emulsion
Antihalation layer
Base

Figure 3-11 Cross section of color film.

layer is sensitive to, and absorbs, the green light, allowing the remaining red light to continue on to the red-sensitive layer.

For "false color" infrared film, the spectral sensitivities of the layers are shifted toward the infrared, so that a layer sensitive to infrared replaces the red-sensitive layer, a red-sensitive layer replaces the green-sensitive layer, and a green-sensitive layer replaces the blue-sensitive layer. This film must be used with a yellow filter so that blue light, which would fog the layers, is not admitted.

Films are classified as either *negative* or *reversal*. A negative film produces an image in which the tone values of the scene are reversed—a light area in the scene will be dark in the image and vice versa. A reversal film produces a positive image directly, in which light areas in the scene are light in the image. In color negative films, colored areas in the scene are shown in their complementary colors: blue regions are shown in yellow, green regions in magenta, and red regions in cyan.

Once film has been exposed to light and a latent image has been formed, it must be processed. Black-and-white film is first put into a *developer,* a solution that reduces each of the silver halide crystals in the latent image to metallic silver, thereby producing a visible image. The amount of time the film is left in the developer must be carefully regulated in order to control the densitometric properties of the film (Section 3.3.2). The film is then placed in a *stop bath* to halt the development reaction, and then into a *fixer* solution, which removes the unexposed silver halide crystals so that the film will no longer be sensitive to light. After this, the film is washed to remove any chemical residue and then dried.

When color negative film is developed, both a negative silver image and a negative dye image are formed. The silver image is formed analogously to the black-and-white film, recording the illumination for each layer of the emulsion for the color to which the layer is sensitive. The dye image is then formed at the same locations as the silver image, translating the density of the silver image into a colored area. After a bleaching step to remove the silver and the yellow filter layer, the film is fixed, washed, and dried, leaving a color negative transparency.

Processing of color reversal film requires a more involved procedure. The first step is developing with a black-and-white developer to form a negative silver image in each layer. The film is then fogged, or illuminated by a bright light, to expose all of the silver grains not originally exposed. The film is then developed again in a color developer, which forms a silver image and a dye image. From this point, the processing is the same as for the color negative: a bleaching step to remove the silver, fixing, washing, and drying.

3.3.2 Sensitometry

Sensitometry is the study of how photographic emulsions respond to light. The properties we are most concerned with are how dark the film becomes in response to

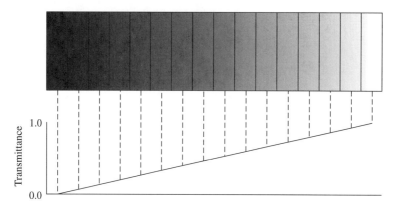

Figure 3-12 Density step wedge.

incoming light and how quickly the film responds. Several measures have been defined to quantify these concepts.

The transparency or *transmittance, T,* of a film is the ratio of the transmitted light to the incident light.

$$T = \frac{I_{\text{trans}}}{I_{\text{incident}}} \qquad (3\text{-}10)$$

The *opacity* of a film is the reciprocal of its transmittance and the *density, D,* is the logarithm of opacity

$$D = \log_{10} \frac{1}{T} \qquad (3\text{-}11)$$

We want to relate the density of an image to the amount of light it has received, in order to determine the optimum exposure to achieve a given density. We define the *exposure, H,* as the product of the illuminance, *E* (in lux or foot-candles), and the time, *t.*

$$H = Et \qquad (3\text{-}12)$$

To evaluate a given photographic material, we expose it for a known time to light of known intensity passed through a series of different known densities, such as the step wedge shown in Fig. 3-12. The resulting image of the step wedge on the film is then measured with a *densitometer,* which records the density or transmittance of the film. Using the known density of each region of the original step wedge and the known exposure time, the exposure, *H,* can be calculated for each region of the film. By plotting the film density against the logarithm of the exposure, we obtain the *D* log *H* or *characteristic curve* for the film. This is also known as the Hurter-Driffield curve, after the researchers who introduced it.

A typical H-D curve for a negative film is shown in Fig. 3-13. We know that we are looking at negative film by the fact that the curve slopes upward to the right, which means that increasing the exposure results in a denser (darker) image.

For analysis purposes, the H-D curve is typically divided into four regions. The beginning of the curve (point 1 in Fig. 3-13) marks the minimum density level. This consists of the density of the base material and fog, which is due to the density of the emulsion and any undeveloped material. Exposures in the region of the toe (between 1 and 2) are underexposed, meaning that the image is very light.

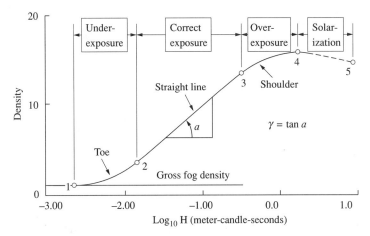

Figure 3-13 Typical Hurter-Driffield (H-D) curve.

We want the film's exposure to fall within the main part of the curve, the straight line from the toe to the shoulder of the curve (from 2 to 3 in Fig. 3-13). In this region a change in exposure results in a linear change in density, giving us an image with good contrast between areas with similar shades. The slope of the curve in this region is known as the *gamma* of the film,

$$\gamma = \frac{\Delta D}{\Delta(\log H)} \tag{3-13}$$

which gives a measure of the *contrast*. A film with a higher gamma will show more density difference between two regions, however, the *dynamic range* of the film will be reduced and some areas may appear as completely black or white.

The shoulder of the curve (from 3 to 4 in Fig. 3-13) is the region of overexposure; in this case, the image will be very dark and no detail will be visible. Past point 4, solarization occurs and the image is destroyed.

A "faster" film, with a numerically higher film speed, gives the same density for less exposure, effectively shifting the H-D curve to the left. Similarly, a "slower" film has an H-D curve shifted to the right.

3.3.3 Electro-optical Sensors

While photographic film uses the incident electromagnetic energy to induce a chemical reaction, electro-optical sensors convert the incident photons into electrons. Depending upon the type of sensor, this electron stream may form the image in an analog manner or be converted to a digital form.

Electro-optical sensors have become much more widely used in recent years, due mainly to the decreased manufacturing costs and increased resolution made possible by advances in semiconductor manufacturing techniques.

There are two main types of electro-optical sensors: imaging tubes and semiconductor sensors. *Imaging tubes,* such as *vidicon* tubes, form an image on a light-sensitive screen and then scan the screen with an electron beam. The varying current of the electron beam is related to the varying intensities of the image. Solid-state *semiconductor sensors* have replaced imaging tubes for most purposes, due to their lower production cost, greater robustness, and, especially for photogrammetric applications, their greater geometric stability.

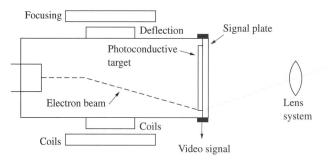

Figure 3-14 Vidicon camera tube.

3.3.3.1   Imaging Tubes

Imaging tubes were the first electronic imaging sensors. Several types of imaging tubes have been developed, but all share the same basic construction: a screen with a coating that varies its electrical properties in response to light and an electron beam that scans the screen and produces the image signal.

For example, the vidicon (Fig. 3-14) consists of a glass vacuum tube. At one end is the target, a transparent plate with a photoconductive coating on the inside. At the other end of the tube is the electron gun, which produces the electron beam. The focusing and deflection coils, which control the electron beam, are mounted around the tube.

The electron beam is accelerated, focused, and deflected to scan the target in a raster pattern. Light from the scene is focused by the lens onto the target screen, which is maintained at a constant charge level. Since the coating on the target screen is photoconductive, the conductivity and therefore the charge on the target varies in proportion to the illumination from the scene. As the beam scans the target, it replenishes the charge lost by the target due to the illumination. This causes the voltage of the beam to change, producing a signal which can be used to generate an image on a display.

Several other types of imaging tubes have been developed, such as the image orthicon. They have been replaced for most applications by semiconductor sensing arrays, especially for photogrammetric applications. The resolution of imaging tubes is limited and they are geometrically unstable, due to the difficulties in precisely controlling the beam deflection.

3.3.3.2   Semiconductor Sensors

The growth of the semiconductor industry has made possible the development and widespread commercial adoption of semiconductor sensors. Such sensors are ubiquitous in consumer video products and are beginning to replace film cameras for amateur photography. The larger arrays and smaller pixel sizes necessary for aerial imaging applications are now becoming available.

Semiconductor sensors are available both as linear arrays, which have one line of pixels, and as two-dimensional area arrays. Linear arrays are easier to manufacture than area arrays, since fewer pixels are involved.

The majority of semiconductor sensors are charge-coupled devices (CCDs). A CCD sensor (Fig. 3-15) consists of a set of photosensitive pixels and an associated shift register, along with the necessary electrodes to apply the appropriate clocking voltages. The photons striking each pixel generate a quantity of electrons proportional to the amount of incident light, due to the *photoelectric effect*. Each pixel location on the array corresponds to a potential well, which attracts the electrons. The position of

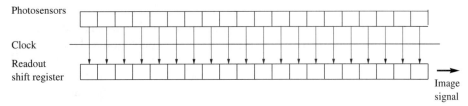

Figure 3-15 Construction of a linear CCD chip.

this well can be changed by manipulating the clock voltages acting upon each pixel; by moving the potential well, the packet of electrons is also moved along. In an imaging sensor, the charge is transferred from each pixel into an adjacent shift register, then the charge is passed along the shift register to an output terminal. The sequence of output electrical charges gives rise to a waveform that can be displayed or digitized for computer processing.

CCD chips are usually characterized in terms of the number of pixels on the chip and the size of each pixel. A chip with more pixels can be used to give a higher-resolution image or to cover a wider field of view at the same resolution. However, a chip with more pixels is more expensive to manufacture since there is a greater probability of bad pixels and therefore lower manufacturing yield. The majority of sensors have square pixels, although some are manufactured with rect-angular pixels.

The chip material and pixel size determine the chip's sensitivity to light. A larger pixel can integrate more light, therefore generating a larger charge and improving the signal-to-noise ratio. However, having a larger pixel size decreases the resolution un-less the optics are changed accordingly.

The spectral sensitivity of the sensor is determined by its materials. The silicon used in most visible-light sensors is actually sensitive into the near-infrared (Fig. 3-1); this infrared capability improves its performance for aerial imaging in hazy condi-tions. Sensor chips designed for other spectral regions use other substances, such as indium-tin (InSb) for 2 to 5 micrometer infrared or mercury-cadmium-telluride (HgCdTe) for 8 to 14 micrometer infrared.

The performance of a CCD chip depends on a number of factors. One of the most important is the *quantum efficiency,* the percentage of incident photons actually con-verted to electrons. For silicon, this is typically from 10 to 20%. Once the electron packets are generated, they must be transferred through the shift register to the outputs at the edge of the chip. The percentage of electrons successfully transferred at each step is known as the *charge transfer efficiency,* and must be very high (> 0.99999) due to the number of transfers involved for large chips.

Noise sources in the imaging process must also be considered. Thermal or quantum noise in the chip material generates the *dark current,* a continual current of electrons independent of incident light intensity and dependent on the temperature. This thermal noise can be reduced by cooling the chip, and this is often done in high-performance applications. Deviations in clock timing also introduce noise in the output signal.

3.3.3.3 Image Recording and Storage

Image storage is not a consideration for film cameras, since the film itself is the storage medium. However, with electro-optical sensors, recording and storing the large amounts of data generated are significant problems.

The signal from an electro-optical sensor is an analog electrical signal, so variations in the radiation from the scene are indicated by variations in the electrical signal. This signal can be displayed on a screen and then photographed on film, although this removes many of the advantages of electronic imaging. The signal can also be recorded in analog form, such as on a videocassette recorder. The best recording method in terms of data fidelity is to perform analog-to-digital conversion on the signal and record it in a digital form. However, the amount of digital data produced requires high-performance processing electronics and high data-rate, large-capacity recording equipment.

Once the data are recorded, the storage and handling requirements are enormous. Consider an area of 100 km^2 imaged at a ground resolution of 0.5 meters. Single image coverage of this area will result in 400 million pixels, while stereo coverage will more than double this. Increasing the ground resolution to 0.25 meters quadruples the storage requirement.

3.4   IMAGE QUALITY

The quality of an image directly affects how well we can measure or interpret the image. For this reason, well-defined criteria have been established for predicting and evaluating the performance of lenses and camera systems.

3.4.1   Resolution

In simplest terms, *resolution* can be thought of as the ability to distinguish two point sources of light on a black background. This is precisely the problem faced by astronomers, who developed some of the first measures of resolution. A point source of light imaged by a lens with a finite aperture will appear as a bright spot surrounded by a set of concentric light and dark rings (Fig. 3-16), due to *diffraction* at the lens aperture.

Diffraction is the spreading of light waves at the edge of an obstruction, a consequence of the wave nature of light. Due to the scattering at the obstruction, the wave propagates into the shadowed region. Diffraction occurs at the lens aperture. If the lens is well-corrected for aberrations, diffraction will become the ultimate limit on resolution and the lens resolution is said to be *diffraction limited*.

The diffraction pattern arising from a point source is known as the *Airy disk*. The radius of the first dark ring in an Airy disk can be calculated from the aperture diameter, d, the focal length, f, and the wavelength, λ, of the incident light.

$$r = 1.22\lambda\frac{f}{d} \tag{3-14}$$

The Rayleigh criterion for judging resolution, one of several possible criteria, states that two point sources are resolved if the bright central spot of one falls in the first dark ring of the second, which is equivalent to saying that the two disks are separated by a distance equal to their radius.

However, we typically deal with images more complex than point light sources, and must define more complicated quality measures. We can still talk about the resolution of an image, but must be more careful in defining what we mean. Perhaps the best known quality measurement is image resolution, expressed in line pairs per mm. (A line pair is one black line and one white line.) However, merely quoting a resolution number by itself means little; a number of factors affect the resolution of an image, each of which must be specified before a resolution measurement can be understood.

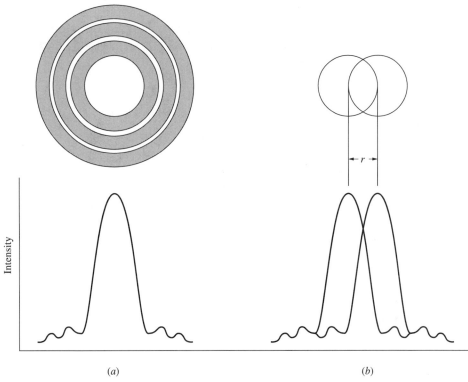

Figure 3-16 (*a*) The Airy disk, formed as a result of the diffraction of light from a point source. (*b*) Resolution measurement based on the Airy disks for two point sources.

The resolution target can consist of alternating white and black bars or intensities varying as a sine wave. One widely-used laboratory standard, the Air Force 1951 target, has sets of three black bars with spacings between the bars varying as a geometric series based on the sixth root of two. A higher-contrast target will give a higher resolution reading, but low-contrast targets are more representative of real-world imaging conditions.

For film cameras, the type of film and the developing procedure must be specified, since film can be developed to emphasize absolute contrast over the amount of gray scale modulation.

The resolution of a lens changes across the field of view and also depends on the orientation of the target. Therefore, resolution is usually measured in several annular zones outward from the principal point, with the target in radial and tangential orientations. The individual measurements in each zone are combined, weighted proportional to the area of the zone, to calculate the area-weighted average resolution (AWAR). This gives a better idea of the overall resolution of the lens.

One of the most important factors in specifying resolution is whether the tests were conducted in laboratory or field conditions. A number of external factors, such as atmospheric haze, camera vibration, and lighting conditions, affect camera performance in field conditions. The resolution measurements obtained by evaluating an actual photograph are therefore seldom as high as those obtained in the laboratory.

Resolution measurements for solid-state sensors are complicated by the fact that the image is formed from discrete samples, the pixels, instead of a relatively continuous medium such as film. The upper limit on resolution for imaging arrays is determined by the *Nyquist sampling theorem,* which specifies that a signal must be sampled at twice its highest *spatial frequency* in order to be reconstructed. Spatial frequency is the reciprocal of the wavelength. For instance, a sine wave with a wavelength of 0.1 mm has a spatial frequency of 10 cycles/mm. A sensor with a pixel size of 10 micrometers could therefore resolve, at most, a sine wave with a wavelength of 20 micrometers, or a spatial frequency of 50 cycles/mm. This theoretical limit is seldom achieved in practice.

3.4.2 Modulation Transfer Function

While the *modulation transfer function* (MTF) is not in itself a measure of image quality, it provides a powerful method for predicting and modeling the quality of images produced by an imaging system under varying environmental conditions. A complete introduction to the modulation transfer function would require the introduction of linear systems theory and Fourier optics, which is beyond the scope of this book. Instead, we will provide a qualitative introduction to the concept and refer the reader to the references for more rigorous details.

If we use a sine wave as the resolution target, the image will also be a sine wave with the same spatial frequency. However, the amplitude of the wave will be reduced, since light from the peaks of the sine wave will be spread (blurred) into the valleys, reducing the overall modulation or difference in amplitude between the peaks and valleys. This is simulated in Fig. 3-17. As the spatial frequency of the sine wave approaches the resolution limit of the system, the modulation will continue to decrease, until at some point the image of the sine wave has a uniform intensity.

Now, suppose that we image a set of sine waves of different spatial frequencies and then measure the ratio of the modulation of the image of each sine wave to its input modulation. These measurements define the modulation transfer function. Figure

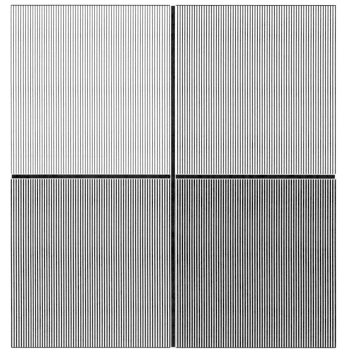

Figure 3-17 Effects of decreasing modulation on sine wave contrast. For 8-bit images (0–255 gray levels), modulation is 127, 64, 32, and 16 gray levels.

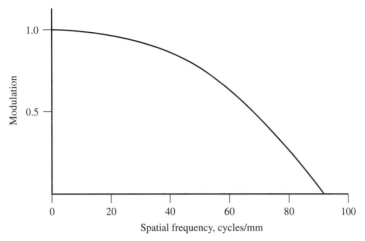

Figure 3-18 Typical MTF curve.

3-18 shows an example of an MTF curve, with modulation (between 0 and 1) plotted as a function of spatial frequency. Note that the modulation decreases as the spatial frequency increases.

The main advantage of MTF characterization is the ability to cascade (multiply in series) the MTF curves of various system components and imaging factors. Since a sinusoidal input yields a (modulated) sinusoidal output, we can use the output of one component as the input for the next component. For example, consider a system with only three components: the atmosphere, the lens, and the film. (Film can be described by its MTF as long as the exposure is within the linear part of the H-D curve.) Atmospheric MTF is determined by the amount of haze present and the length of the path through the atmosphere, and the lens MTF is determined by calibration. We multiply together the MTFs for the three system components, by multiplying the values at each frequency represented, to obtain the system MTF. Now, given an assumed target frequency and modulation, we can multiply this by the system MTF to determine the modulation of the resulting image. If we change our assumptions on any factor, we can simply recalculate the system MTF and compute the effect on the image.

3.5 IMAGING GEOMETRIES

When we view an image, we usually think of it as having been obtained simultaneously, as with a frame camera. However, the introduction of electronic sensors with other geometries and the increase in computer power to support analytical photogrammetry and digital imagery has motivated the development of rigorous photogrammetric models for many nonframe sensors. This section discusses the types of imaging geometries available and some of the considerations in using them.

3.5.1 Point Sensors

A point sensor images only a single point at any one instant of time. This has been a common configuration for electro-optical sensors in nonvisible bands.

One example is the airborne multispectral scanner, in which the pixels within each line of the image are generated by scanning in the cross track direction with mechanical motion, as shown in Fig. 3-19. Each new image line is then generated by the platform motion. This combined side-to-side and forward motion gives rise to the

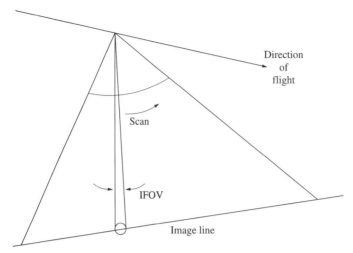

Figure 3-19 Point scanner geometry.

name *whiskbroom sensors.* The scanning speed must be carefully synchronized to the platform velocity, since each scan must be completed in the time required for the platform to move forward one pixel. The internal image geometry is adversely affected by platform motion during the scan.

Since only a single point is imaged, the optics can be much smaller and simpler in a point sensor than in a standard frame camera. This is especially important for non-visible wavelengths, where elaborate reflecting optical trains or special glasses must be used. However, this must be balanced against the additional mechanical complication required for scanning and the decreased integration time compared to linear or area detector arrays.

3.5.2 Line Geometry

In the line imaging geometry, the image is built one line at a time. There are two common configurations, *pushbroom* and *panoramic.*

3.5.2.1 Pushbroom Geometry

In the pushbroom configuration, the line sensor is oriented perpendicular to the platform's motion (Fig. 3-20). This is a popular configuration for solid-state sensors, since linear arrays are easier to build than area arrays and no mechanical scanning is required. This geometry is used in reconnaissance cameras and also in many orbital sensors, such as the SPOT satellite.

Strip cameras use this configuration, with film as the sensing medium. The film is moved across the imaging slit at a speed determined by the platform speed and the scale of the imagery.

The perspective geometry in this case is interesting, since a pushbroom image has perspective only in the direction along the sensor (Fig. 3-21). The image is orthographic in the direction of flight.

The geometric strength of pushbroom images is poor, since each line is an independent image and has its own position and orientation due to platform motion. This can be ameliorated somewhat by the use of navigational sensors such as Global Positioning System (GPS) or Inertial Navigation System (INS), and also by using combi-

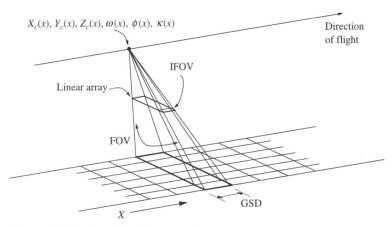

Figure 3-20 Schematic of linear pushbroom sensor.

Figure 3-21 Portion of an image from a linear pushbroom sensor. Courtesy of Lockheed Martin Fairchild Systems.

nations of linear sensors. For instance, the MOMS-02 camera uses three linear sensors, one pointing vertically and two others pointing forward and backward at 21.9 degrees off vertical. The tilted sensors image each ground point in the area of coverage three times, from three different look angles, yielding much stronger geometry for point determination.

   The platform velocity determines the sampling rate of the detector, or the speed with which the film is drawn past the imaging slit. While the sensor can be aimed to

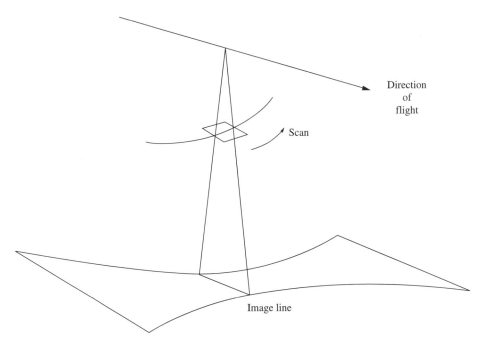

Figure 3-22 Schematic of panoramic camera geometry.

the side, the apparent ground velocity will vary across the sensor and cause over- or under-sampling.

3.5.2.2 Panoramic Geometry

Another way to utilize a linear imaging geometry is in the panoramic configuration. The imaging line is oriented parallel to the direction of flight (Fig. 3-22) and then scanned perpendicular to the flight path. In this way a very wide field of view can be imaged, even horizon to horizon.

Panoramic cameras were first developed for reconnaissance applications, to give high resolution and wide coverage without using a very large and complex lens (Fig. 3-23). A panoramic camera lens needs only a narrow field of view and is therefore easier to design for high resolution and a large aperture. The scanning rate is determined by the platform velocity and altitude and the camera field of view in the direction of flight. The scan must be completed in the time it takes for the platform to cover the field of view in the direction of flight. For stereo coverage, the scan must be completed in half the time; stereo coverage is therefore often difficult to obtain with panoramic cameras.

In some reconnaissance applications, especially with electro-optical sensors, only a small field of view is scanned at depression angles near the horizon. This long-range oblique viewing allows the reconnaissance aircraft to remain at a distance from the target.

The geometry of panoramic cameras is extremely complicated, due to several factors. The photo scale changes drastically across the image, since the object distance changes from the flying height at nadir to infinity at the horizon. A uniform grid imaged by a stationary panoramic camera would therefore appear as in Fig. 3-24. In addition, a panoramic image of a grid on the ground is further distorted by the motion of the camera during the scanning (Fig. 3-25). Another distortion is caused by the image

Figure 3-23 Portions of an image from a panoramic film camera. Courtesy of Recon Optical, Inc.

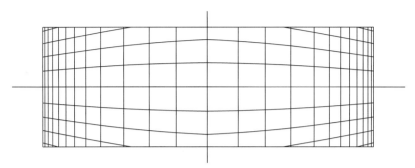

Figure 3-24 Image distortion due to panoramic scan.

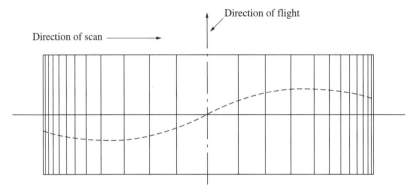

Figure 3-25 Image distortion due to motion of panoramic camera during scan.

motion compensation (IMC) mechanisms used to prevent image blur. As discussed in Section 3.6, these work by moving the lens or the sensor, and in so doing, affect the imaging geometry.

3.5.3 Area (Frame) Imaging Geometry

The frame geometry allows the simultaneous imaging of a large area. The geometry is relatively simple, compared to the dynamic sensors discussed previously, since all points in the image are imaged at the same time and with the same perspective center.

Photogrammetric mapping has for years used the frame mapping camera; a detailed discussion of its construction is given in Section 3.7.

3.6 IMAGE MOTION COMPENSATION

Imaging from a moving platform results in blurred images due to the finite exposure or integration time. To prevent blurring, the film or sensor must be moved at the same (scaled) velocity as the platform. For a vertical photograph with focal length f taken from a platform at altitude H with velocity V, the scaled velocity is

$$v = f\frac{V}{H} \tag{3-15}$$

If we assume a vertical frame photograph, then all parts of the image will have the same velocity. Oblique imaging angles and dynamic sensors pose much more complicated problems (Hotchkiss, 1983).

Image motion compensation adds significant mechanical complication to a sensor. Since other parts of mapping camera systems could be optimized more efficiently, until recently, image motion compensation was mostly applied to reconnaissance systems, which employ high-speed, low-altitude flights and panoramic imaging geometries for which motion compensation is necessary. However, many mapping cameras now have image motion compensation to allow for imaging with slower films (which require longer exposure times), at low altitudes, or under turbulent conditions. Experimental electro-optical area arrays have been developed to implement nonuniform image motion compensation for oblique images.

EXAMPLE 3-5   *Image Motion Compensation*

A mapping camera with a 152 mm focal length is flown at 500 m above terrain at 300 km/hr. Assuming that the camera axis is vertical, what is the apparent image velocity?

$$v_i = 152 \, \frac{300,000 / 3600}{500} = 25.333 \, \text{mm/s}$$

If the exposure time is 1/500 second, how far will the platen have to move to remove any motion blur?

$$25.333 \, \frac{1}{500} = 0.0507 \, \text{mm}$$

3.7   CONSTRUCTION OF A FRAME MAPPING CAMERA

A standard aerial mapping camera has the same four main parts as any frame camera: the *lens,* the *shutter,* the *magazine,* and the *body.* The main differences between standard frame cameras and aerial mapping frame cameras are that the components of a mapping camera are designed to much higher specifications than those of consumer cameras, and a mapping camera is designed to maintain its high geometric precision under a much wider range of operating conditions. For instance, the lens of a mapping camera has extremely high resolution across the format, has almost zero distortion, and is highly color-corrected for optimal image quality; and the lens cone and camera body are designed to resist deformations due to temperature variations or vibration.

A mapping camera installation involves several different components. Figure 3-26 shows the equipment for a standard installation, including the camera in its stabilized mount, the navigation sight used for visually setting and verifying exposure station positions, and the camera control computer. Not shown are the GPS and/or INS navigation sensors, which are now a standard part of mapping camera installations.

3.7.1   Lens

The lens in a modern mapping camera is a complex assembly of multiple optical elements (Fig. 3-10), highly corrected for aberrations and distortion. Current lenses can achieve an AWAR of 120 lp/mm, with radial distortion less than 10 micrometers.

The standard wide-angle mapping lens has a nominal 6 inch focal length which, when used in conjunction with the standard 9 by 9 inch (23 by 23 cm) film format, yields a field of view (FOV) from corner to corner of approximately 90 degrees. Other commonly used focal lengths for aerial cameras are 12 inches (narrow-angle,

Figure 3-26 Frame mapping camera system, including camera, mount, navigation sight, and control computer. Courtesy of LH Systems, LLC.

60-degree FOV), $8\frac{1}{4}$ inches (normal-angle, 75-degree FOV), and $3\frac{1}{2}$ inches (super-wide-angle, 130-degree FOV). Standard mapping lenses are always fixed focus, set at infinity.

Although the filter is not usually part of the lens assembly, aerial cameras are seldom operated without one. Most often this is a yellow (minus blue) filter, to remove the slight blue atmospheric haze. This may also be an anti-vignetting filter, which is slightly darker at the center to even out the illumination across the format of wide-angle lenses (see Section 3.2.6).

The lens is mounted in the *lens cone* (Fig. 3-27), which physically defines the image's *interior orientation,* the focal length and the position of the principal point. The *focal plane* determines the position of the film during exposure for the best image quality.

Within the focal plane are the *fiducial marks,* four or eight well-defined marks located in the corners and/or on the sides of the frame. The positions of the fiducial marks are adjusted during assembly so that the intersection of the lines between opposite marks is as close as possible to the principal point of the lens. The exact locations of the fiducial marks with respect to the principal point are determined during the camera calibration procedure.

In addition to fiducial marks, some cameras have *reseaux,* calibrated markings on a regular grid which are imaged onto the negative itself. These allow the precise cor-

Figure 3-27 Frame mapping camera lens cone. Courtesy of LH Systems, LLC.

rection of film distortion in each small area of the image, instead of only relying on the fiducial marks at the edges, which can give only an estimate of the average distortion across the frame. The reseaux may be etched on a glass plate in front of the film or back-projected onto the film with small lights mounted in the film platen.

The focal plane may also contain the *data strip* (Fig. 3-28), which includes information such as the time, date, altitude, camera serial number, and photo number. This information may be projected by lenses onto the film, or may be directly recorded by LEDs. In modern computer-controlled cameras, this information is also recorded in a computer file for later processing.

The aperture of the camera is controlled by the lens *diaphragm,* which adjusts to control the amount of light entering the lens.

Figure 3-28 Data strips recorded on film borders by a mapping camera. The top strip shows position, altitude, time, and date, while the bottom strip shows exposure and IMC information. Courtesy of LH Systems, LLC.

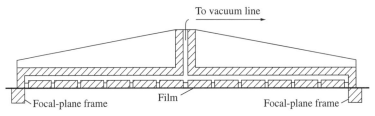

Figure 3-29 Mapping camera vacuum platen.

3.7.2 Shutter

The *shutter* controls the film exposure time. Shutters are designed as either focal-plane shutters, which are located at or near the focal plane, or between-the-lens shutters, which are located within the lens itself. A between-the-lens shutter can expose the whole image simultaneously, since the bundle of rays within the lens is relatively small. A focal-plane shutter has to uncover the whole image format, which takes longer. Since mapping photography is taken from a moving platform, mapping cameras use only high-speed between-the-lens shutters.

3.7.3 Film Magazine

The film is held in a removable *magazine,* which also contains the mechanisms to advance and meter the film and to flatten the film during exposure. Magazines can be interchanged while the camera is mounted in the aircraft, allowing a single mission to involve a large number of exposures or different types of film.

Film flatness is crucial in maintaining the geometric accuracy of the imagery. To accomplish this, the film magazine in a mapping camera contains a vacuum platen. The *platen,* or film backing plate, has a network of grooves in its surface connected to internal tubes, which are connected to an external vacuum source (Fig. 3-29). After the film is in position for exposure, the application of vacuum pulls the film flat against the platen. After the exposure, the vacuum is released, allowing the film to be advanced.

Some aerial cameras have used glass plates to flatten film and avoid film deformation. However, advances in film materials and in platen design have made such cameras unnecessary.

In some cameras, the magazine also implements image motion compensation by moving the film platen during exposure.

3.7.4 Body

The body of a mapping camera holds the other components, the motor for the film advance and shutter, and has the mounting points for attachment to the camera mount. It is designed to isolate the lens cone and magazine from excess vibration.

3.7.5 Camera Control

The exposure (shutter speed and aperture setting) must be set according to the prevailing lighting conditions. Most modern cameras can set this automatically.

The positions of the exposure stations must be carefully controlled to obtain the proper overlap and sidelap between adjacent photos. This can be done using a

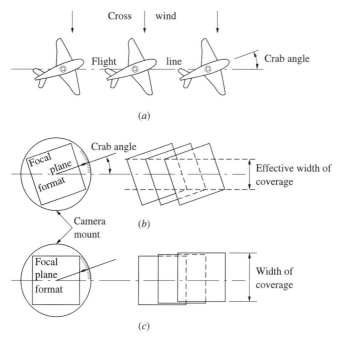

Figure 3-30 "Crab" along flight line.

viewfinder in which a moving cross hair is superimposed on the moving image of the ground. Once the cross hair and the image of the ground are synchronized, the desired overlap can be set and the camera automatically triggered at the proper points.

The most accurate method of controlling exposure station position is the use of the Global Positioning System (GPS, see Chapter 5). With a computer camera controller and a GPS interface, the operator enters the desired coordinates of each exposure station into the control computer, which monitors or controls the position of the aircraft and triggers the camera at the appropriate points.

3.7.6   Camera Mount

The camera mount contains isolation bushings to insulate the camera from vibration and also controls the attitude and heading of the camera. The attitude and heading may be manually adjusted or, in some modern mounts, gyroscopically controlled.

Adjustment of the camera heading is particularly important. Due to crosswinds, the aircraft nose may not be pointing in the direction of flight. If the camera is aligned with the aircraft instead of with the actual flight path, the photos will be rotated with respect to the flight line, or "crabbed." As shown in Fig. 3-30, this reduces the effective width of the coverage, since the photos no longer overlap along their entire width.

Additionally, many modern camera mounts implement angular motion compensation, moving the camera during the exposure to correct for the blurring effects of changes in the aircraft attitude.

3.8   CAMERA CALIBRATION

We can think of camera calibration as having three aspects: geometric calibration, image quality evaluation, and, in some cases, radiometric calibration. Calibration is usually done in a laboratory, although field procedures may also be used. In some cases, the calibration may actually be done as part of the project; this is known as *self-calibration*

(Chapter 5). For electronic sensors, we must think of system calibration, which includes the image digitizer and other data-transfer interfaces.

3.8.1   Geometric Calibration

Geometric calibration establishes the *interior orientation* parameters of the camera. These include the location of the principal point, the focal length, and the radial and tangential distortion. We use the *fiducial marks* imaged on each photo to relate an image point to this calibrated geometry. Geometric calibration determines the locations of the fiducial marks in relation to the principal point and verifies their geometry.

Laboratory calibration can be done using either a *goniometer* or a *multicollimator*, both of which allow the measurement of the angles between the optical axis and light rays entering and leaving the lens. The lens distortion is defined by the change in this angle (Section 3.2.8).

3.8.1.1   Camera Calibration with a Multicollimator

A multicollimator (Fig. 3-31) consists of a set of discrete targets at known angular displacements from a reference direction. Each target is a *collimator,* a light source

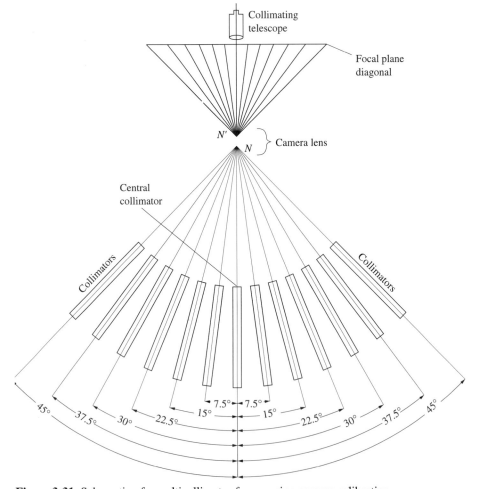

Figure 3-31 Schematic of a multicollimator for mapping camera calibration.

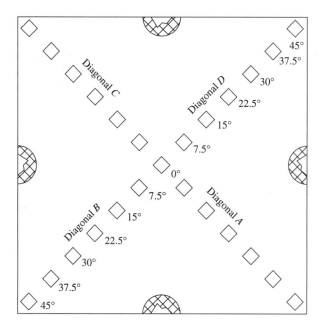

Figure 3-32 Typical image from a camera multicollimator.

beamed through a telescope focused at infinity. The image of the telescope reticle appears to the camera as if it were coming from infinity and is focused onto the focal plane by the lens of the camera being calibrated.

To calibrate a camera using a multicollimator, the camera is first positioned so that its entrance pupil is at the point where the collimator axes converge. Next, the optical axis of the camera is aligned with the reference axis of the multicollimator using autocollimation techniques and a collimator at the reference axis. In autocollimation, a reflecting plate is placed at the camera's focal plane and the camera is adjusted until the autocollimation telescope cross hairs align with their image reflected from the plate.

Once the camera is aligned, a photographic plate is placed in the camera and an exposure made of the collimator array. The resulting image (Fig. 3-32) has lines of targets along the diagonals, along with images of the fiducial marks. Using the measured distance to the first target out from the principal point, d, and the known angle to that collimator, α, the equivalent focal length, f_e, can be calculated.

$$f_e = \frac{d}{\tan \alpha} \tag{3-16}$$

Since four distances are available for each angle (one along each diagonal), an average focal length can be calculated. Given this value of the focal length, we can then calculate the distortion at the other targets by comparing their measured distance from the center d_i averaged over the four targets to the theoretical value d_i' calculated from their angular position, α_i, and the equivalent focal length f_e.

$$r_i = d_i - d_i' = d_i - f_e \tan \alpha \tag{3-17}$$

Note that the radial distortion is positive if the measured radial distance is greater than the calculated distance. If tangential distortion (Section 3.2.8) is present, the sets of collimator images deviate from a straight line. The tangential distortion function is calculated from these deviations.

The principal point of autocollimation (PPA) is located by the image of the central collimator. If the lens and camera were perfectly assembled, the PPA would coincide with the foot of the perpendicular to the focal plane through the rear nodal point of the lens and the distortion curves would be symmetrical around that point. We define the *principal point of symmetry* (PPS) as the point of best symmetry of the distortion curves.

To recover the positions of the principal points, we reference them to the image coordinate axes defined by the fiducial marks. Precisely measured distances between the fiducial marks are used to define the coordinates for each fiducial mark.

3.8.1.2 Camera Calibration with a Goniometer

A *goniometer* (Fig. 3-33) directly measures ray angles at the perspective center, similar to a theodolite. A movable telescope measures the angles around the entrance pupil of the camera with respect to the optical axis of the camera. The camera to be calibrated is mounted in the goniometer with a calibrated grid plate in the focal plane in place of film. The camera is aligned using autoreflection techniques so that the focal plane is perpendicular to the telescope at the zero position. The grid plate is then shifted until the central grid point is aligned with the telescope's cross hairs at the zero position, thereby defining the central grid point as the principal point of autocollimation.

Figure 3-33 Goniometer for camera calibration. Notice camera lens cone hanging vertically above goniometer. Courtesy of LH Systems, LLC.

Once the camera is mounted and aligned, the angles to each intersection on the grid plate are measured. Using the distance to the first grid intersection, d, and the measured angle, α, the equivalent focal length, f_e, can be calculated from Eq. 3-16 in a similar manner as in multicollimator calibration. The distortion at other angular values can then be calculated using the equivalent focal length and the measured angles.

In order to document the location of the principal point of autocollimation with respect to the fiducial marks, the telescope is returned to the zero position and a photographic plate placed in the camera. The image of the telescope reticle is projected onto the plate along with the images of the fiducial marks. These can be measured on the developed plate and their relative coordinates established.

3.8.1.3 Stellar Camera Calibration

Camera calibration requires a number of well-defined targets at infinity whose positions are precisely known. The stars fit this description perfectly, and have in fact been used for camera calibration. A long exposure time is used so that the stars leave trails on the film. Given the camera position and the precise time, the directions of the stars can be calculated. These calculated positions are then corrected for atmospheric refraction and used with the measured coordinates of the stars to compute the distortion functions.

3.8.1.4 Self-Calibration

Another calibration option is *self-calibration*, in which the distortion and camera parameters are included as part of the bundle adjustment solution. This will be discussed in more detail in Chapter 5.

3.8.2 Evaluation of Image Quality

An important part of the camera calibration procedure is the determination of the quality of the images produced by the camera. The basic image properties are fixed by the camera design; the calibration testing verifies that the camera is working correctly.

As discussed in Section 3.4, image quality measurements include the imaging of resolution targets and the performance of MTF measurements on exposed images. Image quality measurements are performed under carefully specified conditions with standard targets and reported on the standard calibration forms.

3.8.3 Radiometric and Spectral Calibration

Radiometric calibration may be either relative or absolute. In relative radiometric calibration, the response of the pixels of an imaging array or a set of arrays to the same input radiance is characterized. This ensures that variations in observed intensity across the image are due to variations in the scene, not to variations in the sensor. In absolute calibration, the relationship between the output signal and the input signal is established so that, given an image of a scene, we can infer the scene radiance. For multispectral sensors, the relative or absolute response in each of the different bands must be calibrated. Additionally, the actual width of each of the spectral bands may be verified.

3.9 ACTIVE SENSORS

The sensor systems discussed thus far are *passive* systems, which record radiation reflected or emitted from the scene that originates from other sources, most often the sun. An *active* sensor, on the other hand, illuminates the scene and then records the returned energy.

The most common active sensor is *radar (radio detection and ranging),* which illuminates a scene with microwave radiation (wavelengths from 0.05 to 1 meter) and uses the reflected energy to estimate the distance and direction to each portion of the scene. Radar can penetrate clouds and haze and can be used at night, but usually has lower spatial resolution and is harder to interpret than optical images. Chapter 11 is devoted to the principles of active sensing systems and the exploitation of data obtained from them.

3.10 PLATFORMS FOR PHOTOGRAMMETRIC SENSING SYSTEMS

The standard platform for photogrammetric mapping has always been the airplane. In recent years, the number of types of platforms has expanded in two different directions. The availability of unmanned platforms, such as radio-controlled model aircraft, provides new opportunities for the acquisition of small-format photography. At the other extreme, the development of reliable satellite platforms and the declassification of high-resolution sensor technology have made 1-meter resolution images available from space.

3.10.1 Unmanned Platforms

Unmanned platforms include kites, blimps, balloons, and radio-controlled model helicopters and aircraft. These allow the acquisition of small-format (35-mm to 70-mm) photography over localized areas of interest. The camera may be mounted on a pendulum mount to keep the axis vertical, or mounted directly to the platform with bushings to isolate it from vibration. The camera may be triggered by remote control or by a timer.

Several considerations must be addressed when using unmanned airborne platforms. Standard flight planning parameters, such as the size of the area of interest, the image scale, and the stereo overlap if stereo is to be used, must be determined. It is difficult to precisely control the location and orientation of exposures, so more photographs than strictly necessary should be taken to avoid gaps in coverage. Resolution and image quality can be an issue; not only are the images being taken from a moving platform, but the combination of motor vibrations and the lightweight platform can produce image blurring. A practical concern is the risk of equipment loss or damage due to accidents during launch or landing.

The military has adopted several semiautonomous platforms, known as unmanned aerial vehicles (UAVs), for use in reconnaissance and mapping. The smaller versions of these are radio-controlled by an operator, but more sophisticated versions can execute preprogrammed mission plans, flying a specified path while imaging at fixed points. Imagery is sent back to the controller by a telemetry link, either by line-of-sight microwave or satellite relay. An example of a military UAV is the Air Force's Predator RQ-1A (Fig. 3-34). The Predator is 27 feet long and has a 48.7-foot wingspan, a top speed of 80 mph, and a range of 400 nautical miles. The Predator is controlled by a ground crew of four, consisting of a vehicle controller and three sensor operators. The sensors include a color camera used by the operator for control, TV and infrared cameras, and Synthetic Aperture Radar (SAR).

3.10.2 Aircraft for Aerial Photography

A wide variety of aircraft are used for mapping purposes, ranging from single-engine propeller craft to multiple-engine jets. The choice of a mapping aircraft is based on both technical and economic considerations.

Figure 3-34 Predator UAV. Imaging sensors are carried in the round pod under the nose of the vehicle. Photo courtesy of General Atomics Aeronautical Systems, Inc.

Technical issues concern the suitability of the aircraft for the acquisition of aerial photography. High-wing aircraft are usually more desirable because they provide better ground visibility and therefore ease of navigation, although most navigation is now done using GPS systems. A stable aircraft is important to reduce camera motion.

The feasibility and cost of modifying the aircraft to mount the camera or cameras must be studied. The camera is most often mounted in a hole cut in the aircraft floor. The hole must be large enough to not vignette any portion of the camera's field of view, allowing for its maximum possible tilt angles. In some aircraft, the maximum size of the hole is limited by the presence of structural members or control cables in the floor. In pressurized aircraft, a window of optical-quality glass must be installed over the camera port.

An alternate camera installation method used in some jet aircraft is to mount the camera in a modified door. The camera-mount door can be easily interchanged with the standard door, allowing a standard aircraft to be used as a part-time mapping platform.

Both the initial cost of the aircraft and the operating cost per hour affect the price of obtaining the imagery, and must be weighed against the productivity of the aircraft. A higher-speed twin-engine aircraft costs more than a single-engine airplane, but will spend less time in transit, cutting crew costs and allowing more area to be covered. The increased operating range and flight time will allow more jobs to be completed.

3.10.3   Satellite Imaging Platforms

Imaging from space presents a number of technical challenges in addition to the considerable difficulty of orbiting large payloads. The hostile environment of space would seem to be incompatible with the requirements for precision imaging, yet high-resolution panchromatic and multispectral imagery from space are routinely available today. The design of a satellite imaging platform begins with its mission. Whether the imagery will be used for mapping, remote sensing, weather, or another application determines the sensor requirements, such as number of bands and the ground resolution.

A first consideration is the satellite's orbit (Appendix F), which determines how much of the Earth's surface can be covered and how often the satellite will be able to revisit a location. A *sun-synchronous* orbit, in which the satellite always passes over any given point on its orbit at the same local solar time, is often used for remote sensing satellites. A satellite in a *geostationary* or *geosynchronous* orbit always remains in the same position relative to the Earth. Obtaining high-resolution imagery using a geostationary satellite is difficult because the satellite must remain at a distance of 35,800 km. Weather satellites are placed in geostationary orbits so that they can obtain continuous imagery of a large portion of the Earth, while communications satellites use geostationary orbits so that ground station antennas will not have to be continuously aimed.

The sensor geometry, as discussed in Section 3.5, can be frame, panoramic, or linear pushbroom. The Russian KOSMOS satellite carries the TK-350 frame film camera, along with the KVR-1000 panoramic camera. Most film sensors used panoramic geometries, as did the early Landsat MSS sensors. Nearly all current imaging satellites use linear pushbroom sensors.

Some satellites are able to steer the sensor in the cross-track and along-track directions, instead of its view direction being fixed relative to the satellite. This has two advantages: It expands the coverage area available from any particular orbit, effectively decreasing the time between overflights, and it enables the acquisition of stereo coverage. Stereo coverage may be obtained in cross-track or along-track configurations. In cross-track stereo, the sensor is steered to image the same area from two adjacent orbits, as shown in Fig. 3-35*a*. In along-track stereo, the sensor first images the scene ahead of the satellite nadir point, then is steered to image the scene again when it is behind the nadir point in a single orbit pass, as shown in Fig. 3-35*b*. In some cases a separate sensor is used for each view angle: one tilted forward, one tilted backward, and one at nadir.

The imagery produced on a satellite imaging platform must somehow be returned to earth. Satellites that use film eject a container containing the exposed film, which is (hopefully) recovered on the ground. Satellites with digital sensors may either beam the imagery to the ground as it is obtained or store it and download it later. Continuous downloading requires a ground station near any area where imagery will be desired, which can be logistically complicated and expensive. Systems that store imagery for later downloading need fewer ground stations but require large amounts of on-board storage; downloading a large amount of imagery in the short amount of time any particular ground station is visible also requires high transmission rates.

The earliest imaging satellites were for weather forecasting. The first TIROS satellites, launched in 1960, carried two video cameras, which transmitted pictures of cloud formations. The first high-resolution imaging satellites were designed for reconnaissance by the United States; their capabilities, and even their existence, were kept secret until 1995, when the CORONA program was declassified. The CORONA satellites, which obtained their first images in 1960, used panoramic film cameras, with the film returned to earth in separate reentry vehicles. The first images returned had resolutions of about 25 feet; later improvements gave resolutions as high as 6 feet. The first CORONA satellites carried a single camera, but later satellites carried dual cameras, which allowed stereo coverage. The final CORONA missions were flown in 1972. The film-based satellite systems have been replaced by reconnaissance satellites using electronic sensors. Current reconnaissance satellite capabilities are highly classified.

The Landsat series of satellites could be considered the first remote sensing satellites. The first Landsat 1 (initially called ERTS or earth resource technology satellite)

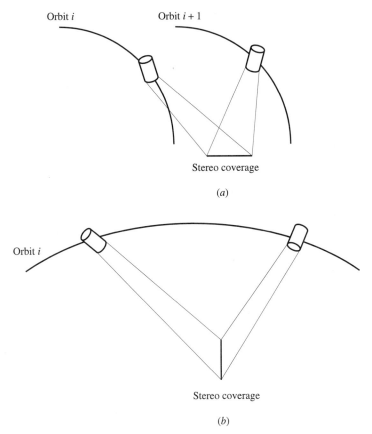

Figure 3-35 Satellite stereo imaging configurations. (*a*) cross-track. (*b*) along-track.

was launched in 1972. It carried a set of three return beam vidicon (RBV) cameras and a four-band multispectral scanner (MSS). While the performance of the RBV cameras and the associated videotape storage system degraded soon after launch, the MSS functioned for several years. The MSS was a panoramic line-scanning system, which scanned six lines simultaneously with each sweep of the imaging mirror. The ground pixel size was 68 meters in the cross-track direction and 83 meters in the along-track direction. Images were stored using on-board tape recorders and downloaded to ground stations. Landsat 1 repeated its coverage every 18 days. Later Landsat satellites (4 and 5) used the thematic mapper, an improved multispectral sensor with seven bands, one of which was an IR band, and a ground resolution of 30 meters.

Landsat was the first of several remote sensing satellites that have revolutionized earth observation. The SPOT satellite, launched in 1986, carried a four-band multispectral sensor with a 20-meter pixel size and a panchromatic sensor with a 10-meter pixel size.

While both Landsat and SPOT were originally funded as government projects, due to the high costs and technological risks involved, recent years have seen the launch of purely commercial high-resolution satellites for remote sensing and mapping purposes. These have been made possible by ongoing developments in satellite and launch vehicle technology, the availability of reliable commercial launch services,

Figure 3-36 One-meter GSD image of Washington, D.C., from the IKONOS satellite. Courtesy of SpaceImaging, Inc.

and continuing improvements in electronic sensors, data transmission links, and computer processing.

High-resolution mapping satellites have raised many issues in national and international security. Currently available imagery gives civilian users or potentially hostile countries access to imagery nearly equal in resolution to that from classified reconnaissance satellites. The ramifications of this are still being worked out. U.S. satellite licensing calls for "shutter controls" over some areas and countries, while other countries have differing policies.

As of this writing, the highest-resolution commercial imaging satellite in orbit is the IKONOS, launched by Space Imaging, Inc. in 1999. IKONOS collects 1-meter resolution panchromatic (Fig. 3-36) and 4-meter resolution multispectral imagery, as frequently as every 2.9 days.

It is not necessary to have a dedicated satellite to obtain imagery from space. For instance, the space shuttle has flown several remote sensing packages, including the large format camera, a calibrated mapping camera with a 9 by 18 inch format. The space shuttle also carried the shuttle radar topography mission package, an interferometric radar sensor designed to provide a digital surface model of 80% of the earth's land surface. The Soviet Mir space station was host to the MOMS-2P sensor, which contains five separate imaging objectives and linear arrays and can be operated in four different modes. By utilizing forward-, backward-, and downward-looking arrays, along-track stereo views can be obtained.

PROBLEMS

3.1 A ray of light in air strikes a plate of glass with index of refraction 1.48 at an incidence angle of 20 degrees. Compute the angle of refraction.

3.2 If the angle of incidence of Problem 3.1 is increased to 40 degrees, what is the corresponding angle of refraction?

3.3 A ray of light in water ($n = 1.333$) is directed toward a plate glass window in the bottom of a boat. The refractive index of the glass is 1.60. Compute the critical angle.

3.4 The apex of an optical wedge is 40 seconds of arc and the refractive index of the glass is 1.560. How far, in millimeters, will a monochromatic beam of light be deflected laterally at a distance of 10 meters after passing through the wedge?

3.5 A plane-parallel glass plate 8 mm thick ($n = 1.52$) is used to shift a ray of light parallel to itself. If the ray strikes the plate normally and passes through undeviated, through what angle must the plate be rotated in order to shift the ray 5 mm?

3.6 A simple lens is made of glass with an index of refraction of 1.55. The front surface radius is $+50$ mm and the back surface is planar. If we treat the lens as a thin lens, what is its focal length? If an object is 100 mm from the lens, where is the image formed? If we take into account the lens thickness of 15 mm, what is the focal length?

3.7 If a lens with focal length 50 mm has an aperture diameter of 25 mm, what is the f number?

3.8 The table below gives the MTFs for the atmosphere, lens, and film for a certain camera. Calculate the MTF for the overall camera system.

Spatial frequency, (cycles/mm)	Atmospheric MTF	Lens MTF	Film MTF
0	1.0	1.0	1.0
10	0.95	0.97	0.97
20	0.90	0.95	0.93
30	0.87	0.93	0.89
40	0.85	0.90	0.84
50	0.80	0.88	0.79

3.9 What are the three primary components of photographic film, and what is the function of each?

3.10 List some advantages and disadvantages of using glass plates instead of film to support the photographic emulsion.

3.11 Explain the difference between contrast and density.

3.12 List the variables that must be taken into consideration in order to obtain a photographic negative with the desired contrast and density.

3.13 List the sensitivity range of wavelengths in the electromagnetic spectrum for orthochromatic, panchromatic, and infrared emulsions.

3.14 Explain how the characteristic curve is established for a photographic film.

3.15 If a CCD array with pixels that measure 0.010 by 0.010 millimeters is flown at 2000 meters above the terrain and images the earth through a 50-mm lens, what is the nominal ground sample distance (GSD)? If the same CCD array is used in a panoramic configuration, what is the GSD in the along-track and cross-track directions at an angle of 30 degrees from nadir?

3.16 What are the differences between a cartographic camera and a reconnaissance camera?

3.17 A 6000-pixel linear pushbroom sensor with 0.010-mm square pixels and a 50-mm focal length lens is flown in an aircraft 3000 meters above the terrain at 900 km/hr. At what line rate must the sensor be read out so that the image is neither over-sampled nor under-sampled (the GSD is exactly equal to the aircraft's forward motion during the line sample time)? What is the width of the coverage strip?

3.18 Describe the effect of crab on a linear pushbroom sensor.

3.19 How much image blur is present in a frame camera with a 152-mm focal length, flown in an aircraft 1000 meters above the terrain at 300 km/hr?

3.20 A panoramic camera with a 12-inch focal length lens contains 5-inch wide film. The flying height is 5000 feet above the ground. If the camera axis swings in a vertical plane from side to side, calculate and plot the ground coverage out to 60 degrees on either side. Neglect earth curvature and the forward speed of the aircraft.

3.21 A cartographic camera contains a lens with a focal length of 152.25 mm and uses a 230 by 230 mm format. If this camera is retrofitted with a reseau plate 8 mm thick ($n = 1.584$), how far toward or away from the focal plane must the lens be moved in order to refocus the image on the focal plane? What is the approximate distortion caused by the glass plate at the corners of the format?

3.22 What is the approximate crab angle if the aircraft ground speed is 150 knots and a crosswind is blowing at 10 knots?

3.23 Assume that the vacuum failed during an exposure and the film buckled out 5 mm from the focal plane at a point whose x and y coordinates with respect to the principal point are 102.45 mm and 78.24 mm, respectively. The camera lens focal length is 151.80 mm. What are the magnitude and direction of the displacement of the point due to the buckling?

3.24 The distances to the 7.5 degree, 15 degree, 22.5 degree, 30 degree, 37.5 degree, and 45 degree marks, measured from the center cross along a diagonal out to the corner of a photographic plate during the calibration process, are 20.22, 41.177, 63.663, 88.726, 117.866, and 153.435 mm, respectively. Compute the focal length based on the distance to the 7.5 degree mark and compute the radial distortion at the remaining angular distances. Plot the distortion curve.

3.25 From the data in Problem 3.24, compute the individual focal lengths based on the distances to the 15, 22.5, 30, 37.5, and 45 degree marks. Compute and plot the individual distortion curves for the corresponding focal lengths.

3.26 A camera is calibrated in a goniometer with the following results:

Known distance to grid point (mm)	Measured angle	Known distance to grid point (mm)	Measured angle
0	0	84.853	28° 58' 31"
14.144	5° 16' 22"	98.995	32° 51' 56"
28.284	10° 27' 11"	113.137	36° 26' 26"
42.426	15° 28' 06"	127.279	39° 42' 40"
56.569	20° 15' 20"	141.421	42° 41' 44"
70.711	24° 46' 04"	155.563	45° 25' 06"

Compute the focal length based on the angle measured to the first grid point. Compute and plot the distortion curve.

REFERENCES

BOLAND, J. 2000. ASPRS camera calibration review panel report executive summary. *Photogrammetric Engineering and Remote Sensing* 66(3):239–245.

BRACEWELL, R. N. 1999. *The Fourier Transform and Its Applications,* 3rd edition. WCB/McGraw-Hill.

BROWN, D. C. 1986. Unflatness of plates as a source of systematic error in close-range photogrammetry. *Photogrammetria* 40(4):343–363.

CHAMPLEBOUX, G., LAVALLÉE, S., SAUTOT, P., and CINQUIN, P. 1992. Accurate calibration of cameras and range imaging sensors: The NPBS method. In *Proceedings of the IEEE International Conference on Robotics and Automation.* pp. 1552–1557. Nice, France: IEEE.

COMER, R., KINN, G., LIGHT, D., and MONDELLO, C. 1998. Talking digital. *Photogrammetric Engineering and Remote Sensing* 64(12):1139–1144.

GOODMAN, J. W. 1996. *Introduction to Fourier Optics,* 2nd edition. San Francisco: McGraw-Hill.

HECHT, E., ZAJAC, A., and GUARDINO, K. 1997. *Optics,* 3rd edition. Reading, MA: Addison-Wesley Publishing Co.

HINCKLEY, T. K., and WALKER, J. W. 1993. Obtaining and using low-altitude/large-scale imagery. *Photogrammetric Engineering and Remote Sensing* 59(3):310–318.

HOTCHKISS, R. N. 1983. Image motion considerations in electro-optical panoramic cameras. In *Airborne Reconnaissance VII Proceedings of the SPIE Symposium.* Vol. 424 pp. 172–181. San Diego, CA: Society of Photo-Optical & Instrumentation Engineers.

JAMES, J. F. 1995. *A Student's Guide to Fourier Transforms: With Applications in Physics and Engineering.* Cambridge: Cambridge Univ. Press.

LEACHTENAUER, J., DANIEL, K., and VOGL, T. 1998. Digitizing satellite imagery: Quality and cost considerations. *Photogrammetric Engineering and Remote Sensing* 64(1):29–34.

LI, R. 1998. Potential of high-resolution satellite imagery for national mapping products. *Photogrammetric Engineering and Remote Sensing* 64(12):1165–1169.

LIGHT, D. L. 1992. The new camera calibration system at the U.S. Geological Survey. *Photogrammetric Engineering and Remote Sensing* 58(2):185–188.

LIGHT, D. L. 1996. Film cameras or digital sensors? The challenge ahead for aerial imaging. *Photogrammetric Engineering and Remote Sensing* 62(3):285–291.

MAAS, H. 1999. Image sequence based automatic multi-camera system calibration techniques. *ISPRS Journal of Photogrammetry and Remote Sensing* 54(5–6):352–359.

MCDONALD, R. A. 1995. CORONA: Success for space reconnaissance, a look into the cold war, and a revolution for intelligence. *Photogrammetric Engineering and Remote Sensing* 61(6):689–720.

MUNJY, R. A. H. 1986. Self-calibration using the finite element approach. *Photogrammetric Engineering and Remote Sensing* 52(3):411–418.

OPPENHEIM, A.V., and SCHAFER, R. 1975. *Digital Signal Processing.* Englewood Cliffs, NJ: Prentice Hall.

PAPOULIS, A. 1991. *Random Variables and Stochastic Processes.* New York: McGraw Hill College Div.

Chapter 4

Mathematical Concepts in Photogrammetry

4.1 FUNDAMENTALS OF PERSPECTIVE GEOMETRY

As stated in the previous chapter, perspective geometry forms the basis of the imaging model for frame cameras, and has proven useful, by extension, for modeling other sensors as well. The human visual system, with its image-forming lens and its image-sensing retina, gives each of us an intuitive feeling for the effects of perspective projection. These effects include the appearance of distant objects as being smaller and the convergence of parallel lines to a vanishing point. Artists of the Renaissance, including Leonardo da Vinci and Albrecht Dürer, sought to incorporate perspective geometry into landscape and still life renderings and even produced mechanical devices to aid in the systematic production of these drawings.

4.1.1 Perspective Projection of Lines: Cross Ratio

The fundamental elements of the perspective relationship between lines are shown in Fig. 4-1. These consist of a point L, known as the *perspective center,* a bundle of lines through point L, and a pair of lines $\overleftrightarrow{a}$ and $\overleftrightarrow{A}$ not containing L. Four lines of the bundle intersect line $\overleftrightarrow{a}$ at points d, e, f, and g and line $\overleftrightarrow{A}$ at points D, E, F, and G. A unique relationship exists between the various line segments of line $\overleftrightarrow{a}$, and the same relationship holds between the segments of line $\overleftrightarrow{A}$. This relationship, the *cross ratio* or *anharmonic ratio*, can be shown to be completely determined by the angles among the four lines, α, β, γ, and δ. For the line $\overleftrightarrow{A}$, the cross ratio is defined as

$$r = \frac{DF}{DG} \times \frac{EG}{EF} \tag{4-1}$$

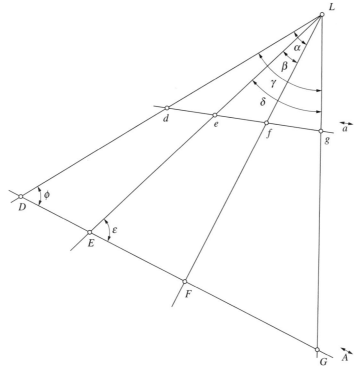

Figure 4-1 Elements comprising the cross ratio of lines.

From Fig. 4-1, by the law of sines

$$DF = LF \frac{\sin \alpha}{\sin \phi} \quad DG = LG \frac{\sin \gamma}{\sin \phi}$$

$$EF = LF \frac{\sin \beta}{\sin \epsilon} \quad EG = LG \frac{\sin \delta}{\sin \epsilon}$$

(4-2)

and therefore

$$r = \frac{LF \sin \alpha}{LG \sin \gamma} \times \frac{LG \sin \delta}{LF \sin \beta} = \frac{\sin \alpha \sin \delta}{\sin \gamma \sin \beta}$$

(4-3)

Any line intersecting these four lines will produce this *invariant* ratio; in particular, line $\overleftrightarrow{a}$ yields an identical value:

$$\frac{df}{dg} \times \frac{eg}{ef} = r = \frac{DF}{DG} \times \frac{EG}{EF}$$

(4-4)

Thus, given three known points on each of two projectively related lines, the location of a fourth point can be transferred from one line to the other. This can also be accomplished by a three-parameter transformation, as introduced in the next section.

4.1.2   Perspective Projection of Lines: Three-Parameter Transformation

Referring to Fig. 4-1, a one-dimensional coordinate system may be attached to each of the two projectively related lines: the x-system to line $\overleftrightarrow{a}$ and the X-system to line $\overleftrightarrow{A}$. The perspective transformation from x to X is given by

$$X = \frac{e_1 x + f_1}{e_0 x + 1}$$

(4-5)

in which e_0, e_1, and f_1 are the three unknown transformation parameters. Given a minimum of three points of known x and X coordinates, these parameters can be estimated. Once the values of the parameters are determined, Eq. 4-5 may be used to compute the coordinate X on line $\overleftrightarrow{A}$ for each additional point of known coordinate x on line $\overleftrightarrow{a}$. It should be clear that this approach is completely equivalent to that using the invariant quantity r described in the preceding section. In fact, these same two possibilities exist for two projectively related planes, as discussed in the following two sections.

EXAMPLE 4-1

In Fig. 4-1 we have the following dimensions: $de = 1.3129$, $ef = 0.6306$, $fg = 1.2484$, $DE = 2.7294$, and $EF = 1.8409$. Determine FG using the cross ratio, and again using the parameter method.

SOLUTION   First, we use the lowercase lengths and Eq. 4-4 to solve for $r = 1.8143$. Then we rewrite Eq. 4-4, using x for the unknown segment length, and solve algebraically for x.

$$r = \left(\frac{DF}{DF + x}\right)\left(\frac{EF + x}{EF}\right)$$

$$x = \frac{DF \times EF - r \times EF \times DF}{r \times EF - DF}$$

Substituting the given values yields $x = FG = 5.5683$.

To determine FG using the parameter method, we rewrite Eq. 4-5 as follows:

$$X = -xXe_0 + xe_1 + f_1$$

We must pick an origin on both lines, call it points d and D. Now for points d, e, and f we write three linear equations in the three unknown parameters and solve.

$$\begin{bmatrix} 0 & 0 & 1 \\ -3.5834 & 1.3129 & 1 \\ -8.8824 & 1.9435 & 1 \end{bmatrix} \begin{bmatrix} e_0 \\ e_1 \\ f_1 \end{bmatrix} = \begin{bmatrix} 0 \\ 2.7294 \\ 4.5703 \end{bmatrix}$$

The solution vector is $(-0.1481, 1.6746, 0)$. Evaluating Eq. 4-5 at point $g (x = 3.1919)$, we obtain $G = 10.1386$. Subtracting the lengths DE and EF, we obtain $FG = 5.5683$.

4.1.3   Perspective Projection of Planes: Two-Dimensional Invariance

Figure 4-2 shows the perspective projection between plane P_x and plane P_X. Any point with coordinates (x, y) in P_x is projected to a corresponding point in P_X with coordinates (X, Y), such that the two points and the perspective center fall on a straight line. The relationship between these two planes is completely specified when four points, no three of which fall on a line, are given with known coordinates (x, y) and known projected coordinates (X, Y). This means that for any additional point with known coordinates in plane P_x, its projected coordinates in plane P_X can be determined.

Denoting by subscripts 1, 2, 3, and 4 the given control points, and by (x_n, y_n) and (X_u, Y_u) the *known* and *unknown* coordinates, respectively, of the additional point to be

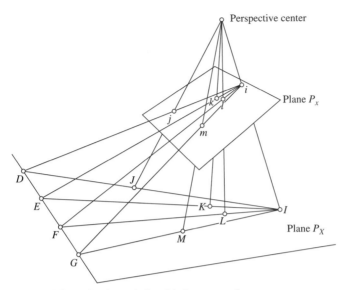

Figure 4-2 Projective relationship between planes.

transferred, the *invariance* property between the two planes is given by:

$$
\frac{\begin{vmatrix} x_1 & x_2 & x_3 \\ y_1 & y_2 & y_3 \\ 1 & 1 & 1 \end{vmatrix}}{\begin{vmatrix} x_1 & x_2 & x_4 \\ y_1 & y_2 & y_4 \\ 1 & 1 & 1 \end{vmatrix}} \times \frac{\begin{vmatrix} x_1 & x_4 & x_n \\ y_1 & y_4 & y_n \\ 1 & 1 & 1 \end{vmatrix}}{\begin{vmatrix} x_1 & x_3 & x_n \\ y_1 & y_3 & y_n \\ 1 & 1 & 1 \end{vmatrix}} = \frac{\begin{vmatrix} X_1 & X_2 & X_3 \\ Y_1 & Y_2 & Y_3 \\ 1 & 1 & 1 \end{vmatrix}}{\begin{vmatrix} X_1 & X_2 & X_4 \\ Y_1 & Y_2 & Y_4 \\ 1 & 1 & 1 \end{vmatrix}} \times \frac{\begin{vmatrix} X_1 & X_4 & X_u \\ Y_1 & Y_4 & Y_u \\ 1 & 1 & 1 \end{vmatrix}}{\begin{vmatrix} X_1 & X_3 & X_u \\ Y_1 & Y_3 & Y_u \\ 1 & 1 & 1 \end{vmatrix}} \tag{4-6}
$$

This represents a linear equation (i.e., of a straight line) in the two unknowns X_u, Y_u. A second equation is obtained by interchanging the positions of any two points, say 1 and 2, in *all* of the determinants in Eq. 4-6. These two independent linear equations may be solved for the unknown coordinates X_u, Y_u of the additional point. This approach can be applied to any points that need to be transferred from one plane (P_x) to the other (P_X).

EXAMPLE 4-2

Four corresponding points in each of two projectively related planes are:

$$(x_1, y_1) = (1, 2), (x_2, y_2) = (2, 7), (x_3, y_3) = (7, 6), (x_4, y_4) = (8, 4)$$

$$(X_1, Y_1) = (1, 1), (X_2, Y_2) = (1, 3), (X_3, Y_3) = (3, 3), (X_4, Y_4) = (3, 1)$$

A fifth point in the *xy* plane has coordinates $(x_5, y_5) = (5.9351, 4.3896)$. Using the invariance property between these two planes, determine the corresponding coordinates in the *XY* plane.

SOLUTION Evaluating the determinants in Eq. 4-6 with number elements, and expanding the two with the unknown coordinates yields the linear equation

$$2.0000\, X_5 - 4.0000\, Y_5 = -2.0000$$

Interchanging the positions of points 1 and 2 and repeating this process yields another linear equation

$$-2.000 X_5 - 1.3333 Y_5 = -6.0000$$

Solving these two equations simultaneously yields the coordinates of the unknown point

$$(X_5, Y_5) = (2, 1.5)$$

Two-dimensional invariance may also be established between two planes on the basis of corresponding lines. Thus, if centrally projected straight lines are expressed in the form of Eq. D-4 in Appendix D,

$$a_i x + b_i y + 1 = 0$$

then $(a, b)_i$, $(A, B)_i$, (a_n, b_n), and (A_u, B_u) replace $(x, y)_i$, $(X, Y)_i$, (x_n, y_n), and (X_u, Y_u), respectively, in Eq. 4-6, as follows:

$$\frac{\begin{vmatrix} a_1 & a_2 & a_3 \\ b_1 & b_2 & b_3 \\ 1 & 1 & 1 \end{vmatrix}}{\begin{vmatrix} a_1 & a_2 & a_4 \\ b_1 & b_2 & b_4 \\ 1 & 1 & 1 \end{vmatrix}} \times \frac{\begin{vmatrix} a_1 & a_4 & a_n \\ b_1 & b_4 & b_n \\ 1 & 1 & 1 \end{vmatrix}}{\begin{vmatrix} a_1 & a_3 & a_n \\ b_1 & b_3 & b_n \\ 1 & 1 & 1 \end{vmatrix}} = \frac{\begin{vmatrix} A_1 & A_2 & A_3 \\ B_1 & B_2 & B_3 \\ 1 & 1 & 1 \end{vmatrix}}{\begin{vmatrix} A_1 & A_2 & A_4 \\ B_1 & B_2 & B_4 \\ 1 & 1 & 1 \end{vmatrix}} \times \frac{\begin{vmatrix} A_1 & A_4 & A_u \\ B_1 & B_4 & B_u \\ 1 & 1 & 1 \end{vmatrix}}{\begin{vmatrix} A_1 & A_3 & A_u \\ B_1 & B_3 & B_u \\ 1 & 1 & 1 \end{vmatrix}} \tag{4-7}$$

Linear estimation of (A_u, B_u) requires four reference lines, no two of which are parallel.

4.1.4 Perspective Projection of Planes: Eight-Parameter Transformation

Another expression of the perspective relationship between planes is given by the eight-parameter transformation. The fundamental elements of this perspective projection between planes consist of a point, known as the *perspective center*, a bundle of lines through this point, and two planes which cut the bundle of lines and do not contain the perspective center. These are illustrated in Fig. 4-2. There is a natural one-to-one correspondence between each line of the bundle and its intersection point in the cutting planes. Of course, this excludes the cases of lines from the bundle parallel to the cutting plane, since such lines would not intersect the cutting plane. The relationship between the Cartesian coordinates of a point (x, y) in plane P_x of Fig. 4-2 and the coordinates of its corresponding point (X, Y) in plane P_X is

$$X = \frac{e_1 x + f_1 y + g_1}{e_0 x + f_0 y + 1}$$
$$Y = \frac{e_2 x + f_2 y + g_2}{e_0 x + f_0 y + 1} \tag{4-8}$$

This is referred to as the *projective transformation between planes* or the *eight-parameter transformation*. It will be derived in Section 4.5.6. It can be inverted by multiplying by the denominators in Eq. 4-8 and solving for x and y,

$$x = \frac{(f_2 - f_0 g_2)X + (f_0 g_1 - f_1)Y + (g_2 f_1 - g_1 f_2)}{(e_2 f_0 - e_0 f_2)X + (e_0 f_1 - e_1 f_0)Y + (e_1 f_2 - e_2 f_1)}$$
$$y = \frac{(e_0 g_2 - e_2)X + (e_1 - e_0 g_1)Y + (e_2 g_1 - e_1 g_2)}{(e_2 f_0 - e_0 f_2)X + (e_0 f_1 - e_1 f_0)Y + (e_1 f_2 - e_2 f_1)} \tag{4-9}$$

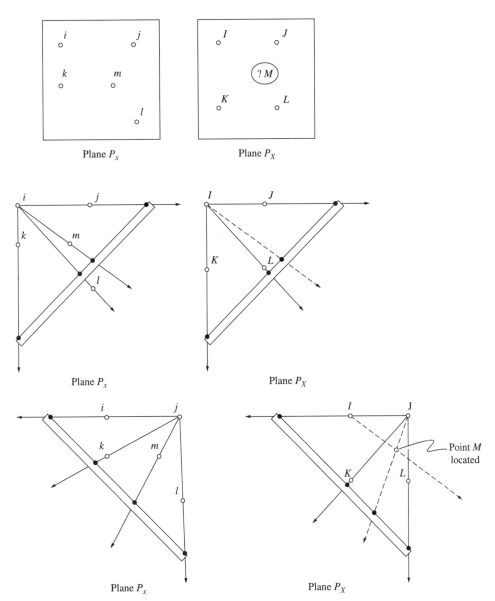

Figure 4-3 Graphical transfer of a point between projective planes using the paper strip method.

Note that in order to estimate the eight unknown transformation parameters $e_0, f_0, \ldots,$ g_2, a minimum of four points of known coordinates (x, y) and known projected coordinates (X, Y) is necessary.

Figure 4-3 illustrates how a point could be transformed graphically between two projective planes if there are four corresponding points given in each plane. First note in Fig. 4-2 that the bundle of lines in plane P_x with point i as the vertex intersects the bundle of lines in plane P_X with point I as the vertex along the line of intersection between the two planes at points D, E, F, and G. Since points D, E, F, and G occupy both planar bundles, the cross ratio from Eq. 4-4 will be the same for both bundles. In other words, the cross ratio is *invariant* under projective transformation of a planar bundle of lines. Given this fact, a graphical process called the *paper strip method* can be used to transfer a fifth point between planes, given four corresponding points in the two planes. See Fig. 4-3. The steps in the process are

1. Select a point from among the four known points and call it the vertex of a planar bundle.
2. Draw lines through the other three known points and through the fifth point.
3. Mark the intersection points of these four lines with an arbitrary cutting line (the paper strip).
4. Construct three lines of the bundle in the second plane using the same point as the vertex.
5. Using the paper strip, establish a point on the fourth line in the second plane. This defines a line through the unknown point position.
6. Repeat steps 1–5 with another point among the four as the vertex.
7. The intersection of the two lines corresponding to the fifth point defines its position in the second plane.

Equation 4-8 represents the classical projectivity equations based on point correspondence. A comparable pair of projectivity equations for two projectively related straight lines is given by

$$
A = \frac{r_1 a + s_1 b + t_1}{r_0 a + s_0 b + 1}
$$
$$
B = \frac{r_2 a + s_2 b + t_2}{r_0 a + s_0 b + 1}
$$

$$(4\text{-}10)$$

where (A, B) and (a, b) are the line parameters (see Eq. D-4 in Appendix D) and $r_0, s_0,$ $\ldots, t_1, t_2$ are the eight projective parameters.

A different form of the projectivity equations that retains the same transformation parameters (i.e., $e_0, \ldots, g_2$) as in Eq. 4-8 is given by

$$
p'[e_1 \cos\alpha' \cos\alpha'' + f_1 \sin\alpha' \cos\alpha'' + e_2 \cos\alpha' \sin\alpha'' + f_2 \sin\alpha' \sin\alpha''
$$
$$
-p''(e_0 \cos\alpha' + f_0 \sin\alpha' + 1)] + g_1 \cos\alpha'' + g_2 \sin\alpha'' = 0
$$

$$(4\text{-}11)$$

$$
e_1 \sin\alpha' \cos\alpha'' - f_1 \cos\alpha' \cos\alpha'' + e_2 \sin\alpha' \sin\alpha'' - f_2 \cos\alpha' \sin\alpha''
$$
$$
-p''(e_0 \sin\alpha' - f_0 \cos\alpha') = 0
$$

$$(4\text{-}12)$$

in which $e_0, b_0, \ldots, g_1, g_2$ are the eight projective parameters and (p', α'') and (p'', α'') are the line descriptors for the line equation of the form $x \cos\alpha + y \sin\alpha = p$, given by Eq. D-2 and depicted in Fig. D-1 in Appendix D. This formulation has the advantage of having the same parameters for points and lines, which allows for combining both features in the same solution.

Combined Point/Line Projectivity

Equations 4-8 and 4-10 can be used together in the same least-squares adjustment to effect projectivity between two planes based on a combination of points, Eq. 4-8, and lines, Eq. 4-10. However, in this case two sets of eight parameters, $e_0, \ldots, g_2$ and $r_0, \ldots, t_2$ are required in the adjustment. Since there are only eight independent parameters in a two-plane projectivity, the following eight constraint equations must be used:

$$r_1 = (f_2 - f_0 g_2)/(e_1 f_2 - e_2 f_1)$$
$$s_1 = (e_0 g_2 - e_2)/(e_1 f_2 - e_2 f_1)$$
$$t_1 = (e_2 f_0 - e_0 f_2)/(e_1 f_2 - e_2 f_1)$$
$$r_2 = (f_0 g_1 - f_1)/(e_1 f_2 - e_2 f_1)$$
$$s_2 = (e_1 - e_0 g_1)/(e_1 f_2 - e_2 f_1) \qquad (4\text{-}13)$$
$$t_2 = (e_0 f_1 - e_1 f_0)/(e_1 f_2 - e_2 f_1)$$
$$r_0 = (f_1 g_2 - f_2 g_1)/(e_1 f_2 - e_2 f_1)$$
$$s_0 = (e_2 g_1 - e_1 g_2)/(e_1 f_2 - e_2 f_1)$$

Alternatively, Eqs. 4-11 and 4-12 could be used for the lines in conjunction with Eq. 4-8 for the points, since they all have the same eight parameters $e_0, \ldots, g_2$.

4.1.5   Perspective Projection Between Three-Dimensional Space and a Plane

In the perspective projection between lines (1-D) and between planes (2-D), there is a one-to-one correspondence between points in the domain and points in the range. However, the perspective projection between three-dimensional space and a plane is unique only in one direction. For any point in space, there is a unique projective point in the plane, but for any point in the plane, there are an infinite number of corresponding points in space. If a frame photograph represents the plane, then we see the necessity for multi-station photogrammetry. At least one additional ray (or other constraint) is needed to resolve the ambiguity inherent in going from 2-D to 3-D. The equations which describe this projection are referred to as the *collinearity equations*, and are derived in Section 4.5.1.

4.2   COORDINATE REFERENCE FRAMES

Two primary reference coordinate systems will be considered here: the *image space coordinate system* and the *object space coordinate system*. These will always be considered to be Cartesian and right-handed. *Right-handed* refers to the placement of the axes such that if the fingers of the right hand curl from $+X$ toward $+Y$, then the thumb points in the $+Z$ direction. Numerous supporting or derived coordinate systems may be useful for certain explanations and applications. These auxiliary systems need not be Cartesian or right-handed, but would have to be transformed into such in order to satisfy the imaging equations. For example, ellipsoidal object space coordinates expressed as latitude, longitude, and height, (ϕ, λ, h), might be transformed into a local space rectangular (LSR) topocentric system or into a universal space rectangular (USR) geocentric system, both of which are Cartesian and right-handed. Screen display coordinate systems of pixel and line are usually left-handed to follow computer graphics practices, but would be transformed into a conventional (x, y) system for use in the imaging equations.

4.2.1   Image Space Coordinate System

The image space coordinate system for frame photographs was described in Section 2.4. The image coordinates (x, y) are referenced to a coordinate system defined by the fiducial marks. Introducing a shift so that the origin is at the principal point yields $(x - x_0, y - y_0)$. Placing the 3-D origin at the perspective center yields $(x - x_0, y - y_0, -f)$, referred to as *sensor coordinates* or *image space coordinates*, as shown in Fig. 2-11.

As an example of an auxiliary image space coordinate system, consider the equivalent vertical photograph associated with any given tilted photograph, as shown in Fig. 4-4. If the image space coordinates of a point are $(x, y, -f)$ and M is the rotation matrix

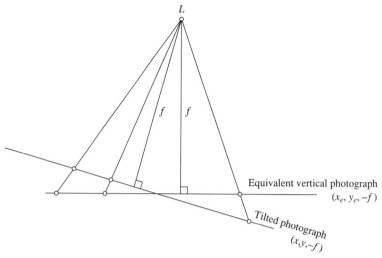

Figure 4-4 Original tilted photograph and equivalent vertical photograph.

relating the object and image space coordinate systems (see Sections 4.4 and 4.5.1 and Appendix A), then we can obtain the coordinates $(x_e, y_e, -f)$ of the point in the equivalent vertical photograph by

$$
\begin{bmatrix} u \\ v \\ w \end{bmatrix} = M^T \begin{bmatrix} x \\ y \\ -f \end{bmatrix}
$$

$$
\begin{bmatrix} x_e \\ y_e \\ -f \end{bmatrix} = \left(\frac{-f}{w}\right) \begin{bmatrix} u \\ v \\ w \end{bmatrix}
$$

(4-14)

4.2.2   Object Space Coordinate System

The *object space* is the three-dimensional region covered by the photograph or image. The *object space coordinate system* is a Cartesian system used to locate features and points shown in the image in the object space (Fig. 4-5). The origin can be the Earth's center of mass (a geocentric system) or some more convenient local point (a topocentric system). The surveyor, geodesist, engineer, or cartographer who is the recipient of photogrammetrically compiled data or the supplier of control point coordinates may prefer to work in a latitude-longitude-height (ϕ, λ, h) system or in a map projection system (x, y, h), but these would have to be transformed into a Cartesian system to serve as the object space coordinate system for photogrammetry.

4.3   SENSOR MODEL (INTERIOR ORIENTATION)

The *sensor model*, or *interior orientation*, defines the sensor or camera characteristics required for the reconstruction of the object space bundle of rays from the corresponding image points. In a frame camera, these characteristics would include at least the focal length or principal distance, the location of the principal point in the image plane, and a description of the lens distortion. For cameras not focused at infinity, the principal distance is the image distance as defined by the lens equations (Chapter 3). For cameras focused at infinity, the principal distance is equal to the focal length. These elements are shown in Fig. 4-6 for both a frame camera and a pushbroom linear sensor.

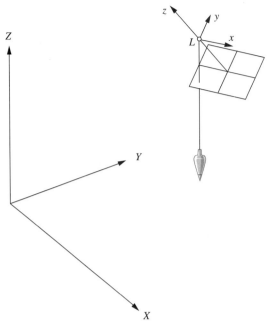

Figure 4-5 Object and image space coordinate systems.

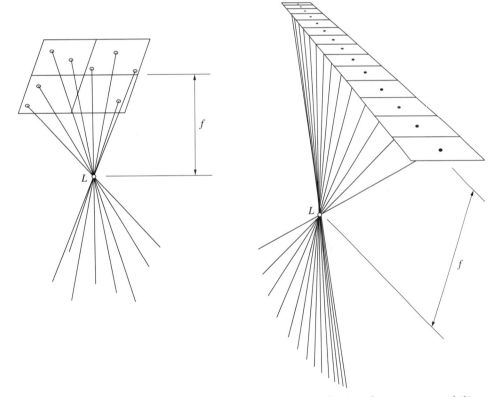

Figure 4-6 Sensor model to reconstruct the object space rays for (*a*) a frame camera and (*b*) a pushbroom linear sensor.

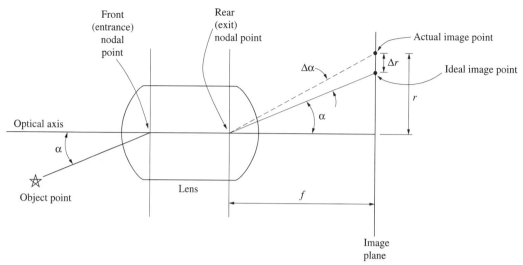

Figure 4-7 Lens distortion.

The principal point, defined as the point of intersection with the image plane of the perpendicular from the perspective center, is usually given with respect to the coordinate axes defined by fiducial marks. In the case of digital sensors, the principal point is given with respect to the image row-column coordinate system. Lens distortion, as shown in Fig. 4-7, is present when the actual image space rays are not parallel to the corresponding object space rays. Figure 4-7 illustrates that in actual (i.e., not "thin") lens systems, the perspective center is split into the *front and rear nodal points*. Other than the small axial displacement, the lens system behaves as if the front and rear nodal points were coincident at a single perspective center. An incident ray through the front nodal point will exit the rear nodal point at the same angle with respect to the optical axis except for the effects of lens distortion. The radial component of lens distortion is given by

$$\Delta r = r - f \tan \alpha \tag{4-15}$$

Tangential components of lens distortion may also be present. Sufficient redundancy and measurement precision may also detect systematic departures from film platen flatness. Some or all of these sensor parameters are determined during camera or sensor calibration (see Chapter 3). These parameters are usually estimated under controlled laboratory conditions, but with sufficient redundancy and proper layout of exposure stations, they may be estimated during block adjustment by *self-calibration* (see Chapter 5).

Knowledge of the parameters of inner orientation allows correction of the raw image measurements for all known systematic errors or displacements, and, with a properly reconstructed bundle, the determination of valid imaging condition equations with the object point coordinates and the exterior orientation parameters.

4.4 PLATFORM MODEL (EXTERIOR ORIENTATION)

Interior orientation, or the sensor model, described in Section 4.3, establishes the bundle of rays from the image points. *Exterior orientation*, or the *platform model*, establishes the position and orientation of the bundle of rays with respect to the object space coordinate system. Each bundle requires six independent elements; three for position and three for orientation. In the case of a frame camera, one bundle represents the entire image. In the case of a linear sensor, each line defines a new bundle, theoretically with its own six elements of exterior orientation. In practice, due to near functional dependency among these numerous parameters, they are usually estimated with a much smaller set of independent parameters.

For a bundle of rays, the three elements of position fix the location of the vertex or center of perspective. This is point L in Fig. 4-5. The coordinates of point L are often referred to as the camera station or exposure station coordinates, and are expressed as

$$L = \begin{bmatrix} X_L \\ Y_L \\ Z_L \end{bmatrix} \tag{4-16}$$

With only this point of the bundle established, the rays themselves can still take any orientation in space. Analytical geometry tells us that three angles, or three independent parameters, are sufficient to describe the orientation or attitude of this bundle in the object space coordinate system. This is equivalent to saying that three independent parameters are necessary to define the rotation matrix that relates the object space and image space systems. See Section 4.5.1 and Appendix A for a summary of the different choices of three parameters that can be used to construct the 3×3 rotation matrix. The exterior orientation defines the relationship between the object and image space coordinate systems by the following equation:

$$\begin{bmatrix} x \\ y \\ -f \end{bmatrix} = k\boldsymbol{M} \begin{bmatrix} X - X_L \\ Y - Y_L \\ Z - Z_L \end{bmatrix} \tag{4-17}$$

in which $(x, y, -f)$ are the image space coordinates, k is a scale factor, $\boldsymbol{M}$ is a 3×3 matrix containing the rotation parameters, and (X, Y, Z) represent the object point.

The standard approach to constructing $\boldsymbol{M}$ is by using three sequential rotations: ω about the X-axis, ϕ about the once-rotated Y-axis, and κ about the twice-rotated Z-axis.

$$\boldsymbol{M}_\omega = \begin{bmatrix} 1 & 0 & 0 \\ 0 & \cos\omega & \sin\omega \\ 0 & -\sin\omega & \cos\omega \end{bmatrix}$$

$$\boldsymbol{M}_\phi = \begin{bmatrix} \cos\phi & 0 & -\sin\phi \\ 0 & 1 & 0 \\ \sin\phi & 0 & \cos\phi \end{bmatrix} \tag{4-18a}$$

$$\boldsymbol{M}_\kappa = \begin{bmatrix} \cos\kappa & \sin\kappa & 0 \\ -\sin\kappa & \cos\kappa & 0 \\ 0 & 0 & 1 \end{bmatrix}$$

The total rotation matrix is then constructed as

$$\boldsymbol{M} = \boldsymbol{M}_\kappa \boldsymbol{M}_\phi \boldsymbol{M}_\omega \tag{4-18b}$$

$$\boldsymbol{M} = \begin{bmatrix} \cos\phi\cos\kappa & \cos\omega\sin\kappa + \sin\omega\sin\phi\cos\kappa & \sin\omega\sin\kappa - \cos\omega\sin\phi\cos\kappa \\ -\cos\phi\sin\kappa & \cos\omega\cos\kappa - \sin\omega\sin\phi\sin\kappa & \sin\omega\cos\kappa + \cos\omega\sin\phi\sin\kappa \\ \sin\phi & -\sin\omega\cos\phi & \cos\omega\cos\phi \end{bmatrix}$$

The selection of the order of rotations, that is, the primary, secondary, and tertiary axes, is arbitrary, but it will affect the attitudes at which singularities occur. More details on the construction of rotation matrices are given in Section A.4 of Appendix A.

4.5 PHOTOGRAMMETRIC CONDITIONS

The functional model of the imaging system will be realized in the *condition equations,* which relate image points, object points, and the imaging system parameters. A given condition equation may be used for different purposes, depending on which variables are considered observables, knowns, or unknowns in the stochastic model. For example, the collinearity condition equation may be used for space intersection, space resection, or relative orientation, as well as for other tasks. This flexibility will force us to allow any variable occurring in the equations to take the roles of observable, known, or unknown, depending on the circumstances.

4.5.1 The Collinearity Equations

Figure 4-8 illustrates the imaging geometry of a single point in a frame photograph. The image plane in Fig. 4-8 is depicted as lying below the perspective center, rather than above it, as would be the case with the film plane of an actual camera. This allows us to work with image geometry as found on a right-reading paper print or film diapositive rather than that found on a photographic negative. The fundamental characteristic of frame imaging is that the perspective center, the image point, and the corresponding object point all lie on a line in space. This line can be expressed as vector components in the image space coordinate system or as vector components in the object space coordinate system. These image and object coordinate systems will be related by three position parameters and three orientation parameters, as described in Section 4.4.

The three orientation parameters may be considered as three sequential rotations, ω, ϕ, κ or α, t, and s; as three nonsequential rotations, t_x, t_y, and H; or as strictly algebraic parameters, a, b, and c. In any case, they will be implicitly expressed in the nine elements of a 3×3 rotation matrix M. This matrix M is orthogonal and will have the sense of being applied to the object space coordinates to produce coordinates parallel to the image space system. Considering only the rotation and neglecting other parameters, such as scale,

$$\begin{bmatrix} x \\ y \\ z \end{bmatrix} = M \begin{bmatrix} X \\ Y \\ Z \end{bmatrix} \tag{4-19a}$$

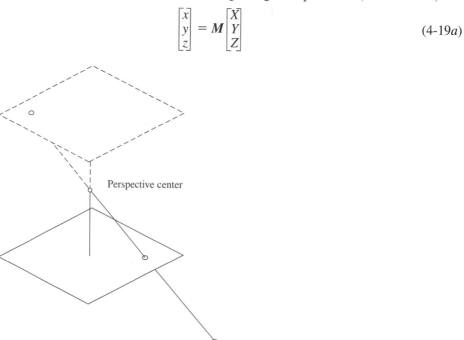

Perspective center

Figure 4-8 Imaging geometry for a single point on a frame photograph.

Since M is orthogonal, its inverse is equal to its transpose, and the order of the previous equation can be reversed:

$$\begin{bmatrix} X \\ Y \\ Z \end{bmatrix} = M^T \begin{bmatrix} x \\ y \\ z \end{bmatrix} \tag{4-19b}$$

In Eqs. 4-19, the origins of the two coordinate systems are assumed to coincide. In fact the origins do not coincide, and shift terms are introduced to place a local origin of object space coordinates at the perspective center. The difference in magnitude between the image space vector and the corresponding object space vector necessitates the introduction of a scale factor, k, into the equation.

$$\begin{bmatrix} x \\ y \\ z \end{bmatrix} = kM \begin{bmatrix} X - X_L \\ Y - Y_L \\ Z - Z_L \end{bmatrix} \tag{4-20}$$

The image space coordinates for a frame camera system will have a z coordinate fixed at the negative of the principal distance, or focal length. In addition, there may be small offsets x_0 and y_0 from a fiducial-based origin to a perspective center origin. These are reflected in the revised image space coordinates:

$$\begin{bmatrix} x - x_0 \\ y - y_0 \\ -f \end{bmatrix} = kM \begin{bmatrix} X - X_L \\ Y - Y_L \\ Z - Z_L \end{bmatrix} \tag{4-21}$$

The matrix M can be expressed in terms of its elements:

$$\begin{bmatrix} x - x_0 \\ y - y_0 \\ -f \end{bmatrix} = k \begin{bmatrix} m_{11} & m_{12} & m_{13} \\ m_{21} & m_{22} & m_{23} \\ m_{31} & m_{32} & m_{33} \end{bmatrix} \begin{bmatrix} X - X_L \\ Y - Y_L \\ Z - Z_L \end{bmatrix} \tag{4-22}$$

Multiplying the matrix and vector on the right-hand side of the equation, we obtain three scalar equations instead of a matrix equation:

$$\begin{aligned} x - x_0 &= k[m_{11}(X - X_L) + m_{12}(Y - Y_L) + m_{13}(Z - Z_L)] \\ y - y_0 &= k[m_{21}(X - X_L) + m_{22}(Y - Y_L) + m_{23}(Z - Z_L)] \\ -f &= k[m_{31}(X - X_L) + m_{32}(Y - Y_L) + m_{33}(Z - Z_L)] \end{aligned} \tag{4-23}$$

The scale factor is something of a nuisance parameter and can be eliminated by dividing the first two equations in 4-23 by the third to obtain the classical form of the collinearity equations:

$$x - x_0 = -f \frac{m_{11}(X - X_L) + m_{12}(Y - Y_L) + m_{13}(Z - Z_L)}{m_{31}(X - X_L) + m_{32}(Y - Y_L) + m_{33}(Z - Z_L)}$$

$$y - y_0 = -f \frac{m_{21}(X - X_L) + m_{22}(Y - Y_L) + m_{23}(Z - Z_L)}{m_{31}(X - X_L) + m_{32}(Y - Y_L) + m_{33}(Z - Z_L)} \tag{4-24}$$

In Eq. 4-21 we can take M to the other side and perform similar steps to eliminate the scale factor k:

$$X - X_L = (Z - Z_L) \frac{m_{11}(x - x_0) + m_{21}(y - y_0) + m_{31}(-f)}{m_{13}(x - x_0) + m_{23}(y - y_0) + m_{33}(-f)}$$

$$Y - Y_L = (Z - Z_L) \frac{m_{12}(x - x_0) + m_{22}(y - y_0) + m_{32}(-f)}{m_{13}(x - x_0) + m_{23}(y - y_0) + m_{33}(-f)} \tag{4-25}$$

These equations can be used for simple applications in their present form when quantities on the right are known and the quantities on the left are unknown. For instance, Eqs. 4-24 may be used when the interior and exterior orientations are known and an object point is known in order to compute the corresponding image coordinates. Equations 4-25 may be used when the interior and exterior orientations are known and an image coordinate is known, along with one component of the object space point coordinate, in order to obtain the remaining two components of the object space point position.

For applications with more unknown variables to be solved for from a series of minimally sufficient or redundant observations, the nonlinearity of the equations forces us to use Taylor series approximations. In this case, we can reorganize Eqs. 4-24 slightly into the following form:

$$F_1 = x - x_0 + f\frac{U}{W} = 0$$
$$F_2 = y - y_0 + f\frac{V}{W} = 0$$

$(4\text{-}26)$

where

$$\begin{bmatrix} U \\ V \\ W \end{bmatrix} = M \begin{bmatrix} X - X_L \\ Y - Y_L \\ Z - Z_L \end{bmatrix}$$

$(4\text{-}27)$

Then the condition equations can be written in linearized form as:

$$\begin{bmatrix} \dfrac{\partial F_1}{\partial l} \\ \dfrac{\partial F_2}{\partial l} \end{bmatrix} v + \begin{bmatrix} \dfrac{\partial F_1}{\partial x} \\ \dfrac{\partial F_2}{\partial x} \end{bmatrix} \Delta = \begin{bmatrix} -F_1(l^o, x^o) \\ -F_2(l^o, x^o) \end{bmatrix}$$

$(4\text{-}28)$

or

$$Av + B\Delta = f$$

$(4\text{-}29)$

The partial derivatives are presented in Appendix C.

The collinearity equations find application as the basis of the multi-station bundle adjustment, in which this pair of condition equations is written for each pass point and control point for every photograph in which the point appears. Thus, for a point appearing in three photographs, six equations of this type would be written. The collinearity equations are also the traditional means of solving the problems of space resection and intersection (see Chapter 5). By judiciously de-weighting the object space point coordinates, and constraining seven of the exterior orientation parameters of a pair of photographs, they can also be used for relative orientation. In short, they can be considered to be the workhorse of analytical photogrammetry.

An interesting variation on the collinearity equations can occur when the object space points are effectively at infinity, as in the case of star photographs. Such a situation can occur when photographing satellites in Earth orbit from the ground, or when undertaking a stellar camera calibration. In these cases, rather than characterizing a point by position, we may characterize it by direction. The equations then take the form:

$$x - x_0 = -f\frac{m_{11}C_x + m_{12}C_y + m_{13}C_z}{m_{31}C_x + m_{32}C_y + m_{33}C_z}$$
$$y - y_0 = -f\frac{m_{21}C_x + m_{22}C_y + m_{23}C_z}{m_{31}C_x + m_{32}C_y + m_{33}C_z}$$

$(4\text{-}30)$

where the C vector components are the direction cosines of the star in the stellar coordinate system:

$$\begin{bmatrix} C_x \\ C_y \\ C_z \end{bmatrix} = \begin{bmatrix} \cos\alpha\cos\delta \\ \sin\alpha\cos\delta \\ \sin\delta \end{bmatrix} \tag{4-31}$$

and α and δ are the right ascension and declination of the star, respectively.

EXAMPLE 4-3

For a photograph with exterior orientation elements $(\omega, \phi, \kappa) = (2, 5, 15)$ degrees and $(X_L, Y_L, Z_L) = (5000, 10{,}000, 2000)$ meters and camera parameters $(x_0, y_0, f) = (0.015, -0.020, 152.4)$ mm, compute via the collinearity equations the coordinates of the ground point $(X, Y, Z) = (5100, 9800, 100)$ in the the fiducial-based image system.

SOLUTION Evaluating the rotation matrix yields

$$\begin{bmatrix} 0.9622 & 0.2616 & -0.0751 \\ -0.2578 & 0.9645 & 0.0562 \\ 0.0871 & -0.0348 & 0.9956 \end{bmatrix}$$

Using Eq. 4-24 to obtain the image coordinates with respect to the principal point yields $(x - x_0, y - y_0) = (15.159, -26.449)$ mm. The coordinates in the fiducial-based system are $(x, y) = (15.174, -26.469)$ mm.

EXAMPLE 4-4

Determine where the ray corresponding to the image point in Example 4-3 intersects the plane in object space defined by $Z = 500$ m.

SOLUTION Evaluating Eq. 4-25 yields $(X, Y) = (5078.95, 9842.10)$ m.

4.5.2 Line-Based Collinearity Condition

The basic collinearity condition for a *point* with image coordinates (x, y) was given by Eq. 4-21, or by Eq. 4-24 after the elimination of the scalar k. The equivalent to these for a straight line feature is given by

$$\begin{vmatrix} \rho_x & A_x & (X - X_L) \\ \rho_y & A_y & (Y - Y_L) \\ \rho_z & A_z & (Z - Z_L) \end{vmatrix} = 0 \tag{4-32}$$

where

$$\begin{bmatrix} \rho_x \\ \rho_y \\ \rho_z \end{bmatrix} = M^T \begin{bmatrix} x_i - x_0 \\ y_i - y_0 \\ -f \end{bmatrix} \tag{4-33}$$

M = rotation matrix that rotates the object coordinate system parallel to the image coordinate system

(x_i, y_i) = image coordinates of a point on the straight line in the image space

(x_0, y_0, f) = elements of interior orientation

(A_x, A_y, A_z) = components of a vector along the line in the object space

(X, Y, Z) = coordinates of a specific point on the line in the object space

(X_L, Y_L, Z_L) = coordinates of the exposure station

Equation 4-32 can be written for each point on the straight line in the image space. Since two points define a straight line, a minimum of two equations are required for each line when determining the redundancy in the triangulation process. Equations from additional points contribute further to the line fitting process. The point N at (X_n, Y_n, Z_n) on the line closest to the origin is selected as the specific point and thus Eq. 4-32 becomes

$$\begin{vmatrix} \rho_x & A_x & (X_N - X_L) \\ \rho_y & A_y & (Y_N - Y_L) \\ \rho_z & A_z & (Z_N - Z_L) \end{vmatrix} = 0 \tag{4-34}$$

An alternative photogrammetric condition using straight line descriptors in both image and object spaces may be derived based on the collinearity equation for point features, Eq. 4-21.

The image and object point coordinates, (x_i, y_i) and (X_i, Y_i, Z_i), respectively, can be written in straight line coordinate systems as (see Appendix D)

$$\begin{bmatrix} x_i \\ y_i \end{bmatrix} = \begin{bmatrix} \cos\alpha & -\sin\alpha \\ \sin\alpha & \cos\alpha \end{bmatrix} \begin{bmatrix} p \\ v_i \end{bmatrix} \tag{4-35}$$

where p and α are the straight line parameters in the image, v_i is a length along the line to the point, and

$$\begin{bmatrix} X_i \\ Y_i \\ Z_i \end{bmatrix} = M_\beta \begin{bmatrix} q \\ 0 \\ W_i \end{bmatrix} \tag{4-36}$$

where q and the three rotation angles β_1, β_2, and β_3 in the matrix M_β are the four line parameters in the object space, and W_i is given in Eq. D-11.

Substituting the above two equations into Eq. 4-21 yields

$$\begin{bmatrix} p\cos\alpha - v_i\sin\alpha - x_0 \\ p\sin\alpha + v_i\cos\alpha - y_0 \\ -f \end{bmatrix} = k_i M \begin{bmatrix} m_{\beta_{11}}q + m_{\beta_{13}}W_i - X_L \\ m_{\beta_{21}}q + m_{\beta_{23}}W_i - Y_L \\ m_{\beta_{31}}q + m_{\beta_{33}}W_i - Z_L \end{bmatrix} \tag{4-37}$$

Let

$$\begin{bmatrix} C_x \\ C_y \\ C_z \end{bmatrix} = \begin{bmatrix} m_{\beta_{11}}q + m_{\beta_{13}}W_i - X_L \\ m_{\beta_{21}}q + m_{\beta_{23}}W_i - Y_L \\ m_{\beta_{31}}q + m_{\beta_{33}}W_i - Z_L \end{bmatrix} \tag{4-38}$$

Equation 4-37 can be written as

$$p\cos\alpha - v_i\sin\alpha - x_0 = k_i(m_{11}C_x + m_{12}C_y + m_{13}C_z) \tag{4-39a}$$

$$p\sin\alpha + v_i\cos\alpha - y_0 = k_i(m_{21}C_x + m_{22}C_y + m_{23}C_z) \tag{4-39b}$$

$$-f = k_i(m_{31}C_x + m_{32}C_y + m_{33}C_z) \tag{4-39c}$$

Eliminating k_i by dividing Eqs. 4-39a and 4-39b by Eq. 4-39c and multiplying both sides of the equations by $-f$ yields

$$p\cos\alpha - v_i\sin\alpha - x_0 = -f\frac{m_{11}C_x + m_{12}C_y + m_{13}C_z}{m_{31}C_x + m_{32}C_y + m_{33}C_z} \tag{4-40a}$$

$$p\sin\alpha + v_i\cos\alpha - y_0 = -f\frac{m_{21}C_x + m_{22}C_y + m_{23}C_z}{m_{31}C_x + m_{32}C_y + m_{33}C_z} \tag{4-40b}$$

Introducing the auxiliary quantity

$$D = p - x_0\cos\alpha - y_0\sin\alpha \tag{4-41}$$

and replacing C_x, C_y, and C_z by their values from Eq. 4-38 yields, after a few manipulation steps, the final form of the equations:

$$\begin{aligned}
&(Dm_{31} + m_{11}f\cos\alpha + m_{21}f\sin\alpha)(m_{\beta_{11}}q - X_L) \\
&+ (Dm_{32} + m_{12}f\cos\alpha + m_{22}f\sin\alpha)(m_{\beta_{21}}q - Y_L) \\
&+ (Dm_{33} + m_{13}f\cos\alpha + m_{23}f\sin\alpha)(m_{\beta_{31}}q - Z_L) = 0
\end{aligned} \tag{4-42a}$$

$$\begin{aligned}
&(Dm_{31} + m_{11}f\cos\alpha + m_{21}f\sin\alpha) \\
&+ (Dm_{32} + m_{12}f\cos\alpha + m_{22}f\sin\alpha) \\
&+ (Dm_{33} + m_{13}f\cos\alpha + m_{23}f\sin\alpha) = 0
\end{aligned} \tag{4-42b}$$

These are the two equations for a straight line in an image, corresponding to the two collinearity equations for a point in an image. The photogrammetric condition equations using straight lines will fail when the straight line in object space is in the same plane as the two (or more) image perspective centers (the *epipolar condition*). When this occurs, the projection planes for the corresponding straight lines from the images are coplanar. Since the planes do not intersect in a single line, the object line cannot be determined. Consequently, in practice, straight lines that are parallel or are close to parallel to the epipolar plane should be avoided.

4.5.3 The Coplanarity Equation

Figure 4-9 illustrates a pair of overlapping frame photographs and a pair of conjugate image points. If these photographs are relatively oriented with respect to each other, then the object space rays defined by the two image points and their respective perspective centers will intersect exactly. This intersection defines the *model space* position of this point. The phrase *model space* refers to the object space, often taken at reduced scale. The two rays and the base vector connecting the two perspective centers constitute the three sides of a triangle and thus implicitly define a plane. The equation that enforces this relationship between the three rays is the *coplanarity condition*. The equation can be constructed by recalling the formula for the volume of a parallelepiped. If we define the base of the parallelepiped by two adjacent edge vectors, and its height by an edge vector not in the plane of the base, then the volume of the parallelepiped is the triple scalar product of these three vectors. These elements are shown in Fig. 4-10. The triple scalar product of three vectors $\vec{a}$, $\vec{b}$, and $\vec{c}$ is given by

$$\vec{a}\cdot(\vec{b}\times\vec{c}) = \begin{vmatrix} a_x & a_y & a_z \\ b_x & b_y & b_z \\ c_x & c_y & c_z \end{vmatrix} \tag{4-43}$$

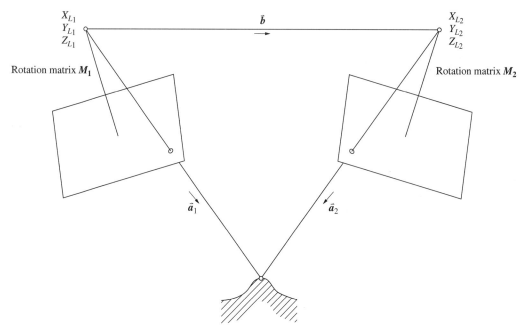

Figure 4-9 Geometry for the coplanarity condition.

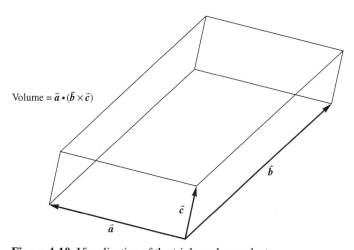

Figure 4-10 Visualization of the triple scalar product.

The volume of the parallelpiped will be zero if the third vector lies in the plane of the base, making it coplanar with the first two vectors.

As illustrated in Fig. 4-9, the base vector $\vec{b}$ represents the displacement between the perspective centers. The vector $\vec{a}_1$ represents the object space vector from the image point on the left photograph, and the vector $\vec{a}_2$ represents the object space vector from the image point on the right photograph. Components of these three vectors may be expressed as

$$\vec{b} = \begin{bmatrix} b_X \\ b_Y \\ b_Z \end{bmatrix} = \begin{bmatrix} X_{L_2} - X_{L_1} \\ Y_{L_2} - Y_{L_1} \\ Z_{L_2} - Z_{L_1} \end{bmatrix} \tag{4-44}$$

$$\hat{a}_1 = \begin{bmatrix} u_1 \\ v_1 \\ w_1 \end{bmatrix} = M_1^T \begin{bmatrix} x - x_0 \\ y - y_0 \\ -f \end{bmatrix}_1 \tag{4-45}$$

$$\hat{a}_2 = \begin{bmatrix} u_2 \\ v_2 \\ w_2 \end{bmatrix} = M_2^T \begin{bmatrix} x - x_0 \\ y - y_0 \\ -f \end{bmatrix}_2 \tag{4-46}$$

We use the triple scalar product to express the coplanarity of these three vectors:

$$\hat{b} \cdot (\hat{a}_1 \times \hat{a}_2) = 0 \tag{4-47}$$

or, in the determinant form using the vector components,

$$F = \begin{vmatrix} b_X & b_Y & b_Z \\ u_1 & v_1 & w_1 \\ u_2 & v_2 & w_2 \end{vmatrix} = 0 \tag{4-48}$$

The partial derivatives of F with respect to each of the variables in Eq. 4-48 are given in Section C.4 of Appendix C. The coplanarity equation is primarily used for computing the relative orientation between pairs of photographs. This is usually done by fixing seven parameters among X_{L_1}, Y_{L_1}, Z_{L_1}, ω_1, ϕ_1, κ_1, X_{L_2}, Y_{L_2}, Z_{L_2}, ω_2, ϕ_2, and κ_2, then solving for the remaining five. Approximate model space coordinates do not need to be generated in this case, since object coordinates do not appear in the equation.

Successive relative orientations performed between the two pairs of photographs in a triplet using only the coplanarity condition do not guarantee that the three rays for any point in the triple overlap area will intersect in a single point, but would only ensure that two of the pairs of rays intersect. We could write another coplanarity equation between the rays from the first and third photographs, but this breaks down when the three planes are coincident or nearly so. Unfortunately, this will often be the case with normal aerial strip geometry. To ensure accurate three-ray intersection, we use the scale restraint equation described in Section 4.5.5.

4.5.4 Alternative Forms of the Collinearity and Coplanarity Equations

The term *homogeneous coordinates* refers to the representation of an n-dimensional coordinate value with an extra coordinate component set to unity (see the related discussion in Section 9.3.5). Any multiples of this expanded representation are considered to be equivalent, since it is always possible to recover the original form by dividing all components by the $(n + 1)$th component. Homogeneous coordinates are used extensively in computer graphics applications. One of the factors which makes them attractive is that translations can be implemented by matrix multiplication.

If we have two vectors

$$\hat{a} = \begin{bmatrix} a_x \\ a_y \\ a_z \end{bmatrix} \text{ and } \hat{b} = \begin{bmatrix} b_x \\ b_y \\ b_z \end{bmatrix} \tag{4-49}$$

and we arrange the elements of $\hat{b}$ into the skew-symmetric matrix

$$K = \begin{bmatrix} 0 & -b_z & b_y \\ b_z & 0 & -b_x \\ -b_y & b_x & 0 \end{bmatrix} \tag{4-50}$$

then the following vector cross product can be formulated as a matrix product:

$$\vec{b} \times \vec{a} = K\vec{a} \tag{4-51}$$

The collinearity condition of Eq. 4-21 can be rewritten as

$$M\begin{bmatrix} X - X_L \\ Y - Y_L \\ Z - Z_L \end{bmatrix} = \frac{1}{k}\begin{bmatrix} x - x_0 \\ y - y_0 \\ -f \end{bmatrix} = \frac{1}{k}\begin{bmatrix} 1 & 0 & -x_0 \\ 0 & 1 & -y_0 \\ 0 & 0 & -f \end{bmatrix}\begin{bmatrix} x \\ y \\ 1 \end{bmatrix} = \frac{1}{k}C\begin{bmatrix} x \\ y \\ 1 \end{bmatrix} \tag{4-52}$$

in which the image coordinate vector is in the form of homogeneous coordinates and the interior orientation elements (x_0, y_0, f) are in the calibration matrix C. Rearranging the previous equation slightly,

$$\vec{a} = \begin{bmatrix} X - X_L \\ Y - Y_L \\ Z - Z_L \end{bmatrix} = \frac{1}{k}M^T C\begin{bmatrix} x \\ y \\ 1 \end{bmatrix} \tag{4-53}$$

For two photographs, the base vector $\vec{b}$ can be expressed as in Eq. 4-44. The coplanarity condition, as expressed in Eq. 4-47, can be reordered as

$$\vec{a}_1 \cdot \vec{b} \times \vec{a}_2 \tag{4-54}$$

and using Eq. 4-51, this may be rewritten as

$$\vec{a}_1^T K_b \vec{a}_2 = 0 \tag{4-55}$$

Using Eq. 4-52, we may rewrite Eq. 4-55 as

$$\frac{1}{k_1}\frac{1}{k_2}[x_1 \quad y_1 \quad 1] C_1^T M_1 K_b M_2^T C_2 \begin{bmatrix} x_2 \\ y_2 \\ 1 \end{bmatrix} = 0 \tag{4-56}$$

Dividing through by the scale factors and collecting the matrices together, we obtain

$$[x_1 \quad y_1 \quad 1] C_1^T E C_2 \begin{bmatrix} x_2 \\ y_2 \\ 1 \end{bmatrix} = 0 \tag{4-57a}$$

or

$$[x_1 - x_{0_1} \quad y_1 - y_{0_1} \quad -f_1] E \begin{bmatrix} x_2 - x_{0_2} \\ y_2 - y_{0_2} \\ -f_1 \end{bmatrix} = 0 \tag{4-57b}$$

The matrix $E = M_1 K_b M_2^T$ is called the *essential matrix*. Its elements are functions of only the exterior orientation elements of the two photographs. In the literature, particularly that involved with image understanding and computer vision, E is used with what are referred to as *calibrated* cameras. For *uncalibrated* cameras, the C matrices are combined with E to form what is termed the *fundamental matrix* F,

$$F = C_1^T M_1 K_b M_2^T C_2 \tag{4-58}$$

In this case the coplanarity condition takes the form

$$[x_1 \quad y_1 \quad 1] F \begin{bmatrix} x_2 \\ y_2 \\ 1 \end{bmatrix} = 0 \tag{4-59}$$

The elements of both E and F are recoverable only to a scale factor since Eqs. 4-57b and 4-59 are quadratic forms equal to zero and are not affected when multiplied by a scalar. Of the remaining eight independent elements in the essential matrix, E, five relative orientation parameters of a stereo pair are recoverable. On the other hand, *seven* independent parameters are recoverable from the fundamental matrix, F, since its rank (and the rank of K_b) is 2, and therefore $|F| = 0$. These seven parameters include the five of relative orientation plus two of the six interior orientation elements (three for each photograph).

4.5.5   The Scale Restraint Equation

Figure 4-11 shows the arrangement of three photographs and an object point visible on all three photographs. We allow a mismatch between the rays from photographs 1 and 2, and between the rays from photographs 2 and 3. The three object space rays corresponding to the three photographs are

$$\vec{a}_1 = \begin{bmatrix} a_{1_x} \\ a_{1_y} \\ a_{1_z} \end{bmatrix} = M_1^T \begin{bmatrix} x_1 - x_0 \\ y_1 - y_0 \\ -f \end{bmatrix} \tag{4-60}$$

$$\vec{a}_2 = \begin{bmatrix} a_{2_x} \\ a_{2_y} \\ a_{2_z} \end{bmatrix} = M_2^T \begin{bmatrix} x_2 - x_0 \\ y_2 - y_0 \\ -f \end{bmatrix} \tag{4-61}$$

$$\vec{a}_3 = \begin{bmatrix} a_{3_x} \\ a_{3_y} \\ a_{3_z} \end{bmatrix} = M_3^T \begin{bmatrix} x_3 - x_0 \\ y_3 - y_0 \\ -f \end{bmatrix} \tag{4-62}$$

The direction of the mismatch between the rays from photographs 1 and 2 is given by the vector mutually perpendicular to the two rays:

$$\vec{d}_1 = \vec{a}_1 \times \vec{a}_2 \tag{4-63}$$

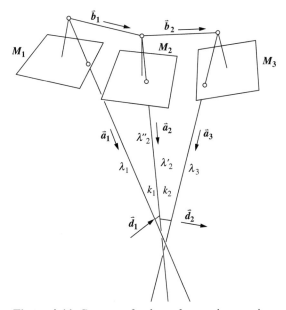

Figure 4-11 Geometry for the scale restraint equation.

The direction of the mismatch between the rays from photographs 2 and 3 is given by the vector mutually perpendicular to the two rays:

$$\vec{d}_2 = \vec{a}_2 \times \vec{a}_3 \tag{4-64}$$

Summing the four vectors around each of the two loops corresponding to the two photo pairs, we obtain the following equations, with the λ's and k's being scale factors:

$$\begin{bmatrix} b_{1_x} \\ b_{1_y} \\ b_{1_z} \end{bmatrix} = \lambda_1 \begin{bmatrix} a_{1_x} \\ a_{1_y} \\ a_{1_z} \end{bmatrix} + k_1 \begin{bmatrix} d_{1_x} \\ d_{1_y} \\ d_{1_z} \end{bmatrix} - \lambda_2' \begin{bmatrix} a_{2_x} \\ a_{2_y} \\ a_{2_z} \end{bmatrix} \tag{4-65}$$

or

$$\begin{bmatrix} b_{1_x} \\ b_{1_y} \\ b_{1_z} \end{bmatrix} = \begin{bmatrix} a_{1_x} & d_{1_x} & -a_{2_x} \\ a_{1_y} & d_{1_y} & -a_{2_y} \\ a_{1_z} & d_{1_z} & -a_{2_z} \end{bmatrix} \begin{bmatrix} \lambda_1 \\ k_1 \\ \lambda_2' \end{bmatrix} \tag{4-66}$$

and

$$\begin{bmatrix} b_{2_x} \\ b_{2_y} \\ b_{2_z} \end{bmatrix} = \lambda_2'' \begin{bmatrix} a_{2_x} \\ a_{2_y} \\ a_{2_z} \end{bmatrix} + k_2 \begin{bmatrix} d_{2_x} \\ d_{2_y} \\ d_{2_z} \end{bmatrix} - \lambda_3 \begin{bmatrix} a_{3_x} \\ a_{3_y} \\ a_{3_z} \end{bmatrix} \tag{4-67}$$

or

$$\begin{bmatrix} b_{2_x} \\ b_{2_y} \\ b_{2_z} \end{bmatrix} = \begin{bmatrix} a_{2_x} & d_{2_x} & -a_{3_x} \\ a_{2_y} & d_{2_y} & -a_{3_y} \\ a_{2_z} & d_{2_z} & -a_{3_z} \end{bmatrix} \begin{bmatrix} \lambda_2'' \\ k_2 \\ \lambda_3 \end{bmatrix} \tag{4-68}$$

Solving each of Eqs. 4-66 and 4-68 for the unknown λ_2 using Cramer's rule, we obtain

$$\lambda_2' = \frac{\begin{vmatrix} a_{1_x} & d_{1_x} & b_{1_x} \\ a_{1_y} & d_{1_y} & b_{1_y} \\ a_{1_z} & d_{1_z} & b_{1_z} \end{vmatrix}}{\begin{vmatrix} a_{1_x} & d_{1_x} & -a_{2_x} \\ a_{1_y} & d_{1_y} & -a_{2_y} \\ a_{1_z} & d_{1_z} & -a_{2_z} \end{vmatrix}} \tag{4-69}$$

and

$$\lambda_2'' = \frac{\begin{vmatrix} b_{2_x} & d_{2_x} & -a_{3_x} \\ b_{2_y} & d_{2_y} & -a_{3_y} \\ b_{2_z} & d_{2_z} & -a_{3_z} \end{vmatrix}}{\begin{vmatrix} a_{2_x} & d_{2_x} & -a_{3_x} \\ a_{2_y} & d_{2_y} & -a_{3_y} \\ a_{2_z} & d_{2_z} & -a_{3_z} \end{vmatrix}} \tag{4-70}$$

Forcing λ_2' and λ_2'' to be equal yields the scale restraint condition equation:

$$F = -\lambda_2' + \lambda_2'' = 0 \tag{4-71}$$

which is expressed in terms of the determinants, with some sign manipulation to be consistent with the previous equation:

$$
F = \frac{\begin{vmatrix} a_{1_x} & d_{1_x} & b_{1_x} \\ a_{1_y} & d_{1_y} & b_{1_y} \\ a_{1_z} & d_{1_z} & b_{1_z} \end{vmatrix}}{\begin{vmatrix} a_{1_x} & d_{1_x} & a_{2_x} \\ a_{1_y} & d_{1_y} & a_{2_y} \\ a_{1_z} & d_{1_z} & a_{2_z} \end{vmatrix}} + \frac{\begin{vmatrix} b_{2_x} & d_{2_x} & a_{3_x} \\ b_{2_y} & d_{2_y} & a_{3_y} \\ b_{2_z} & d_{2_z} & a_{3_z} \end{vmatrix}}{\begin{vmatrix} a_{2_x} & d_{2_x} & a_{3_x} \\ a_{2_y} & d_{2_y} & a_{3_y} \\ a_{2_z} & d_{2_z} & a_{3_z} \end{vmatrix}} = 0
\tag{4-72}
$$

4.5.6   The Plane Projectivity Equation as a Form of the Collinearity Equations

The plane projectivity condition equation and its inverse form are stated in Section 4.1.4 and in Appendix A. The projectivity equation models a perspective projection between two planes, as shown in Fig. 4-12. In this section, it is shown that this projectivity equation is a straightforward extension of the fundamental collinearity equations derived in Section 4.5.1. The derivation presented here begins with the collinearity equations, projects 3-D points into a 2-D plane, then constrains the 3-D points in object space to lie in a plane. Simplifying the resulting equations leads to a form which is identical to the previously stated projectivity equations. Restating Eq. 4-24,

$$
x = -f\frac{m_{11}(X - X_L) + m_{12}(Y - Y_L) + m_{13}(Z - Z_L)}{m_{31}(X - X_L) + m_{32}(Y - Y_L) + m_{33}(Z - Z_L)} + x_0
$$
$$
y = -f\frac{m_{21}(X - X_L) + m_{22}(Y - Y_L) + m_{23}(Z - Z_L)}{m_{31}(X - X_L) + m_{32}(Y - Y_L) + m_{33}(Z - Z_L)} + y_0
\tag{4-73}
$$

Separating the terms, we get

$$
x = \frac{-(fm_{11}X + fm_{12}Y + fm_{13}Z) + (fm_{11}X_L + fm_{12}Y_L + fm_{13}Z_L)}{(m_{31}X + m_{32}Y + m_{33}Z) - (m_{31}X_L + m_{32}Y_L + m_{33}Z_L)} + x_0
$$
$$
y = \frac{-(fm_{21}X + fm_{22}Y + fm_{23}Z) + (fm_{21}X_L + fm_{22}Y_L + fm_{23}Z_L)}{(m_{31}X + m_{32}Y + m_{33}Z) - (m_{31}X_L + m_{32}Y_L + m_{33}Z_L)} + y_0
\tag{4-74}
$$

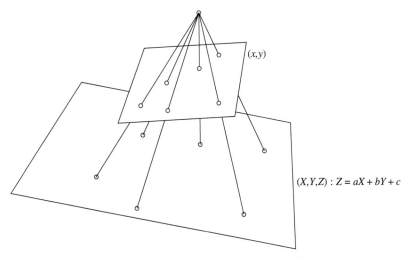

Figure 4-12 Projectivity condition as a special case of collinearity.

Replacing the parenthetical terms on the right by P, Q, and R,

$$x = \frac{-(fm_{11}X + fm_{12}Y + fm_{13}Z) + P}{(m_{31}X + m_{32}Y + m_{33}Z) - R} + x_0$$

$$y = \frac{-(fm_{21}X + fm_{22}Y + fm_{23}Z) + Q}{(m_{31}X + m_{32}Y + m_{33}Z) - R} + y_0$$

(4-75)

Making a common denominator,

$$x = \frac{-fm_{11}X - fm_{12}Y - fm_{13}Z + P + x_0 m_{31}X + x_0 m_{32}Y + x_0 m_{33}Z - x_0 R}{m_{31}X + m_{32}Y + m_{33}Z - R}$$

$$y = \frac{-fm_{21}X - fm_{22}Y - fm_{23}Z + Q + y_0 m_{31}X + y_0 m_{32}Y + y_0 m_{33}Z - y_0 R}{m_{31}X + m_{32}Y + m_{33}Z - R}$$

(4-76)

To enforce the constraint that all points (X, Y, Z) occupy a plane, we can use the equation

$$Z = aX + bY + c$$

(4-77)

Substituting this into the previous expression and collecting coefficients we obtain

$$x = \frac{(-fm_{11}-fm_{13}a+x_0 m_{31}+x_0 m_{33}a)X + (-fm_{12}-fm_{13}b+x_0 m_{32}+x_0 m_{33}b)Y + (-fm_{13}c+P+x_0 m_{33}c-x_0 R)}{(m_{31}+m_{33}a)X + (m_{32}+m_{33}b)Y + (m_{33}c-R)}$$

$$y = \frac{(-fm_{21}-fm_{23}a+y_0 m_{31}+x_0 m_{33}a)X + (-fm_{22}-fm_{23}b+y_0 m_{32}+y_0 m_{33}b)Y + (-fm_{23}c+Q+y_0 m_{33}c-y_0 R)}{(m_{31}+m_{33}a)X + (m_{32}+m_{33}b)Y + (m_{33}c-R)}$$

If numerator and denominator in these two equations are divided by $(m_{33}c - R)$ then it is clear that the equations take the form shown in Eq. 4-8.

4.5.7   Further Consideration of the Orientation Parameters

The orientation matrix of a frame photograph plays an important role in the solution of many photogrammetric problems. Estimating the three independent parameters involved in this matrix is critical, particularly because there exist many geometric situations in practice for which this is quite difficult. In Appendix A, we present several different parameterizations of the rotation matrix, each of which has advantages and disadvantages for particular applications.

4.6   SATELLITE IMAGERY AND ORBITS

In order to develop the imaging equations that relate a terrestrial point position to its corresponding satellite image position, it is necessary to introduce the relevant coordinate systems. The orientation of these coordinate systems is best understood through a presentation of the geometry of *normal* orbits. The phrase *normal orbit* implies a geometric solution to the classical two-body problem, which can be presented in closed form. In practice, orbital perturbations arising from the effects of other celestial bodies, the Earth's oblateness, atmospheric drag, and other sources all contribute to the normal orbit being just a first approximation to the actual satellite trajectory. A review of some elementary mechanics and other pertinent derivations are provided in Appendix F.

4.6.1   Imaging Equations

If we assume that we have a linear sensor on a satellite platform, oriented in a *pushbroom* configuration, the imaging equations can be constructed in the following way:

Each line of imagery, corresponding to one integration period of the sensor, can be considered to be a separate perspective image, with only a single dimension in the y-direction. Each of these lines would therefore have its own perspective center and orientation parameters. Thus, what may appear from a casual inspection to be a static image frame is, in fact, a mosaic of many tiny *framelets*. The near functional dependency among this multitude of orientation parameters makes it impossible to attempt to estimate them all. The usual strategy is to select a few independent parameters, estimate them, and then express all of the others in terms of these few. Figure F-6 shows the instantaneous position and orientation of an orbiting sensor. The equations which relate a terrestrial object point (X_e, Y_e, Z_e), an instantaneous perspective center (X_L, Y_L, Z_L), and a set of image space coordinates are (refer to Appendix F for mathematical details)

$$\begin{bmatrix} 0 \\ y \\ -f \end{bmatrix} = kR_{\mathrm{III}}R_{\mathrm{II}}R_{\mathrm{I}} \left[R_0^T \begin{bmatrix} X_e \\ Y_e \\ Z_e \end{bmatrix} - \begin{bmatrix} X_L \\ Y_L \\ Z_L \end{bmatrix} \right] \tag{4-78}$$

where R_0 performs the Earth-fixed to quasi-inertial rotation, R_{I} performs the quasi-inertial to nominal platform rotation, R_{II} performs the nominal platform to actual platform rotation, and R_{III} performs the platform to sensor rotation. The rotation matrix R_{I} can be constructed in the following way:

$$R_{\mathrm{I}} = R_3(\pi)R_2\left(\frac{\pi}{2} - \omega - f(t)\right)R_1\left(i - \frac{\pi}{2}\right)R_3(\Omega) \tag{4-79}$$

The angles in R_{II} would likely be parameterized as a low-order polynomial function of time, possibly with the x-pixel coordinate serving as a time analog if the line rate is constant. The rotation from platform to sensor, R_{III}, would often be constant during an image formation, changing for example between images as the *off-nadir* viewing angle is modified. The first two equations of 4-78 would be divided by the third, as in the derivation of the collinearity equations, to eliminate the nuisance scale parameter. Of necessity, some of the rotations described above would be held constant for the duration of image acquisition. The perspective center coordinates would probably also be modeled as a low-order polynomial function of time. Specific strategies for parameter modeling depend on the particular sensor being modeled. The arrival of commercially available satellite data in the 1-meter GSD (ground sample distance) range means that this kind of image modeling will become an important component of computational photogrammetry.

PROBLEMS

4.1 In Fig. 4-1, we have the following dimensions: $de = 1.5$, $ef = 1.6$, $fg = 1.8$, $DE = 3.1$, and $EF = 3.5$. Determine FG using the cross ratio. Determine FG again using the three-parameter solution.

4.2 For a photograph with exterior orientation elements $(\omega, \phi, \kappa) = (3, -2, 88)$ degrees, and $(X_L, Y_L, Z_L) = (12{,}000, 14{,}000, 3000)$ meters, and camera parameters $(x_0, y_0, f) = (-0.012, 0.006, 153.000)$ mm, compute via the collinearity equations the coordinates of the ground point $(X, Y, Z) = (11{,}850, 13{,}700, 200)$ m in the fiducial-based image system.

4.3 For the image point determined in Problem 4.2, determine where the corresponding ray intersects the plane in object space defined by $Z = 0$.

REFERENCES

ATKINSON, K. 1996. *Close Range Photogrammetry and Machine Vision.* Scotland: Whittles Publishing.

BATE, R., MUELLER, D., and WHITE, J. 1971. *Fundamentals of Astrodynamics.* New York: Dover Publications.

BOULET, D. 1991. *Methods of Orbit Determination for the Micro Computer.* Richmond, VA: Willmann-Bell, Inc.

FAUGERAS, O. 1993. *Three-Dimensional Computer Vision.* Cambridge, MA: The MIT Press.

GREVE, C. 1996. *Digital Photogrammetry: An Addendum to the Manual of Photogrammetry.* Bethesda, MD: American Society for Photogrammetry and Remote Sensing.

HARTLEY, R., and ZISSERMAN, A. 2000. *Multiple View Geometry.* Cambridge: Cambridge University Press.

HECHT, E., ZAJAC, A., and GUARDINO, K. 1997. *Optics.* 3rd edition. Reading, MA: Addison-Wesley.

HOFFMAN, B. 1966. *About Vectors.* New York: Dover Publications.

KARARA, H., ed. 1989. *Non-Topographic Photogrammetry.* Bethesda, MD: American Society for Photogrammetry and Remote Sensing.

KRAUS, K. 1993. *Photogrammetry.* Bonn: Dummler.

LEICK, A. 1995. *GPS Satellite Surveying.* New York: John Wiley & Sons.

MAKKI, S. 1991. *Photogrammetric Reduction and Analysis of Real and Simulated SPOT Imageries.* Ph.D. Thesis. West Lafayette, IN: Purdue University.

MOFFITT, F., and MIKHAIL, E. 1980. *Photogrammetry.* New York: Harper & Row.

PADERES, F. 1986. *Geometric Modeling and Rectification of Satellite Scanner Imagery and Investigation of Related Critical Issues.* Ph.D. Thesis. West Lafayette, IN: Purdue University.

SCHENK, T. 1999. *Digital Photogrammetry.* Laurelville, OH: TerraScience.

SLAMA, C. 1980. *Manual of Photogrammetry.* Bethesda, MD: American Society for Photogrammetry and Remote Sensing.

WOLF, P., and DEWITT, B. 2000. *Elements of Photogrammetry.* New York: McGraw Hill.

Chapter 5

Resection, Intersection, and Triangulation

5.1   INTRODUCTION

The basic operations of photogrammetry are *resection*, the determination of the position and orientation of an image in space, and *intersection,* the calculation of the object space coordinates of a point from its coordinates in two or more images. These operations are combined in *triangulation*, or *block adjustment*, in which the image orientation parameters and the point coordinates are calculated simultaneously.

The history of photogrammetric triangulation is in many ways the history of computational capability. Although the basic equations of analytical photogrammetry were formulated in the mid-1800s, there was no practical means of performing the amount of calculation required for realistic numbers of images or points. Instead, analog stereoplotters were developed to perform stereo plotting, while strip or block adjustments were performed using analog equipment or graphical methods. Early analytical block adjustment procedures emulated these analog and graphical adjustment methods, due to the limited power of the first available computers. As computers became more powerful, the mathematical models used in triangulation became more rigorous and capable, allowing the solution of blocks of thousands of photos and points with simultaneous determination of residual systematic errors in the data. Photogrammetric triangulation, having begun as a way to reduce control requirements, now rivals or surpasses ground surveying in the accuracy attainable.

We are currently witnessing another revolution in triangulation, brought about by the widespread availability and utilization of Global Positioning System (GPS) receivers and other navigational sensors. Control point requirements are being reduced to an absolute minimum by the continuing development of techniques for exploiting the positioning information available.

This chapter describes the elementary operations of resection and intersection, then discusses the underlying mathematics of block adjustment. An introduction to the use of added parameters in block adjustment is given, as well as a discussion of the utilization of navigational information. Programs implementing many of the operations described in this chapter are described in Appendix G and are included on the accompanying CD-ROM.

5.2   IMAGE COORDINATE REFINEMENT

The mathematical models used in photogrammetry describe an idealized imaging geometry in order to simplify the equations. Factors that cause the light ray to deviate from a straight line or otherwise distort the image coordinates are omitted from the models, with the assumption that they have been corrected in a preprocessing step. These nonrandom deviations from the mathematical model are called *systematic errors*. We attempt to model these errors from their causes and correct for them as much as possible before performing photogrammetric calculations, as when we correct for lens distortion using camera calibration data. In some cases, we can identify systematic trends in the data but not their source, and in these cases we must try to remove the nonrandom effects during the block adjustment solution; this is known as *selfcalibration*, which is described in Section 5.9.

The goal of image coordinate refinement is to ensure that the errors in the image coordinates fit the assumption of the mathematical model that errors are strictly random with no remaining systematic components. Systematic errors are corrected in the order opposite that in which they occur. In the imaging of a point on the ground, the first error might be not allowing for the curvature of the Earth in expressing the point's coordinates. Next, the light ray from the point is refracted by the atmosphere and then distorted as it goes through the lens. The point where the light ray hits the sensor is offset by the principal point displacement and possibly perturbed by film deformation. When we measure the point's position using a comparator or digitize the film using a scanner, mechanical errors affect the measured coordinates of the point.

Of course, not all of these systematic errors occur in every image, and there may be additional effects that must be corrected in some special cases.

5.2.1   Comparator or Scanner Calibration

The design of the measuring or scanning instrument used determines the types of errors introduced into the measured coordinates. (The design and operation of comparators and scanners are described in Chapter 7.) We determine these errors using calibration procedures and correct the measured image coordinates for their effects.

Comparators and scanners move an optical system across the film or, equivalently, move the film under an optical system. There are typically separate systems for motion and measurement; each must be precise and stable. Inaccuracies or instabilities in the measurement system lead to systematic errors in the measured coordinates.

Movement and measurement occur along two perpendicular axes; if the coordinate axes are not perpendicular, the measured coordinates will undergo an affine deformation. Misalignments or distortions in the optical system will also result in incorrect measurements.

Comparators are calibrated by measuring grid plates with precisely located targets. Multiple readings are taken on each target in at least four sets, with the grid rotated ninety degrees between each set. The first step in calibration is determining an affine transformation (Appendix A.4) that models the overall scale along each coordinate axis and the nonorthogonality of the measurement axes. Nonlinear effects are then modeled by polynomials in the x and y directions, using a least squares adjustment. In many computer-controlled instruments, the calibration correction is applied automatically by the controller, so that the output coordinates do not need further processing.

5.2.2   Film and Platen Deformations

Although great improvements have been made in the stability of film bases, there may still be some residual distortions in the developed film. Some older cameras also exhibit deformations in the film platen, but this is seldom significant in modern cameras. These deformations are corrected using the fiducial marks or reseaux (Section 3.7.1) imaged on the film. This also transforms the image coordinates into the fiducial coordinate system.

Fiducial marks may be located in the corners or on the edges of the image; most modern cameras have both. The number and locations of the fiducial marks determines how well film deformation can be corrected and also the type of transformation (Appendix A.4) that can be applied. Enough fiducials should be measured to provide sufficient redundancy to allow bad measurements to be detected. The number required varies with the type of transformation used. For example, if two fiducials are measured, then only an overall scale can be determined (along with rotation and translation into the fiducial coordinate system) and a four-parameter similarity transformation applied. If one of the fiducials is measured incorrectly, the transformation residuals will not indicate this (they will always be zero), and the measurement error will contaminate all measurements from that image. If four fiducials are measured, then an affine (six-parameter) transformation can be used. The measurement of more than four fiducials allows the use of a projective transformation or general polynomial equations.

If camera reseaux are available, those nearest the image point of interest are measured and the transformation between their imaged positions and their calibrated positions is calculated. The same types of transformations used for fiducial marks are employed. If the reseaux are part of the film platen and are not attached to the lens body, the relationship between the reseaux and the fiducials will change each time the magazine is replaced. The transformation from reseau coordinates to fiducial coordinates must therefore be determined by measuring both reseaux and fiducial marks.

5.2.3   Principal Point Displacement

Ideally, the lines between opposite fiducial marks should intersect exactly at the principal point. In practice, there is usually a small offset, determined during camera calibration (Section 3.8), that must be corrected.

5.2.4   Lens Distortion Correction

Once we have obtained image coordinates relative to the principal point, we can apply the corrections for radial and tangential lens distortion, which are determined during calibration. Radial distortion is corrected using a lookup table or a polynomial function, while tangential distortion is corrected by formula (Section 3.2.8).

5.2.5 Atmospheric Refraction

Light rays passing through the interface of two media with different refractive indices are refracted according to Snell's Law (Section 3.2.1). The refractive index of air decreases as the air becomes less dense with increasing altitude; a light ray is therefore bent as it goes from the denser air near the ground to the thinner air at the camera altitude.

Atmospheric refraction always acts radially outward on a vertical aerial image, bending the ray away from vertical (Fig. 5-1). If we define α as the angle the refracted ray makes with vertical, the angular displacement Δd (in microradians) is

$$\Delta d = K \tan \alpha \qquad (5\text{-}1)$$

The constant K can be thought of as the amount of refraction for a ray at an angle of 45 degrees, and is determined by the assumed density profile of the atmosphere. While the exact relationship between density and altitude varies according to meteorological conditions, several standard atmospheric models have been developed. A commonly used one is the Air Force ARDC (Air Research and Development Command) 1959 model, although other models exist (Gyer, 1996). Based on the ARDC 1959 model, the constant K can be computed as

$$K = \frac{2410H}{H^2 - 6H + 250} - \frac{2410h}{h^2 - 6h + 250}\left(\frac{h}{H}\right) \qquad (5\text{-}2)$$

If H, the flying height, and h, the terrain elevation, are both expressed in kilometers, the constant K is obtained in microradians.

The correction for atmospheric refraction is very similar to that for lens distortion. For each image point, the angle α between its corresponding ray and vertical is calculated using the approximate orientation of the image. The constant K is then calculated using the approximate flying height and the terrain elevation. The correction angle, Δd,

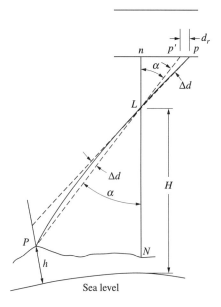

Figure 5-1 Atmospheric refraction calculation.

is then applied to α and new image coordinates obtained. If the exact image orientation is not known at the start of the process, the refraction correction will have to be repeated during the triangulation procedure as the image orientation is refined.

EXAMPLE 5-1   *Atmospheric Refraction Calculation*

Given a vertical frame aerial photograph taken from an altitude of 3000 meters with a 152-mm lens, what is the atmospheric refraction correction (using the ARDC 1959 model) for a point at image coordinates $x = 59.043$ mm, $y = 72.392$ mm? Assume the terrain elevation is 300 meters.

SOLUTION

$$K = \frac{2410(3.0)}{3.0^2 - 6(3.0) + 250} - \frac{2410(0.3)}{0.3^2 - 6(0.3) + 250}\left(\frac{0.3}{3.0}\right) = 29.7088$$

$$r = \sqrt{59.043^2 + 72.392^2} = 93.417 \text{ mm}$$

$$\Delta d = K \tan \alpha = K\frac{93.417}{152.0} = 18.258 \text{ microradians}$$

$$r' = f \tan(\alpha + \Delta d) = 93.421 \text{ mm}$$

$$x' = \frac{r'}{r}x = 59.046 \text{ mm}$$

$$y' = \frac{r'}{r}y = 72.395 \text{ mm}$$

If the imagery involves two different media, as in underwater photography, the refraction correction becomes much more complicated. The exact shape of the interface between the two media must be modeled, along with the angles of the light rays as they traverse the interface.

5.2.6   Curvature of the Earth

The collinearity equations (Section 4.5.1) are written for transformations between two Cartesian coordinate systems. Most triangulation programs therefore work in rectangular coordinate system representations of the Earth's curved surface using either geocentric or local tangent coordinate systems (Section 4.2.2). However, most surveying work is done in coordinate systems defined by a map projection, such as state plane or UTM (Appendix A.6). If these coordinates are not converted into Cartesian coordinates before the triangulation, a correction for the curvature of the Earth must be applied to the image coordinates.

The Earth-curvature correction for vertical photography is straightforward (Fig. 5-2). A point located on the Earth's curved surface at P, with elevation h, is located at P' by the map projection. If the flying height is H, the radius of the Earth is R, the focal length is f (in mm), and the radial distance of the point from the principal point is r (in mm), the correction, in mm, is

$$d_E = \frac{r^3 H}{2f^2 R} \tag{5-3}$$

The correction is toward the nadir. Another approach, short of complete conversion to local Cartesian coordinates, is a tangent plane reduction of heights, with a

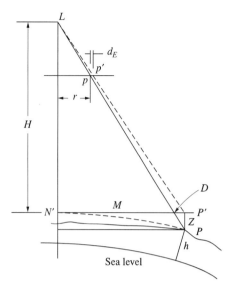

Figure 5-2 Earth-curvature correction.

corresponding inverse correction after the adjustment. If the elevation with respect to sea level is Z_{SL}, the distance from the reference point is D, and R is the Earth's radius, the tangent plane height Z_{TP} is

$$Z_{TP} = Z_{SL} - \frac{D^2}{2R} \tag{5-4}$$

5.3   RESECTION

Resection is the determination of an image's position and orientation parameters with respect to an object space coordinate system. In the standard case, the parameters include the object space coordinates of the perspective center and three angles, ω, ϕ, κ, which describe the orientation of the object space coordinate system with respect to the image coordinate system. Resection also includes the determination of the position and orientation parameters indirectly; for instance, the determination of coefficients of a set of polynomials that describe the position and orientation of a time-varying sensor with respect to time or a set of abstract projective parameters such as the direct linear transform (DLT) model (Chapter 9).

As one of the fundamental photogrammetric problems, resection techniques have been the subject of study since the beginning of photogrammetry. There are many different formulations of the resection problem, optimized for different imaging situations, operational procedures, computational requirements, and accuracy specifications. For instance, both closed-form and iterative solutions have been developed. Closed-form solutions are usually faster, due to their lower computational requirements, and have the advantage of not requiring initial approximations for the exterior orientation parameters. They are therefore popular for computer vision applications that must run without operator input or editing. However, the most accurate solutions require the use of redundant observations and least squares techniques, which cannot be applied to closed-form methods.

Although it may seem odd, in most cases the image parameters themselves are not of interest. The final product is usually the object space point coordinates, and the image parameters are simply auxiliary variables. The calculated image position may not be close to the actual position due to projective compensation, which is the partial compensation for systematic errors by changes in the orientation parameters. For instance, uncorrected radial lens distortion or atmospheric refraction in a vertical image are partially compensated for by changes in the calculated flying height. This is usually not important, the exception to this being when GPS or other navigational information is used to provide an independent estimate of the image position.

A minimum of three noncollinear control points are required for a resection solution of a frame image, assuming that its interior orientation is known. However, even if sufficient control information is available, there are geometric configurations that can result in unsolvable or unstable solutions. If the control points and the perspective center all lie on or close to a cylindrical surface, the resection solution will be unstable or possibly even indeterminate.

Known geometric information within the scene, such as straight lines or circles with known orientations, can also be included in the resection solution. Equations describing the geometry (Appendix D) are added to the collinearity equations and solved as part of the least squares adjustment. Determining the minimum amount of control point or geometric information needed can sometimes be complex.

The standard resection method for photogrammetric applications is based on the collinearity equations (Eqs. 4-24) and can be thought of as a special case of bundle block adjustment, with only one image and no unknown ground points. The six unknowns are, of course, the three perspective center position coordinates and the three orientation parameters, expressed in terms of ω, ϕ, and κ or another system of angles. The observations are the image coordinate measurements of each point. Two collinearity equations are written for each imaged point; with three control points, the resulting six equations allow for the unique solution of the six unknown parameters. If additional points are available, a least squares solution can be performed to obtain better results and to allow for point measurement editing. The collinearity equations must be linearized (Appendix C) and initial approximations must be furnished for the parameters. The initial approximations may come from navigational sensors such as GPS, from flight plans, from reference to maps or previously resected images, or from an initial closed-form solution.

Although resection is a fundamental photogrammetric procedure, it is seldom used in practice except as a first step for generating approximations for a bundle adjustment. The vast majority of photogrammetric applications involve multiple photographs to cover wider areas and to provide stereo measurements. A simultaneous block adjustment is much better in these cases, since it dramatically reduces the control requirements and therefore the cost of control point surveying. Additionally, a block adjustment reestablishes the relative orientations between the photos very accurately, making mapped objects more consistent between stereomodels. Individual resection solutions are sensitive to errors in the control information used for each solution and may not be consistent between photos.

5.4 SINGLE-RAY BACKPROJECTION

Very often we need to determine the object space position of a point appearing in only one image. Since the three-dimensional information about the world is lost when it is projected into a two-dimensional image, additional information must be provided before we can backproject the ray into the world. While we most often think of fixing the elevation coordinate, any one of the three object space coordinates can be provided, or a functional relationship between them specified.

The simplest case is backprojection to a given coordinate plane, for instance, projecting to a fixed elevation (Z coordinate). The solution for X and Y is straightforward, using the inverse form of the collinearity equations (Eqs. 4-25). The equations can also be written for the cases where the X or Y coordinates are known, rather than the Z coordinate.

A more complicated situation occurs when backprojecting the image ray to a digital elevation model (DEM). Elevations in the DEM are indexed by position, but we cannot calculate the position of the point we are backprojecting until we know its elevation. An iterative procedure is therefore required, as shown in Fig. 5-3. Starting with the maximum elevation in the DEM, the backprojected position i is calculated. A step is taken along the ray, the new position and elevation are calculated, and the elevation in the DEM corresponding to that position is retrieved. The DEM elevation is compared to the elevation of the point on the ray to determine if the ray has intersected the terrain surface. Once the surface has been found, the position and elevation can be refined using a smaller step size.

This procedure, known as *ray tracing* in computer graphics, has been extensively studied to improve its efficiency and to avoid potential pitfalls. For instance, Fig. 5-4 shows a ray that intersects a hill. If the initial elevation guess is at the bottom of the hill, the position will be incorrectly calculated to be on the blind side of the hill.

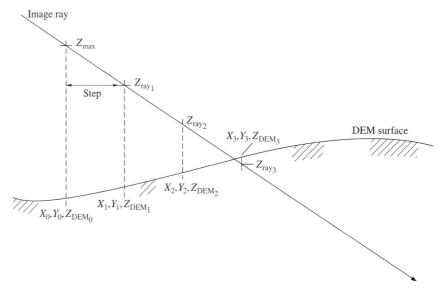

Figure 5-3 Ray tracing calculation for projection from image space to object space.

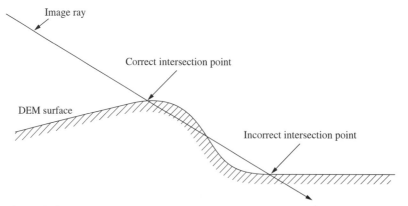

Figure 5-4 Incorrect ray tracing procedure, due to starting at too low an elevation.

5.5 INTERSECTION

Intersection refers to the determination of a point's position in object space by intersecting the image rays from two or more images. The standard method is again the application of the collinearity equations (Eqs. 4-24), with two equations for each image of the point. If two images are available, a total of four equations containing three unknowns, the object space coordinates of the point, are obtained. There is one degree of freedom, and the linearized set of equations can be solved by least squares methods. Adding more images increases the number of degrees of freedom and therefore improves the solution.

Approximate coordinates of the point, calculated either by single-ray projection into the world (Section 5.4) or by approximate intersection calculations, must be supplied to begin the nonlinear solution. For each image, we can write the unit vector from the perspective center through the ground point as

$$\begin{bmatrix} \alpha \\ \beta \\ \gamma \end{bmatrix}_i = M_i^T \begin{vmatrix} x_i - x_0 \\ y_i - y_0 \\ -f_i \end{vmatrix} \tag{5-5}$$

The equations of the line through the perspective center of image i and the object point is therefore

$$\frac{X - X_{L_i}}{\alpha_i} = \frac{Y - Y_{L_i}}{\beta_i} = \frac{Z - Z_{L_i}}{\gamma_i} \tag{5-6}$$

These equations are written for each image, giving four equations in three unknowns, which can be solved for the coordinates of the unknown point.

The observations in the least squares intersection solution are the image coordinate measurements, so performing the least squares adjustment minimizes the weighted sum of the image residuals squared. Intersection solutions minimizing other quantities, such as the perpendicular distance between rays, could be implemented, but the statistical interpretation of the results would be unclear. There is no stochastic information associated with the distances between rays, only with the image measurements, which are directly observed quantities.

5.6 RELATIVE ORIENTATION

Given two images of a scene taken from different viewpoints, we can create a *stereomodel* that can be viewed to obtain a three-dimensional impression of the scene. To form a stereomodel, the projected image rays through conjugate points must intersect in space, thereby reestablishing the original epipolar geometry (Section 2.7) of the pair of images. This procedure is known as *relative orientation*. Relative orientation can be performed manually by adjusting the orientation elements of an analog stereoplotter, as described in Section 7.2.5, or analytically by measuring corresponding image points and calculating the orientation parameters, as described in this section.

Relative orientation involves the determination of five degrees of freedom. If we consider the interior orientation of the images to be known, each image will have six unknown exterior orientation parameters—X_c, Y_c, Z_c, ω, ϕ, and κ—for a total of twelve unknown parameters. The 3-D stereomodel will have an arbitrary uniform scale with respect to the world coordinate system, three unknown rotation parameters (e.g., sequential rotations around each coordinate axis), and three unknown translations (one in each coordinate direction). These seven parameters are determined during *absolute orientation* (Section 5.7), which relates the model space coordinates to

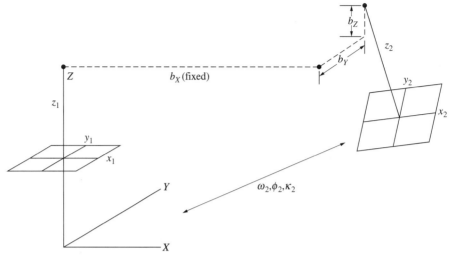

Figure 5-5 Dependent relative orientation.

the object space coordinates. Subtracting these seven parameters from the original twelve leaves five to be determined during relative orientation.

The five parameters to be determined can be chosen in several ways. For instance, all the parameters of one image, along with one position parameter of the second image, can be fixed and the remaining parameters of the second image adjusted (*dependent* relative orientation). As shown in Fig. 5-5, this defines the model coordinate system parallel to the first image's coordinate system, leaving two position parameters and the three orientation parameters of the second image with respect to the model coordinate system to be determined. The model scale is determined by the fixed position parameter of the second image. Alternatively, the κ and ϕ parameters of both images can be adjusted, along with the ω parameter of either image (*independent* relative orientation). This defines the model coordinate system with its X-axis parallel to the base between the two images (Fig. 5-6); the ϕ and κ parameters of each image are then adjusted, along with the ω parameter of one image, to bring the image coordinate systems into alignment with the model system.

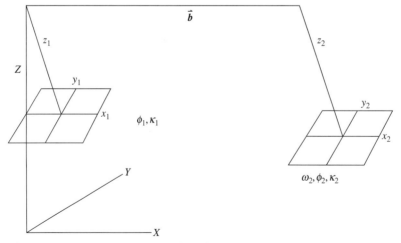

Figure 5-6 Independent relative orientation.

Relative orientation can be performed with more than two images. In a sequential procedure, once the first pair of images is relatively oriented, each successive image is oriented to its preceding image. A simultaneous analytical solution can also be performed, if care is taken to specify the proper degrees of freedom. While the first model has five degrees of freedom, each succeeding model has six, since the scale of succeeding models is no longer arbitrary, having been established by the first model. Regardless of the number of images in the model, absolute orientation always involves seven parameters.

Relative orientation is the basis of independent model block adjustment (Section 5.8.2). Strip relative orientation is the first step of polynomial block adjustments (Section 5.8.1).

5.6.1   Analytical Relative Orientation Using the Coplanarity Condition

As discussed in Section 4.5.3, the coplanarity equation expresses the condition that the two perspective centers and the two conjugate image rays lie in the same plane. All of the twelve exterior orientation elements of the two images appear in the equations, so seven parameters must be fixed for relative orientation. The parameters can be fixed simply by not solving for them, or by using the unified approach to least squares adjustment (Appendix B.9) and setting the variances on the fixed parameters to small values. The latter approach is more flexible, allowing different combinations of parameters to be fixed if desired and also allowing parameters to be held to fixed values. No approximations for the point coordinates are required, since the object space coordinates of the point do not appear in the coplanarity equation. This can be advantageous in many operational situations.

Since the coplanarity condition adds no unknown parameters, each coplanarity equation satisfies one degree of freedom. Five coplanarity equations, written for five points, are therefore sufficient for relative orientation, but it is desirable to have more points to obtain a more precise solution and to allow the detection of bad measurements.

Application of the coplanarity equation to the relative orientation of more than two images is problematic. We can write coplanarity equations for each pair of images, but this will not necessarily ensure that the image rays of a point that appears in more than two images will intersect in a common point, only that pairwise combinations of rays will intersect. Consider the case of three images from a single flight line. Since the perspective centers lie in nearly a straight line, all three of the image rays will be coplanar (Fig. 4-11). The coplanarity condition will be satisfied for any pair of rays, but this does not mean that all three rays intersect in a point. If all three rays do not intersect, the scale is inconsistent between the two models. To enforce a consistent scale, we write a scale restraint equation (Section 4.5.5) for each point that appears on three or more images.

5.6.2   Analytical Relative Orientation Using the Collinearity Equations

Relative orientation using the collinearity equations is a special case of bundle adjustment. Although the use of the collinearity condition requires the generation of approximate model coordinates for the points, the implementation can be simpler since there is only one type of equation to implement. As with the coplanarity approach, unified least squares can be used to fix the values of selected parameters.

Each image of a point contributes two collinearity equations, for a total of four equations per point for a pair of photographs. Since each point adds three unknowns, its model coordinates, there is a redundancy of one. Five points are therefore needed for the five degrees of freedom of relative orientation; this is, of course, the same as for the coplanarity approach.

In strip relative orientation, collinearity equations are again written for each image of each point. Each point that appears on more than two images adds a redundancy of two, since two collinearity equations are added but no more unknown point coordinates are added.

5.7 ABSOLUTE ORIENTATION

Before a relatively oriented stereomodel can be used for mapping, the relationship between the model space and object space coordinate systems must be established. This is known as *absolute orientation*. Seven parameters are involved: a uniform scale, three translations, and three rotations (Fig. 5-7).

When working on an analog stereoplotter, absolute orientation is performed empirically. Two horizontal control points in the model provide the scale, the translations along the X and Y axes, and the κ rotation around the Z axis. Three elevation control points provide leveling information (the Ω and Φ rotations around the X and Y axes) and the Z translation.

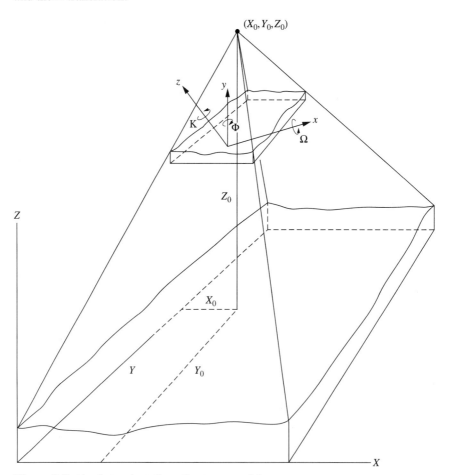

Figure 5-7 Absolute orientation of a stereo model.

Analytical absolute orientation is based on the three-dimensional similarity transform described in Appendix A.4,

$$Y = \mu M X + T \tag{5-7}$$

where Y is the vector of known world coordinates, μ is the scale factor, M is the rotation matrix from model coordinates to the world coordinate system, X is the vector of model coordinates, and T is the translation vector. This set of three scalar equations is written for each point with all coordinates known in both the model and world coordinate systems. For a point with only horizontal coordinates or the vertical coordinate known, only the relevant equations (the first two or the last one, respectively) need to be written. Alternatively, the standard deviations of the unknown coordinates can be set to large values, thereby removing the influence of those coordinates on the solution.

Since the solution is nonlinear, approximations must be generated before the solution can proceed. The model scale can be estimated by calculating the ratio of the distances between a pair of points in model space and in object space. For models constructed from vertical photography, the tilts can be approximated as zero. The approximate rotation around the Z-axis can be determined by calculating the azimuth between two points in both the model space and object space coordinate systems and setting the approximation to be the object space azimuth minus the model space azimuth. The translations can then be calculated by applying the approximate scale and rotation matrix to a point in the model system, then subtracting the transformed coordinates from its object space coordinates. Linear formulations of absolute orientation also exist (Thompson, 1959; Horn, 1987; Horn, 1988).

5.8   BLOCK TRIANGULATION

The development of block triangulation has been a major factor in improving the economic feasibility of photogrammetric mapping. Without triangulation, every stereomodel would need two horizontal and three vertical control points plus additional check points, requiring expensive ground surveys. Additionally, determining the orientations of all images simultaneously yields more accurate and consistent mapping across the entire region.

The development of block triangulation has been driven by the advances in computational power. Before the advent of computers, triangulation methods used analog devices or graphical techniques. Early analytical methods were designed to use the output of existing stereoplotters and to minimize the amount of computation required by combining individual images into larger units. Polynomial strip adjustment methods were modeled after graphical (analog) adjustment methods. Independent model algorithms reduce the number of parameters required by working with models instead of actual images and simplifying the computations required. At the present time, nearly all block triangulation is done using the bundle method, which allows the integration of additional geometric or navigational information. Only a small percentage of block triangulation is still done using independent models.

The availability of accurate navigational information has also advanced block triangulation in recent years. Using the Global Positioning System (GPS) to determine exposure station positions essentially makes a control point of each exposure station, reducing control requirements for the block to minimal configurations. The operational aspects of GPS utilization are still being explored, but the availability of GPS has already had a great impact on block triangulation procedures. Improved attitude determination by inertial navigation systems (INS) is also beginning to have a significant impact on triangulation, especially for nonframe sensors.

This section briefly discusses some of the earlier approaches to block and strip triangulation, then concentrates on the implementation of the bundle adjustment and its extensions.

5.8.1   Polynomial Strip and Block Adjustment

As discussed previously, we can perform a relative orientation of an entire strip of photography by either analog or analytical methods. During the process of tying each successive model in the strip to the one before it, systematic errors in the image coordinates accumulate to warp the strip (Fig. 5-8). In polynomial strip adjustment, the transformation from strip coordinates to world coordinates is modeled as a set of polynomials with terms chosen to correct for the common modes of strip deformation. One set of equations is solved for the planimetric (X, Y) coordinates, while another equation models the Z transformation.

One popular model, the Schut adjustment, developed at the National Research Council of Canada, uses conformal polynomials (Appendix A) for the planimetric adjustment and a separate polynomial for the vertical adjustment.

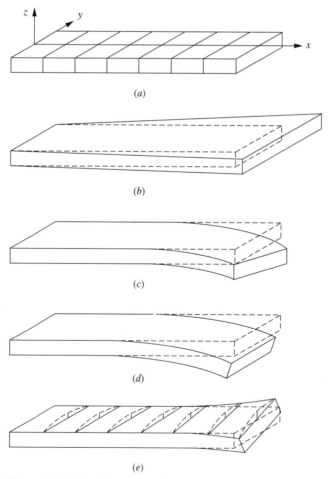

(a)

(b)

(c)

(d)

(e)

Figure 5-8 Strip deformation after strip formation process. Those deformations are corrected by polynomial adjustment methods.

$$x' = x + a_1 + a_3 x - a_4 y + a_5(x^2 - y^2) - 2a_6 xy +$$
$$a_7(x^3 - 3xy^2) - a_8(3x^2 y - y^3) + \cdots$$
$$y' = y + a_2 + a_4 x + a_3 y + a_6(x^2 - y^2) + 2a_5 xy +$$
$$a_7(3x^2 y - y^3) + a_8(x^3 - 3x^2 y) + \cdots \tag{5-8}$$
$$z' = z + b_0 - 2b_2 x + 2b_1 y + c_1 x^2 + c_2 x^3 +$$
$$c_3 x^4 + d_1 xy + d_2 x^2 y + d_3 x^3 y +$$
$$d_4 x^4 y + e_1 y^2 + e_2 xy^2$$

The coefficients of the polynomials are computed after first scaling and translating the strip coordinates into the world coordinate system. If a block adjustment is to be performed, each strip is adjusted sequentially, using adjusted tie points from the previous strip, until the adjustments between strips become negligible. Once the adjustment is completed, the adjusted world coordinates of the tie points are computed, and these points are used to set up stereomodels for map compilation.

Polynomial adjustment is computationally economical, since only the polynomial coefficients must be computed. This method was therefore well suited to the limited computational power available in the early 1960s. However, its accuracy is low since the image geometry is not directly modeled and only gross effects can be corrected. Polynomial adjustment is no longer used in practice, other than occasionally to generate tie point coordinate approximations for a bundle adjustment, since the cost of the computational power required for a more accurate bundle adjustment is much less than that of the additional control points required to obtain satisfactory results with polynomial methods.

5.8.2   Block Adjustment by Independent Models

Instead of grouping the models into strips, we can adjust each model independently as part of an overall block adjustment. This also reduces the number of parameters in the adjustment (seven per model, as opposed to six per image for bundle adjustment), but gives much better results than polynomial adjustment methods. The models can be formed on standard analog stereoplotters or from comparator measurements using analytical relative orientation methods (Section 5.6).

Independent model block adjustment can be thought of as the simultaneous absolute orientation (Section 5.7) of all the models in the block. As discussed previously, if we have two horizontal and three vertical control points in a model, we can relate model coordinates to world coordinates using the three-dimensional similarity transformation. In independent model block adjustment, instead of control points in each model, we have a block of models connected by tie points (including the perspective centers), which among them contain at least two horizontal and three vertical control points (Fig. 5-9).

Inclusion of the perspective centers as tie points is critical for the geometric strength of the solution. In standard block configurations, stereomodels overlap only at the very edges in the along-strip direction. If tie points were chosen only within the edge regions, the tie points would be nearly collinear, allowing the connected models to rotate around the overlap region. Adding the perspective center as a tie point reinforces the connection by preventing this rotation. If an analog stereoplotter is used to form the models, the coordinates of the perspective center must be measured while the model is set in the instrument. If analytical methods are used, the perspective center coordinates are determined as part of the solution.

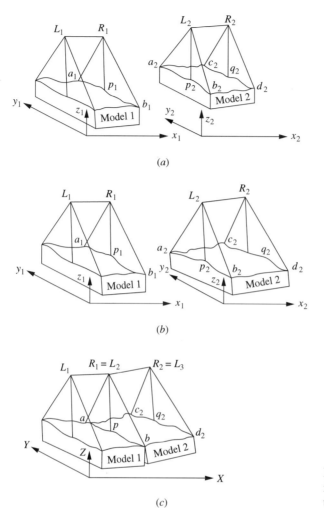

(a)

(b)

(c)

Figure 5-9 Tie points for independent model triangulation.

To implement independent model block adjustment, we write Eq. 5-7 for each time a control or tie point appears in a model, with the absolute orientation parameters for that model as unknowns. The model and world coordinates of the points are treated as observations, with the standard deviations differentiating between known and unknown coordinates. The total number of unknowns for the least squares solution is then seven per model and three per tie point. We can solve the total system of equations and obtain absolute orientation parameters for each model, as well as adjusted world coordinates for the tie points.

Independent model block adjustment was widely adopted, since it could use either existing stereoplotters or comparators for input data generation, the computational expense was well suited to available computers, and the accuracy was much better than that of the polynomial methods.

Several variations of the independent model method were developed. One of the most successful was the implementation in PAT-M-43, developed at the University of Stuttgart. This divided the seven-parameter absolute orientation described above into two separate solutions, a four-parameter planimetric adjustment and a three-parameter height adjustment. The planimetric adjustment incorporated the scale factor, the K rotation around the Z-axis, and the X and Y translations, while the height adjustment incorporated the two leveling rotations, Ω and Φ, and the Z translation. These two adjustments are not independent, since the planimetric adjustment assumes that the model is level and the height adjustment assumes that the points are properly positioned planimetrically. The two adjustments must therefore be repeated alternately until the corrections to the point coordinates are negligible. Since the conditions are usually close to being met at the start of the procedure and the solution is nearly linear, only a few iterations are typically required.

A variation of independent model adjustment is to work with larger units, such as triplets or sub-blocks. Using larger basic units further reduces the number of parameters and simplifies editing of tie points for blunders.

5.8.3 Bundle Block Adjustment

The most elementary geometric unit in photogrammetry is the image ray, which connects an object space point, the perspective center of an image, and the projection of the point on the image. A single image can be thought of as a *bundle* of such rays converging at the perspective center with an unknown position and orientation in space. A bundle block adjustment establishes the position and orientation of each bundle, using the rays in each bundle and the given ground control information.

Bundle adjustment is the most accurate and flexible method of triangulation currently in use. The accuracy levels attainable in aerial triangulation allow its use for geodetic control extension, while in close-range applications, precision of 1:500,000 of the largest dimension of the measured object has been reported. Numerous extensions of the basic method have increased both flexibility and accuracy. The addition of self-calibration parameters (Section 5.9) corrects for remaining systematic errors and increases the overall accuracy of the adjustment. Geometric relationships within the scene, such as collinearity or coplanarity, can be incorporated (Section 4.5.2) to add information and thereby increase the precision. High-quality navigational data from GPS receivers (Section 5.10) can be rigorously included to reduce control requirements dramatically.

Bundle block adjustment is based on the collinearity equations (Section 4.5.1):

$$x - x_0 = f\,\frac{m_{11}(X - X_L) + m_{12}(Y - Y_L) + m_{13}(Z - Z_L)}{m_{31}(X - X_L) + m_{32}(Y - Y_L) + m_{33}(Z - Z_L)}$$
$$y - y_0 = f\,\frac{m_{21}(X - X_L) + m_{22}(Y - Y_L) + m_{23}(Z - Z_L)}{m_{31}(X - X_L) + m_{32}(Y - Y_L) + m_{33}(Z - Z_L)} \tag{5-9}$$

For the least squares adjustment, the collinearity equations are linearized (Appendix C) in the form

$$v + \dot{B}\dot{\delta} + \ddot{B}\ddot{\delta} = f \tag{5-10}$$

The unknowns in Eq. 5-10 are the six exterior orientation parameters of the image (the elements of $\dot{\delta}$) and the three object space coordinates of the point (the elements of $\ddot{\delta}$). The observations are the image coordinate measurements, with residuals v. Bundle adjustment is most often implemented using unified least squares (Appendix B.9), which allows the assignment of realistic standard deviations to each point's object space coordinates. Since bundle block adjustment involves more unknowns than other triangulation methods, efficient algorithms for forming, storing, and solving the normal equations are particularly important. These algorithms are discussed in depth in Section 5.11.

5.9   BLOCK ADJUSTMENT WITH ADDED PARAMETERS (SELF-CALIBRATION)

Camera calibration was formerly performed as a separate procedure from block adjustment. The interior orientation parameters of the camera were determined and then held fixed during the triangulation procedure. Image coordinate measurements were assumed to be corrected for systematic effects, such as lens distortion and atmospheric refraction, before the adjustment. This enabled the use of simplified geometric and stochastic models, since image measurements were assumed to be uncorrelated and to contain only random errors.

However, there are many situations in which we would like to determine the systematic errors under operational conditions as part of the solution. The examination of residuals from block adjustments often shows clear trends attributable to uncorrected systematic errors, due to differences in environmental conditions between the calibration laboratory and the operational environment. In some cases, it is not feasible to completely calibrate the camera in a laboratory, as in close-range applications that use nonmetric cameras. Many navigational sensors, such as inertial navigation systems or statoscopes, have systematic drift errors that increase as a function of time. While the overall characteristics of the errors are known, the exact amount of drift cannot be determined beforehand and must be part of the solution. For these reasons, block adjustment with self-calibration parameters was developed.

Examination of the collinearity equations (Eqs. 4-24) shows that the principal point coordinates and the focal length appear directly in the equations and could be treated as additional unknowns in the solution. It would seem that the refined image coordinates could be replaced by unrefined image coordinates and the correction equations for lens distortion, so that the correction coefficients can be obtained from the solution.

The problem with naively solving for the interior orientation parameters in this way is that in many cases there is not enough geometric information to separate the effects of the parameters so that they can be determined. In this case, we say that the parameters are *correlated,* since choosing a value for one parameter fixes the value of the other. For example, take the case of a perfectly vertical aerial image of flat terrain. There are an infinite number of combinations of flying height and focal length that will yield the same image. If we are given the measurements of an object in the image and on the ground, we cannot determine both the focal length and flying height, since their effects cannot be separated. Exactly the same thing can happen in an aerial triangulation with added parameters—the geometry is often such that the effects of the self-calibration parameters cannot be distinguished from each other or from the effects of the orientation parameters.

Note that parameters that are not perfectly correlated can still cause problems with the solution. In the previous example, if we are given the measurements of two objects at different elevations, we can theoretically determine both the flying height and focal length. However, in a practical sense, we will still have trouble determining both parameters unless the difference in elevation is significant with respect to the flying height. If the difference in scale for the two objects is not greater than the effects of measurement or computational roundoff errors, the parameters determined will not be accurate.

Therefore, to apply self-calibration by added parameters, we must carefully choose the parameters we add to the solution, we must design the image acquisition geometry to minimize correlation, and we must pay careful attention to correlations when we perform the solution.

Several formulations of self-calibration parameters have been proposed. Generally, they are based on first modeling known systematic errors, such as lens distortion, principal distance error, or principal point offset. Additional terms are then added to cover general deformations that might be present. A typical set of added parameters are those proposed by Brown (1976):

$$\Delta x = a_1 x + a_2 y + a_3 xy + a_4 y^2 + a_5 x^2 y + a_6 xy^2 + a_7 x^2 y^2$$
$$+ \frac{x}{c}\left[a_{13}(x^2 - y^2) + a_{14}x^2y^2 + a_{15}(x^4 - y^4)\right]$$
$$+ x[a_{16}(x^2 - y^2)^2 + a_{17}(x^2 + y^2)^4 + a_{18}(x^2 + y^2)^6] + a_{19} + a_{21}\left(\frac{x}{c}\right)$$
$$\Delta y = a_8 xy + a_9 x^2 + a_{10}x^2y + a_{11}xy^2 + a_{12}x^2y^2 \qquad \text{(5-11)}$$
$$+ \frac{y}{c}[a_{13}(x^2 - y^2) + a_{14}x^2y^2 + a_{15}(x^4 - y^4)]$$
$$+ y\left[a_{16}(x^2 + y^2)^2 + a_{17}(x^2 + y^2)^4 + a_{18}(x^2 + y^2)^6\right] + a_{20} + a_{21}\left(\frac{y}{c}\right)$$

where x and y are the image coordinates, c is the principal distance, a_{1-21} are the added parameters, and Δx and Δy are the corrections.

The terms in these equations were designed to handle film deformation, uncorrected lens distortion, and platen unflatness. *Orthogonality*, or statistical independence, of the terms is important. When the terms are orthogonal, no two terms describe the same effect. This reduces parameter correlations.

The self-calibration parameters need not apply to the entire block; they may apply only to individual images, flight lines, or sub-blocks. This may be done if more than one camera was used, the flights were made on different days or under different conditions, etc.

Proper project design, using highly redundant photo coverage, is crucial for the successful use of self-calibration. In close-range applications, one can use convergent photography from multiple orientations and positions. While convergent photography is not an option for aerial mapping, using flight lines with 60% sidelap gives at least four image rays on each point. Another strong configuration is the use of cross strips. Tie points should be well-distributed, arranged in a regular 3 by 3, or preferably a 5 by 5, grid on each image.

Added parameter solutions typically use the unified least squares method, so that *a priori* weights can be applied to the added parameters to control and stabilize the solution. Parameters which are known by experience to be correlated in a given configuration can be constrained to zero. An initial adjustment is performed with the added parameters constrained to small values to obtain good values for the exterior orientation parameters while keeping the solution stable. The parameter correlations are then analyzed and the weights on the appropriate coefficients are adjusted to allow them to contribute to the final solution. The correlations and standard deviations must be carefully monitored, as discussed in Section 5.12, to ensure a valid solution.

5.10   UTILIZATION OF NAVIGATIONAL DATA

Several attempts have been made in the past to utilize navigational data in order to improve accuracy and to decrease the number of control points required. However, the limited accuracy of available sensors was not suitable for photogrammetric applications, and available systems were often too expensive and hard to use for widespread application.

The development of the Global Positioning System (GPS) by the U.S. Department of Defense and its subsequent widespread adoption by commercial users has made highly accurate and relatively inexpensive means of determining positions available. The Global Positioning System is widely used within photogrammetry for surveying ground control points, controlling photo acquisition, and especially for determining image positions. It is typically used in conjunction with inertial navigation systems for the determination of image positions.

5.10.1 The Global Positioning System

The Global Positioning System was originally designed by the U.S. Department of Defense as a means for combatants to determine their position quickly and accurately. Further developments by the Department of Defense and civilian users have improved its accuracy for civilian applications and, in the process, led to its widespread adoption by users in many fields. The Russian government is also fielding a similar system, GLONASS (Global Navigation Satellite System).

5.10.1.1   GPS Operating Principles

The Global Positioning System consists of 21 active satellites (plus some spares) in orbit at an altitude of about 20,000 kilometers. The satellites are placed in orbits such that at least five satellites are always visible from any place on Earth. Each satellite contains four atomic clocks and continuously broadcasts two carrier signals, L1 (1.57542 GHz) and L2 (1.2276 GHz). The L1 carrier transmits the navigation message, which contains the clock and ephemeris data for the satellite. Each carrier is modulated by a pseudo-random noise (PRN) sequence, which is specific to a given satellite. Two PRN sequences are employed: the coarse acquisition (C/A) code modulates the L1 carrier and repeats every millisecond, while the precise (P) code modulates both carriers and repeats every seven days. When anti-spoofing (AS) is activated, the P code is encrypted, necessitating the use of a military receiver with decryption capabilities.

Position determination with GPS is based on intersecting the ranges to at least four satellites. The ranges are calculated by receiving the modulated carriers and correlating the known signal patterns against the transmitted PRN patterns. The time offset of the correlation peak with respect to the start time of the PRN pattern indicates the distance of the receiver from the satellite (Fig. 5-10). If the clocks in the receiver and satellite were synchronized, we could multiply the time offset by the propagation speed of the signal and determine the distance from the receiver to the satellite. In reality, the clocks are different and we can only calculate a pseudo-range, which we must adjust after determining the clock errors.

We can calculate the satellite's position at the time the signal was broadcast from the satellite ephemeris, which is broadcast as part of the navigation message. The pseudo-ranges to each satellite define a sphere around each satellite's position; the four spheres thus defined would intersect in a point if there were no clock errors. By algebraically requiring the spheres to intersect at a point, we can solve for the scaling factor between the receiver and satellite clocks.

Since the GPS was initially conceived as a military system, an important consideration in its design was preventing hostile forces from using it. Two positioning modes are therefore implemented: the Standard Positioning System for civilian users and the Precise Positioning System for military applications. In the Standard Positioning System, the clock signals are intentionally dithered (randomly delayed) to reduce

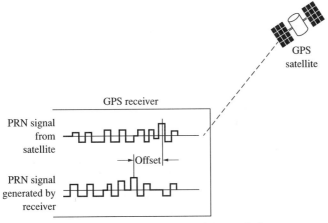

Figure 5-10 Pseudo-Random Noise (PRN) correlation, to determine pseudo-range from the GPS satellite.

the attainable accuracy to about 100 meters. Access to the Precise Positioning System requires a receiver equipped with a cryptographic key to decipher the correction signals, and allows 22 meters horizontal accuracy. The intentional degradation of the signals, known as *selective availability* (SA), has been recently deactivated by the Department of Defense.

Differential GPS techniques have been developed to remove the effect of selective availability, along with many of the other bias errors present in GPS signals. In *code-phase* differential GPS surveying, two or more receivers are used (Fig. 5-11), one or more of that remain stationary. Since the receivers are relatively close together, the biases that affect the satellite signal will be nearly the same for all receivers. Any apparent changes in the position of the stationary receiver are due to changing bias errors in the pseudo-ranges, which can then be detected and corrected. The pseudo-range corrections are transmitted by radio to the moving receiver, which applies them before calculating its own position. Accuracies of 1 to 10 meters can be obtained with code-phase differential correction.

Another method for high-accuracy GPS surveys is carrier-phase differential positioning, in either *static* or *kinematic* operation. In this method, the stationary

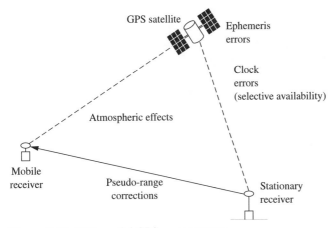

Figure 5-11 Differential GPS measurement.

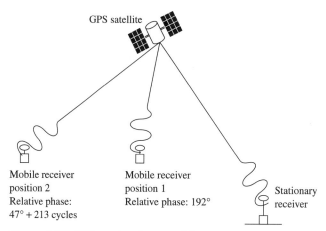

Figure 5-12 GPS carrier-phase tracking.

or mobile receivers both track the phase of the carrier waves from the satellite. Since the wavelength of the carrier is known (about 19 centimeters for the L1 carrier), a change in the relative phase between the receivers indicates a change in distance (Fig. 5-12). The wavelength of the carrier is smaller than that of the transmitted code so positioning can be done much more accurately using this method. Tracking multiple satellites gives multiple phase differences and allows the position to be determined to a small fraction of the wavelength. The greatest difficulties in kinematic positioning are determining the ambiguity in the number of wavelengths between the receivers and continuously tracking the satellite signal so that the established phase differences are not lost.

The accuracy available from GPS signals is affected by a number of biases present in the system. Along with the intentional errors introduced into the transmissions by selective availability, errors in the satellite ephemeris, inaccuracies in calculating atmospheric transmission delays, multipath reflections, and other effects reduce the accuracy of the GPS data. The precision of the calculated position is a function of the number of satellites observed and the geometry of the range intersections. For this reason, the satellites used should be high above the horizon and well-distributed across the sky.

The geometric strength of the satellite configuration is described by the geometric dilution of precision (GDOP), which represents the volume of the ellipsoid defined by the unit vectors from the receiver to the satellites. The GDOP is computed from the geometry of the receiver and satellite positions, assuming equal errors in pseudoranges and times. The GDOP can be divided into separate terms describing the horizontal, vertical, three-dimensional, or time precision. The appropriate terms of the GDOP are multiplied by the estimated range errors to obtain an estimate of the actual precision.

5.10.1.2   Incorporation of GPS Positions into Block Adjustment

Integration of GPS positions into block adjustment first requires careful integration of the GPS receiver, the data collection equipment, and the camera to ensure the collection of usable data. The GPS antenna position on the aircraft is the first consideration. The GPS position is actually calculated for the antenna; the offset between the antenna and the external nodal point of the camera must be taken into account in the solution. A standard technique is to install the antenna directly over the camera mount after allowing for the aircraft's attitude during flight. This minimizes, but does not entirely eliminate, the horizontal offset between the antenna and the camera. If the camera is in a stabilized mount that moves to keep the camera axis vertical, the situation is further complicated. The antenna-camera offset is usually carried as a parameter in the block adjustment, taking into account the aircraft and camera orientations.

Although pseudo-range positioning can be used for navigation and exposure positioning, only differential methods are accurate enough to contribute to a block adjustment. Differential code-phase methods may be used, but differential carrier-phase positioning is the more accurate procedure. The main problem is continuously maintaining the satellite signal reception so that the phase offset is not lost or the number of cycles miscounted (*cycle slip*). To prevent this, initialization is performed both before and after the flight to establish the correct cycle offsets, which can be propagated forward or backward to points where cycle slips occurred. Alternatively, since loss of contact is most likely during the turns at the end of each flight line, each strip may be treated as an independent entity in terms of its GPS positions.

The loss of cycle lock manifests itself as an offset and apparent linear drift in the GPS position. Any systematic errors in synchronization between the camera shutter and the GPS timing will also appear as a constant offset. These parameters can be carried as unknowns in the adjustment and applied either across the entire block or on a strip-by-strip basis.

The condition equation for a GPS observation of an exposure station position is

$$
\begin{bmatrix} X_{\text{GPS}} \\ Y_{\text{GPS}} \\ Z_{\text{GPS}} \end{bmatrix} = \begin{bmatrix} X_L \\ Y_L \\ Z_L \end{bmatrix} + M^T \begin{bmatrix} d_x \\ d_y \\ d_z \end{bmatrix} + \left(\begin{bmatrix} a_X \\ a_Y \\ a_Z \end{bmatrix} + \begin{bmatrix} b_X \\ b_Y \\ b_Z \end{bmatrix} (t - t_0) \right) \tag{5-12}
$$

where X_{GPS}, Y_{GPS}, and Z_{GPS} describe the position of the GPS antenna and X_L, Y_L, and Z_L are the exposure station coordinates. Since the GPS receiver can take position readings only at fixed intervals, which may not coincide with the exposures, the position used in the equations must be interpolated between adjacent readings. The offsets between the GPS antenna and the external node of the camera, measured in the camera coordinate system, are d_x, d_y, and d_z and M is the rotation matrix from the world to the camera coordinate system (Appendix A). If the camera is rigidly mounted to the aircraft, M is a function of the aircraft's attitude; if the camera is in a stabilized mount, then both the mount and aircraft orientations must be taken into account.

The terms a_X, a_Y, and a_Z and b_X, b_Y, and b_Z model the constant offset and linear drift errors of the GPS positioning and are calculated in the solution. The linear drift is a function of the time t since some reference time, t_0.

Another consideration is the relationship between the GPS coordinates, calculated with respect to the WGS84 ellipsoid, and the local datum in which the control points are surveyed. The transformation for horizontal coordinates is usually well defined. However, vertical control points are typically defined with respect to the local geoid, and the relationship between the geoid and the ellipsoid in many areas is not well known. Datum transformation inaccuracies should not be allowed to distort the geometry of the block. Parameters of the datum transformation can be included within the block adjustment, or the block adjustment can be done using only the GPS information with the block adjusted to the control as a unit to prevent internal deformations due to datum errors.

Optimal control point configurations for GPS-controlled blocks are still being studied. At least theoretically, no control points at all are necessary since each exposure station is being treated as a control point. However, some number of control points is needed to provide datum information and extra reliability against blunders.

5.10.2 Inertial Navigation Systems (INS)

Inertial navigation systems contain precise accelerometers to record linear and angular acceleration. Integrating the output signals from the three orthogonal linear accelerometers gives estimates of velocity and position, and integrating the output signals from the three angular accelerometers provides orientation information.

There have been several experiments in the past in using inertial navigation systems for photogrammetric applications, but INS have not been widely applied. This is due in part to the cost of high-accuracy INS, but also to the inherent accuracy limitations of the systems. Inertial systems tend to drift over time, with the error increasing the longer the system has been operating. This drift behavior can be modeled using added parameters in the block adjustment, but still limits the accuracy obtainable from the sensor.

The integration of INS and GPS information has a number of advantages over the use of either system alone. Since the INS provides continuous positioning, while the GPS is read at discrete intervals, the INS acts to smooth out random errors. The INS serves as a backup system in case the satellite signal is lost due to aircraft maneuvers, since its drift will not be significant over short periods of time. The INS can provide orientation data along with position, providing additional information for use in a block adjustment.

5.11 NUMERICAL ASPECTS OF BUNDLE ADJUSTMENT

One of the most important factors in the adoption of bundle adjustment has been the development of efficient algorithms for formation and solution of the normal equations. Detailed analysis of the structure of the B and N matrices (Appendix B) greatly reduces both the computational and storage requirements for the solution.

5.11.1 Normal Equation Formation

The fundamental equation of bundle adjustment is the collinearity equation, which describes the basic unit of photogrammetry, the image ray. The use of the collinearity condition makes equation formation and solution a very efficient process. To begin, we write the linearized collinearity equations (Appendix C) for each image i and point j,

$$\underset{2 \times 1}{v_{ij}} + \underset{2 \times 6}{\dot{B}_{ij}} \; \underset{6 \times 1}{\dot{\delta}_i} + \underset{2 \times 3}{\ddot{B}_{ij}} \; \underset{3 \times 1}{\ddot{\delta}_j} = \underset{2 \times 1}{f_{ij}} \tag{5-13}$$

where $\dot{\delta}$ contains the corrections to the parameters of image i and $\ddot{\delta}$ contains the corrections to the coordinates of point j.

If we now form the least-squares normal equations associated with one image of one point, we obtain

$$\begin{bmatrix} \dot{B}_{ij}^T W_{ij} \dot{B}_{ij} & \dot{B}_{ij}^T W_{ij} \ddot{B}_{ij} \\ \ddot{B}_{ij}^T W_{ij} \dot{B}_{ij} & \ddot{B}_{ij}^T W_{ij} \ddot{B}_{ij} \end{bmatrix} \begin{bmatrix} \dot{\delta}_i \\ \ddot{\delta}_j \end{bmatrix} = \begin{bmatrix} \dot{B}_{ij}^T W_{ij} f_{ij} \\ \ddot{B}_{ij}^T W_{ij} f_{ij} \end{bmatrix} \tag{5-14}$$

where W_{ij} is the 2 x 2 weight matrix (the inverse of the covariance matrix) associated with the image coordinates of point j on image i. Equation 5-14 can be rewritten as

$$\begin{bmatrix} \dot{N}_i & \overline{N}_{ij} \\ \overline{N}_{ij}^T & \ddot{N}_j \end{bmatrix} \begin{bmatrix} \dot{\delta}_i \\ \ddot{\delta}_j \end{bmatrix} = \begin{bmatrix} \dot{t}_i \\ \ddot{t}_j \end{bmatrix}$$

(5-15)

Note that $\dot{N}$ contains coefficients of the image parameters, $\ddot{N}$ contains coefficients of the point parameters (coordinates), $\overline{N}$ and refers to both.

We now make the pivotal assumption that the errors in each point's image coordinates are uncorrelated with those of any other image coordinates, both other points imaged on the same image and other images of the same point, although error in the x and y coordinates of any single image point may be correlated. If image coordinate errors are correlated, they are affected by the same systematic errors instead of being purely random; when we assume that the image coordinate errors are uncorrelated, we are assuming that all systematic errors have been removed. This is seldom completely true in practice and is addressed by the use of added parameters (Section 5.9).

This assumption means that the total covariance, Σ, and weight, W, matrices, containing the covariance and weight information for all the image coordinates, are block diagonal. Each block is a 2×2 matrix corresponding to one set of image measurements for one point, as shown in Fig. 5-13. Since the measurements are independent of each other, and each set of collinearity equations refers to only one image and one point, we can sum the contributions to the normal equations from each set of collinearity equations. The total $\dot{N}$ matrix is block diagonal, with each 6×6 block referring to a separate image. Each $\dot{N}_i$ is the sum of the $\dot{N}_{ij}$ submatrices, formed from the $\dot{B}_{ij}$ and W_{ij} matrices from each set of collinearity equations that refer to image i. Similarly, $\ddot{N}$ has 3×3 blocks on the diagonal, each referring to the coordinates of an individual point. Each $\ddot{N}_j$ is formed from the $\ddot{B}_{ij}$ and W_{ij} of the collinearity equations referring to point j. The structure of the total $\overline{N}$ matrix is determined by which points occur on which images. Each $\overline{N}_{ij}$ submatrix is the product of the $\dot{B}_{ij}$, W_{ij}, and $\ddot{B}_{ij}$ from one set of collinearity equations, so $\overline{N}_{ij}$ submatrices corresponding to a point j that does not occur on image i are zero. A typical normal equation structure is shown in Fig. 5-14.

The advantages of this method of equation formation are obvious, since the alternative would be to first form the entire B matrix and then calculate N by brute force multiplication. For realistic numbers of images and points the total B matrix will be mostly zeros, meaning that large amounts of storage would be used to store zeros and that the majority of the effort in forming N would be in multiplying by zero.

Bundle adjustment is most often implemented using unified least squares (Appendix B.9), which does not change the basic formation algorithm significantly. The weight matrices of the parameter observations for each image and point, $\dot{W}_i$ and $\ddot{W}_j$, are added to the respective $\dot{N}_i$ and $\ddot{N}_j$ submatrices. The total corrections to the original parameter observations, $\dot{f}_i$ and $\ddot{f}_j$, multiplied by the corresponding parameter weight matrices, are subtracted from the $\dot{t}_i$ and $\ddot{t}_j$ vectors.

A simple formation algorithm is summarized in the pseudo-code given in Algorithm 1.

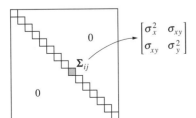

Figure 5-13 Image coordinate covariance matrix, showing 2×2 block diagonal structure.

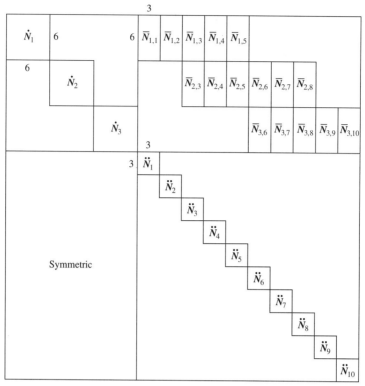

Figure 5-14 Unreduced normal equation structure.

Algorithm 1 Normal equation formation (unified least squares)

```
/* add in initial parameter weights */
for each point j
```
$$\dot{N}_j = \dot{W}_j$$
$$\dot{t} = -\ddot{W}_j\ddot{f}_j$$
```
end point j loop
for each image i
```
$$\dot{N}_i = \dot{W}_i$$
$$\dot{t}_i = -\dot{W}_i\dot{f}_i$$
```
    for each point j
        if point j is on image i
            calculate Ḃ_ij , B̈_ij , and f_ij
```
$$\dot{N}_i = \dot{N}_i + \dot{B}_{ij}^T W_{ij} \dot{B}_{ij}$$
$$\dot{t}_i = \dot{t}_i + \dot{B}_{ij}^T W_{ij} f_{ij}$$
$$\ddot{N}_j = \ddot{N}_j + \ddot{B}_{ij}^T W_{ij} \ddot{B}_{ij}$$
$$\ddot{t}_j = \ddot{t}_j + \ddot{B}_{ij}^T W_{ij} f_{ij}$$
$$\overline{N}_{ij} = \dot{B}_{ij}^T W_{ij} \ddot{B}_{ij}$$

```
      end point j loop
end image i loop
```

5.11.2 Normal Equation Reduction and Solution

The total normal equations are a large system of equations with a well-defined structure. We can take advantage of this structure to reduce the size of the matrices we have to solve by eliminating one set of variables, in this case the point coordinate corrections. After solving for the remaining image parameter corrections, the point coordinate corrections are obtained by back-substitution.

We can think of the total normal equations as two sets of equations:

$$\dot{N}\hat{\delta} + \overline{N}\ddot{\delta} = \dot{t}$$
$$\overline{N}^T\dot{\delta} + \ddot{N}\ddot{\delta} = \ddot{t}$$

(5-16)

Solving the second set of equations for $\ddot{\delta}$ yields

$$\ddot{\delta} = \ddot{N}^{-1}(\ddot{t} - \overline{N}^T\dot{\delta})$$

(5-17)

Substituting this into the first set of equations and simplifying gives

$$(\dot{N} - \overline{N}\ddot{N}^{-1}\overline{N}^T)\dot{\delta} = t - \overline{N}\ddot{N}^{-1}\ddot{t}$$

(5-18)

These are referred to as the *reduced normal equations*. Notice that this solution involves computation of the inverse of $\ddot{N}$. Since $\ddot{N}$ is block diagonal, its inverse is just the inverse of each on-diagonal 3×3 block, an inexpensive operation (Appendix A). Algorithm 2 implements the reduction process.

Algorithm 2   Normal Equation Reduction

```
for each image i
  for each point j
    if point j is on image i
```

$$\dot{N}_{ii} = \dot{N}_{ii} - \overline{N}_{ij}\ddot{N}_j^{-1}\overline{N}_{ij}^T$$

$$\dot{t}_i = \dot{t}_i - \overline{N}_{ij}\ddot{N}_j^{-1}\ddot{t}_j$$

```
      for each other image k point j is on
```

$$\dot{N}_{ik} = \dot{N}_{ik} - \overline{N}_{ij}\ddot{N}_j^{-1}\overline{N}_{jk}^T$$

```
      end image k loop
    end point j loop
  end image i loop
```

The off-diagonal submatrix $\overline{N}_{ij}$ is zero if image i and point j do not appear together in a collinearity equation, that is, if point j is not imaged on image i. This fact determines the structure of the reduced normal equations, which we exploit to efficiently solve the equations.

Analysis of Algorithm 2 shows where the contributions from each point are subtracted from the reduced normal equations. Before reduction, $\dot{N}$ is a block diagonal matrix. Each block corresponds to one image, and the dimensions of each block are determined by the number of parameters for each image (i.e., six for the standard frame camera). If a point appears on only one image, its contributions are subtracted only from the diagonal block corresponding to that image ($\dot{N}_{ii} = \dot{N}_i$). However, if a point appears on both image i and image k, its contributions are subtracted from the on-diagonal blocks of the normal equations corresponding to image i and image k and also from the off-diagonal submatrix $\dot{N}_{ik}$ (Fig. 5-15). This nonzero off-diagonal submatrix changes the structure of the matrix and directly affects the cost of the solution.

For standard aerial block adjustment, the reduced normal equation coefficient matrix is a *sparse* matrix, meaning that most of its entries are zero. It is also a *banded* matrix, since all its nonzero coefficients are within a fixed distance (the *bandwidth*) from the main diagonal. A banded matrix can be solved much more efficiently than a full matrix, since the zero elements can be ignored in the solution.

The bandwidth is determined by the configuration of the block and by how the photos are ordered (numbered) in the normal equations. The farther apart in the normal equations two photos i and k imaging the same point are located, the greater the bandwidth, due to the $\dot{N}_{ik}$ term that is formed during the reduction phase. To reduce the bandwidth, we try to find the method of numbering the photos that will minimize the distance between all photos containing the same points. To illustrate the bandwidth reduction possible, we will now compare down-strip and cross-strip numbering.

Figure 5-16 shows the reduced normal equation structure for photos numbered along the strip. Each photo has tie points in common with the images before and after it in the strip, and also with the images in the strips on either side. A tie point at the corner of an image will also appear on up to three images in the next strip; if there are n images in the strip, the off-diagonal submatrices generated in the reduction will be a distance of $n + 3$ blocks from the main diagonal. Contrast this to the structure resulting from cross-strip numbering shown in Fig. 5-17. If there are m strips, the maximum distance between photos containing the same point will be $2m + 2$ blocks. As long as there are at least twice as many photos per strip as there are strips, cross-strip numbering will yield a smaller bandwidth and therefore a more efficient solution.

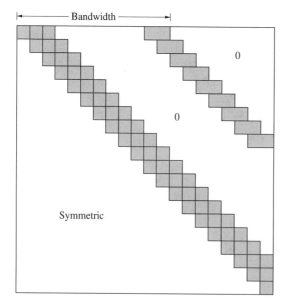

Figure 5-15 Reduced normal equation structure.

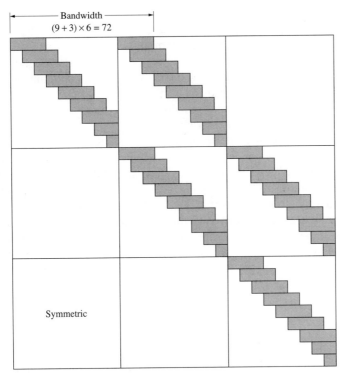

Figure 5-16 Reduced normal equations using down-strip numbering, with three flightlines and nine photos/strip.

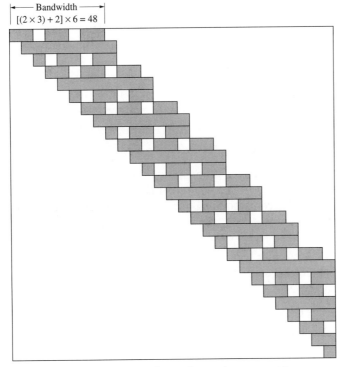

Figure 5-17 Reduced normal equations using cross-strip numbering, with three flightlines and nine photos/strip.

This example assumes a regular strip configuration with more photos in each strip than there are strips. The optimum photo numbering scheme will change if 60% sidelap is used, if the block is square so that the number of photos per strip is equal to the number of strips, if cross strips were flown, or if the photos or strips are not regularly aligned. Determining the photo numbering scheme to obtain the minimum bandwidth is an extremely difficult problem in the most general case, but is such an important determinant of solution efficiency that many triangulation programs have a preprocessing step that attempts to find an optimum ordering.

Again, we should stress that the formation and reduction algorithms presented here were designed for clarity, not necessarily efficiency; for instance, the formation and reduction phases can be combined. There are trade-offs possible between the amount of storage required and the amount of computation required, depending, for instance, on whether the $\bar{N}$ submatrices are calculated once and stored or recalculated each time they are needed. The order of the loops within the algorithms may be changed, depending on whether there are more points or photos. In addition, as discussed in the following sections, the use of self-calibration parameters, geometric constraints, or navigational information necessitates changes in the normal equation structure and in the equation formation and solution procedure.

5.11.3 Normal Equation Formation with Self-Calibration Parameters and/or Constraint Equations

The previous derivation of the reduced normal equations described traditional block adjustment, using only the collinearity equations. However, there are many cases in which we want to add other parameters or equations to the solution. For instance, we may want to determine calibration parameters for the camera (Section 5.9), utilize navigational or GPS information (Section 5.10), or incorporate known geometric properties of the scene. If properly designed and implemented, the additional parameters and/or the additional equations can be added to the solution without adversely affecting its efficiency.

We again start with the collinearity equations, augmented with added parameters. A second set of constraint equations is introduced with the same parameters. The general linearized form of these equations is:

$$v + \dot{B}\dot{\delta} + \hat{B}\hat{\delta} + \ddot{B}\ddot{\delta} = f$$
$$v_c + \dot{C}\dot{\delta} + \hat{C}\hat{\delta} + \ddot{C}\ddot{\delta} = f_c \tag{5-19}$$

where $\dot{\delta}$ and $\ddot{\delta}$ contain the image and point coordinate corrections, as in Eq. 5-13, and $\hat{\delta}$ contains the corrections to any additional parameters for self-calibration, geometric constraints, etc. The corresponding normal equation structure is then

$$\begin{bmatrix} \dot{N} & \breve{N} & \overline{N} \\ \breve{N}^T & \hat{N} & \widetilde{N} \\ \overline{N}^T & \widetilde{N}^T & \ddot{N} \end{bmatrix} \begin{bmatrix} \dot{\delta} \\ \hat{\delta} \\ \ddot{\delta} \end{bmatrix} = \begin{bmatrix} \dot{t} \\ \hat{t} \\ \ddot{t} \end{bmatrix} \tag{5-20}$$

where, assuming the use of the unified approach to least squares,

$$\dot{N} = \dot{B}^T W \dot{B} + \dot{C}^T W_c \dot{C} + \dot{W}$$
$$\breve{N} = \dot{B}^T W \hat{B} + \dot{C}^T W_c \hat{C}$$
$$\overline{N} = \dot{B}^T W \ddot{B} + \dot{C}^T W_c \ddot{C}$$
$$\hat{N} = \hat{B}^T W \hat{B} + \hat{C}^T W_c \hat{C} + \hat{W}$$

$$\bar{N} = \ddot{B}^T W \ddot{B} + \ddot{C}^T W_c \ddot{C}$$
$$\ddot{N} = \ddot{B}^T W \ddot{B} + \ddot{C}^T W_c \ddot{C} + \ddot{W}$$
$$\dot{t} = \dot{B}^T W f + \dot{C}^T W_c f_c - \dot{W} \dot{f}$$
$$\hat{t} = \hat{B}^T W f + \hat{C}^T W_c f_c - \hat{W} \hat{f}$$
$$\ddot{t} = \ddot{B}^T W f + \ddot{C}^T W_c f_c - \ddot{W} \ddot{f}$$

Depending on the types of equations being processed, some submatrices of the total N matrix may be **0**. Assume, for instance, that the constraint equations include only image orientation parameters and parameters related to them, such as the offset between the GPS antenna and the camera; in this case, $\ddot{C}$, and thus $\bar{N}$, are **0**, since there are no equations that involve both ground point coordinates and the constraint parameters. In this case, $\hat{B}$ is also **0**, since none of the constraint parameters will appear in the collinearity equations.

On the other hand, suppose that we are performing self-calibration; if no constraint equations are present, $\dot{C}$, $\hat{C}$, and $\ddot{C}$ will be 0; $\bar{N}$ will contain the calibration parameters; $\ddot{N}$ will be the cross term of the calibration parameters and the orientation parameters, and $\bar{N}$ will be the cross term between the calibration parameters and the ground coordinates. Equation formation proceeds in a manner similar to that described earlier. The procedure shown below assumes that all the matrices exist; for any specific case, matrices that are **0** can be ignored. We begin by writing out the equations represented by the matrices in Eq. 5-20:

$$\dot{N}\dot{\delta} + \ddot{N}\hat{\delta} + \bar{N}\ddot{\delta} = t \tag{5-21a}$$

$$\ddot{N}^T\dot{\delta} + \hat{N}\hat{\delta} + \tilde{N}\ddot{\delta} = \hat{t} \tag{5-21b}$$

$$\bar{N}^T\dot{\delta} + \tilde{N}^T\hat{\delta} + \ddot{N}\ddot{\delta} = \ddot{t} \tag{5-21c}$$

The reduced normal equations are again obtained by eliminating the $\ddot{\delta}$ parameters, the corrections to the point coordinates. Solving Eq. 5-21c for $\ddot{\delta}$ gives

$$\ddot{\delta} = \ddot{N}^{-1}(\ddot{t} - \bar{N}^T\dot{\delta} - \tilde{N}^T\hat{\delta}) \tag{5-22}$$

Substituting this into Eqs. 5-21a and 5-21b and simplifying gives

$$(\dot{N} - N\ddot{N}^{-1}N^T)\dot{\delta} + (\ddot{N} - N\ddot{N}^{-1}\tilde{N}^T)\hat{\delta} = \dot{t} - N\ddot{N}^{-1}\ddot{t}$$
$$(\ddot{N}^T - \tilde{N}\ddot{N}^{-1}\bar{N}^T)\dot{\delta} + (\hat{N} - \tilde{N}\ddot{N}^{-1}\tilde{N}^T)\hat{\delta} = \hat{t} - \tilde{N}\ddot{N}^{-1}\ddot{t} \tag{5-23}$$

If we examine the structure of these equations shown in Fig. 5-18, we notice that the reduced $\dot{N}$ has the same banded structure as before, with the addition of a border

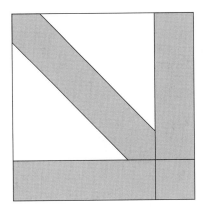

Figure 5-18 Structure of reduced normal equations with added parameters or constraints.

corresponding to the added parameters. This is referred to as a *banded-bordered* matrix.

Algorithm 3 shows a simplified version of the normal equation formation process for a solution involving self-calibration parameters and also additional constraint equations. As with Algorithm 1, the procedure steps through each equation, calculates the partial derivative matrices and discrepancy vectors, and adds their contributions to the corresponding portions of the normal equation matrices. The addition of parameters to the collinearity equation and the inclusion of constraint equations add more steps to the procedure, but do not change the fundamental principles.

Algorithm 3 Formation of Banded-Bordered Normal Equations with Added Parameters (Unified Least Squares)

```
for each point j
```
$$\ddot{N}_j = \ddot{W}_j$$
$$\ddot{t}_j = -\ddot{W}_j \ddot{f}_j$$
```
end point j loop
for each added parameter or constraint equation k
```
$$\widehat{N}_k = \widehat{W}_k$$
$$\widehat{t}_k = -\widehat{W}_k \widehat{f}_k$$
```
end added parameter k loop
/* add the contributions of the collinearity equations,
possibly with self-calibration parameters */
for each image i
```
$$\dot{N}_i = \dot{W}_i$$
$$\dot{t}_i = -\dot{W}_i \dot{f}_i$$
```
    for each point j
        if point j is on image i
            calculate
```
$\dot{B}_{ij}$, $\ddot{B}_{ij}$, and f_{ij}
$$\dot{N}_i = \dot{N}_i + \dot{B}_{ij}^T W_{ij} \dot{B}_{ij}$$
$$\dot{t}_j = \dot{t}_j + \dot{B}_{ij}^T W_{ij} f_{ij}$$
$$\ddot{N}_j = \ddot{N}_j + \ddot{B}_{ij}^T W_{ij} \ddot{B}_{ij}$$
$$\ddot{t}_j = \ddot{t}_j + \ddot{B}_{ij}^T W_{ij} f_{ij}$$
$$\bar{N}_{ij} = \bar{N}_{ij} + \dot{B}_{ij}^T W_{ij} \ddot{B}_{ij}$$
```
            if self-calibration parameter k is used
                calculate
```
$\widehat{B}_{ijk}$
$$\check{N}_{ik} = \check{N}_{ik} + \dot{B}_{ij}^T W_{ij} \widehat{B}_{ijk}$$
$$\widehat{N}_k = \widehat{N}_k + \widehat{B}_{ijk}^T W_{ij} \widehat{B}_{ijk}$$
$$\widehat{t}_k = \widehat{t}_k + \widehat{B}_{ijk}^T W_{ij} f_{ij}$$

```
      end point j loop
   end image loop i
   for each constraint equation
```
$\quad$ calculate $\hat{C}_k$, $\dot{C}_{ik}$, $\ddot{C}_{jk}$ and f_{c_k}

$$\hat{N} = \hat{N} + \hat{C}^T W_c \hat{C}$$
$$\hat{t} = \hat{t} + \hat{C}^T W_c f_c$$
$$\breve{N} = \breve{N} + \dot{C}^T W_c \hat{C}$$
$$\dot{N} = \dot{N} + \dot{C}^T W_c \dot{C}$$
$$\dot{t} = \dot{t} + \dot{C}^T W_c f_c$$

$$\bar{N} = \bar{N} + \hat{C}^T W_c \ddot{C}$$
$$\ddot{N} = \ddot{N} + \ddot{C}^T W_c \ddot{C}$$
$$\ddot{t} = \ddot{t} + \ddot{C}^T W_c f_c$$
$$\breve{\bar{N}} = \breve{\bar{N}} + \dot{C}^T W_c \ddot{C}$$

```
end constraint equation loop
```

The reduction process for normal equations with added parameters presented in Algorithm 4 is similar to that presented in Algorithm 2, with the contributions of the point unknowns being subtracted from the submatrices of $\hat{N}$ and $\breve{N}$ as well as from $\dot{N}$. Once the reduction is finished, both the image and added parameters are determined.

Algorithm 4   Normal Equation Reduction with Added Parameters and/or Constraints

```
for each image i
  for each point j
    if point j is on image i
```
$$\dot{N}_{ii} = \dot{N}_{ii} - \bar{N}_{ij}\ddot{N}_j^{-1}\bar{N}_{ij}^T$$
$$\dot{t}_i = \dot{t}_i - \bar{N}_{ij}\ddot{N}_j^{-1}\ddot{t}_j$$

$\qquad$ for each other image k point j is on

$$\dot{N}_{ik} = \dot{N}_{ik} - \bar{N}_{ij}\ddot{N}_j^{-1}\bar{N}_{jk}^T$$

$\qquad$ end image k loop
$\quad$ if point j occurs in an equation with added parameter m

$$\hat{N}_m = \hat{N}_m - \bar{N}_{mj}\ddot{N}_j^{-1}\bar{N}_{mj}^T$$
$$\breve{N}_{im} = \breve{N}_{im} - \bar{N}_{ij}\ddot{N}_j^{-1}\bar{N}_{mj}^T$$
$$\hat{t}_m = \hat{t}_m - \bar{N}_{mj}\ddot{N}_j^{-1}\ddot{t}_j$$

$\qquad$ for any other parameter n occuring in the equation

$$\hat{N}_{mn} = \hat{N}_{mn} - \tilde{N}_{mj}\ddot{N}_j^{-1}\tilde{N}_{jn}^T$$

$\qquad$ end parameter n loop
```
```

```
      end parameter m loop
    end point j loop
  end image i loop
```

The most important consideration in adding equations to a bundle adjustment is writing the equations so that the structure of the normal equations is not affected. For instance, we may want to write a constraint on the horizontal distance d between two points:

$$\sqrt{(X_1 - X_2)^2 + (Y_1 - Y_2)^2} - d = 0 \qquad (5\text{-}24)$$

This equation will be linearized as

$$\begin{bmatrix} \ddot{C}_1 & \ddot{C}_2 \end{bmatrix} \begin{bmatrix} \ddot{\delta}_1 \\ \ddot{\delta}_2 \end{bmatrix} + \hat{C}\hat{\delta} = f_c \qquad (5\text{-}25)$$

where $\ddot{\delta}_1$ and $\ddot{\delta}_2$ contain the corrections to the coordinates of points 1 and 2, and $\hat{\delta}$ contains the corrections to the known distance d. When the $\ddot{N}$ portion of the normal equations is formed, three submatrices will be affected:

$$\begin{aligned} \ddot{N}_1 + \ddot{C}_1^T W_c \ddot{C}_1 \\ \ddot{N}_2 + \ddot{C}_2^T W_c \ddot{C}_2 \\ \ddot{N}_{12} + \ddot{C}_1^T W_c \ddot{C}_2 \end{aligned} \qquad (5\text{-}26)$$

where $\ddot{N}_1$ and $\ddot{N}_2$ are the on-diagonal 3×3 submatrices that refer to the corrections on the point coordinates. However, $\ddot{N}_{12}$ is an off-diagonal 3×3 submatrix, appearing since points 1 and 2 are referenced in the same equation. Recall that the elimination of the point coordinate unknowns, Eq. 5-22, is based on the assumption that the $\ddot{N}$ matrix is block-diagonal and therefore easy to invert; adding off-diagonal submatrices locally destroys this structure. If constraints between points are necessary, then the two-point structure must be handled as a unit, rather than reducing by individual points. This approach provides flexibility in the utilization of equations at the expense of added complexity in the solution implementation. In some cases, it may be more efficient to initially eliminate the image parameters, taking advantage of the block-diagonal structure of the $\ddot{N}$ matrix (as long as there are no constraints between the image parameters). In other situations, it may be possible to add additional parameters to the constraint equations so that no two points appear in the same equation.

Assume that we are constraining points on the shore of a lake to lie at the same unknown elevation. The straightforward way to write the equations would be

$$Z_1 - Z_2 = 0 \qquad (5\text{-}27)$$

which would introduce off-diagonal terms into $\ddot{N}$. Alternatively, the equations could be written as

$$\begin{aligned} Z_1 - Z_W = 0 \\ Z_2 - Z_W = 0 \end{aligned} \qquad (5\text{-}28)$$

with the added parameter Z_W as the elevation of the lake surface. Corrections to Z_W would then appear in the $\hat{\delta}$ vector, forming a border in the reduced normal equations.

5.11.4 Solution Algorithms

Once we have formed the reduced normal equations, we solve for the $\dot{\delta}$ (image) parameters and any added ($\hat{\delta}$) parameters. This is not done by inverting the normal equation coefficient matrix and multiplying by the constant term, as is usually written. Instead, direct or iterative solution methods are employed that do not explicitly form the inverse, since calculating the inverse is much more computationally expensive than just solving the equations.

The most straightforward direct solution method is Gaussian elimination. Unknowns are progressively eliminated by adding or subtracting scaled rows of the coefficient matrix from one another. When only one unknown is left, its value is directly determined and the remaining unknowns determined by back-substitution. The coefficient matrix is essentially transformed into a triangular matrix, making the set of equations easier to solve.

Several more efficient matrix solution techniques exist, based on factoring the coefficient matrix into triangular matrices. The Cholesky method is particularly well-suited for solving least squares normal equations, since it is predicated on having positive-definite symmetric matrices. The coefficient matrix is factored into a triangular matrix L which yields the original matrix when multiplied by its transpose.

$$(L^T L) \, x \; = b \tag{5-29}$$

Efficient variants of the Cholesky technique exist for both banded and banded-bordered matrices.

5.11.5 Back-Substitution

Once the reduced normal equations are solved for the corrections to the photo parameters, the corrections to the point coordinates are calculated by back-substitution. This can be done on a point-by-point basis; for point j, $\ddot{\delta}_j$ is

$$\ddot{\delta}_j = \ddot{N}_j^{-1}\left(\ddot{t}_j - \sum_i \overline{N}_{ij}^T \dot{\delta}_i\right) \tag{5-30}$$

5.11.6 Error Propagation

To evaluate the results of the solution we need the parameter covariance (Appendix B.10), which is the inverse of the normal equation coefficient matrix. However, as we have seen (Section 5.11.4), direct solution methods such as the Cholesky do not produce the inverse of the normal equation coefficient matrix. Instead, we can make the observation that calculating the inverse of a matrix A can be thought of as solving n equations, one for each column x_i of A^{-1}, each of which has as its right-hand side a column of the identity matrix I_i

$$A x_i = I_i \tag{5-31}$$

where I is an identity matrix with the same dimensions as A. If a factorization method like the Cholesky solution has been used to solve the equations, the matrix does not have to be refactored to solve equations with different right-hand sides. If we keep the last factorization of the coefficient matrix after the solution has converged, we can calculate the inverse with a minimum of extra work. There are also methods to calculate only the portion of the inverse lying within the band of the original matrix, since the inverse of a banded matrix is not, in general, a banded matrix.

Once the covariance of the $\dot{\delta}$ parameters, and the $\hat{\delta}$ parameters if present, is calculated by inversion of the reduced normal equations coefficient matrix, we need to calculate the covariances of the point coordinates, the $\overset{..}{\delta}$ parameters. This can be done by applying error propagation techniques to Eq. 5-17.

$$\overset{..}{\delta} = \overset{..}{N}{}^{-1}(\overset{..}{t} - \overline{N}^T\overset{.}{\delta}) \tag{5-32}$$

There are two stochastic quantities in the equation, $\overset{.}{\delta}$ and the original observations, which are present in $\overset{..}{t}$ (recall that $\overset{..}{t} = \overset{..}{B}{}^TWf$ and that $f = d - l$). The partial derivative of Eq. 5-32 with respect to the observations l is

$$\frac{\partial\overset{..}{\delta}}{\partial l} = \overset{..}{N}{}^{-1}\overset{..}{B}{}^TW \tag{5-33}$$

while the partial derivative with respect to $\overset{.}{\delta}$ is

$$\frac{\partial\overset{..}{\delta}}{\partial\overset{.}{\delta}} = -\overset{..}{N}{}^{-1}\overline{N}^T \tag{5-34}$$

The propagated point covariance is

$$\overset{..}{Q} = \overset{..}{N}{}^{-1}\overset{..}{B}{}^TWQW\overset{..}{B}\overset{..}{N}{}^{-1} + \overset{..}{N}{}^{-1}\overline{N}^T\overset{.}{Q}\overline{N}\overset{..}{N}{}^{-1} \tag{5-35}$$

The first term of Eq. 5-35 reduces to $\overset{..}{N}{}^{-1}$, since $WQ = I$, $\overset{..}{B}{}^TW\overset{..}{B} = \overset{..}{N}$, and $\overset{..}{N}\overset{..}{N}{}^{-1} = I$.

5.12 EVALUATION OF THE BLOCK ADJUSTMENT

After running a block adjustment, we must evaluate its results to be sure that they meet the project specifications and requirements and that the results are valid—that is, not contaminated by bad measurements or assumptions. Evaluation begins in the planning stage, with the verification of the suitability of the block design by prior experience or by simulation. After the block adjustment is run, a first step is often a qualitative evaluation, in which the operator examines graphical representations of the adjustment's output in order to understand broad trends and to catch obviously bad inputs. Finally, statistical analysis techniques are used to protect against bad observations and to quantify the quality of the adjustment. Robust estimation methods, such as iteratively reweighted least squares, L1 estimation, or least median of squares, can be part of the evaluation process, by identifying questionable observations, although the latter two do not allow the use of the derived normal equations.

5.12.1 Block Adjustment Planning by Simulation

The characteristics of a block adjustment are determined by the geometry of the images, the number and arrangement of the tie and control points, and the quality of any additional navigational information. In most mapping projects, the flight lines are laid out in a regular pattern and the flight parameters can be based on past experience to give the desired results. However, for novel situations or cases with irregular flight lines or control point arrangements, a simulation of the block adjustment should be run prior to the actual data acquisition to verify that the project will meet the design criteria and to guide possible changes in the project design.

Simulation is made possible by the fact that any measurement errors in the individual observations have very small effects on the overall geometry, which determines the precision and reliability of the block. We can therefore generate simulated observations from the nominal image geometry and use these to predict the characteristics of the final block adjustment.

A simulated block adjustment requires only one iteration of the solution, since the initial approximations are exact. Only the normal equation coefficient matrix must be formed, so that the error propagation can be performed and the parameter standard deviations and correlations analyzed in the same manner as for a real solution. This allows the detection and correction of any weak geometric conditions in the block design, which can be remedied by additional control or tie points or by changing the flight line arrangement.

5.12.2 Qualitative Evaluation

The purpose of qualitative evaluation is to allow the operator to understand the properties of the adjustment and to diagnose any problems. It is difficult to recognize trends or outliers by looking at pages of numbers; instead, since this is a geometric problem, it is best addressed by graphical displays of the solution outputs.

The plots most often used for qualitative evaluation of results of a block adjustment are of the image residuals, as in Fig. 5-19. The sizes of the residuals are usually exaggerated to emphasize the trends. The image residuals should point in random directions and have comparable sizes. Residuals all pointing in the same direction (either parallel or radially with respect to the center of the image) indicate the presence of a systematic effect. These effects may indicate uncorrected errors, such as atmospheric refraction, or an image parameter that has been weighted too highly and not allowed to adjust. A residual that is larger than its neighbors or points in the opposite direction to those near it indicates a bad measurement. It should be remembered that bad measurements are not always evident in the residuals, depending upon the number of measurements of the point, the imaging geometry, and the size and direction of the error.

If sufficient check points are available, a check point error plot (Fig. 5-20) can show the deformation of the block and indicate any problems with control

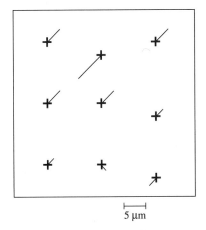

5 µm   **Figure 5-19** Image residual plot.

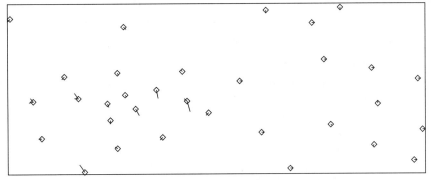

Figure 5-20 Plot of check point error along strip.

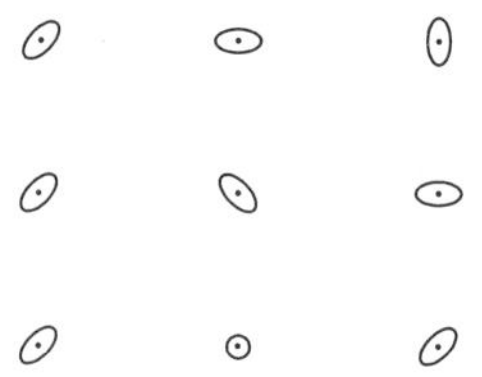

Figure 5-21 Precision plot.

point coordinates. Again, the errors should be randomly distributed and of comparable sizes.

Plotting the calculated precision of points across the block (Fig. 5-21) can also be instructive. Precision should be consistent across the block; variations can indicate points which are not well-determined.

5.12.3 Statistical Evaluation

As with any other least squares solution, statistical methods are used to evaluate the results of a block adjustment (Appendix B.10). Three different aspects, the precision, the accuracy, and the reliability, must be evaluated, each with the appropriate statistical techniques, to ensure that the solution is valid.

5.12.3.1 Precision

We evaluate the *precision* of the solution by examining the covariances of the parameters. The variance of a parameter can be thought of as specifying the spread of its value; a parameter with a small variance has its value known within small bounds, while a large variance indicates that a parameter is not well-determined in the solution. A perfectly known parameter would have a zero variance, while a completely unknown parameter would have an infinite variance. A least squares solution produces parameter variances, but we usually prefer to speak in terms of the *standard deviation*, which is the square root of the variance and has the same units as the parameter.

One of the products of a least squares adjustment is the covariance matrix of the parameters. The diagonal elements of the covariance matrix are the parameter variances, and the off-diagonal elements are the covariances between the parameters. For example, a covariance matrix for a point's X, Y, and Z ground coordinates might be

$$\overset{..}{\Sigma}_j = \begin{bmatrix} \sigma_X^2 & \sigma_{XY} & \sigma_{XZ} \\ \sigma_{YX} & \sigma_Y^2 & \sigma_{YZ} \\ \sigma_{ZX} & \sigma_{ZY} & \sigma_Z^2 \end{bmatrix} = \begin{bmatrix} 1.69 & 0.728 & 1.638 \\ 0.728 & 1.96 & 2.058 \\ 1.638 & 2.058 & 4.41 \end{bmatrix} \tag{5-36}$$

In this example, the variances of the X and Y coordinates are 1.69 and 1.96 meters, respectively, while the variance of the Z coordinate is 4.41 meters. To make the numbers easier to interpret, we often calculate the standard deviation correlation matrix from the covariance matrix. The diagonal elements of this matrix are the standard deviations of the parameters (the square roots of the diagonal elements of the covariance matrix). The off-diagonal terms are the *correlation coefficients*, ρ, calculated for each pair of parameters from their covariance and individual variances. For parameters i and j, the correlation coefficient ρ_{ij} is

$$\rho_{ij} = \frac{\sigma_{ij}}{\sigma_i \sigma_j} \tag{5-37}$$

Parameters with a correlation coefficient of 1 or -1 are linearly related—if one parameter is known, the other is determined. The presence of high correlations implies that the solution is not geometrically well-determined or that the solution is over-parameterized. The standard deviation correlation matrix for the point above is

$$\begin{bmatrix} \sigma_X & \rho_{XY} & \rho_{XZ} \\ \rho_{YX} & \sigma_Y & \rho_{YZ} \\ \rho_{ZX} & \rho_{ZY} & \sigma_Z \end{bmatrix} = \begin{bmatrix} 1.30 & 0.4 & 0.6 \\ 0.4 & 1.40 & 0.7 \\ 0.6 & 0.7 & 2.1 \end{bmatrix} \tag{5-38}$$

One case where the correlations must be carefully analyzed is when self-calibration parameters are included in an adjustment, as discussed in Section 5.11.3.

5.12.3.2   Accuracy

A fundamental requirement for a block adjustment is that it be *accurate*, so that the results from the block adjustment reflect the actual state of the world. Accuracy cannot be determined by examining the solution, since a solution can be very consistent within itself (very precise, as discussed in the previous section), but still not be accurate. An external comparison is required to determine accuracy. For example, a set of measurements made with a 100-meter tape that is only 99 meters long could be very precise, in that the standard deviation of the mean measurement would be small, but the accuracy would be very poor. Rigorously evaluating the accuracy of a block adjustment against external standards verifies that uncorrected systematic errors (such as a short measuring tape) do not contaminate the results and that possible weak block geometry does not prevent the recognition of mistakes in control point or tie point measurements.

Block adjustment accuracy is evaluated using *check points* whose world coordinates are known but which were not used as control in the solution. The root mean square of the differences between the computed coordinates and the known values provides a measure of the solution accuracy. If the check points are not surveyed with much higher accuracy than the expected accuracy of the block adjustment, their accuracy must be taken into account in the evaluation.

If only a few check points are available, it may be possible to repeat the adjustment using some control points as check points and some check points as control points. If the statistics do not change significantly between runs, the operator can be more confident of a valid solution.

It is important to realize that the accuracy cannot be meaningfully evaluated using only control point residuals. Unless the solution has very large redundancy, any measurement errors in the control points will be absorbed into the calculated orientation parameters and decrease the accuracy of the solution.

5.12.3.3   Reliability

Another important consideration is the *reliability* of the solution, which is its resistance to gross errors, or blunders, in the input data. If bad inputs are not detected, the solution results may be compromised or completely invalidated. Bad data can be detected only by having redundant observations, within a strong geometric configuration that allows them to serve as a check on each other. Blunders can be detected only by examining the residuals for each observation. However, the imaging geometry directly

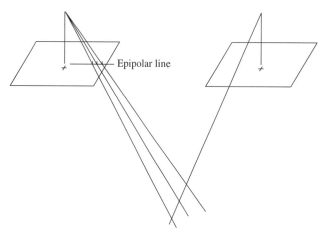

Figure 5-22 Measurement error along the epipolar line.

affects how much of the blunder is reflected by the residual and how much is absorbed into the calculated parameters. Robust estimation methods, used in conjunction with conventional least squares, can be a valuable tool for editing large data sets which may contain blunders.

Consider the case of a point with unknown ground coordinates measured on only one photo; its residual will always be zero, no matter how big an error is made in its measurement, and all of the error in its measurement will be absorbed into the calculated ground coordinates. If the point is measured on two photos (Fig. 5-22), then moving the point in depth results in the images of the point moving along the epipolar lines (Section 2.7). This implies that any measurement error parallel to the epipolar line will have no effect on the calculated residuals and will only cause the depth of the point to be calculated incorrectly. Only if the point is visible on a third image are measurement errors along the epipolar line detectable.

In block adjustment, the situation is more complicated, due to the multiple image rays and more complicated overall geometry. This makes the detection of weakly-determined configurations by inspection difficult. The inspection of individual residuals is also complicated by the number of observations present. Instead, statistical measures are used to quantify the susceptibility of each residual to measurement errors and to establish criteria for rejection of individual measurements based on their residuals.

To use residuals to test for blunders, we must separate the effects of the geometry determining the residuals from the effects of a blunder in the corresponding observation. The geometry determining each individual residual is summarized in the Q_{vv} matrix, the cofactor matrix of the residuals (Appendix B):

$$Q_{vv} = W - BN^{-1}B^T \qquad (5\text{-}39)$$

Residual testing should be done on the *standardized residual,* which is the residual divided by its standard deviation (the square root of the corresponding diagonal element of Q_{vv}). If the redundancy of the solution is large, the *t* test statistic can be used; otherwise, the τ statistic is appropriate (Appendix B). Robust estimation models can also be beneficial here.

PROBLEMS

5.1 A photograph with side fiducials is placed in an *x-y* comparator and the following measurements are made:

Point	x (mm)	y (mm)
1	21.017	144.369
2	255.924	148.918
3	140.762	29.255
4	136.186	264.073
a	146.226	68.824
b	198.828	150.505
c	88.060	163.622

Compute the image coordinates of points *a*, *b*, and *c* with respect to the origin at the indicated principal point (the intersection of the lines between the fiducials).

5.2 The distances between fiducials 1 and 2 and between fiducials 3 and 4 in Problem 5.1 are both 235 mm. Correct the coordinates of points *a*, *b*, and *c* for linear film deformation.

5.3 A photograph with side fiducials is placed in an *x-y* comparator and the following measurements are made:

Point	x (mm)	y (mm)
1	29.103	133.547
2	263.854	143.231
3	151.336	21.079
4	141.627	255.740
d	118.082	37.414
e	38.388	101.008
f	242.102	216.214

Compute the image coordinates of points *d*, *e*, and *f* with respect to the origin at the indicated principal point (the intersection of the lines between the fiducials).

5.4 The distances between fiducials 1 and 2 and between fiducials 3 and 4 in Problem 5.3 are both 235.252 mm. Correct the coordinates of points *d*, *e*, and *f* for linear film deformation.

5.5 Given the following comparator measurements and calibrated coordinates of four corner fiducial marks along with the measurements of three image points:

Point	Measured x (mm)	Measured y (mm)	Calibrated x (mm)	Calibrated y (mm)
1	32.108	41.262	0.000	0.000
2	243.772	29.159	212.000	0.000
3	255.925	240.810	212.010	211.995
4	44.210	252.895	0.002	212.006
a	151.446	101.818	——	——
b	83.062	221.240	——	——
c	206.585	71.476	——	——

If the calibrated principal point coordinates are (106.011, 105.990), convert the measurements for points *a*, *b*, and *c* to the photographic coordinate system with the origin at the principal point using a two-dimensional linear conformal transformation (Appendix A).

5.6 Solve Problem 5.5 using a six-parameter affine transformation.

5.7 Solve Problem 5.5 using an eight-parameter projective transformation.

5.8 Given the following comparator measurements and calibrated coordinates of four corner fiducial marks along with the measurements of three image points:

Point	Measured x (mm)	Measured y (mm)	Calibrated x (mm)	Calibrated y (mm)
Principal point	——	——	0.000	0.000
1	23.061	144.057	−117.148	0.000
2	258.038	145.245	117.142	0.000
3	141.163	27.258	0.015	−117.410
4	139.936	262.155	−0.014	117.451
d	200.515	46.462	——	——
e	106.440	120.162	——	——
f	108.086	236.200	——	——

Using a two-dimensional linear conformal transformation, convert the measurements for points d, e, and f into the photographic coordinate system with the origin at the principal point.

5.9 Solve Problem 5.8 using a six-parameter affine transformation.

5.10 Solve Problem 5.8 using an eight-parameter projective transformation.

5.11 Given the following table of radial lens distortion values:

Radial distance (mm)	Distortion (mm)
0	0.000
20	−0.001
40	−0.003
60	−0.003
80	−0.001
100	0.000
120	0.002
140	0.004
160	0.005
180	0.007

correct the following image coordinates (given relative to the principal point) for radial distortion:

Point	x (mm)	y (mm)
1	−110.420	3.432
2	45.507	−42.65
3	3.222	22.466
4	113.129	−106.905

5.12 Using the ARDC 1959 Model Atmosphere, compute and plot a set of curves for values of $H = 5, 10, 15, 20$ and 25 km to represent the values of displacement d_r due to atmospheric refraction expressed in micrometers, for a vertical photograph with a focal length of 152 mm, and for radial distances in 20-mm intervals out to 160 mm. Assume $h = 0$.

5.13 Compute and plot the curves of Problem 5.12 for a focal length of 88 mm.

5.14 Compute and plot the curves of Problem 5.12 for a focal length of 305 mm.

5.15 Compute and plot a set of curves for values of $H = 5, 10, 15, 20$, and 25 km to represent the values in μm of the displacement d_E due to the curvature of the Earth for a vertical photograph with a focal length of 152 mm, and for radial distances in 20-mm intervals out to 160 mm.

5.16 Compute and plot the curves of Problem 5.15 for a focal length of 88 mm.

5.17 Compute and plot the curves of Problem 5.15 for a focal length of 305 mm.

5.18 A strip of three photographs contains 15 points. Photo 1 contains points 1–9; photo 2 contains points 4–12; photo 3 contains points 7–15. Show the structure of the normal equations and of the reduced normal equations.

5.19 For the following flight-line configurations, determine whether down-strip or cross-strip numbering will produce the smaller bandwidth:

Number of strips	Number of photos per strip
2	12
10	10
20	5

5.20 Convert the following covariance matrix to a standard deviation correlation matrix:

$$\begin{bmatrix} 4.95 & 3.56 & 2.02 \\ 3.56 & 6.23 & 2.52 \\ 2.02 & 2.52 & 10.44 \end{bmatrix}$$

5.21 Convert the following standard deviation correlation matrix to a covariance matrix:

$$\begin{bmatrix} 2.50 & 0.85 & 0.62 \\ 0.85 & 1.96 & 0.69 \\ 0.62 & 0.69 & 3.33 \end{bmatrix}$$

REFERENCES

ACKERMANN, F., EBNER, H., and KLEIN, H. 1973. Block triangulation with independent models. *Photogrammetric Engineering* 39(9):967–981.

ACKERMANN, F. 1980. Block adjustment with additional parameters. In *International Archives of Photogrammetry and Remote Sensing*. Vol. XXIII No. B3. Hamburg: International Society for Photogrammetry and Remote Sensing.

BERTRAM, S. 1966. Atmospheric refraction. *Photogrammetric Engineering* 32(1):76–84.

BERTRAM, S. 1969. Atmospheric refraction in aerial photogrammetry. *Photogrammetric Engineering*, 35(6):560.

BLAIS, R. 1973. *Program SPACE-M Users Manual*. Ottowa, Canada: Surveys and Mapping Branch, Dept. of Energy, Mines, and Resources.

BROWN, D.C. 1955. *A Matrix Treatment of the General Problem of Least Squares Considering Correlated Observations*. Ballistic Research Laboratory Report No. 960. Aberdeen, Maryland: Aberdeen Proving Ground.

BROWN, D.C. 1958. *A Solution to the General Problem of Multiple Station Analytical Stereotriangulation*. RCA Technical Report No. 34. RCA Missile Test Project. Cape Canaveral, FL: Patrick AFB.

BROWN, D.C. 1964. *Research in Mathematical Targeting, the Practical and Rigorous Adjustment of Large Photogrammetric Nets*. RADC-TDR-353. Rome, NY: Rome Air Development Center.

BROWN, D.C. 1968. A unified lunar control network. *Photogrammetric Engineering* 34(12):1272–1292.

BROWN, D.C. 1973. Accuracies of analytical triangulation in applications to cadastral surveying. *Surveying and Mapping* 33(3):281–302.

BROWN, D.C. 1976. The bundle adjustment—progress and prospects. In *International Archives of Photogrammetry and Remote Sensing*. Vol. XIII No.3. Helsinki: International Society for Photogrammetry and Remote Sensing.

BROWN, D. C. 1994. New developments in photogeodesy. *Photogrammetric Engineering and Remote Sensing* 60(7):877–894.

BROWN, L. G. 1992. A survey of image registration techniques. *ACM Computing Surveys* 24(4):325–376.

CHEN, H. H. 1991. Pose determination from line-to-plane correspondences: Existence condition and closed-form solutions. *IEEE Transactions on Pattern Analysis and Machine Intelligence* 13(6):530–541.

CHURCH, E. 1945. *Revised Geometry of the Aerial Photograph*. Bulletin 15. Syracuse, NY: Syracuse University Press.

CHURCH, E. 1948. *Theory of Photogrammetry*. Bulletin 19. Syracuse, NY: Syracuse University Press.

CHURCH, E. 1950. *Illustrative Solutions of Analytic Problems in Aerial Photogrammetry*. Technical Paper No. 97. Columbus, OH: Ohio State University Research Foundation.

DOYLE, F. J. 1964. The historical development of analytical photogrammetry. *Photogrammetric Engineering* 30(2):259–265.

EBNER, H., KORNUS, W., and OHLHOF, T. 1992. A simulation study on point determination using MOMS-O2/D2 imagery. In *International Archives of Photogrammetry and Remote Sensing.* Vol. XXIX No. B4.

EBNER, H. 1976. Self-calibrating block adjustment. In *International Archives of Photogrammetry and Remote Sensing.* Vol. XIII No. 3. Helsinki: International Society for Photogrammetry and Remote Sensing.

FISCHLER, M., and BOLLES, R. 1981. Random sample consensus: A paradigm for model fitting with applications to image analysis and automated cartography. *Communications of the ACM* 24(6):381–395.

FÖRSTNER, W. 1985. The reliability of block triangulation. *Photogrammetric Engineering and Remote Sensing* 51(8):1137–1149.

FÖRSTNER, W. 1987. Reliability analysis of parameter estimation in linear models with applications to mensuration problems in computer vision. *Computer Vision, Graphics, and Image Processing* 40(3):273–310.

FÖRSTNER, W. 1994. Diagnostics and performance evaluation in computer vision. In *Proceedings, NSF/ARPA Workshop on Performance versus Methodology in Computer Vision.* Seattle, WA: IEEE.

FRASER, C. S. 1997. Digital camera self-calibration. *ISPRS Journal of Photogrammetry and Remote Sensing,* 52(4):149–159.

FRITZ, L. W. 1973. *Complete Comparator Calibration.* NOAA Technical Report NOS 57. Rockville, MD: National Ocean Survey, National Oceanic and Atmospheric Administration.

GABET, L., GIRAUDON, G., and RENOUARD, L. 1997. Automatic generation of high resolution urban zone digital elevation models. *ISPRS Journal of Photogrammetry and Remote Sensing* 52(1):33–47.

GOAD, C. C., and YANG, M. 1997. A new approach to precision airborne GPS positioning for photogrammetry. *Photogrammetric Engineering and Remote Sensing* 63(9):1067–1077.

GRUEN, A. 1982. The accuracy potential of the modern bundle block adjustment in aerial photogrammetry. *Photogrammetric Engineering and Remote Sensing,* 48(1):45–54.

GUPTA, R., and HARTLEY, R. I. 1997. Linear pushbroom cameras. *IEEE Transactions on Pattern Analysis and Machine Intelligence* 19(9):963–975.

GYER, M. S. 1967. *The Inversion of the Normal Equations of Analytical Aerotriangulation by the Method of Recursive Partitioning.* RADC-TR-67-69. Rome, NY: Rome Air Development Center.

GYER, M. S. 1996. Methods for computing photogrammetric refraction corrections for vertical and oblique photographs. *Photogrammetric Engineering and Remote Sensing* 62(3):301–310.

HARALICK, R. M., LEE, C. N., OTTENBERG, K., and NOLLE, M. 1994. Review and analysis of solutions of the three point perspective pose estimation problem. *International Journal of Computer Vision* 13(3):331–356.

HEIPKE, C., KORNUS, W., and PFANNENSTEIN, A. 1996. The evaluation of MEOSS airborne three-line scanner imagery: Processing chain and results. *Photogrammetric Engineering and Remote Sensing* 62(3):293–299.

HEIPKE, C., ed. 1997. Special issue: Automatic image orientation. *ISPRS Journal of Photogrammetry and Remote Sensing* 52(3).

HOFMANN-WELLENHOF, B., LICHTENEGGER, H., and COLLINS, J. 1997. *Global Positioning System: Theory and Practice.* Berlin: Springer Verlag.

HORN, B. K. P. 1990. Relative orientation. *International Journal of Computer Vision* 4(1):59–78.

HORN, B. K. P. 1987. Closed-form solution of absolute orientation using unit quaternions. *Journal of the Optical Society of America* A-4:629–642.

HORN, B. K. P., HILDEN, H. M., and NEGAHDARIPOUR, S. 1988. Closed-form solution of absolute orientation using orthonormal matrices. *Journal of the Optical Society of America* A-5:1129–1135.

JEYAPALAN, K. 1972. Calibration of a comparator. *Photogrammetric Engineering* 38(5):472–478.

JIYU, L., XIAOMING, C., DEREN, L., GUANG, W., JINGNIAN, L., WEI, L., and JINXIANG, Z. 1996. GPS kinematic carrier phase measurements for aerial photogrammetry. *ISPRS Journal of Photogrammetry and Remote Sensing* 51(5):230–242.

KRATKY, V. 1989. Rigorous photogrammetric processing of SPOT images at CCM Canada. *Photogrammetria* 44:53–71.

KRUPNIK, A., and SCHENK, T. 1997. Experiments with matching in the object space for aerotriangulation. *ISPRS Journal of Photogrammetry and Remote Sensing* 52(4):160–168.

KUBIK, K., MERCHANT, D., and SCHENK, T. 1987. Robust estimation in photogrammetry. *Photogrammetric Engineering and Remote Sensing* 53(2):167–169.

LEE, C., THEISS, H., BETHEL, J., and MIKHAIL, E. 2000. Rigorous mathematical modeling of airborne pushbroom imaging systems. *Photogrammetric Engineering and Remote Sensing* 66(4):385–392.

LIU, Y., and HUANG, T. S. 1990. Determination of camera location from 2-D to 3-D line and point correspondences. *IEEE Transactions on Pattern Analysis and Machine Intelligence* 12(1):28–37.

McGLONE, J. C., and MIKHAIL, E. M. 1981. *Photogrammetric Analysis of Aircraft Multispectral Scanner Data.* Technical Report CE-PH-81-3. West Lafayette, IN: Purdue University School of Civil Engineering.

McGLONE, C., and MIKHAIL, E. M. 1982. Geometric constraints in multispectral scanner data. In *Proceedings of the 48th Annual Meeting.* Denver, CO: American Society for Photogrammetry and Remote Sensing.

McGLONE, C. 1995. Bundle adjustment with object space constraints for site modeling. In *Proceedings of the SPIE: Integrating Photogrammetric Techniques with Scene Analysis and Machine Vision.* Vol. 2486 pp. 25–36.

McGLONE, C. 1996. Bundle adjustment with geometric constraints for hypothesis evaluation. In *International Archives of Photogrammetry and Remote Sensing.* Vol. XXXI(B3) pp. 529–534. Vienna, Austria: ISPRS.

McGLONE, C. 1998. Block adjustment of linear pushbroom imagery with geometric constraints. In *International Archives of Photogrammetry and Remote Sensing.* Vol. XXXII(B2) pp. 198–205. Cambridge, United Kingdom: ISPRS.

MIKHAIL, E. M., and PADERES, F. C., JR. 1991. *Photo-grammetric Modeling and Reduction of SPOT Stereo Images (Phase II)*. Technical Report CE-PH-913. West Lafayette, IN: Purdue University.

MIKHAIL, E. M. 1970. Relative control for extraterrestrial work. *Photogrammetric Engineering* 36(4):381–389.

MIKHAIL, E. M. 1988. Photogrammetric Modeling and Reduction of SPOT Stereo Images. Technical Report CE-PH-884. West Lafayette, IN: Purdue University.

MIKHAIL, E. M. 1993. Linear features for photogrammetric restitution and object completion. In *Proceedings of the SPIE: Integrating Photogrammetric Techniques with Scene Analysis and Machine Vision*. Vol. 1944 pp. 16–30. Orlando, FL: SPIE.

MOLANDER, C., MORACO, A., and HOUCK, D. 1987. A numerical photogrammetric model for software applications. In *Proceedings of the 1987 ASPRS-ACSM Annual Convention*, pp. 284–292. Baltimore MD: American Society for Photogrammetry and Remote Sensing.

MOLENAAR, M. 1978. Essay on empirical accuracy studies in aerial triangulation. *ITC Journal* 1:81–103.

MORGADO, A., and DOWMAN, I. 1997. A procedure for automatic absolute orientation using aerial photographs and a map. *ISPRS Journal of Photogrammetry and Remote Sensing* 52(4):169–182.

NOERDLINGER, P. D. 1999. Atmospheric refraction effects in Earth remote sensing. *ISPRS Journal of Photogrammetry and Remote Sensing*, 54(5–6):360–373.

NOVAK, K., and TUDHOPE, R., eds. 1993. Special issue: GPS photogrammetry. *Photogrammetric Engineering and Remote Sensing* 59(11).

OHLOFF, T. 1995. Block triangulation using three-line images. In D. Fritsch and D. Hobbie, eds. *Photogrammetric Week*, pp. 197–206. Stuttgart: Wichmann.

ORUN, A. B., and NATARAJAN, K. 1994. A modified bundle adjustment software for SPOT imagery and photography: Tradeoff. *Photogrammetric Engineering and Remote Sensing* 60(12):1431–1437.

PADERES, F. C., MIKHAIL, E. M., and FÖRSTNER, W. 1984. Rectification of single and multiple frames of satellite scanner imagery using points and edges as control. In *NASA Symposium on Mathematical Pattern Recognition and Image Analysis*. Houston, TX: NASA.

PARKINSON, B.W., JR., SPILKER, J. J., AXELRAD, P., and ENGE, P., eds. 1996. *The Global Positioning System: Theory and Application*. Washington, DC: American Institute of Aeronautics and Astronautics.

POPE, A. J. 1975. *The Statistics of Residuals and the Detection of Outliers*. Technical Report NOS 65 NGS 1. Rockville, MD: National Oceanic and Atmospheric Administration.

PRESS, W. H., FLANNERY, B. P., TEUKOLSKY, S. A., and VETTERLING, W. T. 1989. *Numerical Recipes in C*. Cambridge: Cambridge University Press.

SCHMID, H. 1953. *An Analytical Treatment of the Orientation of a Photogrammetric Camera*. Report No. 880. Aberdeen, MD: Ballistic Research Laboratories.

SCHMID, H. 1955. *An Analytical Treatment of the Problem of Triangulation by Stereophotogrammetry*. Report No. 961. Aberdeen, MD: Ballistic Research Laboratories.

SCHMID, H. 1959. *A General Analytical Solution to the Problem of Photogrammetry*. Report No. 1065. Aberdeen, MD: Ballistic Research Laboratories.

SCHUT, G. 1969. Photogrammetric refraction. *Photogrammetric Engineering* 35(16):79–86.

SCHWARZ, K. P., CHAPMAN, M. A., CANNON, M. E., GONG, P., and COSANDIER, D. 1994. A precise positioning/attitude system in support of airborne remote sensing. In *International Archives of Photogrammetry and Remote Sensing*. Vol. XXX(B2) pp. 191–199. Ottawa: International Society for Photogrammetry and Remote Sensing.

STEFANOVIC, P. 1978. Blunders and least squares. *ITC Journal* 1:122–154.

STRUNZ, G. 1992. Feature based image orientation and object reconstruction. In *International Archives of Photogrammetry and Remote Sensing*. Vol. XXIX(B3) pp. 113–118. Washington, DC: International Society for Photogrammetry and Remote Sensing.

TANG, L., and HEIPKE, C. 1996. Automatic relative orientation of aerial images. In *Photogrammetric Engineering and Remote Sensing* 62(1):47–55.

THOMPSON, E. H. 1958. An exact linear solution of the problem of absolute orientation. *Photogrammetria* 15(4):163–179.

THOMPSON, E. H. 1959. A rational algebraic formulation of the problem of relative orientation. *The Photogrammetric Record* 3(14):152–159.

THOMPSON, E. H. 1968. The projective theory of relative orientation. *Photogrammetria* 23(2):67–75.

THOMPSON, E. H. 1971. Space resection without interior orientation. *The Photogrammetric Record* 7(37):39–45.

THOMPSON, E. H. 1975. Resection in space. *The Photogrammetric Record* 8(4):333–334.

TOUTIN, T., and CARBONNEAU, Y. 1992. MOS and SEASAT image geometric corrections. *IEEE Transactions on Geoscience and Remote Sensing* 30(3):603–609.

TSAI, V. J. D. 1995. Automatic photo reordering in a simultaneous bundle adjustment. *Photogrammetric Engineering and Remote Sensing* 61(7):899–908.

WENG, J., HUANG, T. S., and AHUJA, N. 1992. Motion and structure from line correspondences: Closed-form solution, uniqueness, and optimization. *IEEE Transactions on Pattern Analysis and Machine Intelligence* 14(3):318–336.

WESTIN, T. 1990. Precision rectification of SPOT imagery. *Photogrammetric Engineering and Remote Sensing* 56(2):247–253.

ZENG, Z., and WANG, X. 1992. A general solution of a closed-form space resection. *Photogrammetric Engineering and Remote Sensing* 58(3):327–338.

Chapter **6**

Digital Photogrammetry

From its start, photogrammetry has used photographic emulsions on film or glass bases for dimensional stability. Precise mechanical analog instruments were constructed to measure the imagery. In recent years, with the rapid development of digital technology, photogrammetry has also been applied to digital images, which may be generated directly from electronic cameras or scanned from film imagery.

The projective geometry that is the basis of photogrammetry is the same whether it is applied to analog or digital images. The use of images in digital form offers distinct advantages in processing and automation. Digital image processing can be used to alter the image in very precise ways and make it more useful. If the image is "examined" by a computer, many of the operations we think of as requiring a human can be performed automatically.

This chapter describes the properties of digital images and gives a brief overview of *digital image processing*. Since digital images are stored in a computer, we can take advantage of this for image processing. We can also imagine having the computer perform tasks that an operator would normally perform by examining the image. An overview of *computer vision* is given, in terms of the low-level feature processing and vision system design. Normally, photogrammetrists try to model a scene by processing an image. Computers can be used to automate these tasks and also to generate a digital image given a model of the scene. This chapter concludes with a brief discussion of this application of *computer graphics*.

Each of these topics is in itself a major field of research. This chapter is meant to provide only a brief overview. The reader is encouraged to investigate the references for further details.

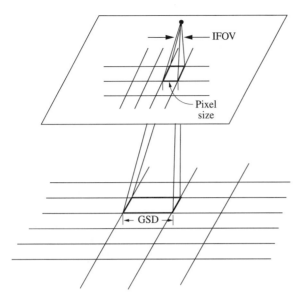

Figure 6-1 Instantaneous field of view (IFOV) and ground sample distance (GSD).

6.1   DIGITAL IMAGERY

A *digital image* is a regular array of *pixels,* or picture elements, and is described in terms of its *geometry* and its *radiometry.*

6.1.1   Digital Image Geometry

Each pixel of a digital image is a direct or indirect sample of the continuous incident light distribution. A digital sensor samples incident light directly. An example of indirect sampling is scanning a piece of film and converting it into a digital image.

Pixels may be arranged in any regular pattern, but in practice most sampling grids are square or rectangular. Hexagonal grids have some theoretical advantages but have seldom been implemented. The ratio between pixel width and pixel height is known as the *aspect ratio;* a square pixel has an aspect ratio of one.

Digital images are often described in terms of their pixel size, but can also be specified in terms of the *instantaneous field of view* (IFOV), which is the angle determined by the pixel size and the focal length. As shown in Fig. 6-1, as the focal length decreases or the pixel size increases, the IFOV increases. A larger IFOV means that one pixel covers a larger portion of the scene or, equivalently, that the image will have lower spatial resolution.

Another measure related to pixel size is the *ground sample distance* (GSD), which is the projection of the pixel size onto the ground plane (Fig. 6-1). This is often erroneously used as a synonym for resolution. In fact, the resolution of a digital image is determined by both the sensor geometry and by factors external to the sensor, such as atmospheric conditions, platform motion, etc. (see Chapter 3). The actual resolution may be as good as indicated by the GSD, but in practical applications it is often lower.

EXAMPLE 6-1   *IFOV and GSD Calculation*

A satellite imaging system has pixels that are 0.015 mm square and a focal length of 1000 mm. What is the IFOV? What is the GSD at nadir, assuming an orbital altitude of 650 km?

SOLUTION   The IFOV is

$$IFOV = 2\tan^{-1}\left(\frac{0.015/2}{1000}\right) = 0.000015 \text{ rad}$$

and the GSD is

$$GSD = IFOV(650{,}000) = 9.75 \text{ m}$$

Digital images sample a continuous analog input image at discrete points and assign a discrete intensity value at each point. How can we determine whether all the information in the original image is present in its sampled digital form or, equivalently, what spatial sampling rate is required to fully capture the input? The answer to this question relies on *sampling theory*. We will not attempt to include the mathematics required to derive sampling theory here, but will instead try to give an intuitive feel for the considerations. Students should consult standard references, such as Bracewell (1986) and Oppenheim and Schaefer (1975), for detailed derivations.

The sampling rate required to accurately reconstruct an input signal depends on how rapidly the signal varies. To quantify the speed at which a signal varies, we recall that the Fourier transform (Bracewell, 1986) expresses a function as the sum of scaled and shifted cosine and sine functions. The more rapidly the signal varies, the higher the frequency of the sinusoids required to model the function. Slowly varying image features have low spatial frequencies; sharp features, such as a strong edge, contain high-frequency information, since short-wavelength sinusoids are required to form them. The range of frequencies in a signal is known as its *bandwidth*; a signal that varies rapidly (i.e., contains many high-frequency components) has a high bandwidth. A *band-limited* function is one which has no frequency components above a certain frequency.

Sampling theory states that a band-limited signal with a highest frequency of ω can be exactly reconstructed with a sampling rate of 2ω. In terms of wavelength, a signal with a minimum wavelength of λ must be sampled with a sample spacing of no greater than $\lambda/2$. The theoretical minimum sampling rate required is known as the *Nyquist rate*. Figure 6-2 shows a signal, its Fourier transform, and the Nyquist sampling points.

What happens if we don't sample at a high enough rate? The higher-frequency components of the signal, which correspond to the fine details in the image, are incorrectly interpreted as lower-frequency components. These erroneous components confuse the overall signal reconstruction, a phenomenon known as *aliasing*. Spatial aliasing often occurs in digital images of sharp edges and in improperly resampled digital images. Figure 6-3 shows a cosine function with a period of 12 units, sampled at every 9 units instead of at the Nyquist rate of every 6 units. Notice the change in frequency that results from the undersampling.

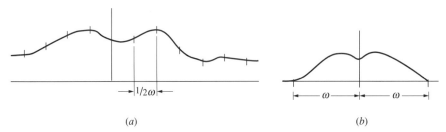

(a)   (b)

Figure 6-2 Sampling of analog signal with bandwidth ω in (a) spatial and (b) frequency domains.

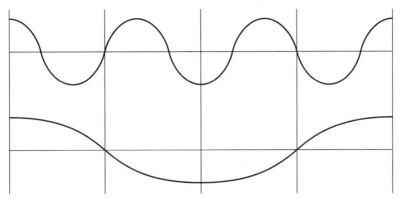

Figure 6-3 Cosine function aliased due to improper sampling.

Sampling theory assumes a perfect sampling function (an impulse with zero width) and that no noise is present in the signal. In real life, the presence of noise requires us to sample more frequently than the theoretical minimum. Also, instead of sampling with an infinitesimal aperture, which would correspond to a perfect sampling function, we must sample with an aperture of finite width that introduces some blurring into the imaging process. This blurring, or *low-pass filtering* (Section 6.2.1.1), reduces the higher-frequency components present and thus alleviates the dangers of aliasing. As an example of this, some sensors are purposely defocused to eliminate high frequencies that would be aliased by the detectors.

For digital images, the lens and the pixel both act as apertures, limiting the maximum detail visible in the image. For most aerial cameras, the lens (and film, if a scanner is used) usually have much higher resolution than the commonly used pixel sizes can record, so the pixel size determines the effective aperture.

When performing measurements on digital imagery, we must keep in mind their discrete nature. A digital image is an array of values on a grid. When we say that a point is located at row 202, column 400, do we mean that it is at that grid line intersection or in the middle of the corresponding grid cell (Fig. 6-4)? Where is point (202.5, 400.75) located? This is primarily a question of definition, but the chosen definition must be consistently adhered to throughout the image acquisition and processing chain. In digital image acquisition, the pixel at (202, 400) covers a finite area and its value is a function of the energy incident on the pixel, with the center of the sample located at the center of the pixel. The definition of pixel location used in programs for display and measurement of images must be understood, along with the definitions used in edge detectors or other feature extractors. For instance, many gradient edge detectors use a 3 by 3 window and calculate an approximation to the image gradient around the pixel of interest. Some operators, however, use a 2 by 2 window and calculate the gradient at the "crack" between pixels.

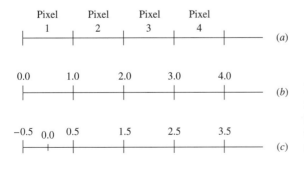

Figure 6-4 (*a*) Discrete (integer) image coordinates. (*b*) Continuous image coordinates defined at pixel edge. (*c*) Continuous image coordinates defined at pixel center.

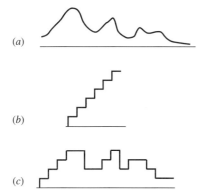

(a)

(b)

(c)

Figure 6-5 Quantization of analog signal. (*a*) Original signal. (*b*) Quantization levels. (*c*) Quantized signal.

6.1.2   Digital Image Radiometry

Each pixel in a digital image represents the intensity value of the image at its array position. This value is the result of *quantization* of the original continuous analog value, which may be the charge accumulated in one detector of a CCD chip due to incident light or the amount of light passing through a piece of film mounted in an image scanner. In the process of conversion from analog to digital, all analog values within one quantization step are assigned to the same digital value, as shown in Fig. 6-5. When the digital image is later displayed, pixels at array positions which initially had different analog values within this range will be displayed with the same intensity value.

Quantization unavoidably introduces noise into a digital image. The amount of quantization noise is determined by the number of quantization levels used (the greater the number of levels, the less noise is introduced) and by the spacing of the quantization levels. Optimum spacing of the quantization levels is determined by the probability distribution of the input values. If the input values are uniformly distributed, so that all values are equally likely, a uniform quantizer is optimal. If more values are likely to occur in a particular portion of the range, then more quantization levels should be used in that region.

Figure 6-6 shows a portion of a digital image and the pixel values corresponding to the area outlined in white. In this case, each pixel is represented by an 8-bit number; the pixel values therefore range from 0 for black to 255 ($2^8 - 1$) for white. The greater the number of bits, the larger the number of gray levels that can be represented. The number of bits used may differ for different applications, going up to 12 or even 16 bits for applications such as radiography where fine shades of gray must be distinguished. A 1-bit image can represent only black and white.

A pixel can also have multiple values associated with it. For instance, each pixel in a digital color image will have three values, which typically represent red, green, and blue intensity values or some other color encoding scheme. A *multispectral* image may have 4 to 40 *bands*, or colors, while *hyperspectral* sensors may have 200 or more bands.

6.2   DIGITAL IMAGE PROCESSING

One advantage of digital imagery is the ability to perform digital processing on the image, in order to modify or emphasize certain aspects of the image. For instance, we may brighten the image to see more detail in shadows or change the frequency content to emphasize edges. Digital image processing is strictly concerned with the image values at each pixel. There is no consideration of the content of the image—an edge detector doesn't know whether the image contains mountains or cows. Digital

(a)

```
102 101  82  84  79  83  94  96  89 153 178 181 141 120 164 175 181 182 184 182 185 191 177  97  81  83  83
148 129 106  86  74  76  96 106  95 147 171 176 182 158 174 178 176 177 177 184 187 186 164  85  80  86  78
163 167 168 152 120  95  91  97  98 154 176 183 178 172 175 180 180 187 186 184 190 186 127  80  75  67  69
167 165 169 172 184 200 180 135 108 169 179 186 184 189 184 175 181 183 188 191 186 185 101  76  83  78  71
168 167 165 164 195 210 204 196 200 194 183 175 181 187 184 191 187 181 186 193 192 169  93  84  80  77  94
136 167 173 173 199 200 193 203 203 197 197 199 194 195 198 198 185 187 191 186 184 134  86  82  83  80  83
 89  86 102 154 196 202 194 198 206 199 202 200 184 194 195 186 188 177 185 183 178 109  84 103  94  83  77
 71  87  83  85  85 116 179 199 199 189 198 195 187 194 188 183 185 189 190 188 159  88  77  87  86  75  79
 85  83  81  87  83  76  82  87 131 183 194 188 190 192 186 187 188 195 185 192 187 147  87  90  75  89  81
 76  71  67  76  78  74  92  83  86 108 165 189 190 187 188 197 188 178 127  98  91 132 179 180 137  91  80
 78  77  70  69  76  68  80 107 105  88 135 184 184 185 187 191 185 172  92 189 181 177 190 188 189 184 181
 68  76  72  67  74  80  77  78  94  91 131 185 187 180 182 178 173 176 191 193 187 186 181 192 187 184 191
 74  68  77  78  73  72  84  77  85  88 146 187 189 190 187 184 178 189 165  95 128 170 180 185 191 190 185
 71  70  72  82  86  80  80  88  87  98 169 181 192 189 184 173 176 181 156  94  84  84 104 131 177 184 187
 73  73  66  73  77  79  86  95  93 127 179 183 174 178 178 180 179 186 119  94  87 102 100  92  86  88 130
 71  71  64  66  74  76  90  91  89 150 177 178 179 183 168 178 174 173  98  94  88  88  81  95  89  93  91
 68  71  67  68  75  71  87  91 107 164 167 169 175 176 176 182 182 158 112  88  81  76  77  82  79  81  79
 69  68  68  65  76  77  80  93 121 163 172 169 172 181 175 181 180 142  99  82  74  68  73  70  80  87  81
 69  65  69  66  70  76  92  93 141 162 162 168 170 175 183 174 169 107  99  82  67  71  75  70  72  78  72
 79  60  61  69  74  74  92  98 161 173 169 168 178 181 175 185 175 118  99  76  72  71  80  68  64  70  71
 65  60  65  66  70  76 103 114 172 181 172 174 172 180 177 182 161 108  90  78  90  86  69  74  75  73  70
 58  57  64  75  74  94  92 151 182 176 174 178 181 180 180 179 162 113  92  79  70  81  67  65  74  77  69
 61  63  68  73  76  83  94 167 169 171 172 175 176 178 174 183 142 103  87  75  83  79  77  78  73  70  68
 61  67  75  79  85  80 108 170 178 182 182 176 172 181 180 183 144 102  83  78  85  80  74  84  79  71  68
 65  61  74  76  85  90 138 176 172 176 173 178 179 184 174 172 123 103  89  79  75  82  71  84  74  71  80
 70  71  79  89  83  76 154 178 167 172 175 179 179 175 183 177 122  82  88  84  90  75  81  84  70  76  76
 64  77  74  88  86 117 168 171 177 175 187 179 186 179 132 114  68  83  82  75  80  87  75  78  74  68  75
 69  73  76  90  84 139 174 176 166 175 184 171 177 183 154 173 115  99  89  92  87  83  81  82  74  68  72
```

(b)

Figure 6-6 (a) Digital image, and (b) the pixel values for the region outlined in white.

image processing can be *radiometric*, in which individual pixel values are modified based on the pixel value distribution across the image or within a local neighborhood, or *spatial*, in which the values of certain neighboring pixels or the location of the pixel within the image is used in the procedure. Spatial image processing should be distinguished from *geometric* processing, in which the image geometry is changed.

6.2.1   Radiometric Processing

Radiometric processing changes pixel values based on global considerations. The global properties of the pixel values in an image are given by a *histogram* (Fig. 6-7) showing the number of pixels that have each possible value. Figure 6-7 shows a histogram for a

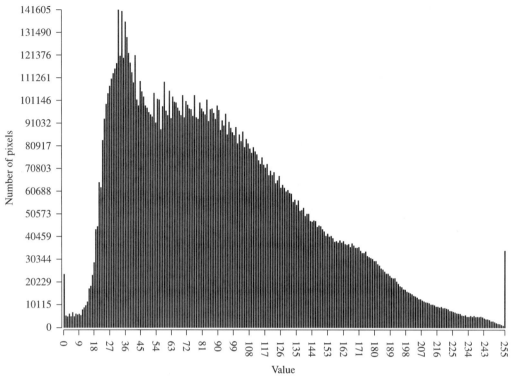

Figure 6-7 Typical image histogram.

gray-scale image with 8 bits per pixel, with pixel values varying from 0 to 255. Notice that few pixels have values near the extremes; most pixels have values near the middle of the range. No information on pixel location is given by a histogram. An image with one half black and the other half white would have exactly the same histogram as an image with black and white pixels arranged in a checkerboard pattern.

In histogram processing, we define a function that maps the input histogram to another histogram with the desired properties. The choice of remapping function is determined by the intended modification to the appearance of the image. Remapping functions are typically shown as in Fig. 6-8, with the input values along the X-axis and the output values along the Y-axis. A diagonal straight line sloping upwards to the right at a 45-degree angle (positive slope) indicates that no change takes place. If we wanted to invert the image (turn a positive into a negative), the remapping function would be a diagonal straight line sloping downwards (negative slope).

If a remapping function has a slope greater than 45 degrees, the contrast of the image is increased, since a given range of input gray levels is mapped to a greater output range. Since the total number of gray levels is constant, this implies that contrast must be decreased elsewhere in the image, compressing the range of some input values. This is most often done by specifying that values near the extremes, for instance, the minimum and maximum 5% of pixels, are mapped to the extreme values.

Suppose that the image is too dark overall, or that we want to see more contrast in the dark areas. If a logarithmic remapping function is applied, pixel values in the dark areas will now cover an increased range, thereby increasing their contrast. Figure 6-9*a* shows an image partially obscured by a cloud shadow and Fig. 6-9*b* shows the same image after a logarithmic remapping. On the other hand, if portions of an image are

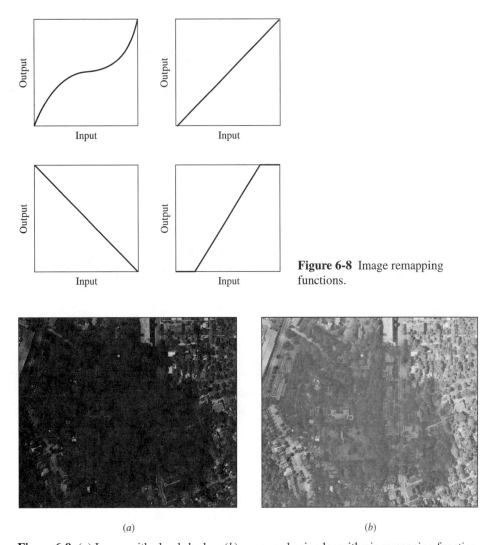

Figure 6-8 Image remapping functions.

Figure 6-9 (*a*) Image with cloud shadow (*b*) remapped using logarithmic remapping function.

too bright to distinguish details, as in Fig. 6-10*a*, an exponential remapping function brings out the details, as shown in Fig. 6-10*b*.

We often want to increase the contrast across the whole dynamic range of the image. In this case we use *histogram equalization*, which transforms the image into one which has a uniform histogram, thereby spreading the pixel values out across the image dynamic range. Figure 6-11 shows a low-contrast image and the histogram-equalized image.

Radiometric processing of color images is more complicated. Simply processing the histograms of the red, green, and blue bands independently will result in an image with shifted colors, since the ratios of the color band values for each pixel will be changed. The RGB (red, green, blue) representation of the image color must be transformed into another representation better suited to digital intensity processing. Several transformations have been defined which separate the RGB values into an intensity component and two other components that encode the color information (Shih, 1995). A common representation is intensity, hue, and saturation (IHS). In the RGB cube

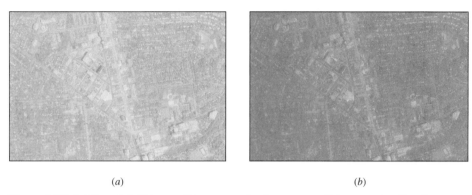

(a)   (b)

Figure 6-10 (a) Over-bright image (b) remapped using exponential remapping function.

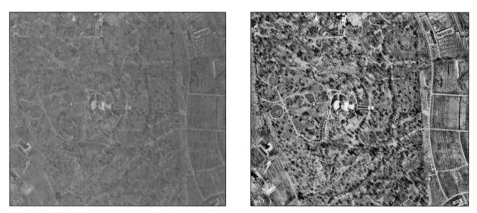

Figure 6-11 Histogram equalization.

shown in Fig. 6-12, the line running from black (0, 0, 0) to white (1, 1, 1) is defined as the intensity axis. The hue and saturation color axes are defined relative to this intensity axis. The hue, H, describes the color as an angle. The saturation, S, can be thought of as describing the amount of the color present, on an axis ranging from white (no color present) to the pure color.

The transformation between RGB and IHS has been defined in a number of ways. One standard transformation is defined as (Pratt, 1991; Shih, 1995)

$$\begin{bmatrix} I \\ V_1 \\ V_2 \end{bmatrix} = \begin{bmatrix} \dfrac{1}{3} & \dfrac{1}{3} & \dfrac{1}{3} \\ -\dfrac{\sqrt{6}}{6} & -\dfrac{\sqrt{6}}{6} & \dfrac{\sqrt{6}}{3} \\ \dfrac{\sqrt{6}}{6} & -\dfrac{\sqrt{6}}{6} & 0 \end{bmatrix} \begin{bmatrix} R \\ G \\ B \end{bmatrix}$$ (6-1)

$$H = \arctan\frac{V_2}{V_1} \quad , \qquad V_1 \neq 0$$

$$S = \sqrt{V_1^2 + V_2^2}$$

Representing colors in IHS allows independent processing of each component. For instance, when merging higher-resolution panchromatic satellite imagery with

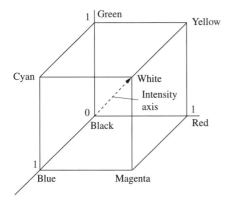

Figure 6-12 RGB (red, green, blue) coordinates.

lower-resolution multispectral imagery, the panchromatic image is used as the intensity component of the fused image, along with the color information from the multispectral image. The RGB representation is then regenerated from the fused image. The transformation from IHS back into RGB is

$$V_1 = S\cos H$$
$$V_2 = S\sin H$$

$$\begin{bmatrix} R \\ G \\ B \end{bmatrix} = \begin{bmatrix} 1 & -\dfrac{\sqrt{6}}{6} & \dfrac{\sqrt{6}}{2} \\[2mm] 1 & -\dfrac{\sqrt{6}}{6} & -\dfrac{\sqrt{6}}{2} \\[2mm] 1 & \dfrac{\sqrt{6}}{3} & 0 \end{bmatrix} \begin{bmatrix} I \\ V_1 \\ V_2 \end{bmatrix} \tag{6-2}$$

EXAMPLE 6-2 *RGB-IHS Conversion*

A pixel in a 24-bit color image has RGB values (150, 150, 75). The corresponding IHS values are

$$\begin{bmatrix} I \\ V_1 \\ V_2 \end{bmatrix} = \begin{bmatrix} 0.333 & 0.333 & 0.333 \\ -0.408 & -0.408 & 0.816 \\ 0.408 & -0.408 & 0 \end{bmatrix} \begin{bmatrix} 150 \\ 150 \\ 75 \end{bmatrix} = \begin{bmatrix} 125.0 \\ -61.237 \\ 0.0 \end{bmatrix}$$

$$H = \arctan\frac{V_2}{V_1} = \arctan\frac{0.0}{-61.237} = -180°$$

$$S = \sqrt{V_1^2 + V_2^2} = 61.237$$

To make the pixel brighter we increase I, the intensity, to 200. The corresponding RGB values are now

$$\begin{bmatrix} R \\ G \\ B \end{bmatrix} = \begin{bmatrix} 1 & -0.408 & 1.225 \\ 1 & -0.408 & -1.225 \\ 1 & 0.816 & 0 \end{bmatrix} \begin{bmatrix} 200 \\ -61.237 \\ 0.0 \end{bmatrix} = \begin{bmatrix} 225.0 \\ 225.0 \\ 150.0 \end{bmatrix}$$

6.2.2 Spatial Processing

In some cases we want to perform *spatial processing* on an image, which we will define as processing a pixel based on some function of its neighboring pixels. Spatial processing can be linear or nonlinear and can be based on *convolutions* or other approaches.

6.2.2.1   Filtering by Convolution

One of the most common image processing operations is *convolution*, a basic operation in linear systems theory. It can be thought of as processing a signal through some function; for example, a recorded image is the convolution of the signal at the entrance pupil of the lens with the transfer function describing the lens.

The convolution of two one-dimensional discrete signals f and g, written $f \otimes g$, is defined as

$$f(m) \otimes g(m) = \sum_{i=-\infty}^{\infty} f(i)g(m-i) = \sum_{i=-\infty}^{\infty} f(m-i)g(i) \tag{6-3}$$

The convolution output at position m is the sum of the products of the corresponding function elements. While convolution is commutative $f \otimes g = g \otimes f$, we usually think of the function with the greater spatial extent as the signal and the smaller one as the filter.

A graphical representation of convolution is shown in Fig. 6-13 for two discrete funtions $f(m)$ and $g(m)$, which are both nonzero only for values $m = 1, 2, 3, 4$. An output value $h(m)$ is obtained by summing the products of the values of the two functions over the index i, as given in Eq. 6-3. While the index i theoretically goes from minus infinity to plus infinity, the functions need only be evaluated at values of i for which

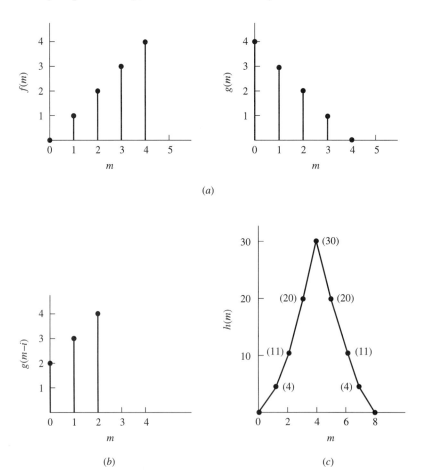

Figure 6-13 Discrete convolution of two functions.

Table 6-1   Convolution of $f(m)$ and $g(m)$ in Fig. 6-13 for $m = 2$

i	$f(i)$	$g(m-i)$	$f(i)\,g(m-i)$
-2	0	0	0
-1	0	1	0
0	0	2	0
1	1	3	3
2	2	4	8
3	3	0	0
		$h(2)$	11

both are nonzero. For example, consider the calculation of $h(2)$, shown in Table 6-1. In this case, the values of i for which $g(m - i)$ is nonzero are from -1 to 2.

A graphical representation of convolution makes clear the operations performed. As shown in Fig. 6.13b, the evaluation of $g(m - i)$ is equivalent to "flipping" the function. The flipped function $g(m - i)$ is aligned with $f(m)$ at the position of interest, in this case $m = 2$, and the corresponding elements are multiplied. The sum of these products is the value of the output at $m = 2$. The result of the convolution operation, $h(m)$ is shown in Fig. 6-13c.

Convolution is a linear operation. Specifically,

$$af \otimes bg = ab(f \otimes g) \tag{6-4}$$

Convolution is also an associative operation:

$$(f \otimes g) \otimes h = f \otimes (g \otimes h) \tag{6-5}$$

Perhaps the most important property of the convolution operation, both in terms of linear system analysis and for visualizing its effects, is the fact that convolution in the spatial domain is equivalent to multiplication in the frequency domain,

$$f \otimes g = FG \tag{6-6}$$

where F and G are the Fourier transforms of f and g, respectively. This property gives us the option of implementing a convolution either in the spatial domain or the frequency domain, depending on the properties of the two signals and available computational resources.

The convolution theorem gives us an easy way to visualize the effects of convolution with a particular filter. If the Fourier transform of a particular filter is zero at higher frequencies, convolution with that filter will remove high-frequency components from the output. This is usually known as a *low-pass filter*. Conversely, a filter with zeros at lower frequencies will accentuate the fine detail in the image. This is known as a *high-pass filter*.

Figure 6-14b shows a perfect low-pass filter in frequency space, and the corresponding spatial filter in Fig. 6-14a. The function in the spatial domain is a *sinc* function, $(\sin x)/x$, which is the Fourier transform of a rectangle function. The sinc function has an infinite extent, an illustration of the fact that a function and its transform cannot both be limited in extent.

When would we use a low-pass filter? As discussed in Section 6.1, when resampling an image we must make sure that we are sampling at a rate at least twice that of the highest image frequency. By removing higher frequencies from the image with a low-pass filter, we can ensure that spurious image detail will not be introduced. Another application

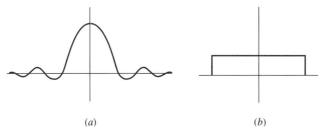

(a)   *(b)*

Figure 6-14 Low-pass filter in (*a*) spatial and (*b*) frequency domains.

is image noise reduction, since low-pass filtering reduces the noise at the expense of fine image detail. Averaging is a form of low-pass filtering.

High-pass filters are used for image *sharpening*. Extracting high-frequency details and adding them back to the original image enhances the detail in the image and makes it appear sharper.

6.2.2.2   Nonlinear Processing

There are several useful nonlinear image processing operations. One of the most common is *median filtering* (Pratt, 1991). Median filtering is a one-dimensional or two-dimensional processing operation in which the value of the pixel of interest is replaced by the median of the pixel values in some neighborhood around it. This is often used to remove image noise, since impulse or *shot* noise (pulses with values very different from the actual image) can be completely removed with minimal image degradation. Noise pulses with a width of less than half the median window will be completely removed. Of course, if the image has legitimate features that are pulse-like, these will also be removed. Step and ramp edges (Fig. 6-15) are not affected by median processing.

6.3   DIGITAL IMAGE RESAMPLING

We often need to resample a digital image into a different geometry, for example, to create an orthophoto. Any type of geometric transformation, including similarity, affine, projective, polynomial, etc. (see Appendix A), can be applied to a digital image in exactly the same way as to any two-dimensional set of coordinates. However, once we have calculated the new image coordinates, which are often fractional, we must decide which gray value to assign to them. A new value must be interpolated from the pixels neighboring the coordinates; a number of mathematical techniques are available (Wolberg, 1990).

The biggest issue is *aliasing*. When an image containing fine detail is sampled at too low a rate, the output image will contain spurious artifacts, such as jagged edges. In Fig. 6-16, an image has been reduced by taking alternate pixels, producing the patterns shown. If the input image is first smoothed to reduce the amount of detail, the sampling is successful.

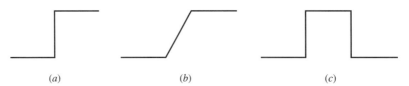

(a)   *(b)*   *(c)*

Figure 6-15 Some typical edge profiles: (*a*) step, (*b*) ramp, and (*c*) line.

Figure 6-16 Subsampled image showing aliasing on straight lines.

The simplest and least expensive resampling method is *nearest neighbor* interpolation, which simply uses the value of the pixel closest to the transformed coordinates. If the scale and geometry change from the input image to the output image is not too great, nearest neighbor interpolation may be satisfactory. If, however, the output image is scaled relative to the input, the appearance will be degraded. Nearest neighbor sampling is very susceptible to aliasing.

Bilinear interpolation is a better solution. To determine an image value w_0 at (x_0, y_0) which lies between pixels (x_i, y_j), (x_{i+1}, y_j), (x_{i+1}, y_{j+1}), and (x_i, y_{j+1}), as shown in Fig. 6-17, we first calculate the auxiliaries t and u and then the desired value w_0

$$t = \frac{x_0 - x_i}{x_{i+1} - x_i}$$

$$u = \frac{y_0 - y_j}{y_{j+1} - y_j} \tag{6-7}$$

$$w_0 = (1 - t)(1 - u)w_{i,j} + t(1 - u)w_{i+1,j} + tu w_{i+1,j+1} + (1 - t)u w_{i,j+1}$$

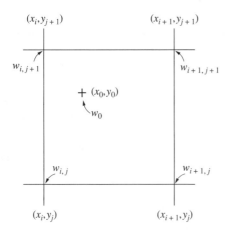

Figure 6-17 Bilinear interpolation.

EXAMPLE 6-3   *Nearest Neighbor Interpolation and Bilinear Interpolation*

A digital image has the following values:

Row	Column	Intensity
10	12	168
10	13	162
11	12	164
11	13	160

Calculate the intensity for a point at row 10.4, column 12.8 by nearest neighbor interpolation and bilinear interpolation.

SOLUTION   The nearest row coordinate is 10 and the nearest column coordinate is 13, so the intensity for the point is 162 by nearest neighbor interpolation.
Using bilinear interpolation (Eq. 6-7), the intensity is

$$t = \frac{10.4 - 10}{11 - 10} = 0.4$$

$$u = \frac{12.8 - 12}{13 - 12} = 0.8$$

$$w_0 = (0.6)(0.2)168 + (0.4)(0.2)164 + (0.4)(0.8)160 + (0.6)(0.8)162 = 162.24$$

More sophisticated interpolation schemes, using higher-order polynomial or spline functions, may give better results at the cost of more computation.

One of the most common applications of resampling in digital photogrammetry is to reproject a stereo pair of digital images into epipolar alignment for stereo viewing or for stereo matching, as discussed in Section 6.6.4.8 and in Chapter 7.

6.4   DIGITAL IMAGE COMPRESSION

Digital images require large amounts of storage space. A 9 by 9 inch aerial photo scanned at 8 bits per pixel with a 0.050 mm pixel size requires 20.9 megabytes (MB) of storage; if scanned at a 0.0125 mm pixel size, it will require 324.4 MB of storage. A color image would require three times that amount of storage. Fortunately, we can take advantage of three characteristics of digital images to reduce the storage requirements.

First, images contain redundant data. Most pixels aren't very different from their neighbors—there is spatial redundancy in the image. Color images will also have spectral redundancy, since the color bands will tend to have similar values. Image sequences (such as video) will have temporal redundancy, since each image will be similar to the one before it.

Second, not all pixel values are equally likely. Examining an image histogram (Fig. 6-7) reveals that most pixel values occur near the middle of the dynamic range. Instead of coding each value with a full eight bits, we can use variable-length codes to represent the pixel values and assign the shortest codes to the most probable values.

Third, very small changes in images are usually hard to notice. Instead of compressing an image using techniques that allow us to exactly reconstruct an image (*lossless* compression), we can allow small, and presumably insignificant, errors to be introduced into the reconstructed image (*lossy* compression).

Evaluation of the quality of images compressed with lossy techniques is very much dependent upon the intended use of the image. A number of theoretical statistics, such as root mean square (RMS) error between the reconstructed and original images, are used to describe lossy techniques, but these may not accurately reflect the impressions of the human observer. For instance, the human eye has much more tolerance for compression error artifacts in a rapidly changing low-resolution video sequence than in a high-resolution static image display that will be examined closely. Photogrammetric applications have especially high requirements, since the precision of image measurements is dependent upon the presence of high-frequency detail in the image and it is this very detail which is sacrificed in lossy compression schemes. The use of lossy compression techniques, such as that developed by the Joint Photographic Experts Group (JPEG), has in fact been shown to affect target measurement (Mikhail et al., 1984; Tempelman et al., 1995) as well as multispectral classification and automated stereo matching (Gupta et al., 1993).

We can use the similarities between adjacent pixels to reduce the amount of storage space required. Very seldom do adjacent pixels have values differing by large amounts, even at edges. This immediately suggests a simple image compression algorithm; instead of storing each pixel's value, store the value of the first pixel in each row and then store only the difference between each pixel and the one before it. The difference between pixels is typically much smaller than the total dynamic range of the image, so we could use fewer bits per pixel.

For example, suppose we know that the maximum difference between adjacent pixels in a row is always less than 32 (5 bits). We could then store the value of the first pixel in each row and then for each subsequent pixel i, store only the difference between pixels i and $i-1$. In this way, we could reduce the storage required to only slightly more than 5 bits per pixel. Since there is no loss of image information (we could reconstruct the image exactly from its compressed form), this would be a *lossless* compression algorithm, as long as the difference between adjacent pixels in a row did not exceed 5 bits.

This is a very simplified example of *predictive coding,* a compression method that predicts the value of a pixel from earlier pixels. Prediction can be based only on earlier pixels in the same row (one-dimensional prediction), or can include pixels in earlier rows (two-dimensional prediction); the pixel at row r, column c could be predicted by pixels $(r-1, c), (r-1, c-1)$, and $(r, c-1)$. The prediction coefficients can be solved by regression calculations over the image or, more often, by using a set of nominal coefficients from similar images.

EXAMPLE 6-4   *Predictive Coding*

The pixels in a section of a typical row of a digital image have the following values:

$$145, 148, 150, 152, 155, 158, 161, 150, 149, 151, 150\ldots$$

We want to implement a one-dimensional predictive coder for image compression. For a zeroth-order predictor, we predict that each pixel will have the same value as the preceding pixel. The transmitted data would therefore be:

$$145, 3, 2, 2, 3, 3, 3, -11, -1, 2, -1\ldots$$

The first value transmitted gives the base value for the line and each successive value gives the difference between the pixel and the preceding pixel. A first-order predictor would predict each pixel using the two preceding pixels. This can be thought of as adding the change in pixel values between the two preceding pixels (the first derivative) to the value of the preceding pixel.

$$x_i = x_{i-1} + (x_{i-1} - x_{i-2})$$

The transmitted values would then be:

$$145, 3, \ -1, 0, 1, 0, 0, \ -14, 10, 3, -3 \ldots$$

In this case, the first transmitted value gives the base value for the line and the second transmitted value is based on a zeroth-order prediction. After two values have been transmitted, the first-order predictor can be used.

Another way to model the redundancy between pixels is in terms of basis functions. The Fourier transform (Bracewell, 1986) uses cosine and sine basis functions of varying phase and frequency. A more commonly used basis function set for image compression is the cosine transform, which uses shifted and scaled cosine curves as the representation. The discrete cosine transform (DCT) is the basis of the JPEG compression standard.

A recent development is the *wavelet transform* (Mallat, 1998), which is used for image compression and many other applications. The basis functions for a wavelet transform are not restricted to trigonometric functions as for the Fourier transform, but can be any of an infinite set of functions. The basis functions are specified according to their desired characteristics, in terms of localization, smoothness, and other properties.

Performing a transform by itself provides no compression—since no information is lost, the transformed image will require as much storage as the input image. However, many of the transform coefficients will be very small. If we are willing to accept some amount of degradation in the reconstructed image, we can discard coefficients that do not contribute significantly to the reconstructed image. The smallest coefficients are at the higher spatial frequencies and represent the finer details in the image. Discarding the high-frequency coefficients causes the reconstructed image to appear blurred compared to the input image.

A useful side-effect of transform coding is the ability to perform *progressive transmission*. By transmitting the lowest-frequency components first, a blurred version of the image can be reconstructed and the user can decide if he wants to see the complete image, thereby saving time and bandwidth. Progressive transmission technology is often used by web browsers when JPEG images are transmitted.

6.5 DIGITAL IMAGE MEASUREMENT

As photogrammetrists, we need to make measurements on digital images and to understand the precision of these measurements. Our ability to make manual measurements is determined by the characteristics of the image display, mainly its contrast and resolution. For automated measurement techniques, the quality of the measurements is determined by the algorithm used, be it *template matching*, *correlation*, or more sophisticated techniques.

The most basic determinants of measurement accuracy in digital images are the characteristics of the digital image itself. Digital images are *discrete* representations, quantized spatially by the pixel sampling geometry and in intensity by quantization into some number of intensity levels. This dual quantization, along with characteristics of the imaged objects, imposes a limit on measurement precision (Havelock, 1989; Havelock, 1991).

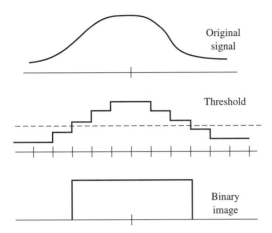

Original
signal

Threshold

Binary
image

Figure 6-18 Measurement of a
quantized binary target.

For simplicity, consider a one-dimensional target in a binary image (Fig. 6-18).
The image of the target is blurred by the imaging process, and then sampled at each
pixel; if the amount of signal captured at each pixel is above the threshold, a '1' is recorded. We can move the target laterally without affecting its binary image until we
move the target far enough that the input seen by any of the pixels drops below the
threshold or rises above it. This positional uncertainty limits the precision with which
we can measure this target in this image, no matter what method is used to analyze the
resulting digital image.

In the absence of noise, we could improve the precision by decreasing the
pixel size. However, consider what happens when noise is present. As the pixel
size decreases, the amount of signal received in each pixel will be smaller and the
signal-to-noise ratio will decrease. As the signal-to-noise ratio decreases, the
noise will cause pixel values to randomly vary over or under the threshold, changing the image and the computed target position.

Improving the precision by adding more quantization levels is similarly limited
by the effects of noise. Without noise, the distance we can move the target without affecting the quantized gray level is reduced by more finely quantizing the input. No
matter how fine the quantization, however, there is still some distance that the target
can move without affecting the quantized gray levels. When noise is present, any
quantization with steps smaller than the noise level will add no information. We
would be trying to position based on the random noise fluctuations.

The combined influence of pixel size and quantization has been borne out by empirical studies (Unruh and Mikhail, 1982; Trinder, 1989; Trinder et al., 1995; Mikhail
et al., 1984).

Scene content also influences measurement precision, particularly in the case of
image correlation. In the same way that a human operator has trouble viewing stereo
on featureless terrain, a digital correlation operator will have trouble making accurate
matches in areas where there is little detail in the image (Förstner, 1982).

6.6   COMPUTER VISION

Computer vision was one of the first applications of *artificial intelligence*. Artificial
intelligence can be defined as the ability of a computer to be programmed to perform tasks normally performed by a human. In computer vision, the computer uses
an image or images of a scene to do the things a human might do, such as recognize
a particular object or navigate through the scene.

Why isn't computer vision a solved problem? Why can't a computer "look at" an aerial image and produce a line map showing the roads and buildings? A number of factors combine to make the computer vision problem difficult. An image is a projection of the three-dimensional world onto a two-dimensional plane. The size and shape of three-dimensional objects are lost. Effects such as perspective and occlusion further complicate the mapping. Even separating the image into regions corresponding to different objects in the scene is a complicated task. Separating objects requires knowing what each object is, but to recognize each object, the computer needs to separate it from the background. Not all objects can be identified by their appearance, especially those which are functionally defined. Contextual or functional knowledge from additional sources is required to locate such objects in the image.

Computer vision can be divided into a number of sub-areas. *Machine vision* typically refers to computer vision used in close-range industrial applications, where environmental conditions such as lighting and camera positions can be controlled and a relatively limited number of objects are to be recognized or positioned. Applications include dimensional or cosmetic inspection of parts, and robot positioning and control. *Image understanding* refers to attempting to build a geometric model of the scene using primitive features to segment and recognize objects. In *active vision*, less emphasis is placed on explicit scene models than on vision algorithms to perform specific tasks, such as obstacle avoidance or target tracking. The video cameras are actively controlled to point, focus, and zoom at points of interest in the scene.

We tend to think of computer vision as always being applied to single black-and-white frame images, but computer vision techniques have been applied to color images, ultraviolet/infrared images, radar, or 3-D "images" from 3-D scanners (Chapter 9) or LIDAR (laser radar, Chapter 11).

This section will discuss some typical computer vision problems, focusing on cartographic applications, to give an understanding of the issues involved and some of the approaches that have been taken. Such a large and dynamic field as this cannot be covered in a single chapter; a broader understanding can be gained by consulting the books listed in the references (Nalwa, 1993; Horn, 1986; Ballard and Brown, 1982; Marr, 1982). Current developments are described in the journals and conference literature.

Although progress has been made in many applications of computer vision, it is still an active area of research. No one algorithmic approach has been shown to work for all applications; indeed, many algorithms perform poorly when run on images or scenes different from those on which they were developed. Therefore, instead of presenting "the" computer vision algorithm, this section will first present some of the topics that must be considered when designing an algorithm for a particular application, then describe some of the technologies used, and finally discuss a current computer vision system.

6.6.1 Computer Vision Algorithm Design and Development

A computer vision algorithm begins with an application goal, such as to recognize a specific object (e.g., a machined part) or a class of objects (e.g., buildings or roads), to acquire precise measurements of an object, or to enable an autonomous robot to navigate through obstacles. If we are interested in object recognition, for instance, we must first construct a model of the object and also of the scene, which gives the object context.

Suppose that we are trying to extract roads from aerial imagery, a task that on the surface seems almost trivial. First, we must decide which types of roads the algorithm will extract; rural two-lane roads have different characteristics than urban multi-lane freeways or city streets. For rural roads in open terrain, we may assume a simple road model with a constant road width, constant intensity across and along the road, and no major occlusions. However, a more realistic and comprehensive model must take into account occlusions due to trees and buildings; nonuniform intensities due to shadows, traffic, road markings, and road surface changes; and varying road geometry due to intersections or entrance and exit ramps.

The appropriate model must be designed into the system from the start. If we focus on one type of road, the system may be optimized for better performance on that road type, but we will be left with the problem of implementing detectors for other types of roads and the meta-problem of deciding which of the extraction systems to run on any given image.

A basic consideration in system design is the type of imagery available. In our example, the resolution of the imagery determines the geometry and amount of detail shown. On a SPOT panchromatic image with 10-meter pixels, a typical rural road may be only two or three pixels wide, while high-resolution mapping imagery will show cars and road markings. The spectral characteristics of the imagery also determine the types of features that can be used for recognition. In panchromatic imagery, only shades of gray are available, but color imagery gives better discrimination between the road surface and possible backgrounds. If multispectral imagery is available, spectral classification techniques (Chapter 10) may be used to help identify asphalt or concrete road surfaces.

When using aerial imagery, we usually have little control over imaging conditions other than to specify sun angles and photo scale. In industrial situations, however, a great deal of effort is applied to designing lighting, targets, and fixtures in order to simplify the computer vision task.

Once we have determined the image characteristics, we can select the image features to be used in the algorithm. A number of features are available, as discussed in the next section. We must select the features that are most salient to the object or task of interest. In our road example, we know that roads are typically elongated linear features. We could use edges and look for long parallel edges, or we could use region-based methods and look for elongated regions. At low resolutions, roads have fairly uniform gray levels, but at higher resolutions cars and road markings will complicate the appearance of the road.

System design begins with top-level decisions concerning the type of algorithm to use, its inputs and outputs, and its expected performance. One fundamental choice is whether the system is to be completely automated or semi-automated, allowing operator inputs to guide the processing and edit the output. Humans can quickly examine a scene and indicate where the algorithm should focus its work, while the computer does the final delineation. Alternatively, the computer can generate a number of hypotheses and have the operator accept, edit, delete, or add to the initial hypotheses. However operator interaction is utilized, the system must be designed so that the end result is more efficient than purely manual methods. If too much time is required for initial setup or excessive editing of the results is required, the system will not be more efficient than purely manual methods.

Overall operational flow is also a design decision. We may use a data-driven approach, starting from image features and generating object hypotheses, or we may use a *top-down approach*, starting from objects of interest and trying to find them within the scene.

The expected performance of the system and how that performance is to be measured are important design criteria that are often neglected. If we can generate a reference model (or "ground truth") for the scene, which contains all of the features of interest, we can compare the system's output against the perfect data. For systems with metric output, we can use standard statistical covariance measures to compare dimensions or positions output from the system with the reference model. If the system output is discrete objects, we can characterize performance in terms of:

True positives: objects successfully detected by the system

False positives: objects detected by the system which do not exist in the world

True negatives: objects the system did not detect where no object exists

False negatives: objects not detected by the system which exist in the world

Each type of error has a correction cost associated with it. For instance, a false positive produced by a building extraction system requires only that the operator delete it; a false negative may require the operator to manually delineate the building, a much more expensive operation. In many cases, the number of false positives and the number of false negatives are related and can be controlled somewhat by adjusting system parameters. If a building hypothesis must obtain a certain score to be output by the system as a building, then lowering the threshold score will produce more building hypotheses. While more of these additional hypotheses may be wrong, buildings which were missed before may be found.

Another important aspect of evaluation is the range of test imagery and test scenes. A standard cliche of computer vision is that "every image has its algorithm," in other words, any algorithm can be tuned to work on a specific image. Although it would seem obvious, systems should not be evaluated on the same imagery used for algorithm development. Instead, imagery from different flights, over different areas, and over scenes with differing characteristics must be used if any meaningful performance statistics are to be generated.

6.6.2   Image Representation

A number of algorithms make use of *image pyramids,* a series of reduced-resolution versions of an image formed by smoothing and subsampling of the original image (Fig. 6-19).

Examining reduced-resolution images allows us to focus on the important structures in the image, since only strong edges and large regions are visible. Random noise is reduced by the inherent averaging in the resolution reduction.

A number of algorithms are based on multiple resolution image pyramids (Terzopoulos, 1986). In edge detection (Canny, 1986), edges that appear at coarse scales are usually more significant in interpreting the scene; tracking the edges up to the finer scales allows more precise location. Working with a reduced-resolution image reduces the search space for stereo image matching (Section 6.6.4.8). A solution obtained at a low resolution can be used to initialize a solution at the next higher resolution (Witkin et al., 1987), making the algorithm both robust and accurate. There is biological evidence that mammalian visual systems also do a form of pyramid processing, using receptor cells at differing scales.

Image pyramids are an example of *scale space* (Witkin, 1983; Koenderink, 1984). A scale space consists of a series of representations of a signal, at resolutions ranging from the original down to some arbitrary minimum, parameterized by a scale factor. As the scale is decreased, the amount of detail in the signal is reduced.

An important property of scale space is that no new details are introduced into the signal as the scale is reduced. It has been shown (Yuille and Poggio, 1986) that the only operator which does not introduce false details is the Gaussian and its derivatives. For *Gaussian filtering*, the scale factor is the standard deviation of the kernel.

Figure 6-19 Image pyramid.

Note that an image filtered with the Gaussian is still the same size, only with less visible detail. To form an image pyramid, the filtered images are subsampled after each filtering operation to reduce the number of pixels. Due to the reduction in image information from the filtering, the subsampling does not discard information.

Although Gaussian filtering is the most theoretically acceptable method, in most practical systems image pyramids are formed by simple averaging and subsampling.

6.6.3   Object Representation

In order to recognize an object in a scene, we must be able to represent three-dimensional objects. A large number of representations exist, developed for applications such as computer-aided design (CAD). Computer vision has some special requirements. We need representations that can be generated from image data and can be easily matched against some reference model. Some objects are precisely defined by geometry, such as industrial parts; others have a generic geometry, with the same basic shape but different proportions or parts.

Object representations can be divided into three basic classes (Ballard and Brown, 1982):

1. Surface or boundary representations
2. Sweep representations
3. Volumetric representations

6.6.3.1   Surface or Boundary Representations

This class of representations consists of a set of surfaces and the boundaries between them. This type of representation is not suitable for all objects, since not all objects have well-defined faces or edges. These are natural object representations for computer vision, since most algorithms extract edges and surface shapes.

This representation is best suited for polyhedral objects, where the surfaces are planar and the edges between them are straight. While curved surfaces can be used, based on polynomials, splines, or other functions, recovery of such surfaces can be difficult.

6.6.3.2   Sweep Representations (Generalized Cylinders)

Sweep representations, usually called generalized cylinders, are defined by a cross section of an arbitrary shape and an axis along which the cross section is swept (Fig. 6-20). The cross section may vary during the sweep according to some parametric rule. Generalized cylinders are a very useful representation for elongated objects, objects with a well-defined central axis, or combinations of parts with those properties.

Several object recognition systems have been built around generalized cylinder representations (Marr and Nishihara, 1978; Zerroug and Nevatia, 1996). An estimate of the cylinder axis can often be extracted from an image by finding the axis of symmetry of an image region corresponding to the object. If more than one view is available, the radius of the cylinder may be determinable by intersecting the rays tangent to the object.

6.6.3.3   Volumetric Representations

Volumetric representations represent an object as a combination of elementary spatial primitives. In their simplest form, space is divided into a regular array of *voxels*

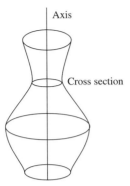

Figure 6-20 A generalized cylinder.

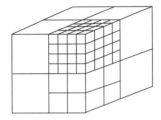

Figure 6-21 An octree representation, showing the recursive subdivision scheme.

(volume elements) and the occupied voxels are noted. Instead of dividing the whole space into uniform cells, a more efficient approach is to use an *octree* representation. Each cell in an octree representation can be divided into eight smaller cells, each of which can also be divided, down to the final spatial resolution (Fig. 6-21). Once a cell is completely occupied or completely empty at any given resolution, it need not be further subdivided, thereby saving storage space.

This type of representation is often used for navigation applications, in which only the presence or absence of some obstacle is important. A variant of this representation is the inclusion of the certainty that a specific cell is occupied, allowing an autonomous robot to navigate only through areas it is certain are unobstructed or to focus its sensors on uncertain areas.

Another type of volumetric representation is *constructive solid geometry* (CSG), in which rotated, translated, and scaled geometric primitives are combined according to the rules of set theory. Typical geometric primitives include rectangular or triangular prisms and cylinders, while operations may include addition (placing one primitive on top of another) or subtraction (subtracting a cylinder from a rectangular prism leaves a hole).

6.6.4 Low-Level Vision: Feature Extraction

Very few vision algorithms operate directly on the entire array of pixels. Instead, low-level processing is performed to extract image *features* for later reasoning. This reduces the amount of data to be processed and also extracts the most salient information from the image. For instance, a line drawing can convey much of the same information as a photograph, but with much less data. While it can be dangerous to use the human visual system as a justification for a computer vision system, there is physiological evidence that the human visual processing system contains, among other detectors, feature detectors "hardwired" to detect edges or motion in a set of directions.

A number of image features have been used in computer vision, including the following:

- edges
- regions
- textures
- interest points
- stereo disparity
- color
- motion
- shading

Extracted features may be used to segment the image into regions corresponding to objects or parts of objects, or to determine the shapes of objects. Known as the *shape-from-x* problem, shape recovery algorithms have been developed for shading, texture, illumination, and motion, among others.

Some of these features are briefly discussed in the following sections; more detailed descriptions of how they are extracted and used are available in the literature.

6.6.4.1 Edge Detection

Many processes in the world give rise to edges in an image:

- Depth discontinuities between objects, which correspond to object boundaries.
- Surface reflectance discontinuities at markings on an object's surface.
- Discontinuities in surface orientation at the boundaries between object surfaces.
- Changes in illumination due to shadows or reflections.

These situations all produce intensity discontinuities, which may be of several different types. Common edge profiles (Fig. 6-15) include the *step* edge, an abrupt transition between two discrete levels; the *ramp* edge, a more gradual transition; and the *line*, which may be thought of as parallel step edges with their gradients in opposite directions.

Edge detection begins as an image processing operation, but can be carried out in various ways depending on assumptions about the imaging process and how the edges will be used in the algorithm.

The first step is the edge operator itself. If an edge is defined as a change in image intensity, a gradient operator is the obvious choice. If the differences between pixels in two orthogonal image directions (usually, but not necessarily, the row and column directions) are δ_1 and δ_2, the gradient magnitude is

$$\sqrt{\delta_1^2 + \delta_2^2}$$

and the gradient direction is

$$\arctan \frac{\delta_2}{\delta_1}$$

Unfortunately, gradient operators have the property of emphasizing noise in the image. The standard way to combat random noise is by averaging, or calculating the gradient using image values averaged over a region. Although using larger support areas gives better noise resistance, the operator will confuse edges spaced more closely than the

0 1	1 0		−1 0 1	1 1 1		−1 0 1	1 2 1
−1 0	0 −1		−1 0 1	0 0 0		−2 0 2	0 0 0
			−1 0 1	−1 −1 −1		−1 0 1	−1 −2 −1

Roberts   Prewitt   Sobel

Figure 6-22 Roberts, Prewitt, and Sobel edge detector kernels.

size of the support regions. The use of larger regions for averaging also causes problems at edge corners or junctions.

Commonly used gradient operators include the *Roberts, Prewitt*, and *Sobel* operators, shown in Fig. 6-22. Notice that the Prewitt and Sobel operators include adjacent pixels to provide better resistance to noise.

Another popular edge operator is the *Canny* operator (Canny, 1986), which was designed to simultaneously optimize both the signal-to-noise ratio and edge localization, while giving only one response to an edge. In two dimensions, the Canny operator finds the maximum partial derivative of the image perpendicular to the edge direction and smooths the image perpendicular to the edge direction. If we denote the image by I and the Gaussian operator by G, this procedure can be written as

$$\frac{\partial^2}{\partial \bar{n}^2}(G \otimes I) \tag{6-8}$$

where G is a Gaussian function, I is the image function, and $\bar{n}$ is the unit vector in the direction of the (smoothed) image gradient,

$$\bar{n} = \frac{\nabla G \otimes I}{|\nabla G \otimes I|} \tag{6-9}$$

This is usually implemented by smoothing with a Gaussian, then taking partial derivatives ∇x and ∇y in the horizontal and vertical directions. The gradient direction is then

$$\bar{n} = \begin{bmatrix} \nabla x \\ \nabla y \end{bmatrix} \tag{6-10}$$

$$|\bar{n}| = \sqrt{(\nabla x)^2 + (\nabla y)^2}$$

The edge detector is designed to work at different scales, defined as using varying widths of Gaussians, in order to extract edges with differing signal-to-noise ratios (essentially, weak versus strong edges). The detector starts with edges at the finest resolution (smallest operator width), then searches for evidence of the edges at larger widths. Another variation introduced by Canny is the use of hysteresis thresholding. Instead of thresholding the edge operator output at a single value, once an edge pixel has been determined to be above the first threshold, connected pixels are also labeled as edges as long as their values are above a second, lower threshold. This produces edges without the gaps that often occur.

The *Marr-Hildreth edge detector* is based on the second derivative (Laplacian) function.

$$\nabla^2 I = \frac{\partial^2 I}{\partial x^2} + \frac{\partial^2 I}{\partial y^2} \tag{6-11}$$

Edges, which are peaks in the output of first-derivative operators, become zero crossings in the output of second-derivative operators. Unfortunately, the second derivative is even more sensitive to noise than the first derivative. We therefore smooth the image with a Gaussian filter before taking the Laplacian, in order to remove some of the noise. Since convolution is a linear operator, taking the Laplacian of the image convolved with a Gaussian is equivalent to convolving the image with the Laplacian of the Gaussian:

$$\nabla^2(G \otimes I) = (\nabla^2 G) \otimes I \qquad (6\text{-}12)$$

Edges are located at the zero crossings of the operator output (Fig. 6-23). The Laplacian of the Gaussian can be approximated as the difference of two Gaussians with a 1.6:1 ratio of their standard deviations, for further efficiency.

Given the large number of edge detectors available (only a fraction of which have been discussed here), one may wonder how to select the best one. Several, such as the

Figure 6-23 Zero-crossing edges from a Laplacian-of-Gaussian edge detector.

Canny operator, are described as optimal, but they are optimal only relative to the assumptions made in their derivation. In many cases, edge detectors give essentially equivalent results; evaluating edge detectors in the context of the specific application and imagery to be evaluated may show that one has a slight advantage.

In general, an application whose performance depends on very high edge quality is unlikely to be successful. Even an optimal edge detection does not give a wireframe depiction of the scene, but instead produces broken pieces of object boundaries along with a number of edges that may correspond to shadows, texture, surface markings, image noise, etc. Notice in Fig. 6-24 the edges corresponding to shadows

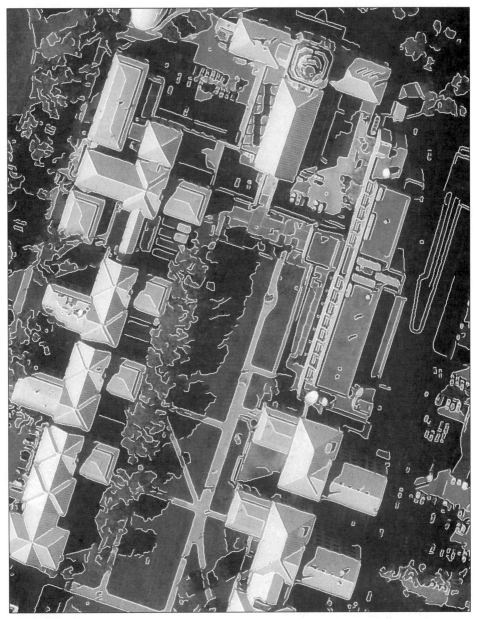

Figure 6-24 Typical edges from edge detection. Notice the broken and missing edges and also the edges due to shadows.

and random patterns in the trees and grass, the missing edges on the building roofs due to lack of contrast, and the number of broken edges.

6.6.4.2 Sub-Pixel Edge Location

For photogrammetric applications, we are often concerned with the precise location of edges. Standard edge detectors typically work at integer pixel spacings. In order to get sub-pixel edge locations, we must interpolate the position of the maximum gradient or second derivative zero crossing from the image data. However, as discussed in the previous subsection, the gradient can be very noisy. In order to get accurate edge locations, most sub-pixel edge location algorithms apply a fitting operation over the edge neighborhood to obtain a better estimate.

One of the earliest examples of a sub-pixel edge detector was formulated by Hueckel (1971; 1973). This edge detector used moments calculated within a circular window to determine the location, orientation, and intensity of a straight edge. The moment calculations were used for computational efficiency.

In the facet model (Haralick, 1984), a surface described by orthogonal polynomials is fit to the neighborhood surrounding the point of interest. The edge is defined to be at the zero crossing of the second directional derivative in the direction of the gradient. The derivative is calculated in terms of the polynomial coefficients, to improve noise resistance.

Another approach is to fit an edge model to the image. Mikhail et al. (1984) used least squares to fit an edge model, along with a parameterized model of the image's spread function, to the image intensity data. Steger (1998) first detects line points and links them into lines, and then fits an asymmetrical line model to the image to determine the line width. Scale space processing (Section 6.6.2) is used to allow the detector to work for lines of different widths.

Another edge localization approach is the use of *snakes* (Kass, et al., 1988; Gruen and Li, 1997). A snake optimizes the fit between image properties, such as edge strength, and its own internal model, usually related to geometry. For example, in the initial formulation of snakes (Kass, et al., 1988), two energy terms were minimized: the internal energy, E_I, which modeled the bending of the snake; and the external energy, E_E, which was the negative of the edge gradient intensity:

$$E_I(i) = \frac{\alpha_i |v_i - v_{i-1}|^2}{2h^2} + \frac{\beta_i |v_{i-1} - 2v_i + v_{i+1}|^2}{2h^4} \tag{6-13}$$

Equation 6-13 gives the expression for the internal energy at each point i in discrete form, where v_i is the coordinate along the snake, h is the spacing between snake points, and α_i and β_i control the stiffness of the snake. The external energy is evaluated from the image at each snake point. The snake solution begins from an approximate initial position established by operator inputs or from another operator; an iterative discrete differential equation solution then proceeds until equilibrium is reached. The solution reached may be a local minimum instead of a global minimum; for instance, the snake may get stuck on a very bright image point instead of the edge adjacent to it. In this case, operator interaction may be used to guide the snake. A number of variations of snake solutions have been implemented, using solution methods such as dynamic programming and B-splines (Gruen and Li, 1997).

6.6.4.3 Edge Thresholding, Thinning, and Linking

Edge operators produce an output at each pixel. To produce discrete edges, this output must be thresholded and the pixels over the threshold labeled as edge pixels.

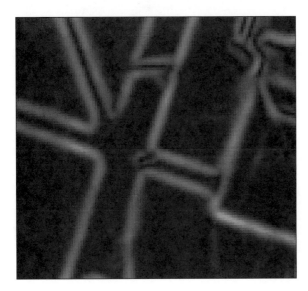

Figure 6-25 Edge detector results prior to thinning.

Real image edges usually cause a region of pixels around the edge to be over the threshold (Fig. 6-25). A *thinning* operation is therefore employed so that only the maximum response in the direction perpendicular to the edge is called an edge.

Extracted edges are seldom continuous, due to image noise and scene artifacts. A *linking* step fills in small edge gaps based on the examination of pixels in the gap (they may have been just under the threshold) or on heuristics such as gap size versus edge length.

Edge detectors extract a large number of edges, many of which may not be useful or relevant. For instance, the objects we want to recognize may consist mostly of straight lines. If we can concentrate on just the straight lines, our processing requirements will be reduced.

The *Hough transform* (Illingworth and Kittler, 1988) is used to extract geometric shapes from an image. The Hough transform works in the parameter space of the curve, with each axis of the accumulator array corresponding to a quantized curve parameter.

As an example, suppose we want to extract straight lines in an image. A set of line parameters determines the coordinates of the points on the line; conversely, a point determines a family of lines (parameter sets) passing through it. For a straight line, the Hough space is therefore two-dimensional. To use the Hough transform to extract straight lines, we write the line equation in the form

$$x\cos\theta + y\sin\theta = \rho \tag{6-14}$$

Use of the ρ-θ formulation avoids the singularities in the slope-intercept form of the line equation, as discussed in Appendix D.2. Each point has an infinite number of lines passing through it, each with a different ρ-θ representation determined by Eq. 6-14. To apply the Hough transform, we first set up an accumulator which divides the ρ-θ space into discrete cells. For each point (Fig. 6-26) we then step through the values of ρ for each cell and calculate the corresponding θ (or vice versa). Each point generates a sinusoidal line in parameter space, as shown in Fig. 6-27 for the points in Fig. 6-26. For each of these ρ-θ pairs, we increment the corresponding accumulator cell. Cells in the accumulator with the largest values indicate the straight lines with the largest number of points.

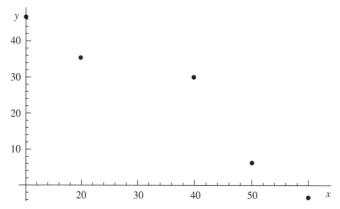

Figure 6-26 Points for Hough transform.

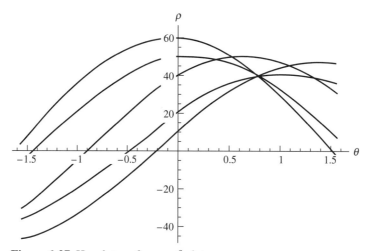

Figure 6-27 Hough transform ρ-θ plot.

| EXAMPLE 6-5 | *Hough Transform* |

Given the following points (shown in Fig. 6-26), use the Hough transform to find which points lie on straight lines.

Point	x	y
1	10.0	46.6
2	20.0	36.0
3	40.0	30.7
4	50.0	6.6
5	60.0	−3.4

For the sake of this example, assume that we can quantize the θ axis with just three values: 30, 45, and 60 degrees. For each point, we then calculate the value of ρ corresponding to the values of θ:

θ	ρ, point 1	ρ, point 2	ρ, point 3	ρ, point 4	ρ, point 5
30	32.0	35.3	50.0	46.6	50.3
45	40.0	39.6	50.0	40.0	40.0
60	45.4	41.2	46.6	30.7	27.1

When plotted, as shown in Fig. 6-27, the strongest line is seen to be at $\theta = 45$ degrees, $\rho = 40.0$.

Although the Hough transform is a powerful technique, it does have limitations. The finer the ρ-θ space is quantized, the more computation is required. Too coarse a quantization, however, will not distinguish lines with similar parameters. Noise effects also tend to spread the correct maximum across adjacent accumulator cells, possibly causing lines to be missed or incorrectly parameterized. Note that the Hough transform does not extract line end points; this must be done in a separate step.

6.6.4.4   Region Segmentation

Instead of trying to directly find the edges in an image, region segmentation relies on trying to group together similar pixels. The exact definition of a region is determined by the algorithm that will use the regions. In nearly all cases, pixels are uniquely assigned to a single region. Regions may or may not have holes, and there may be requirements for the boundary shape.

The simplest way to form regions is to simply threshold the image at some intensity level that separates object and background. Selection of a useful threshold is difficult, if not impossible, and segmented pixels will usually not be spatially connected to form valid regions. This method works only for very simple scenes, such as the binary images often used in industrial applications.

More sophisticated region segmentation schemes define a region by some *similarity measure*. This may be a range of gray levels, a maximum gray level difference between adjacent pixels, maximum deviation from some neighborhood intensity function, such as a plane, color similarities, allowable boundary shapes, image texture characteristics (as described in Section 6.6.4.5), or any number of other possibilities.

Once region similarity criteria are chosen, the next step is to actually form the regions. Common approaches include split-and-merge, region growing, and relaxation.

Split-and-merge is a recursive process. Each region is examined to see if it is a valid region according to the similarity measure. To start the process, the entire image is defined as the first region. If the region is consistent, then it is accepted. If not, the region is split into halves or quarters and each new region is examined (Fig. 6-28). The splitting and examination process continues until all regions are consistent. At this point, separate adjacent regions may exist which should actually be the same region. The merging step therefore examines adjacent regions and combines those that are consistent.

In *region growing*, initial region seeds are scattered across the image, either in a regular or random pattern. Neighboring pixels are added to each region if they meet the criteria for region membership (Fig. 6-29). This method has drawbacks, in that the segmentation can depend on the location of the starting points and that regions begun first tend to collect more of the pixels. Growing regions simultaneously can alleviate some of these problems.

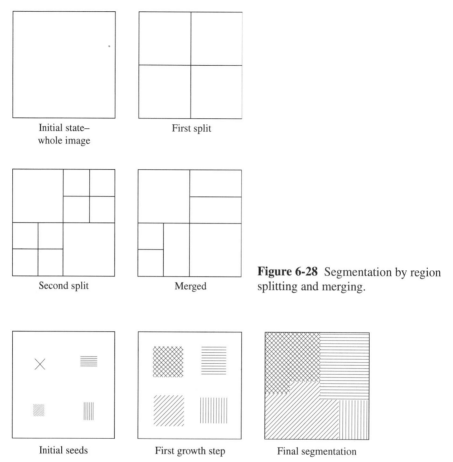

Initial state–
whole image

First split

Second split

Merged

Figure 6-28 Segmentation by region splitting and merging.

Initial seeds

First growth step

Final segmentation

Figure 6-29 Segmentation by region growing.

Relaxation methods (Davis and Rosenfeld, 1981; Hummel and Zucker, 1983) have also been applied to image segmentation. The process begins with a set of labels, and each pixel is assigned one of the labels, either based on initial information or randomly. The neighbors of each pixel are examined and the compatibility between their labels evaluated according to some model. This may involve comparisons of intensity levels or geometric criteria such as boundary shape. The process is iterated until pixel labels quit changing.

There is the potential that the relaxation process will converge to a local minimum instead of a global minimum; in other words, small groups of locally consistent labels instead of a globally consistent labeling. Stochastic relaxation methods (Geman and Geman, 1984) have been developed to prevent this. Instead of the updating process being done in a deterministic fashion, the probability of a label changing is based on compatibility with its neighbors and also the global *temperature*. This temperature parameter varies with the state of the solution. At the start of the process, the temperature is high and changes are mostly random. As the solution nears convergence, the temperature is lowered and compatibility becomes the determining factor. Stochastic relaxation is often called *statistical annealing*, since it is analogous to the annealing process for metals in which the temperature is slowly lowered to reach a minimum energy state.

6.6.4.5 Image Texture

Many natural and man-made surfaces in the world are covered with repetitive patterns. These may be regular, such as the pattern of bricks in a wall, or irregular, such as grass in a lawn. These patterns in the world produce image patterns; these image patterns are called *textures*, since they are not part of the underlying geometry of the scene, but rather a property of the imaged surfaces. The repeated patterns that make up the texture are often called *texels*, or texture elements.

Textures can give useful visual cues to understanding a scene. The changes in the appearance of the texture across the image give cues as to the orientation of the surface, while changes in texture indicate object boundaries. Figure 6-30 shows some examples of textures commonly seen in images.

Before textures can be used for image understanding, they must be described, a difficult task given the wide range of possible textures. Two main approaches have been studied: statistical and structural (Tomita and Tsuji, 1990; Haralick, 1979).

Statistical methods describe the intensity variations of the texture using statistical measures. The simplest of these methods use first-order statistics, which work only with properties of single pixels, describing histograms of image intensity in terms of symmetry and spread. Second-order statistics methods attempt to capture the spatial properties of the texture by using the properties of pairs of pixels. One such statistic is the *autocorrelation*, which is calculated by taking a window in the image, shifting it across the image, and calculating the correlation. If the contents of the window match

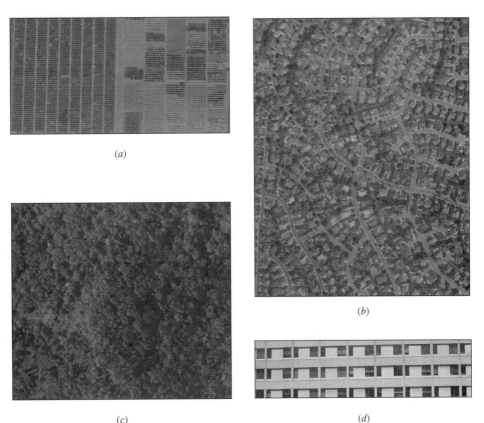

(a)

(b)

(c)   (d)

Figure 6-30 Some typical image textures: (*a*) orchard, (*b*) subdivision, (*c*) forest, and (*d*) windows.

the selected region, the correlation will be high; a periodic structure in the autocorrelation indicates regularly repeated image elements.

A related statistical measure is the *co-occurrence matrix*, which describes the relationship between the values of pixels at specified distances and orientations from each other. The co-occurrence matrix is square, with the number of rows and columns determined by the number of bins into which the gray levels of the image are grouped. The co-occurrence matrix for a distance of 1 pixel and a rotation angle of 0 would be generated by looking at each pixel and its neighbor to the right, (pixels i and $i + 1$) within a defined window. The matrix entry corresponding to the gray levels for pixels i and $i + 1$ would then be incremented. A complete set of co-occurrence matrices is generated for varying distances and rotations, and then examined to determine peaks indicating structure and direction. This method can obviously generate a large amount of data, making analysis complicated. If the window size or gray level ranges are badly chosen, the structure may not be evident in the co-occurence matrices.

Edges can also be used as a statistical texture measure. One possibility is to measure edge density per unit area; another is to examine the histogram of edge directions over an image window for peaks, which indicate structure.

Structural methods of texture extraction first try to identify texture primitives, such as edges or regions of constant gray level, and then extract the relationship between adjacent texture primitives. Recognition of texture elements can be difficult and unreliable.

Texture measures can be used for image segmentation by calculating texture measures over the image, and then grouping regions with similar texture characteristics. Texture-based segmentation typically has problems with the boundaries between regions, since most texture measures are calculated over a window that will overlap the boundary and confuse the texture measures.

Perspective effects present in texture elements can be used to estimate approximate surface orientations. Several methods for shape recovery from texture patterns have been developed. Each shape recovery method requires different assumptions about the texture properties and the imaging model. One approach to the determination of surface orientation is to first extract the texture elements and then look at the variations in texel density or distribution (Ikeuchi, 1984; Aloimonos and Swain, 1988; Blostein and Ahuja, 1989). The main difficulty in this approach is reliably extracting the texels. If we assume that the statistical properties of the texture are isotropic (do not vary with location or orientation), then any variations in the texture statistics across the image can be used to determine the surface orientation (Witkin, 1981). However, since most textures have oriented elements (e.g., brick walls, agricultural fields) the statistical properties will vary with respect to orientation. A less restrictive assumption is that the texture is homogeneous throughout the scene (Kanatani and Chou, 1989).

6.6.4.6   Interest Point Operators

An *interest point operator* is used to detect well-defined points or corners for use in matching or measurement in the same way that a human operator selects tie points for triangulation. In fact, a main application of interest point operators is in automated triangulation (Heipke, 1997*b*). A related application is in their use as a first step in stereo matching (Hannah, 1989). After well-defined points are found, they are matched to determine the basic shape of the surface. Surrounding points can then be matched using the approximate surface, enabling more efficient matches and helping to prevent false matches.

An image window containing a strong point or a corner will have a high variance among its pixel values. Unfortunately, the variance contains no directional information, so that an edge across the window and a randomly distributed set of intensities can have the same variance as a well-defined point. Interest operators therefore use statistics with spatial information to determine how well-defined the position of a particular window is.

The first interest point operator was developed by Moravec (1981). In his implementation, the image was divided into 4 by 4 or 8 by 8 windows. The sums of the squares of the differences between adjacent pixels in the horizontal, vertical, and both diagonal directions were calculated. The score for the window was the minimum of these four sums. Windows with scores above a set threshold were determined to be interest points.

Another interest operator is the Förstner operator (Förstner and Gülch, 1987), which is based on least squares fitting of the gradient information in each window. Once promising windows are found, a final solution optimizes the position of the point. Figure 6-31 shows the points selected by a Förstner operator in a typical aerial scene.

6.6.4.7 Target Measurement

Photogrammetry is often concerned with locating and measuring predefined targets. These targets may be fiducial marks or reseau crosses on an image or control points in the scene. A number of different techniques are used to locate predefined targets, depending on the type and shape of the target, whether its image will be affected by perspective effects, and how much its average gray level differs from the background.

Finding the target in the image may be a difficult task. In some cases, such as for fiducial marks, the approximate location of the target in the image is known. If good

Figure 6-31 Förstner interest points.

approximations for the image location and orientation and the target object space co-ordinates are known, then an approximate image position can be established by projection. Hough transform techniques (Section 6.6.4.3) can also be used.

Once a target is identified in the image, its exact coordinates must be determined. If the target gray levels are sufficiently different from the background, simple thresholding can be used to identify target pixels. If the target is symmetrical, the centroid of the pixels above threshold can then be calculated and used as the target measurement. This method works better for large, well-defined targets, since image noise may result in an uneven target boundary. This also does not allow for changes in target shape due to viewpoint.

If the edge of the target can be reliably identified and extracted, a function describing the target edge shape can be fit to the edge pixel coordinates. For instance, if the target is circular, a circle would be fit and the center point of the circle used as the target location. If edges can be extracted to sub-pixel precision, the precision of the target location is improved. Least squares edge fitting is equivalent to the use of snakes (Section 6.6.4.2).

Template matching may be used to measure the target, if the shape and size of the target image are known. If the target is thresholded, counting the common pixels between the template and image is sufficient. Otherwise, correlation procedures may be used.

Least squares matching (discussed in Section 6.6.4.8) may be performed between the target template and the image. If geometric transformation parameters are included in the least squares solution, the template can be corrected for scale or perspective effects. The use of least squares matching gives a very accurate measurement for the target, since all image information is integrated into the solution. However, the approximate target location must be known for the solution to converge.

6.6.4.8 Stereo Disparity

Stereo image matching can be an application in itself, as when it is used to generate a *digital elevation model* (DEM), or the result of stereo matching may be used as an input to another system. Like other computer vision techniques, the implementation of an accurate and robust image matcher is much harder than would be indicated by the ease with which humans can generate a stereomodel of a scene.

Image matching is an intrinsically difficult problem. No matter which features of an image are used as inputs for matching, ambiguous matches will exist for most features. Some features will have no correct match due to occlusion or to noise effects. Scene structure occludes parts of the scene differently from each viewpoint. Surfaces in the scene will be viewed at varying aspects due to the different viewing angles, so that a line which is 20 pixels long in one image may be only 10 pixels long in the other. A successful image matching algorithm must deal with all these problems in an efficient and robust manner.

In theory, matching could proceed with no knowledge of the relative geometry between the images. However, because of the ambiguities in matching and computational efficiency requirements, the *epipolar geometry* (Chapter 2), defined by the relative orientation of the images, is often used to reduce the two-dimensional search space to a one-dimensional search along the epipolar line.

Epipolar matching can be accomplished in two ways. The matching program can calculate the position of the epipolar line in each image and interpolate the values of pixels to be matched as required. Alternatively, the images can be resampled so that the epipolar lines are parallel to the rows of the image. The matching can then proceed along the image rows and no additional computation is required during the matching process. The latter approach is most often used, since the resampling can be done in a separate batch process and the epipolar-aligned resampling will be required if an operator is to view the stereo pair.

Image matching algorithms can be classified into two basic approaches: area matching and feature matching. In *area matching*, the gray levels within a window are matched against the gray levels within windows in the other image. In *feature matching*, extracted features such as interest points, line segments, or closed contours are matched between images. *Relational matching* is a variation of feature matching in which groups of features and the relationships among them are matched.

Area matching is typically done using the *normalized correlation* between windows in the two images to be matched. The normalized correlation, N, takes into account differences in brightness and variance between the two images. This is shown in Eq. 6-15, which includes the expected values, $E(w)$, of the pixel values in both windows, as well as the expected values of the products of the pixels values in the two windows, divided by the product of the standard deviations of the two windows. This normalizes each image by subtracting the mean and then dividing by the standard deviation. Without normalization, image regions with larger average pixel values (such as a very bright or overexposed region) or with a higher variance among the pixels will tend to have higher correlation values than possibly correct matching regions with more normal gray levels.

$$N = \frac{E(w_1 w_2) - E(w_1)E(w_2)}{\sigma(w_1)\sigma(w_2)} \tag{6-15}$$

A pixel in the master image is correlated by selecting a window around it, then calculating the correlation between this window and a window in the slave image. This calculation is repeated as the window in the slave image is moved along the epipolar line through the *search range* (Fig. 6-32). The window with the highest correlation value within the search range is selected as the match. The window size and the search range are parameters that must be set.

EXAMPLE 6-6 *Image Correlation*

Calculate the normalized correlation of the signal (10, 20, 30, 10) against the signal (10, 15, 20, 10) using Eq. 6-15.

SOLUTION

$$E(w_1 w_2) = (10 \cdot 10 + 20 \cdot 15 + 30 \cdot 20 + 10 \cdot 10)/4 = 275$$

$$E(w_1) = (10 + 20 + 30 + 10)/4 = 17.5$$

$$E(w_2) = (10 + 15 + 20 + 10)/4 = 13.75$$

$$\sigma^2(w_1) = [(10 - 17.5)^2 + (20 - 17.5)^2 + (30 - 17.5)^2 + (10 - 17.5)^2]/4 = 68.75$$

$$\sigma^2(w_2) = [(10 - 13.75)^2 + (15 - 13.75)^2 + (20 - 13.75)^2 + (10 - 13.75)^2]/4 = 17.1875$$

$$N = \frac{275 - 17.5 \cdot 13.75}{\sqrt{68.75}\sqrt{17.1875}} = 1.0$$

The normalized correlation is perfect, since the second signal is just a scaled version of the first.

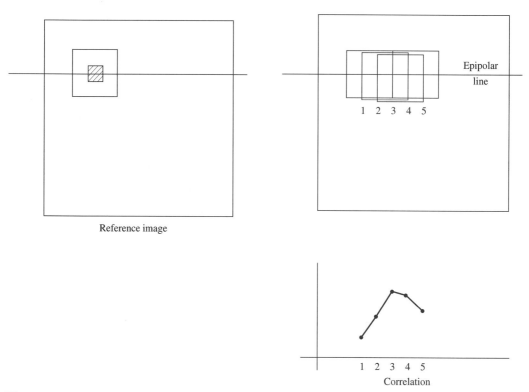

Figure 6-32 Image correlation windows matched along epipolar line, and the resulting correlation function.

A larger window makes the computation more resistant to noise but causes problems with matching across sudden changes in disparity, since the window will be at two different disparities. Small features will tend to be averaged into the background. A larger window is also more computationally expensive than a smaller one.

The search range required depends on the expected disparity, which is determined by the depth of the scene and the imaging geometry (Chapter 4). The search range is chosen as small as possible to reduce the computational requirements, but if the search range is too small, the correct match may be missed.

To address this problem, some stereo matching systems use previous matches to predict the disparity at the next point. This works well as long as there are no large jumps in disparity (buildings, cliffs, etc.). Prior information for the scene depth, if available, can be used to determine search ranges at each point. These techniques can be applied iteratively; an initial matching is used to get a rough idea of the terrain shape and this approximate depth map is then used to set the search ranges for the next iteration.

One approach to solving the trade-offs involving window size and search range is to perform the matching at multiple resolutions or scales (Section 6.6.2). If we reduce the image size by averaging pixels together, we effectively create an image with a shorter focal length. This reduces the apparent disparity and therefore the search range. With fewer pixels to work on, we can perform the matching relatively quickly. Although the resulting disparity map does not have the detail required for the final result, it can be used as a starting point for matching at the next resolution level. This process continues until the final matching is performed at full resolution.

Since surfaces are seen from different viewpoints in the two images, their apparent sizes will differ between the two images. Large differences in scale will cause problems for the correlation since we will be trying to match nonidentical signals. To address this problem, some systems reshape or scale the correlation window using the disparities determined in the neighborhood. A variation on this is iterative orthophoto refinement (Norvelle, 1992), in which the calculated disparities are used to generate orthoimages. The resulting orthoimages are then matched again, and any remaining disparities are used to correct the orthoprojection. The process continues until a convergence criterion is met.

The quality of a correlation match is determined by the image characteristics. As the image signal-to-noise ratio decreases, the probability of getting a wrong match increases, as the random fluctuations of the noise overwhelm the actual image information. The precision of the match is determined by the frequency content (the amount of detail present) in the image (Förstner, 1982). The quality of the match is also degraded by the presence of systematic errors, such as uncorrected lens distortion or atmospheric refraction.

Area-based correlation can be reformulated as a least squares adjustment problem, either in image space or in object space. In image space (Gruen, 1996) the condition equations to be minimized in the adjustment express the difference in gray levels between patches in the master and slave images. A gray level offset and a gain factor are included to model overall radiometric differences between the two images. These parameters are determined as part of the solution. A geometric transformation between the image patches, such as a polynomial, affine, or projective transformation, is included in the solution to allow for the difference in viewpoint between the two images. The matching solution then tries to minimize the radiometric differences between the image patches while calculating the relative geometric and radiometric parameters. The epipolar geometry between images can also be used to constrain the matching.

Least squares matching is very accurate but can be subject to the same problems as other matching algorithms, such as difficulty matching areas without sufficient texture or variation. However, a major advantage of least squares matching is that the quality of the results can be evaluated using standard least squares techniques. The precision of the estimated transformation parameters indicates how well the match has been performed. If the precision is too low, some of the parameters may be removed from the solution (by switching to a simpler geometric model, for instance) and the match redone, or the match may just be flagged as unreliable. The success of least squares matching is dependent on having good initial approximations for the match parameters. The maximum deviation from the correct parameters that still leads to a stable least squares solution is sometimes referred to as the *pull-in range*.

In object-space least squares matching (Wrobel, 1987; Helava, 1988; Ebner and Heipke, 1988), the scene is divided into elements (*groundels*) that model the geometry of the scene (location, elevation, and slope) as well as the radiometry of the area. The position of the groundel in each image is determined by its location and slope. The radiometry is determined by the reflectance of the ground patch and the relative radiometry (gain and offset) of each image. Additionally, the surface described by the groundels is assumed to be continuous and smooth. Least squares condition equations are written to embody these properties. A solution for the entire scene is typically impractical, due to the number of unknowns involved; instead, the solution is performed over areas of interest, such as tie points.

Least squares matching can also be used for template matching, as in the measurement of fiducial marks. The pattern to be located, such as the fiducial mark shape, is used as one of the images in the matching.

Because of the limitations of area matching, many stereo matching systems use feature matching. Features may be interest points (Hannah, 1989), edges (Förstner, 1986; Hsieh et al., 1992), or image contours. Although feature properties, such as edge contrast or interest point neighborhood, may be used in the matching, constraints must be used to obtain unambiguous matches.

The epipolar constraint is used in feature matching for matching interest points or the intersection of the epipolar line and an edge. Only edges that are not parallel to the epipolar lines are usable. Edge matching has an additional matching constraint, since the edge matches must be continuous between epipolar lines. The order in which edges are matched must also stay the same, unless an object in the scene is floating in front of the background.

Image pyramids may also be used in feature matching, as they were for area-based methods. Edges extracted from lower-resolution versions of the images will be significant edges and will tend to be easy to match. The disparity map obtained at low resolution can then be used to guide matching at higher resolutions.

Feature-based matching systems usually provide better matches in discontinuous scenes, such as urban scenes with numerous buildings, since they match the edges usually visible at the tops of buildings. Figure 6-33 shows typical matching results for both methods on several buildings. Note how the area matcher blurs the building edges, while the feature matcher produces a sharper edge, if the edge is detected.

One disadvantage of feature-based methods for producing DEMs is the sparsity of the output disparity map. Intermediate points must be interpolated to provide the dense elevation model usually required. If the feature-based matcher misses a feature, the disparity map will be inaccurate over a large area due to the sparsity of the matching.

Image matching systems are currently in widespread production use, even though they still have significant limitations. Many types of scenes, such as trees, low-contrast fields, or tall buildings, cannot be reliably matched due to the inherently limited information present or the abrupt changes in disparity. Trying to match such areas is a waste of computation and may degrade the results on adjacent areas, which would otherwise have been acceptably solved. Some production systems therefore try to automatically recognize such areas by analyzing texture or autocorrelation, or allow the operator to outline areas where no matching is to be performed.

In production situations, an operator must still examine and edit the automated results, to guard against matching blunders. For most applications, the operator must

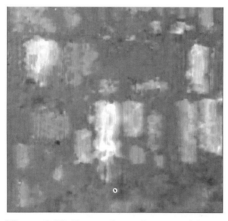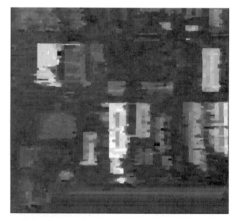

Figure 6-33 Comparison of area- and feature-based matching.

also edit the extracted elevations to obtain the required model of the terrain surface, instead of building roofs, tree tops, etc.

6.6.5   PIVOT: A Typical Computer Vision Application

To give an appreciation of how a typical computer vision algorithm is constructed, this section will describe the PIVOT (Perspective Interpretation of Vanishing points for Objects in Three dimensions) system, built by Dr. Jefferey Shufelt in the Digital Mapping Laboratory in the Computer Science Department of Carnegie Mellon University (Shufelt, 1996*a*; 1996*b*; 1999*a*). PIVOT is typical of the state of the art and illustrates many of the aspects of computer vision described in this chapter. It also has a strong photogrammetric component, making extensive use of camera and scene geometry.

The goal of PIVOT is to extract 3-D building models from individual aerial photographs. It begins with edges extracted from the scene, combines the edge data to form geometric primitives, and then generates building hypotheses. Buildings are described as combinations of 3-D geometric primitives, specifically, rectangular and triangular prisms, as shown in Fig. 6-34.

Edges are labeled with hypothesized direction information using vanishing point information. Vanishing points are determined using the Gaussian sphere (Barnard,

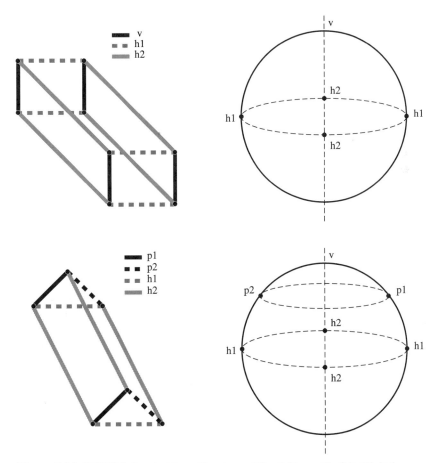

Figure 6-34 PIVOT 3-D primitives. Courtesy of Dr. Jefferey Shufelt, Digital Mapping Laboratory, Carnegie Mellon University.

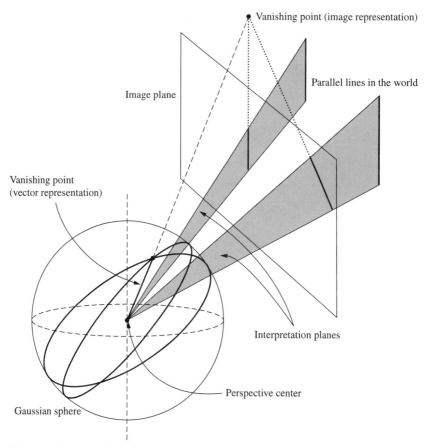

Figure 6-35 Gaussian sphere and vanishing point geometry. Courtesy of Dr. Jefferey Shufelt, Digital Mapping Laboratory, Carnegie Mellon University.

1983), as shown in Fig. 6-35. The Gaussian sphere is a unit sphere centered at the perspective center of the image. Each straight line in the scene defines an *interpretation plane* through the perspective center, which cuts the sphere in a great circle. Just as parallel lines in the scene converge to a point in the image, the great circles produced by the intersection of their interpretation planes with the sphere intersect in two points, which represent the vanishing points for lines in that direction. A line from the perspective center through the vanishing point on the sphere intersects the image plane in the image vanishing point for lines in that direction.

To determine vanishing points, the Gaussian sphere is divided into accumulator cells. The interpretation plane corresponding to each edge is calculated and each cell it passes through on the sphere is incremented, similar to the Hough transform described in Section 6.6.4.3. After all of the edges are processed, peaks in the accumulator array indicate vanishing points. Unfortunately, this procedure is subject to noise problems, since any noise in the image orientation or edge location will tend to spread the peaks. To lessen this problem, PIVOT calculates the location of the vertical vanishing point from the image orientation. The horizon line, on which the vanishing points of horizontal lines must lie, is then defined by the plane perpendicular to the line between the vertical vanishing points.

Since we are looking for predefined geometric primitives, we know either the orientations of lines in which we are interested or the relative orientations of sets of lines.

For instance, the rectangular prism primitive contains vertical lines and horizontal lines in two orthogonal directions. If we search along the horizon line of the Gaussian sphere for peaks corresponding to orthogonal directions, we can identify vanishing points of orthogonal horizontal lines in the scene.

This vanishing point information is now used to label edges with plausible object space orientations. These labels are not infallible or unique, due to the ambiguities of imaging geometry; a horizontal edge may coincidentally pass through the vertical vanishing point, or a nonhorizontal edge may happen to pass through a horizontal vanishing point. Lines may therefore have more than one orientation label.

After the lines are labeled, corners are formed from edges with compatible labels. For instance, two edges labeled vertical cannot form a corner, but two horizontals or a horizontal and a vertical can. The formation of corners in this way greatly reduces the number of features that need to be considered.

The next step is the formation of 2-corners, which consist of two corners that share a common line segment and typically represent a portion of a primitive face. The same geometric constraints on combining edge types still apply.

The 2-corners indicate the types of geometric primitives potentially present. For each 2-corner extracted, PIVOT hypothesizes 3-D primitives which could contain the 2-corner and searches the image for evidence of additional edges and corners. The search for consistent edges usually results in several possibilities for each edge slot in the primitive; a different interpretation is generated for each consistent set of edges.

6.6.5.1   Building Hypothesis Generation

PIVOT can now project the set of hypothesized 3-D primitives into object space. The floor points are projected first, using the camera model and a digital elevation model. Building heights are calculated using the vertical edges.

Building hypotheses are now formed by joining primitives. Adjacent primitives of the same type, formed due to edge fragmentation, are merged if the adjacent faces are of the same shape. If a triangular prism does not have a supporting rectangular prism, a search is performed to instantiate one. PIVOT tries to find vertical edges by searching in the direction of the vertical vanishing point starting from the corner of the prism. Shadow information is also used by iteratively extruding the triangular prism from the ground and comparing the shadow extent calculated from the sun angle with the image intensities.

6.6.5.2   Building Hypothesis Verification

The set of building hypotheses at this point contains every possible building detected, to ensure that no valid building hypotheses are prematurely deleted. The verification stage examines the image information supporting each hypothesis and computes a score describing the quality of each hypothesis. The score has three components: *shadow-to-ground consistency*, which quantifies the evidence of a shadow for the building hypothesis; *surface illumination consistency*, which describes building illumination evidence; and *image gradient fit*, defined as how well intensity edges in the image agree with edges in the building hypothesis.

Hypotheses are sorted by score and PIVOT begins by selecting the hypothesis with the highest score. Any other hypotheses which were formed by merging this hypothesis with others or which were merged to form this building hypothesis are removed from the list. If hypotheses overlap in object space, only the one with the best score is retained. The remaining hypothesis with the next highest score is examined next and the process is repeated until all hypotheses have been verified or deleted.

6.6.5.3 Performance Evaluation

Performance evaluation is an important but often-overlooked aspect of algorithm development. PIVOT was evaluated extensively to understand its performance characteristics and to compare its performance to other systems (Shufelt, 1999*b*).

In this case, the results were evaluated against 3-D building models manually generated from multiple images, which were then backprojected into the single images used for PIVOT testing. Evaluation was done both in 2-D (comparing buildings as delineated on the image) and 3-D (comparing 3-D object space models). Evaluation was extensive, using over 80 images from a variety of test areas.

Typical performance on representative images is shown in Figs. 6-36 and 6-37. Basic evaluation statistics included the true and false positives and negatives, as discussed earlier, calculated on a per-pixel basis for 2-D evaluation and on a per-voxel basis for 3-D evaluation. Some additional statistics gave additional insight into algorithm performance. The *branching factor*, defined as the number of false positives divided by the number of true positives, describes the rate of false positives. The *miss factor,* the number of false negatives divided by the number of true positives, is the proportion of buildings missed by the system. The quality percentage, 100 times the number of true positives divided by the sum of true positives, false positives, and false negatives, describes how likely a building hypothesis produced by the system is to be a real building. As an example of the statistics generated, performance figures for the scene shown in Fig. 6-36 are given in Table 6-2.

6.7 DIGITAL IMAGE SIMULATION: COMPUTER GRAPHICS

Computer graphics can be thought of as artificial photography, or inverse vision. Computer graphics tries to artificially generate an image of a given scene from its geometry, reflectance, and illumination. The basic problem is deceptively simple; determine the path,

Figure 6-36 PIVOT results on test image FLAT_L. Courtesy of Dr. Jefferey Shufelt, Digital Mapping Laboratory, Carnegie Mellon University.

Figure 6-37 Orthoimage texture-mapped onto a digital elevation model. Courtesy of TerraSim, Inc.

Table 6-2   PIVOT Evaluation Results for Test Image FLAT_L

Evaluation	Building detection percentage	Branch factor	Miss factor	Quality percentage
2-D	67.6%	0.56	0.48	49.0%
3-D	54.9%	0.82	0.82	37.9%

intensity, and color of each ray of light entering the camera lens. However, there are myriad complications to this simple statement, related to geometric modeling, surface reflectance, radiosity, and computational expense.

To generate a computer graphic image, we must first choose a geometric model. How can we best represent the geometry of the scene? Man-made objects can often be modeled by geometric primitives, but modeling the geometry of a cloud or natural terrain can be very difficult. Nonrigid objects, such as cloth or skin also present problems.

Next, surface reflectance must be considered. How does a surface reflect light? The simplest assumption is Lambertian reflectance, where the reflectance is independent of surface angle. In real life, reflectance may involve specular highlights, changes in color properties across the object, or a rough surface that reflects in an irregular manner. Objects may have surface markings or textured patterns.

Third, the light sources must be modeled. One or two point sources of light may be relatively simple to model, if we do not have significant inter-reflection between objects. If there is diffuse lighting, from the sky or a window, or complex shadowing and inter-reflections between objects, computing realistic lighting becomes a complex problem.

Finally, computational limitations affect the generation of computer graphics. Many graphics algorithms are relatively well understood, but implementing them within realistic computational budgets is extremely difficult. Computational requirements are often measured in terms of *polygons*, the planar regions which make up the scene.

Computer-generated images cover a wide range of applications. The simplest involve renderings of simple geometric objects, such as in computer aided design/computer aided manufacturing (CAD/CAM) systems. Such views will have hidden line removal and realistic shading, but will not involve complicated backgrounds or lighting.

Graphics becomes much more complicated and the computational expense more significant when animation or real-time graphics is considered. The graphics processing used in video games is highly optimized to allow efficient processing of the types of objects and scenes. For instance, instead of trying to model the detailed geometry of a character's face, an approximate geometric model such as a cylinder is used and an image of a face is *texture-mapped* onto it. While some 3-D effects are lost, such as a realistic nose, the inconsistencies are not noticed in the midst of rapid action and motion.

In texture mapping, a pattern is mapped onto an object, usually through projection or warping. When the object is imaged, the geometric model is used to calculate the position and visibility of the pixel, while the color of the pixel is determined by the appropriate part of the texture map.

A common application of texture mapping in photogrammetry is the production of DEMs (digital elevation models) draped with an aerial image. In this case, each pixel of the digital image is projected onto a corresponding part of the terrain, as represented by the elevation model. When the terrain is rendered for a fly-through, the geometry is taken from the DEM while the appearance is taken from texture-mapped image. Again, this is a convincing effect as long as the viewer is at a distance or is moving. At ground level it becomes obvious that the buildings and trees do not rise above the ground, as shown in Fig. 6-37.

Along with the production of texture-mapped DEMs, computer graphics have found other uses in photogrammetric applications. Proposed structures may be superimposed over an image to illustrate a planned project in relation to its surroundings. An increasingly common application is in *visual simulation*, where cartographic features are rendered in 3-D for training or visualization purposes (See Fig. 8-14 in Chapter 8).

PROBLEMS

6.1 A digital frame sensor has pixel dimensions of 0.010 mm by 0.010 mm and a principal distance of 300 mm. What is the instantaneous field of view (IFOV) in microradians? What is the ground sample distance (GSD) when imaging vertically at 3000 meters above terrain?

6.2 Edge detectors have varying responses to different types of edges. Suppose we are given two one-dimensional edge detectors:

$$1/2 \ [-1 \ 0 \ 1] \ \text{ and } \ 1/4 \ [-1 \ -1 \ 0 \ 1 \ 1 \ ]$$

What is their response on the following edge types? Explain.

    (a) step edge 100, 100, 100, 100, 100, 100, 110, 110, 110, 110, ...
    (b) line 100, 100, 100, 100, 100, 100, 110, 110, 110, 100, 100, 100, ...

(c) parallel step edges 100, 100, 100, 100, 110, 110, 110, 120, 120, 120, ...

(d) ramp 100, 100, 100, 102, 104, 106, 108, 110, 110, 110, 110, ...

6.3 Given the following six points, find the straight line through the largest subset of points using the Hough transform. Assume that θ is known to be either 115, 135, or 155 degrees.

x	y
10	34.0
16	29.9
25	49.0
33	57.0
40	60.3
47	71.0

6.4 We want to detect circles with a known radius r in a digital image using the Hough transform. If the equation of a circle is

$$(x - x_0)^2 + (y - y_0)^2 = r^2$$

where x_0 and y_0 are the coordinates of the center of circle and x and y are image coordinates, what are the parameters of the axes of the Hough space? What is the equation that specifies which cells in the accumulator are incremented for each point?

6.5 Two pixels in a digital color image have RGB values of (68, 144, 102) and (102, 216, 153). Which pixel has the higher intensity? Are they the same hue? Which has the more saturated color?

6.6 Calculate the normalized correlation (Eq. 6-15) between the signal windows (32, 45, 50, 52, 36) and (40, 50, 53, 54, 39).

6.7 Calculate the normalized correlation between the signal windows (32, 45, 50, 52, 36) and (40, 27, 22, 20, 36).

6.8 Convolve the signal (95, 97, 96, 94, 120, 118, 97, 94, 92) with the filter $(\frac{1}{8}, \frac{1}{4}, \frac{1}{4}, \frac{1}{4}, \frac{1}{8})$. What effect does this have on the signal? (Hint: What has happened to the spike with value 120?) Is this a low-pass or high-pass filter?

6.9 Convolve the signal (95, 97, 96, 94, 102, 103, 102, 93, 92, 90) with the filter $(-1, -1, 1, 1)$. What effect does this have on the signal? (Hint: Is the output smoother or more jagged than the original signal?)

6.10 Given the four pixel coordinates and intensity values below, calculate the intensity values at points a, b, and c using both nearest-neighbor and bilinear interpolation.

Point	Row	Column	Intensity
1	596	1010	145
2	596	1011	140
3	597	1011	135
4	597	1010	141
a	596.5	1010.2	?
b	596.7	1010.9	?
c	596.2	1010.0	?

REFERENCES

ALOIMONOS, Y., and SWAIN, M. J. 1988. Shape from patterns: Regularization. *International Journal of Computer Vision* 2(2):171–187.

BALLARD, D. H. 1981. Generalizing the Hough transform to detect arbitrary shapes. *Pattern Recognition* 13(2):111–122. See also Fischler and Firschein, 1987.

BALLARD, D. H., and BROWN, C. M. 1982. *Computer Vision*. Englewood Cliffs, NJ: Prentice Hall.

BARNARD, S. 1983. Interpreting perspective images. *Artificial Intelligence* 21:435–462.

BAUMGARTNER, A., STEGER, C., MAYER, H., ECKSTEIN, W., and EBNER, H. 1999. Automatic road extraction based on multi-scale, grouping, and context. *Photogrammetric Engineering and Remote Sensing* 65(7):777–786.

BLOSTEIN, D., and AHUJA, N. 1989. Shape from texture: Integrating texture-element extraction and surface estimation. *IEEE Transactions on Pattern Analysis and Machine Intelligence* 11(12):1233–1251.

BRACEWELL, R. N. 1986. *The Fourier Transform and Its Applications*. 2nd edition. New York: McGraw-Hill.

BRUNN, A., GÜLCH, E., LANG, W., and FÖRSTNER, W. 1998. A hybrid concept for 3D building acquisition. *ISPRS Journal of Photogrammetry and Remote Sensing* 53(2):119–129.

CANNY, J. 1986. A computational approach to edge detection. *IEEE Transactions on Pattern Analysis and Machine Intelligence* 8(6):679–698. See also Fischler and Firschein, 1987.

DAVIS, L. S., and ROSENFELD, A. 1981. Cooperating processes for low-level vision: A survey. *Artificial Intelligence* 17:245–263.

EBNER, H., and HEIPKE, C. 1988. Integration of digital image matching and object surface reconstruction. In *International Archives of Photogrammetry and Remote Sensing*. Vol. XXVII No. B11 pp. 534–545. Tokyo: International Society for Photogrammetry and Remote Sensing.

EL-HAKIM, S., and FÖRSTNER, W., eds. 1998. Special issue: Imaging and modelling for virtual reality. *ISPRS Journal of Photogrammetry and Remote Sensing* 53(6).

FISCHLER, M., and FIRSCHEIN, O., eds. 1987. *Readings in Computer Vision: Issues, Problems, Principles, and Paradigms*. Los Altos, CA: Morgan Kaufmann.

FÖRSTNER, W., and GÜLCH, E. 1987. A fast operator for detection and precise location of distinct points, corners and centres of circular features. In *Proceedings, ISPRS Intercommission Workshop on Fast Processing of Photogrammetric Data*. pp. 281–305. Interlaken, Switzerland: International Society for Photogrammetry and Remote Sensing.

FÖRSTNER, W., and GÜLCH, E. 1999. Automatic orientation and recognition in highly structured scenes. *ISPRS Journal of Photogrammetry and Remote Sensing* 54(1):23–34.

FÖRSTNER, W. 1982. On the geometric precision of digital correlation. In *International Archives of Photogrammetry and Remote Sensing*. Vol. XXIV No. 3 pp. 176–189. Helsinki: International Society for Photogrammetry and Remote Sensing.

FÖRSTNER, W. 1986. A feature based correspondence algorithm for image matching. In *International Archives of Photogrammetry and Remote Sensing* Vol. XXVI No. 3 pp. 150–166. Rovaniemi, Finland: International Society for Photogrammetry and Remote Sensing.

GEMAN, S., and GEMAN, D. 1984. Stochastic relaxation, Gibbs distributions, and the Bayesian restoration of images. *IEEE Transactions on Pattern Analysis and Machine Intelligence* 6(6):721–741.

GRUEN, A., and LI, H. H. 1997. Linear feature extraction with 3-D LSB-snakes. In *Proceedings of the International Workshop on Automatic Extraction of Man-Made Objects from Aerial and Space Images (II)* pp. 287–298. Basel, Switzerland: Birkhauser Verlag.

GRUEN, A. 1996. Chapter 8: Least squares matching: A fundamental measurement algorithm. In *Close Range Photogrammetry and Machine Vision*. Caithness, Scotland: Whittles.

GRUEN, A. 1998. TOBAGO—a semi-automated approach for the generation of 3-D building models. *ISPRS Journal of Photogrammetry and Remote Sensing* 53(2):108–118.

GUPTA, R., BLOOMER, J., ABDEL-MALEK, A., and ZINSER, R. 1993. Effect of on-board compression on stereo and classification. In *Proceedings of the SPIE: Integrating Photogrammetric Techniques with Scene Analysis and Machine Vision*. Vol. 1944 pp. 196–207. Orlando, FL: SPIE.

HANNAH, M. J. 1989. A system for digital stereo image matching. *Photogrammetric Engineering and Remote Sensing* 55(12):1765–1770.

HARALICK, R. M. 1979. Statistical and structural approaches to texture. *Proceedings of the IEEE* 67(5):786–804.

HARALICK, R. M. 1984. Digital step edges from zero-crossings of second directional derivatives. *IEEE Transactions on Pattern Analysis and Machine Intelligence* 6(1):58–68.

HAVELOCK, D. I. 1989. Geometric precision in noise-free digital images. *IEEE Transactions on Pattern Analysis and Machine Intelligence* 11(10):1065–1075.

HAVELOCK, D. I. 1991. The topology of locales and its effects on position uncertainty. *IEEE Transactions on Pattern Analysis and Machine Intelligence* 13(4):380–386.

HEIPKE, C. 1992. A global approach for least-squares image matching and surface reconstruction in object space. *Photogrammetric Engineering and Remote Sensing* 58(3):317–323.

HEIPKE, C. 1997a. Automation of interior, relative, and absolute orientation. *ISPRS Journal of Photogrammetry and Remote Sensing* 52(1):1–19.

HEIPKE, C., ed. 1997b. Special issue: Automatic image orientation. *ISPRS Journal of Photogrammetry and Remote Sensing* 52(3).

HELAVA, U. V. 1988. Object-space least-squares correlation. *Photogrammetric Engineering and Remote Sensing* 54(6):711–714.

HORN, B. K. P. 1986. *Robot Vision*. Cambridge, MA: MIT Press.

HSIEH, Y., MCKEOWN, D., and PERLANT, F. 1992. Performance evaluation of scene registration and stereo matching for cartographic feature extraction. *IEEE Transactions on Pattern Analysis and Machine Intelligence* 14(2):214–238.

HSIEH, Y. 1995. *Design and Evaluation of a Semiautomated Site Modeling System*. Technical Report CMU–CS–95–195. Pittsburgh, PA: School of Computer Science, Carnegie Mellon University.

HUECKEL, M. H. 1971. An operator which locates edges in digitized pictures. *Journal of the Association for Computing Machinery* 18(1):113–125.

HUECKEL, M. H. 1973. A local visual operator which recognizes edges and lines. *Journal of the Association for Computing Machinery* 20(4):634–647.

HUMMEL, R. A. and ZUCKER, S. W. 1983. On the foundations of relaxation labeling processes. *IEEE Transactions on Pattern Analysis and Machine Intelligence*, 5(3):267–287.

IKEUCHI, K. 1984. Shape from regular patterns. *Artificial Intelligence* 22(1):49–75.

ILLINGWORTH, J. and KITTLER, J.V. 1988. A survey of the Hough transform. *Computer Vision, Graphics, and Image Processing* 44(1):87–116.

KANATANI, K. I. 1993. *Geometric Computation for Machine Vision*. Oxford: Clarendon Press.

KANATANI, K. I. and CHOU, T. C. 1989. Shape from texture: General principle. *Artificial Intelligence*, 38(1):1–48.

KASS, M., WITKIN, A., and TERZOPOULOS, D. 1988. Snakes: Active contour models. *International Journal of Computer Vision* 1(4):321–331.

KOENDERINK, J. J. 1984. The structure of images. *Biological Cybernetics* 50:363–370.

KOFLER, M., REHATSCHEK, H., and GRUBER, M. 1996. A database for a 3D GIS for urban environments supporting photo-realistic visualization. In *International Archives of Photogrammetry and Remote Sensing*. Vol. XXXI No. B2 pp. 198–206, Vienna: International Society for Photogrammetry and Remote Sensing.

MALLAT, S. G. 1998. *A Wavelet Tour of Signal Processing*. San Diego, CA: Academic Press.

MARR, D., and NISHIHARA, H. K. 1978. Representation and recognition of the spatial organization of three-dimensional shapes. *Philosophical Transactions of Royal Society of London*, B 200:269–294.

MARR, D. 1982. *Vision: A Computational Investigation into the Human Representation and Processing of Visual Information*. San Francisco: W. H. Freeman.

MCGLONE, C. 1989. Automated image-map registration using active contour models and photogrammetric techniques. In *Proceedings of the SPIE*. Vol. 1070 pp. 109–115. Los Angeles, CA: SPIE.

MIKHAIL, E. M., Akey, M. L., and Mitchell, O. R. 1984. Detection and sub-pixel location of photogrammetric targets in digital images. *Photogrammetria* 39(3):63–84.

MORAVEC, H. 1981. Rover visual obstacle avoidance. In *Proceedings of the International Joint Conference on Artificial Intelligence*. pp. 785–790. Vancouver, BC: International Joint Conference on Artificial Intelligence.

NALWA, V. S. 1993. *A Guided Tour of Computer Vision*. Reading, MA: Addison-Wesley.

NORVELLE, F. R. 1992. Stereo correlation: Window shaping and DEM corrections. *Photogrammetric Engineering and Remote Sensing* 58(1):111–115.

OPPENHEIM, A. V., and SCHAFER, R. W. 1975. *Digital Signal Processing*. Englewood Cliffs, NJ: Prentice-Hall.

PAPOULIS, A. 1977. *Signal Analysis*. New York: McGraw-Hill.

PARKER, J. R. 1996. *Algorithms for Image Processing and Computer Vision*. Wiley Computing Publishing.

PRATT, W. K. 1991. *Digital Image Processing*. 2nd edition. New York: John Wiley & Sons, Inc.

ROSENFELD, A., and KAK, A. C. 1982. *Digital Picture Processing*. New York: Academic Press.

SALEH, R. A., ed. 1996. Special issue: Softcopy Photogrammetry. *Photogrammetric Engineering and Remote Sensing* 62(6).

SHIH, T. Y. 1995. The reversibility of 6 geometric color spaces. *Photogrammetric Engineering and Remote Sensing* 61(10):1223–1232.

SHUFELT, J. 1996a. Exploiting photogrammetric methods for building extraction in aerial images. In *International Archives of Photogrammetry and Remote Sensing*. Vol. XXXI No. B6 pp. 74–79. Vienna: ISPRS.

SHUFELT, J. 1996b. *Projective Geometry and Photometry for Object Detection and Delineation*. Ph.D. Thesis, Pittsburgh, PA: School of Computer Science, Carnegie Mellon University. Available as Technical Report CMU–CS–96–164.

SHUFELT, J. A. 1999a. *Geometric Constraints for Object Detection and Delineation*. Kluwer International Series in Engineering and Computer Science. Vol. SECS 530. Boston: Kluwer Academic Publishers.

SHUFELT, J. A. 1999b. Performance evaluation and analysis of monocular building extraction from aerial imagery. *IEEE Transactions on Pattern Analysis and Machine Intelligence* 21(4):311–326.

STEGER, C. 1998. An unbiased detector of curvilinear structures. *IEEE Transactions on Pattern Analysis and Machine Intelligence* 20(2):113–125.

TEMPELMAN, U., Nwosu, Z., and ZUMBRUNN, R. 1995. An investigation into the geometric consequences of processing substantially compressed images. In *Proceedings of the SPIE: Integrating Photogrammetric Techniques with Scene Analysis and Machine Vision*. Vol. 2486 pp. 48–58. Orlando, FL: SPIE.

TERZOPOULOS, D. 1986. Image analysis using multigrid relaxation methods. *IEEE Transactions on Pattern Analysis and Machine Intelligence* 8(2):129–139.

TOMITA, F., and TSUJI, S. 1990. *Computer Analysis of Visual Textures*. Boston, MA: Kluwer.

TRINDER, J. C., JANSA, J., and HUANG, Y. 1995. An assessment of the precision and accuracy of methods of digital target location. *Photogrammetria* 50(2):12–20.

TRINDER, J. C. 1989. Precision of digital target location. *Photogrammetric Engineering and Remote Sensing* 55(6):883–886.

UNRUH, J. E., and MIKHAIL, E. M. 1982. Mensuration tests using digital images. *Photogrammetric Engineering and Remote Sensing* 48(8):1343–1349.

WITKIN, A., TERZOPOULOS, D., and Kass, M. 1987. Signal matching through scale space. *International Journal of Computer Vision* 1(2):133–144. See also in Fischler and Firschein, 1987.

WITKIN, A. P. 1981. Recovering surface shape and orientation from texture. *Artificial Intelligence* 17:17–45.

WITKIN, A. P. 1983. Scale-space filtering. In *Proceedings of the International Joint Conference on Artificial Intelligence.* pp. 1019–1022. Karlsruhe, Germany: William Kaufmann, Inc. See also in Fischler and Firschein, 1987.

WOLBERG, G. 1990. *Digital Image Warping.* Los Altos, CA: IEEE Press.

WROBEL, B. 1987. Facet stereo vision (FAST vision). In *Proceedings, ISPRS Intercommission Workshop on Fast Processing of Photogrammetric Data.* pp. 231–258. Interlaken, Switzerland: ISPRS.

YUILLE, A. L., and POGGIO, T. A. 1986. Scaling theorems for zero-crossings. *IEEE Transactions on Pattern Analysis and Machine Intelligence* 8(1):15–25.

ZERROUG, M., and NEVATIA, R. 1996. Three-dimensional descriptions based on the analysis of the invariant and quasi-invariant properties of some curved-axis generalized cylinders. *IEEE Transactions on Pattern Analysis and Machine Intelligence* 18(3):237–253.

Chapter 7

Photogrammetric Instruments

7.1 INTRODUCTION

Until about 1980, the field of photogrammetry was dominated by and reliant upon many special-purpose instruments and devices. These included stereoplotters, mono-comparators, stereocomparators, point marking instruments, auto-dodging printers, simple rectifiers, and differential rectifiers (ortho projectors), as well as a variety of hardcopy vector plotting devices. This equipment was often manufactured by established companies with decades of experience in the unique aspects of modeling frame photograph geometry. Many of these photogrammetric instruments were marvels of optical and mechanical design and craftsmanship. Owners and operators would regard and treat them as one would a fine watch. It was not unusual for a well-maintained mechanical stereoplotter to put in 25 or 30 years of service, with minimal upgrades except perhaps for data collection peripherals. However, these instruments are now gradually being supplanted by general-purpose digital computers, special-purpose photogrammetric software, and high-resolution raster display screens, all operating on digital imagery. The photogrammetric film scanner and computer-compatible stereo viewing systems are emerging as the only special-purpose hardware components necessary for the photogrammetric workstation. Even the scanner may have a limited period of usefulness as the transition is gradually made to direct digital cameras and sensors.

Software-based systems have indisputable advantages in flexibility, potential for automation, accommodation of new sensors and sensing modalities, and in cross-application functionality and integration with other data types within the GIS framework. Software-based systems are less permanent than the steel and glass instruments that formerly dominated photogrammetry. They have a compressed life cycle and tend to be more transient, with new versions and updates issued at frequent intervals. This is probably inevitable. On the whole, the transition to digital systems must be regarded as an exciting prelude to a future of productivity growth and the accomplishment of tasks that would otherwise have been impossible.

This transition from hardcopy-based systems is relatively recent and, in many cases, still ongoing. The fundamental designs and operating principles of optical/mechanical/analytical systems are still relevant and are often directly applicable to their digital successors. We will therefore present a review of both photogrammetric equipment and systems based on traditional hardcopy photographs (Section 7.2) and those based on digital imagery (Section 7.3).

7.2 HARDCOPY-BASED INSTRUMENTS

7.2.1 Monocomparator

The term *comparator* connotes a device to measure two-dimensional point positions on a hardcopy image placed on a measuring stage. As the name implies, a *monocomparator* is limited to observing a single image at a time. They are primarily used for discrete point data collection for block adjustment or aerial triangulation. Bundle block adjustment uses the refined image coordinates directly, whereas an independent model adjustment first requires that the image coordinates be combined with conjugate points from adjacent photographs to form models and model coordinates. Figure 7-1 shows an instrument that can be configured as a monocomparator or as a stereocomparator.

An important consideration in comparator design is adherence to the *Abbe comparator principle,* which stipulates that, for each axis, the measuring scale and the point being measured should lie along the same line. Fulfillment of this design goal ensures that the system is minimally sensitive to any small rotations of the stage as it traverses the guide system. Conversely, systems that do not adhere to this principle are sensitive to such rotations, and such sensitivity cannot be compensated by calibration. Noncompliant systems for which other design goals take precedence can nevertheless achieve satisfactory performance if stable mechanical constraints are implemented to prevent rotation during the stage motion.

Comparator stages are usually constructed as a two-axis cross slide system. A measuring microscope of from $5\times$ to $25\times$ magnification is used to observe the points to be measured on the photograph. Analogous to the comparator principle, the optical measuring mark should be located as near as possible to the point being measured. Often such a measuring mark is projected into the viewing optics via a beam-splitter prism or half-silvered mirror at a point near the photograph plane. Early systems used lead screws and rotary encoders to perform the position determination. Later systems used linear encoders to perform this task. A linear encoder consists of a glass scale with an accurate but coarse set of graduations.

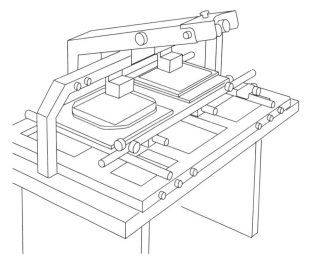

Figure 7-1 Instrument that can serve as a monocomparator or a stereocomparator.

Moiré effects can be exploited to interpolate or subdivide this coarse graduation down to the requisite one-micrometer resolution, the expected least count for an instrument to be used for aerial triangulation. Early systems also relied exclusively on human observation and pointing skills, whereas later hybrid systems employ video view capture and automated target detection and location for the actual measurements.

Calibration of an *XY* stage is a fundamental operation that has relevance across many diverse instrument types. Ideally, any comparator coordinate readings would directly result in Cartesian coordinate data. This implies a known, uniform, isotropic scale and strictly orthogonal axes. In practice, due to manufacturing limitations, such perfect instruments cannot be built. Calibration then determines the equations necessary to transform raw stage coordinates into calibrated Cartesian coordinates with the required characteristics. This calibration is usually carried out by observing points from a calibrated grid plate whose grid cross locations are known to within a micrometer. The mathematical model relating raw stage coordinates (x_s, y_s) and calibrated grid plate coordinates (x_c, y_c) depends on the characteristics of the particular *XY* stage. If nonuniform scale and axis nonorthogonality are the only errors to be modeled, then the six-parameter transformation would be appropriate. Although it may seem more geometrically appropriate to apply the transformation parameters to the stage coordinates, in practice they are more often applied to the calibrated coordinates, as shown in the following equations:

$$x_s = a_0 + a_1 x_c + a_2 y_c$$
$$y_s = b_0 + b_1 x_c + b_2 y_c \tag{7-1}$$

In matrix form this is

$$\begin{bmatrix} x_s \\ y_s \end{bmatrix} = \begin{bmatrix} a_1 & a_2 \\ b_1 & b_2 \end{bmatrix} \begin{bmatrix} x_c \\ y_c \end{bmatrix} + \begin{bmatrix} a_0 \\ b_0 \end{bmatrix} \tag{7-2}$$

If the measuring stage has curvature in the guide rails, then a higher-order transformation may be needed, for example the following 12-parameter polynomial:

$$x_s = a_0 + a_1 x_c + a_2 y_c + a_3 x_c^2 + a_4 y_c^2 + a_5 x_c y_c$$
$$y_s = b_0 + b_1 x_c + b_2 y_c + b_3 x_c^2 + b_4 y_c^2 + b_5 x_c y_c \tag{7-3}$$

Note that there is not an equivalent matrix form of this system of equations since they are nonlinear in the coordinates. With the ability of modern digital computers to implement these transformations with ease, it becomes more important that the comparator stages be stable, repeatable, and consistent than that the stages be free of geometric errors. As mentioned earlier, if stage errors originate from poorly constrained carriage movement or other such factors, then attempts to compensate by calibration will prove unsuccessful.

If the comparator stage is under computer control, then the target position is in the calibrated system and the actual stage position is determined by forward evaluation of Eq. 7-1 or 7-3. If, on the other hand, a manually driven stage is used, coordinates are generated in the stage coordinate system, and the transformations in Eqs. 7-1 and 7-3 must be applied in the inverse direction. Equation 7-2 is directly invertible as

$$\begin{bmatrix} x_c \\ y_c \end{bmatrix} = \begin{bmatrix} a_1 & a_2 \\ b_1 & b_2 \end{bmatrix}^{-1} \left(\begin{bmatrix} x_s \\ y_s \end{bmatrix} - \begin{bmatrix} a_0 \\ b_0 \end{bmatrix} \right) \tag{7-4}$$

Equation 7-3 is not directly invertible. A numerical iteration approach such as Newton's method is required. The iteration equations are

$$\begin{bmatrix} x_{c_{i+1}} \\ y_{c_{i+1}} \end{bmatrix} = \begin{bmatrix} x_{c_i} \\ y_{c_i} \end{bmatrix} - \begin{bmatrix} \dfrac{\partial F_x}{\partial x} & \dfrac{\partial F_x}{\partial y} \\ \dfrac{\partial F_y}{\partial x} & \dfrac{\partial F_y}{\partial y} \end{bmatrix}^{-1} \begin{bmatrix} F_x(x_{c_i}, y_{c_i}) \\ F_y(x_{c_i}, y_{c_i}) \end{bmatrix} \tag{7-5}$$

where

$$\begin{aligned} F_x(x_c, y_c) &= x_s - a_0 - a_1 x_c - a_2 y_c - a_3 x_c^2 - a_4 y_c^2 - a_5 x_c y_c \\ F_y(x_c, y_c) &= y_s - b_0 - b_1 x_c - b_2 y_c - b_3 x_c^2 - b_4 y_c^2 - b_5 x_c y_c \end{aligned} \tag{7-6}$$

7.2.2   Stereocomparator

The stereocomparator permits the measurement of two photograph transparencies, or diapositives, simultaneously while viewing in stereo. Like the monocomparator, its primary use is in aerial triangulation or block adjustment. The significant advantage of the stereocomparator over the monocomparator is that not all of the points used in the triangulation need to be marked on every photograph. Points used in conventional triangulation fall into three categories: (a) control points, (b) pass points for *along-strip* connection, and (c) tie points for *cross-strip* connection. For horizontal control points, the point on the ground is either signalized, for example, a painted white cross on asphalt, or it is a photo-ID point, for example, the base of a utility pole or the well-defined corner of a structure. These can be readily seen and measured on every photograph on which they occur, and the stereocomparator offers no great advantage over the monocomparator. For pass points and tie points, it is another story. These points, collectively, constitute by far the largest share of points carried in the block adjustment. It is advantageous to mark them on only one photograph (using a point marking instrument described in the next section) and use the stereo transfer capability of the stereocomparator to measure them on each photograph in which they occur. The comparator in this case works as a point transfer instrument and eliminates the necessity of marking the point on multiple photographs. This saves time and reduces the opportunities for errors in the stereo point marking operation, which are not easily correctable. See Section 7.2.3 for a discussion of point marking instruments.

A typical strip triangulation control and pass point layout is illustrated in Fig. 7-2. If a stereocomparator is used, the pass points would only have to be marked on the

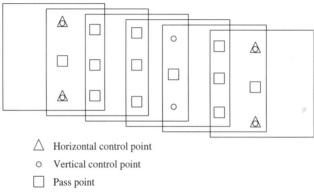

△   Horizontal control point

○   Vertical control point

▢   Pass point

Figure 7-2 Control and pass point layout for strip triangulation.

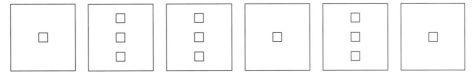

Figure 7-3 Point marking locations for the strip in Fig. 7-3.

diapositives at the positions indicated by squares in Fig. 7-3. A sketch of a typical stereocomparator is shown in Fig. 7-1. If individual zoom control of the magnification is available for the left and right channels, then stereo point transfer between photographs of different scales is possible. The typical 60% forward overlap found in mapping photography yields pass points falling on three adjacent photographs. Thus a three-stage comparator (two viewable at a time) could in principle increase the efficiency of measurement. Such instruments were never widely adopted in commercial practice but were employed by some government agencies. Calibration of a multistage comparator simply represents multiple applications of the single-plate calibration methods outlined in the previous section.

7.2.3 Point Marker

In the triangulation of hardcopy imagery, physically marking pass point and tie point locations on the diapositive is an essential step. Point making is necessary (a) to be able to unambiguously transfer the location of the point to overlapping photographs, (b) to have an unambiguous point to observe for subsequent absolute orientation in a stereoplotter, and (c) to have a point to return to in case any remeasurement should become necessary. A point marker typically allows for point selection on the diapositive under high magnification, following which a small drill (from 25 to 100 micrometers in diameter) physically removes the emulsion at this point so it becomes a permanently marked location. For reasons unknown to the authors, this operation is referred to as *pugging* and the instruments themselves are also called *pugs*. Point markers usually are made with stereo viewing and marking ability, making it possible for all pass points and tie points to be located and marked on every photograph on which they occur, thus allowing measurement to take place on a monocomparator. The stereo transfer of pass points has more commonly been accomplished using a stereocomparator or analytical plotter, thus minimizing the need to physically mark the valuable diapositives. A sketch of a point marker is shown in Fig. 7-4.

7.2.4 Rectifier

The major departures of frame imagery from the corresponding map geometry are the displacements due to tilt and relief. An optical rectifier is a special-purpose photographic enlarger with either a tilting projector or a tilting easel, or both, designed to eliminate displacements due to tilt. To rectify an image, a map sheet with at least four widely spaced horizontal control points is prepared and the projected image and the control plot are aligned so that the imagery matches the plotted positions of the control points. Since the image and object planes of this optical system are no longer perpendicular to the optical axis, a special condition, the *Scheimpflug condition*, must be enforced to ensure that the focus is sharp in the projected image. The Scheimpflug condition requires that the negative plane, the lens plane, and the easel plane all intersect in a single line. This is often enforced by mechanical means. In very flat terrain,

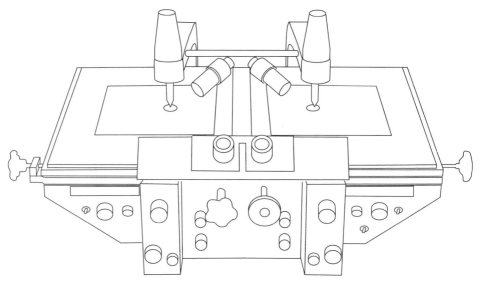

Figure 7-4 Point marking instrument, or pug, for hardcopy diapositives.

such rectifications could be used as image maps, but more commonly these rectified enlargements are used to prepare mosaics. The benefit of the rectification step is to greatly reduce the image mismatches along the seams between adjacent enlargements. A sketch of a typical rectifier is shown in Fig. 7-5.

In order to take the next step and remove both tilt and relief displacement from an enlargement, the orthorectifying projector was developed. In principle this is similar to the simple rectifier just described, with the addition of a curtain and a scanning slit which traverses the enlargement area on the easel. If, while the slit is profiling through the enlargement area, the height of the easel is made to follow the corresponding terrain height, then relief displacement will be eliminated. The necessity of having a terrain model makes this differential rectification process much more expensive than simple rectification. There have been many instrument designs to accomplish this task. Most utilize optical projection as described, but some use video image capture of small patches which are then transferred to light-sensitive material via image display tubes. Needless to say, digital methods of image rectification and orthorectification (Section 8.2.3) have brought enormous benefits and efficiencies to this task.

7.2.5   Stereoplotter

The stereoplotter is the workhorse of the map compilation business. Early instruments based on optical projection had either full-scale or reduced-scale projectors to create an enlarged stereo model in the *model space* above a plotting surface. Left/right channel separation was achieved by the anaglyph technique (red and blue filters on the two projectors, with corresponding filters worn by the viewer), the image alternator technique (alternate display and supression of each image, with synchronized control for the viewer), or the polarization technique (different polarization of left and right images, with corresponding viewing spectacles). The enlargement ratio from photograph to map was variable within a very small range. A *tracing table* with an illuminated *floating mark* was fitted with Z-motion control, an elevation counter, and a pencil for tracing map features. Later instruments could be fitted with guide rails and digitizing

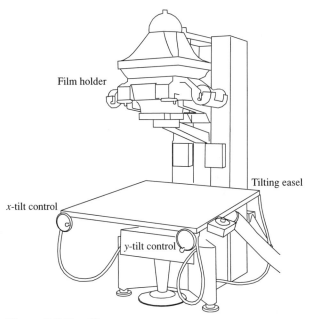

Figure 7-5 Rectifier.

electronics for digital data capture into a CAD system rather than manual plotting of the map directly. A sketch of a typical optical stereoplotter is shown in Fig. 7-6.

The *C-factor* is an attempt to quantify the accuracy of a stereoplotter. This number multiplied by the desired contour interval yields the flying height above the terrain needed to meet conventional map accuracy standards. A typical optical projection plotter might have a C-factor of 1500, whereas a high-performance analytical plotter might have a C-factor of 2500. Thus, when using an analytical plotter, the imagery could be obtained with higher flights (at smaller scale and thus more economically), and the same level of accuracy could be achieved. For the optical plotter, small

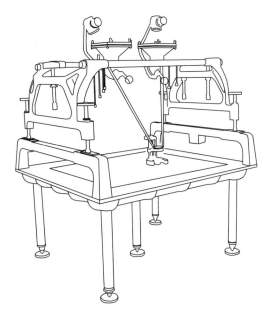

Figure 7-6 Kelsh-type optical stereoplotter.

changes in principal distance could be easily introduced, but changing from a six-inch (152.4-mm) lens to an $8\frac{1}{4}$-inch (209.5-mm) lens requires changing major components of both projectors. Orientation procedures are, in principle, similar for all analog stereoplotters and will be summarized only once.

Inner orientation consists of (1) centering the photograph transparency in the photo carrier by aligning the fiducial marks in the photograph with the registration marks on the carrier and (2) setting the projector principal distance equal to the camera focal length. Cam systems are occasionally available to implement lens distortion correction. *Relative orientation* (Section 5.6) can be done in many ways, some dictated by the projection motions available. A widely used technique utilizes the following steps, associated with the point distribution shown in Fig. 7-7. Note that the outlined steps describe *independent* relative orientation, rather than a *dependent* approach in which only one projector is moved.

1. Clear y-parallax at point 1 with κ_2 rotation.
2. Clear y-parallax at point 2 with κ_1 rotation.
3. Clear y-parallax at point 3 with ϕ_2 rotation.
4. Clear y-parallax at point 4 with ϕ_1 rotation.
5. Clear y-parallax at point 5 with ω_1 or ω_2, then overcorrect by 50% to speed convergence.
6. Repeat steps 1–5 until no further parallax is detected.
7. Check at point 6, and then check everywhere else in the model, to confirm that the entire model is free of y-parallax.

The relatively oriented model could be used for some kinds of data collection, for example, model coordinates for aerial triangulation, but for most map compilation an *absolute orientation* also has to be performed. Absolute orientation brings the model to a fixed scale and levels the model, requiring a minimum of two horizontal control points and three vertical control points. Scaling the model involves lengthening or shortening the base. Leveling the model requires common rotation motions or individual projector rotation motions together with selected base component adjustments.

The mechanical stereoplotter uses a steel *space rod* and a *gimbal* with a sliding *cardan* joint to model the projection of the rays from conjugate image points. The space rods guide viewing optics, which contain the floating mark; channel separation can be achieved by having a separate optical train for each photograph. Physical separation of the plate carriers can be achieved by clever devices such as the *Zeiss parallelogram*. *First order* instruments have a *base-in/base-out* capability, which allows the left and right projectors to interchange roles, via an optical switch. This allows successive models to be set up while retaining the orientation of the common photograph. All stereoplotters can be used both as data collectors for independent model aerial triangulation and for map compilation. A drawing of a typical mechanical

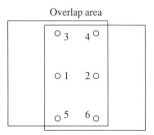

Figure 7-7 Overlap area of photograph pair with point locations for relative orientation.

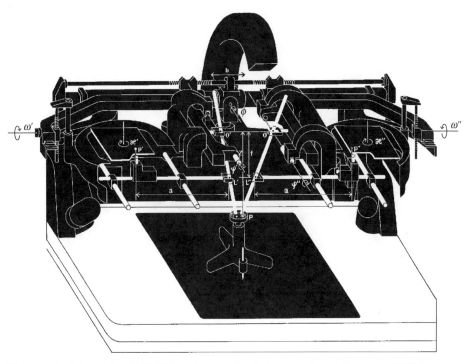

Figure 7-8 Schematic of Wild B8 analog mechanical stereoplotter. Courtesy of LH Systems.

stereoplotter is shown in Fig. 7-8. A photograph of a representative mechanical stereoplotter is shown in Fig. 7-9.

The analytical stereoplotter began to bring the advantages of the digital age into the formation and navigation of stereomodels. In these instruments, the projection is done synthetically by a real-time computer program that tracks operator position requests and maintains a parallax-free stereo view. Thus no physical rotations or base component motions are needed for the analytical stereoplotter. The analytical stereoplotter could, in theory, work with hardcopy imagery from any sensor, since the sensor model is part of the real-time program that controls the stage positions. A schematic diagram of coordinate transformations needed for the analytical stereoplotter is shown in Fig. 7-10. A photograph of such an instrument is shown in Fig. 7-11. In addition to accommodating different sensor models, steps that were previously difficult, such as changing the principal distance, become trivially easy with this type of instrument. Thus the analytical stereoplotter made possible the new development of *close-range photogrammetry* with small-format cameras and with nonmetric cameras. Late-model plotters of this type include video data capture of the stereo view for automated target detection and pointing, and also for semiautomatic digital elevation model (DEM) generation.

Map compilation from stereoplotters has progressed from direct drawing, to pantograph-controlled drawing, to mechanical table linkages, to direct control of electronic pen plotters, to the current stage of digital vector data collection into a computer-aided design (CAD) file. CAD data can be collected in 2-D or in 3-D depending on the application.

7.2.6 Scanner

The photogrammetric scanner is currently a necessary component of large-scale digital mapping systems. The high information content, strong geometry, large dynamic

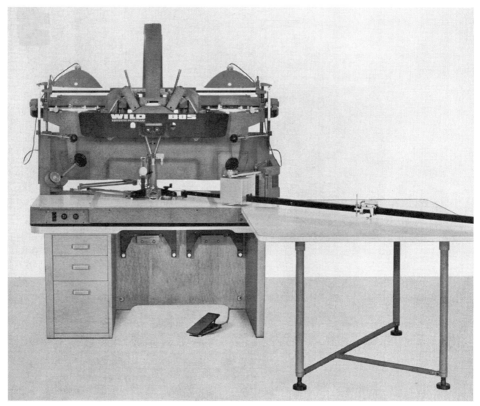

Figure 7-9 Wild B8 analog mechanical stereoplotter. Courtesy of LH Systems.

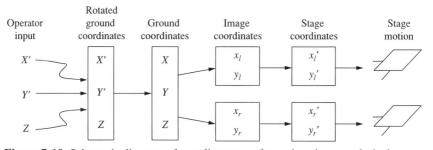

Figure 7-10 Schematic diagram of coordinate transformations in an analytical plotter.

range for radiometry, and compact storage characteristics currently make conventional film-based aerial cameras the most successful means to capture image data at large scale. Recent changes in government policy have allowed digital imagery from space to be commercially acquired at spatial resolutions of about one meter. This will undoubtedly change the economics for medium-scale to small-scale mapping, but for map scales of 1:600 or 1:1200, film-based imaging remains the only viable medium. Digital airborne cameras could, in theory, acquire data at the needed spatial resolution, but current digital sensors, whether linear or area array based, suffer from some combination of size limitations, geometric weakness, and dynamic range limitations. Thus, at least for the foreseeable future, film will continue to be the medium of choice for large-scale mapping. However, the flexibility and

Figure 7-11 Analytical plotter. Courtesy of Z/I Imaging.

economics of digital image manipulation for the actual mapping process dictate that the film image be converted to digital form. This must be done in such a way as to preserve those aspects of the film image that made the use of film compelling in the first place; that is, resolution, dynamic range, and geometry. Thus arises the current need for photogrammetric film scanners.

Film scanners are generally based on three architectures:

1. Rotating drum holding the film transparency with a photomultiplier tube (PMT) performing the actual intensity detection
2. Flatbed stage holding the film transparency with a linear CCD array scanning the image area
3. Flatbed stage holding the film transparency with an area CCD array stepping and capturing a sequence of frames to cover the image area

A brief description will be given for each of these architectures.

PMT-based scanners offer the largest dynamic range for capturing imagery. This may not be an issue when scanning dodged, low-contrast diapositives, but it may be a significant advantage when working directly with film negatives. Mounting the transparency on a drum requires physically cutting apart the film roll, a disadvantage for automation of film roll scanning. Color data capture can be implemented either by a multi-pass arrangement with RGB filters or by three PMTs simultaneously observing the same spot using splitting optics. Maintaining good geometry requires careful attention to the mechanical and electronic details of the scanning motion and the detection timing. Pixel sizes are determined by a mask and by the translation speed of the detector moving along the drum. The PMT is capable of supporting more than eight-bit digitization. Lookup tables can be used to remap the gray level response to another response function.

The linear array CCD scanner either moves a glass stage over the CCD array or moves the array over the stage. One or more passes may be required, depending on the width of the CCD array. Some have tried to create a large effective linear array by aligning several smaller arrays end to end. For the best signal quality and noise minimization, CCDs should be cooled. If the CCD array is not cooled, then capturing the full range of gray tones or colors in a high-contrast negative may not be possible. This architecture is capable of achieving very good geometric accuracy, and can be very fast. The base level pixel size is fixed by the optics. Other pixel sizes may be produced by aggregation or resampling. Color can be captured by multiple passes over the image with RGB filters. Integrated CCDs with three linear arrays, each with its own color filter, could potentially reduce the scanning time for color imagery by a factor of three.

The area array CCD approach is similar in many ways to the scanning linear array approach. Depending on the relative alignment of the CCD array and the photograph, individual frames may need to be resampled so that the rows and columns are fixed with respect to the photograph, rather than with respect to the sensor. A creative approach to the geometric accuracy issue has been demonstrated by embedding grid crosses into the stage plate, and then alternately making them visible and invisible by illumination. These grid crosses can then be used to determine geometric resampling parameters for the image acquired when the crosses are not visible (Leberl et al., 1992). Color acquisition options are similar to those outlined for the linear CCD array scanners.

7.3 SOFTCOPY-BASED SYSTEMS

The capabilities of the photogrammetric software used in softcopy-based systems are the main points of interest for photogrammetry. The ergonomics of the stereo viewing system and the performance of the graphics display system while panning, zooming, polygon rendering, or texture mapping can have an important impact on productivity, but these are factors that are not unique to photogrammetric workstations. The principal mathematical models and algorithms used in photogrammetry have been covered in earlier chapters. The following sections will describe how these models may be integrated and sequenced to provide a powerful and flexible working environment within the photogrammetric workstation. Section 7.3.2 describes softcopy stereo and includes a derivation and an example of the important topic of pairwise rectification or normalization, which is a necessary preprocessing step for most softcopy stereo viewing systems. A photograph of a typical softcopy photogrammetric workstation is shown in Fig. 7-12. A screen print from such a system is shown in Fig. 7-13.

7.3.1 Single Image Environment

The simplest softcopy equivalent of the monocomparator is an image display with a cursor for pointing and collecting image coordinates from raw imagery. If one limits the softcopy system to manual measurement on a screen, then it is hard to argue that this represents much of an advance over the hardcopy comparator. If, on the other hand, software is included to scan the image for targets of interest, perform measurements automatically to sub-pixel accuracy, archive the results in a database, and combine the new measurements with others to estimate parameters of interest, then the softcopy environment becomes a compelling tool for productivity growth.

Figure 7-12 Softcopy photogrammetric workstation. Note special cursor for 3-D measuring mark motion, and active stereo glasses. Courtesy of Z/I Imaging.

The softcopy equivalent of point marking during multi-image triangulation is saving and labeling a small subimage around the pass point or tie point of interest. Single image displays can present enhanced capabilities in a number of ways. If the image displayed is rectified or orthorectified, then ground point planimetric positions can be displayed and recorded directly. This setup can be used to implement a *heads-up* digitizing system, in which one extracts point, line, or area features directly from a digital orthophoto. An equivalent functionality can be implemented using a *mono-plotter* approach, in which one views raw imagery, but employs the image orientation parameters to track the intersection of the ray defined by the current cursor position with a digital elevation model, thereby obtaining rigorous 3-D ground coordinates from a single image. Intersecting a ray with a complex 3-D surface can be slow. Intersecting a ray with a 2.5-D surface (a DEM) is faster, but can still present some problems if occlusions are not handled properly (Section 5.4).

7.3.2 Stereo Environment

One of the difficult aspects of classical analog and analytical stereoplotters has always been the training of new operators. Hardcopy stereoplotter viewing optics are configured for a single individual, and adding a second set of training optics is expensive and inconvenient, if possible at all. This is an area where softcopy stereo

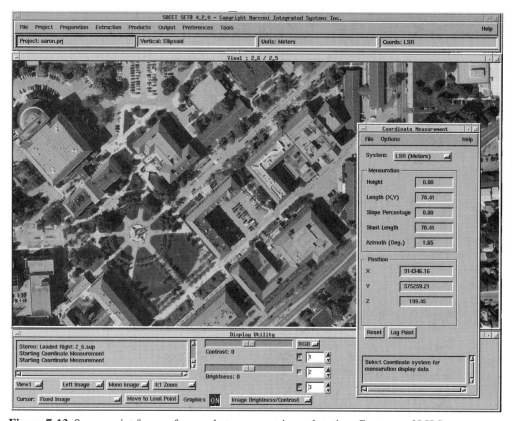

Figure 7-13 Screen print from softcopy photogrammetric workstation. Courtesy of LH Systems.

has a distinct advantage. Three of the four technologies for presenting softcopy stereo, which will be discussed later, permit multiple viewers to observe the stereo model simultaneously. This opportunity for discussion and collaboration while viewing the same 3-D model is a significant value of the softcopy stereo environment. Training new operators in the use of stereo, in particular, benefits from this possibility. Other ambiguous measurement and interpretation tasks also benefit in the same way.

Analog and analytical stereoplotters with oriented imagery and image rotation optics present parallax-free views of the model in the vicinity of the two projected floating half-marks. Softcopy stereo viewers can either pan and scroll the imagery over fixed floating marks, or pan and scroll the floating marks over fixed imagery. Real-time image rotation is not usually provided by image display hardware. Therefore some preprocessing is usually necessary to align the photo base with the viewing setup. Softcopy systems use the method of image normalization or pairwise rectification, which guarantees that conjugate points have zero y-parallax. This permits the left and right images to be presented with only the height component, or x-parallax, unresolved. Images that are taken with nearly parallel view directions can be pairwise rectified or normalized. This process is also called *epipolar resampling*. Figure 7-14 illustrates the geometry of the image normalization process. For clarity the sketch shows photographs that are not nadir looking, but the mathematical approach is general.

In Fig. 7-14, P_1 and P_2 represent the original photograph pair and N_1 and N_2 represent the normalized photograph pair. The transformation between the two will be

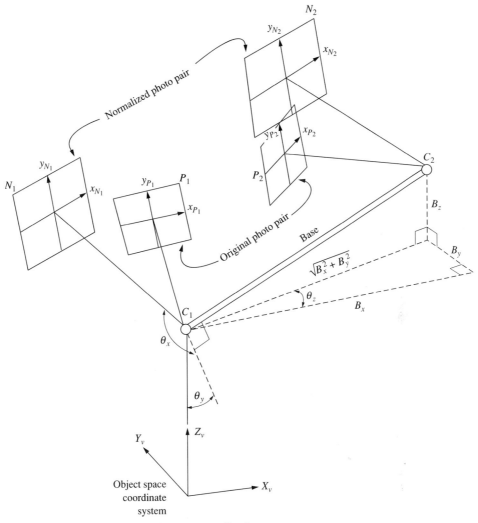

Figure 7-14 Geometry of the image normalization process.

derived. The rotation matrices that relate the object space (axes parallel to X_v, Y_v, Z_v) and the two original photographs are M_1 and M_2. First rotate X_v to lie in the vertical plane through the base.

$$\theta_z = \tan^{-1}\left(\frac{B_y}{B_x}\right) \qquad (7\text{-}7)$$

Then make the once rotated X'_v parallel to the base.

$$\theta_y = \tan^{-1}\left(\frac{-B_z}{\sqrt{B_x^2 + B_y^2}}\right) \qquad (7\text{-}8)$$

Make the twice rotated Z''_v close to the original direction of view (this step is not unique).

$$\theta_x = \frac{\omega_1 + \omega_2}{2} \qquad (7\text{-}9)$$

The ω terms in this equation are interpreted as tertiary rather than primary x-rotations in the sequential formation of M_1 and M_2.

The rotation matrices corresponding to the θ_x, θ_y, and θ_z rotations just derived are M_x, M_y, and M_z, respectively. Their product is

$$M_B = M_x\, M_y\, M_z \tag{7-10}$$

The matrix M_B relates the object space coordinate system and the system parallel to the normalized image coordinate system. The composite rotation matrices between the original photographs and the normalized photographs are

$$\begin{aligned} M_{N_1} &= M_B\, M_1^T \\ M_{N_2} &= M_B\, M_2^T \end{aligned} \tag{7-11}$$

Coordinates (x_P, y_P) in the original photographs are transformed into their normalized counterparts (x_N, y_N) by

$$\begin{aligned} x_N &= -f\frac{m_{N_{11}}x_P + m_{N_{12}}y_P + m_{N_{13}}(-f)}{m_{N_{31}}x_P + m_{N_{32}}y_P + m_{N_{33}}(-f)} \\[2mm] y_N &= -f\frac{m_{N_{21}}x_P + m_{N_{22}}y_P + m_{N_{23}}(-f)}{m_{N_{31}}x_P + m_{N_{32}}y_P + m_{N_{33}}(-f)} \end{aligned} \tag{7-12}$$

Note that for simplicity we are assuming that $(x_0, y_0) = (0, 0)$. Coordinates in the normalized photographs are transformed into their counterparts in the original photographs by

$$\begin{aligned} x_P &= -f\frac{m_{N_{11}}x_N + m_{N_{21}}y_N + m_{N_{31}}(-f)}{m_{N_{13}}x_N + m_{N_{23}}y_N + m_{N_{33}}(-f)} \\[2mm] y_P &= -f\frac{m_{N_{12}}x_N + m_{N_{22}}y_N + m_{N_{32}}(-f)}{m_{N_{13}}x_N + m_{N_{23}}y_N + m_{N_{33}}(-f)} \end{aligned} \tag{7-13}$$

Equations 7-12 can be used to find the limits of coverage in the normalized system. A grid of appropriate spacing is then defined in the normalized system and the resampling is carried out using Eq. 7-13 to actually create the normalized image. Fractional values of x_P and y_P will necessitate interpolation in the original image to obtain a gray value or color components.

EXAMPLE 7-1   *Normalization Computations*

Two photographs have the following exterior orientation parameters:

$$(X_L,\ Y_L,\ Z_L)_1 = (5000,\ 5000,\ 610)\,\text{meters}$$
$$(\omega,\ \phi,\ \kappa)_1 = (1.4145001,\ 1.414070,\ 44.982543)\,\text{degrees}$$
$$(X_L,\ Y_L,\ Z_L)_2 = (5260,\ 5260,\ 630)\,\text{meters}$$
$$(\omega,\ \phi,\ \kappa)_2 = (-0.707143,\ -0.707089,\ 44.995636)\,\text{degrees}$$

The camera parameters are $(x_0, y_0, f) = (0, 0, 152.4)$ mm. A ground point has coordinates $(5080, 5180, 50)$ meters.

Compute the transformation to normalize each of the photographs. Then verify that it is correct by computing the photo coordinates of the given point in each photograph, transforming each one to the normalized plane, and then showing that the normalized y coordinates are equal.

SOLUTION

$$M_1 = M_{\kappa_1}M_{\phi_1}M_{\omega_1} = \begin{bmatrix} 0.707107 & 0.707107 & 0 \\ -0.706676 & 0.706676 & 0.034899 \\ 0.024678 & -0.024678 & 0.999391 \end{bmatrix}$$

$$M_2 = M_{\kappa_2}M_{\phi_2}M_{\omega_2} = \begin{bmatrix} 0.707107 & 0.707107 & 0 \\ -0.706999 & 0.706999 & -0.017452 \\ -0.012341 & 0.012341 & 0.999848 \end{bmatrix}$$

$$\begin{bmatrix} B_X \\ B_Y \\ B_Z \end{bmatrix} = \begin{bmatrix} X_{L_2} - X_{L_1} \\ Y_{L_2} - Y_{L_1} \\ Z_{L_2} - Z_{L_1} \end{bmatrix} = \begin{bmatrix} 260 \\ 260 \\ 20 \end{bmatrix}$$

Extract the ω angles from the two rotation matrices under the assumption that the rotation order is κ (primary), ϕ (secondary), ω (tertiary). Note that the usual assumed order in this text is just the reverse of this. The extraction is done by expressing the matrix symbolically and solving for the unknown angles.

$$\phi = -\sin^{-1}(m_{13})$$

$$\omega = \tan^{-1}\left(\frac{m_{23}/\cos(\phi)}{m_{33}/\cos(\phi)}\right)$$

Therefore, $\omega_1 = 2.0$ degrees, $\omega_2 = -1.0$ degrees. Next, obtain the angles for the base matrix from Eqs. 7-7 through 7-9.

$$\theta_z = 45.0°$$

$$\theta_y = -3.113412°$$

$$\theta_x = 0.5°$$

Compute the base matrix and the two transformation matrices using Eqs. 7-10 and 7-11.

$$M_B = M_x(\theta_x)M_y(\theta_y)M_z(\theta_z) = \begin{bmatrix} 0.706063 & 0.706063 & 0.054312 \\ -0.707415 & 0.706745 & 0.008714 \\ -0.032233 & -0.044574 & 0.998486 \end{bmatrix}$$

$$M_{N_1} = \begin{bmatrix} 0.998524 & 0.001895 & 0.054279 \\ -0.000474 & 0.999657 & -0.026190 \\ -0.054310 & 0.026125 & 0.998182 \end{bmatrix}$$

$$M_{N_2} = \begin{bmatrix} 0.998524 & -0.000948 & 0.054304 \\ -0.000474 & 0.999657 & 0.026164 \\ -0.054310 & -0.026151 & 0.998181 \end{bmatrix}$$

The original photo coordinates of the given ground point are (49.843, 13.860) and (−48.418, 21.285). The normalized coordinates, from Eq. 7-12, are (40.968, 17.585) and (−57.530, 17.585). Note that the y coordinates are equal as desired.

Just as in the case of the optical projection plotter, channel separation for stereo viewing within a computer environment can be achieved in several basic ways.

1. The anaglyph approach displays the left image in blue and the right image in red (or vice versa), and the viewer uses a corresponding set of glasses. A disadvantage of this approach is that color imagery cannot be presented since the color component is used for the separation. Multiple viewers are possible.

2. The split-screen approach puts the left image on the left side of the computer screen and the right image on the right side of the computer screen and forces the viewer to look through an optical stereoscope to perceive stereo. This is a low-cost solution, and allows color imagery, but the field of view is cut in half, and, most importantly, it restricts the viewer's head motion and precludes having multiple simultaneous viewers.

3. Polarization can be used, just as in the case of the optical projection plotter. A liquid crystal panel in front of the monitor alternates between two polarization states and the viewer wears a pair of spectacles with corresponding polarizations. Each image of the stereo pair corresponds to one of the polarization states; the images are alternately displayed on the monitor and the polarization changed accordingly. In this case the active element is the screen and the passive element is the viewer glasses. Color imagery can be used and multiple simultaneous viewers can be accommodated.

4. High-frequency flicker is very similar to the image alternator approach used in the early optical projection instruments. In this case a high-frequency (120 Hz) flicker between left and right images is presented on the monitor. Synchronized spectacles worn by the viewer function as the active viewing element, alternately making the left and right lens opaque, thus effecting the desired channel separation. Because of the synchronization required, the spectacles must communicate with the video driver, either via a tether cable, radio, or infrared optical communication. This system allows color imagery and permits multiple viewers.

7.3.3 General Photogrammetric Workstation Capabilities

Another advantage of the softcopy stereo environment is the ease of including what used to be called *superimposition*. It is often desirable to display collected vector feature information overlaid on the source imagery, in stereo if possible. Such a capability was available only at the cost of enormous complexity in the optical/mechanical/analytical plotter environment, yet comes almost for free in the softcopy environment. This is very important for checking accuracy and completeness in a mapping project.

The most important capabilities of the digital photogrammetric workstation are for the performance of tasks that were not possible on earlier platforms. Furthermore, these disparate tasks can be done in a manner such that they fit into a unified project workflow, and can interact with each other. Figure 7-15 shows a block diagram of the major components of a softcopy stereo workstation. Figure 7-16 shows an outline of the major functional capabilites of such a workstation. Although there are differences between vendors, many of these elements will be present in some form in all photogrammetric workstations. Following is a general discussion of these capabilities.

7.3.3.1 Import

The import function of photogrammetric software must be able to read and interpret images from a variety of sensors, and in a variety of image formats. Sensor options always include frame, and may include others such as pushbroom, scanner, and SAR.

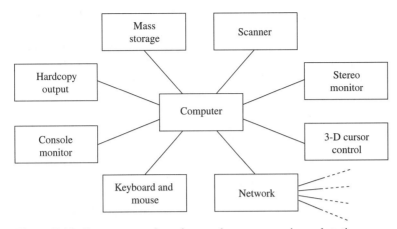

Figure 7-15 Components of a softcopy photogrammetric workstation.

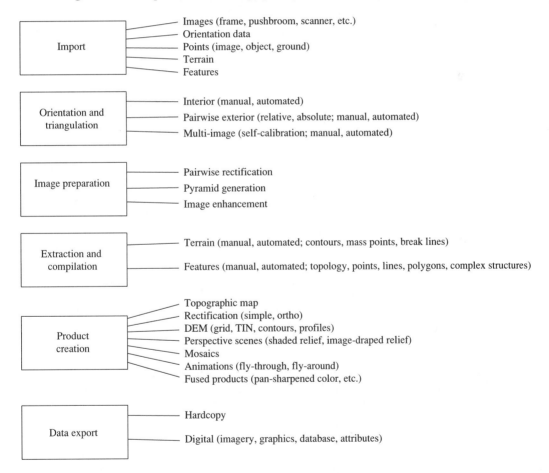

Figure 7-16 Major functions performed by a softcopy photogrammetric workstation.

An ideal system would also be able to accommodate a variable number of color or spectral planes and a variable bit depth per plane. Also desirable is the related ability to read different file formats, for example, TIF, JPEG, RAW, BMP, etc. The import function is also used to read in point data. This includes both image and object coordinates, names or identifiers, and weights or sigmas for image points, ground points, control points, pass points, and tie points. Ideally this module would also include a transformation and projection capability to merge and convert data from different sources, datums, ellipsoids, map projections, etc. Depending on the application, one may also need to import terrain elevation data as a DEM or DTED. These might be grids in any coordinate systems, or TIN organized data (Section 8.2.1). Feature data, vectors, polygons, or wireframes might be needed and therefore must be imported. This raises the issue of multiple CAD formats and user-defined data formats. Lastly, orientation data must be imported if the triangulation data is produced outside of the system or if there is auxiliary orientation data.

7.3.3.2   Orientation and Triangulation

The system should be self-contained in the sense that all necessary orientation computations can be carried out. These include interior orientation, relative orientation, absolute orientation, and multi-image triangulation. Orientation computations may be done via manual measurement or an automated process. There are limitless possibilities to introduce automation to find and measure fiducial marks, pass points, tie points, and control points. Robust estimation or blunder detection is usually necessary to compensate for the usual human or machine frailties when dealing with large quantities of data. Variable parameter selection, such as is used in self-calibration, is a powerful capability for dealing with new or unusual sensors. Ideally, the system can process any number of images in any configuration, including images from multiple cameras or sensors.

7.3.3.3   Image Preparation

Image preparation can include a variety of processes, such as digital image enhancement by contrast modification or sharpening via digital filtering (Section 6.2). It could also include the generation of an image pyramid to avoid repetitive computations when zooming in and out. For a stereo environment, image preparation would include a pairwise rectification capability as outlined earlier in this chapter.

7.3.3.4   Extraction and Compilation

Extraction should encompass terrain as well as features. For terrain, collection modes are necessary to extract contours, profiles, cross sections, mass points, spot heights, break lines, and form lines. The manual mode should be augmented by some capability for terrain generation by image matching. The range of these capabilities is very wide, from simple two-photo correlation-based schemes to hierarchical, feature-based, multi-image, multi-spectral, multi-sensor solutions. As with all of the system functions described in this section, practicalities and market considerations may force niche-oriented systems to address only some of the suggested functions. Feature extraction in the softcopy environment is also primarily a manual task augmented by automated tools. Manual methods must be provided to collect 2-D and 3-D vectors to describe the usual features: roadways, buildings, bodies of water, and elements of the civil infrastructure. If the data are destined for a GIS system, then additional tools must be present for topology and consistency checking, both in 2-D and 3-D. Attribution is a very important and time-consuming aspect of feature extraction, and flexible tools are needed to accommodate each organization's unique needs for attributes, linkages, and formats. As in the case of terrain extraction, autonomous tools and semiautomated functions should be present to assist the user. Completely automated feature extraction is a formidable task and is not a reality at this point, but aspects of the prob-

Table 7.5   Selected Acronyms and Standard File Formats

ASCII	American Standard Code for Information Interchange
AVI	Audio Video Interleaved
BMP	Bitmap
DEM	Digital Elevation Model
DGN	Design
DTED	Digital Terrain Elevation Data
DWG	Drawing
DXF	Drawing Exchange Format
GEOTIF	Extension of TIF file format with georeferencing tags
JPEG	Joint Photographic Experts Group
MPEG	Motion Picture Experts Group
RAW	Raw image
SHP	Shape
TIF	Tagged Image File
TIN	Triangulated Irregular Network
VRML	Virtual Reality Modeling Language

lem can be solved (Section 6.6). With manual cuing and user guidance of model parameters, some types of roadways and buildings can be successfully extracted.

7.3.3.5   Product Creation

The creation of products is closely linked to the previously described functions of data extraction. The list of products available from a photogrammetric workstation should include at least some of the following: topographic maps, GIS data layers, 3-D site models, rectified and orthorectified imagery, mosaics, DEM data in various formats, synthetic images and perspective views, visibility or line-of-sight maps, fly-through and fly-around animations, thematic maps from multispectral analysis, and fused products such as pan-sharpened color imagery.

7.3.3.6   Data Export

The data products just described may need to be actually delivered in a variety of hardcopy and digital formats. Vendors typically support a reasonable number of well-defined interchange formats for digital data, which the customers can import into particular applications. Imagery and image products should be exportable in standard formats. Vector and feature data should be exportable in standard CAD formats. Some standard image and CAD file formats are listed in Table 7-1. Attribute data should be exported in standard database formats as well as ASCII, with well-defined linkages to associated features or image fragments.

In summary, the power and the flexibility of software-based functionality in photogrammetric workstations means that nearly anything is possible. Vendors must make difficult choices as the software capabilities continue to evolve, to include ambitious and promising tools while still ensuring that everything is at a maturity and reliability level that is realistic for production purposes.

PROBLEMS

7.1 Two photographs have the following exterior orientation parameters:

$$(X_L, Y_L, Z_L)_1 = (8000, 4000, 3000) \text{ meters}$$
$$(\omega, \phi, \kappa)_1 = (1, 2, 29) \text{ degrees} (\omega \text{ primary})$$
$$(X_L, Y_L, Z_L)_2 = (9593, 4920, 2985) \text{ meters}$$
$$(\omega, \phi, \kappa)_2 = (-2, 1, 31) \text{ degrees} (\omega \text{ primary})$$

The camera parameters are $(x_0, y_0, f) = (0.010, -0.015, 152.4)$ mm. A ground point has coordinates (9046, 4027, 150) meters. As in Example 7-1, compute the transformation to normalize each of the photographs. Verify that it is correct by computing the coordinates of the given point in each photograph, transforming each one to the normalized plane, and then showing that the normalized y coordinates are equal.

7.2 For the data in Problem 7.1, the photograph limits (in image coordinates) are $(x_{min}, y_{min}) = (-115, -115)$ mm, and $(x_{max}, y_{max}) = (+115, +115)$ mm. What are the limits in the normalized space for photo 1 and photo 2?

REFERENCES

ACKERMANN, F. 1996. Digital photogrammetry: Challenge and potential. *Photogrammetric Engineering and Remote Sensing* 62(6):679.

BURNSIDE, C. 1996. The Photogrammetric Society Analogue Instrument Project: A seventh extract. *The Photogrammetric Record* 15(88):527–544.

CUCURULL, J. 1994. A brief history of color imposition devices and review of current offerings. *Photogrammetric Engineering and Remote Sensing* 60(8):965–969.

DOWMAN, I., EBNER, H., and HEIPKE, C. 1992. Overview of European developments in digital photogrammetric workstations. *Photogrammetric Engineering and Remote Sensing* 58(1):51–56.

GRUEN, A. 1996. Digital photogrammetric stations revisited. In *International Archives of Photogrammetry and Remote Sensing*. Vol. 31, Comm. 2, pp. 127–134. Vienna: International Society for Photogrammetry and Remote Sensing.

KOLBL, O., and BACH, U. 1996. Tone reproduction of photographic scanners. *Photogrammetric Engineering and Remote Sensing* 62(6):687–694.

KRAUS, K. 1993. *Photogrammetry*. Bonn: Dummler.

LEACHTENAUER, J., DANIEL, K., and VOGL, T. 1998. Digitizing satellite imagery: Quality and cost considerations. *Photogrammetric Engineering and Remote Sensing* 64(1):29–34.

LEBERL, F., BEST, M., and MEYER, D. 1992. Photogrammetric scanning with a square array CCD camera. In *International Archives of Photogrammetry and Remote Sensing*. Vol. 29, Comm. 2, pp. 358–363. Washington, DC: International Society for Photogrammetry and Remote Sensing.

MIKHAIL, E. 1996. From the Kelsh to the digital photogrammetric workstation, and beyond. *Photogrammetric Engineering and Remote Sensing* 62(6):680.

MILLER, S., HELAVA, U., and DEVENECIA, K. 1992. Softcopy photogrammetric workstations. *Photogrammetric Engineering and Remote Sensing* 58(1):77–83.

MOFFITT, F., and MIKHAIL, E. 1980. *Photogrammetry*. New York: Harper & Row.

SCARPACE, F., and SALEH, R. 1996. Investigation of aerial triangulation and surface generation using a softcopy photogrammetric system. In *International Archives of Photogrammetry and Remote Sensing*. Vol. 31, Comm. 2, pp. 340–344. Vienna: International Society for Photogrammetry and Remote Sensing.

SCHENK, T. 1999. *Digital Photogrammetry*, Vol. I. Laurelville, OH: TerraScience.

SLAMA, C., ed. 1980. *The Manual of Photogrammetry*. Bethesda, MD: American Society for Photogrammetry and Remote Sensing.

SMITH, M., and SMITH, D. 1996. Operational experiences of digital photogrammetric systems. In *International Archives of Photogrammetry and Remote Sensing*. Vol. 31, Comm. 2, pp. 357–362. Vienna: International Society for Photogrammetry and Remote Sensing.

THORPE, J. 1993. Aerial photogrammetry: State of the industry in the US. *Photogrammetric Engineering and Remote Sensing* 59(11):1599–1604.

TOUTIN, T., and BEAUDOIN, M. 1995. Real-time extraction of planimetric and altimetric features from stereo SPOT data using a digital video plotter. *Photogrammetric Engineering and Remote Sensing* 61(1):63–68.

WALKER, S., and PETRIE, G. 1996. Digital photogrammetric workstations, 1992–96. In *International Archives of Photogrammetry and Remote Sensing*. Vol. 31, Comm. 2, pp. 384–395. Vienna: International Society for Photogrammetry and Remote Sensing.

WALKER, S. 1995. Analogue, analytical, and digital photogrammetric workstations: Practical investigations of performance. *The Photogrammetric Record* 15(85):17–25.

WOLF, P., and DEWITT, B. 2000. *Elements of Photogrammetry*. New York: McGraw Hill.

Chapter 8

Photogrammetric Products

The traditional photogrammetric product is a paper topographic map showing elevation contours and selected terrain and man-made features. Paper maps are now being supplemented and even replaced by digital products. Some of these digital products are simply digital representations of standard hardcopy products, but others contain entirely new types of information specifically designed for computer display and manipulation. The best example of the new applications of digital data is the geographic information system (GIS), which allows the user to display a map generated to his own requirements, using stored data or information derived from stored data. This chapter describes the basic types of standard photogrammetric products and discusses some common examples of each.

8.1   HARDCOPY PHOTOGRAMMETRIC PRODUCTS

Large numbers of paper maps are still printed and distributed, but nearly all are now produced from digital source material. Map data was formerly drawn by a stereoplotter operator, using a plotting table attached to the stereoplotter. Rough plots of contours and planimetric features were done in pencil, then edited and converted to a reproducible form by a draftsman. This process has been replaced by computer methods. The computer can perform many editing functions automatically. For instance, the computer can square buildings on command from the stereoplotter operator or check for interference and consistency between different data layers.

8.1.1   Contour Maps

The familiar *contour map* (Fig. 8-1) indicates elevation by showing lines of equal elevation (isolines). The elevation difference between adjacent contour lines is the

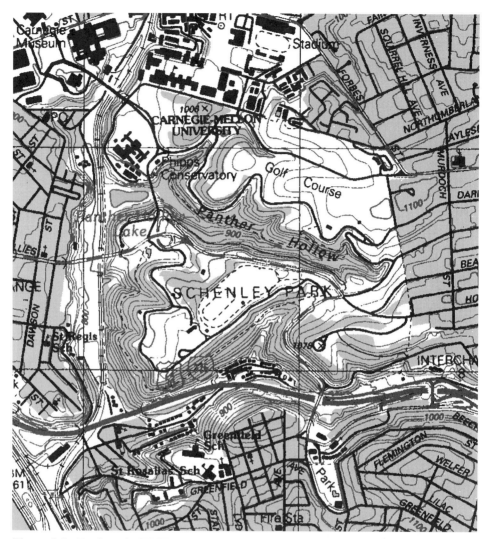

Figure 8-1   Portion of a USGS 1:24,000 topographic map, showing contours and planimetric features.

contour interval, which is chosen as a function of the map scale and the local terrain relief. The smaller the map scale, the larger the contour interval used, since contour lines will tend to be closer together. Areas of steep relief also require a larger contour interval, so that the contour lines are not too close together. The contour interval is also determined by the accuracy of the elevation determination. For instance, the U.S. National Map Accuracy Standards require that 90% of all points have elevations accurate to within half the contour interval. By this standard, to map with a 10-meter contour interval, the elevations must be determined with a standard deviation of 3 meters. (Since 90% of the values in a normal distribution lie within ±1.645 times the standard deviation, to be accurate to within half the contour interval would require a standard deviation of 5/1.645 = 3 meters.)

To estimate the maximum flying height that can be used to produce maps with a given contour interval, photogrammetrists use the *C-factor*, which is the ratio of the flying height and the contour interval. The C-factor is influenced by all parts of the mapping system, including the camera, the quality of the imagery, the stereo-plotter, and the operator. However, it can provide a rough guide for standard production situations. Older direct-projection analog plotters were rated at a C-factor of 700 to 1500, while analytical plotters have C-factors of 2000–2200. For softcopy (digital) systems using scanned aerial imagery, the C-factor is affected by the same geometric factors as for analog plotters and also by the resolution at which the image is scanned. C-factors between 800 and 2100 have been estimated for softcopy systems, depending on the scanning resolution (Light, 1999).

Contours are well-suited to photogrammetric production. To produce a contour map, the operator sets the floating mark (Chapter 7) at the desired elevation in an absolutely-oriented stereomodel and guides it along the ground surface, tracing each contour in turn. The raw contours produced by the operator must be edited to ensure that no contours cross and that the contours are consistent with the linear features on the map. For instance, contours should cross roads perpendicular to the direction of the road, and the bends of the contours should align with the drainage.

Contour maps usually include spot elevations at the tops of hills, in road intersections, and at other points as required to fully describe the topography.

8.1.2   Planimetric Maps

Planimetric maps show features such as roads, buildings, and power lines. The types and amount of detail shown are determined by the scale of the map as well as its purpose. A road may be depicted on a small-scale map by the centerline or as the two edges. On a large-scale map, the median strip, curbs and shoulders, drainage, and manhole covers may be shown. Planimetric features are usually shown with contours, as in Fig. 8-1. Planimetric maps may also show abstract features, such as political boundaries, which are, of course, not generally derived from photogrammetric information.

8.1.3   Rectified Images

Rectified images are another standard hardcopy photogrammetric product. A rectified image is produced by reprojecting the original tilted aerial photograph into an equivalent vertical photograph (Chapter 4). The reprojected image is geometrically identical to a vertical photograph taken from the same point as the original photograph. Since the rectified image is vertical, it shows horizontal planes at a uniform scale (Chapter 2). Rectification is done using a rectifier (Chapter 7), which mechanically recovers the projective geometry of the original and rectified photographs.

If the terrain is relatively flat, the rectified image can be enlarged to a nominal scale and used as a low-precision map. However, the more the terrain deviates from a plane, the more inaccurate such a map will be. Even though rectified prints are relatively inexpensive to produce, they have been replaced in most applications by digital orthophotos (Section 8.2.3).

8.1.4   Photo Mosaics

Photo mosaics are used as low-accuracy maps of areas larger than the coverage of a single photo. Mosaics are manually produced by cutting out central sections of photos and matching them together to cover the area of interest. Only central sections are used, to minimize errors due to relief displacement. Rectified prints are sometimes used rather than the original photographs to improve the accuracy and consistency of the mosaicking.

Manually produced image mosaics are seldom used now, except as indices to locate specific images within a set of flight lines. Most mosaics are now generated automatically from digital orthophotos (Afek and Brand, 1998; McGwire, 1998).

8.2 DIGITAL PHOTOGRAMMETRIC PRODUCTS

Nearly all mapping is currently done using digital systems, even when the final product is a paper map. There is an important distinction, however, between a digital graphical representation and a true digital cartographic data structure. A graphical representation only displays a particular set of data, and lacks the ability to access the underlying semantics of the data set. A cartographic data structure can be manipulated on the basis of the information it represents. Briefly, a digital graphical representation is a picture of a map, while a digital data structure can be used and understood like a map.

For instance, one current product of the U.S. Geological Survey (USGS) is the digital raster graphic (DRG), a 1:24,000 topographic map scanned into digital image form. This is distributed as a TIFF image file, which can be displayed and manipulated with standard image manipulation programs. However, the user cannot specify that only the roads and drainage be displayed or have the computer calculate the total mileage of two-lane roads shown on the map. In contrast, given the digital elevation model and the feature layers associated with the map in a true cartographic data structure, the user can generate alternative representations of the data or process the data layers to extract new information.

Another example of the difference between a simple graphical representation and a digital data structure is the usefulness of different representations of elevation data. Contours may be represented digitally as connected strings of point coordinates with attached attributes indicating their elevation. In some sense a digital file of contours is a digital elevation model. However, contours cannot be easily or efficiently used to determine the elevation of an arbitrary point. Instead, digital contours are primarily a display representation.

8.2.1 Digital Elevation Models (DEMs)

One of the most common photogrammetric products is the *digital elevation model.* DEMs have a number of uses, both as raw data and as input to orthoimage production. The development of automated procedures to generate DEMs has greatly increased their production, and therefore also their utilization.

A digital elevation model represents the Earth's surface elevation digitally as an array of points. Figure 8-2 shows a DEM (containing the area shown in Fig. 8-1) in two common depictions. In Fig. 8-2*a*, the elevation is indicated by the image intensity, with higher points having brighter values. Figure 8-2*b* is an image simulated by assuming a solar azimuth and elevation and then rendering the appearance of the terrain surface based on its slope and aspect relative to the incident solar illumination.

The most common DEM format is the *raster* grid, with elevations given at regularly spaced points, or *posts*. The horizontal positions may be relative to any coordinate system; for instance, the USGS supplies DEMs in UTM (Universal Transverse Mercator) coordinates with the post spacings given in meters, while National Imagery and Mapping Agency (NIMA) DEMs are produced with respect to geodetic (latitude-longitude) coordinate systems and the post spacing is specified in arc-seconds. DEMs are often classified by their post spacing, for example, a 1-arc-second or approximately 30-meter DEM. A smaller horizontal spacing usually implies more accurate elevation values, but in practice elevation accuracy is determined by the production method and the product specifications.

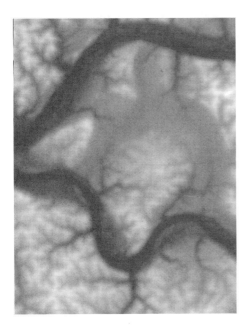
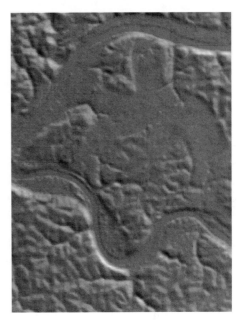

(a)   (b)

Figure 8-2 Two depictions of a DEM of Pittsburgh, PA, (a) with elevation encoded by intensity values, with brighter values being higher, and (b) a shaded relief representation, showing the DEM as if the surface were illuminated by a light source in the upper right corner.

The fidelity with which the Earth's surface can be represented is determined by the horizontal spacing of the DEM and the vertical accuracy of the elevation determination. Because DEMs are discrete representations of the Earth's continuous surface, they are subject to the same sampling theorem considerations as digital images (Chapter 6). Sudden elevation changes, such as cliffs, cannot be represented well by gridded DEMs (Fig. 8-3). Interpolation between DEM posts on either side of the cliff will be misleading, since slopes on the Earth's surface are not continuous. The same problem exists when interpolating across drainage features, since the surface is not continuous between the adjacent posts. To alleviate these problems, some gridded DEM formats include *breaklines*, which indicate cliffs, drainage lines, and other significant discontinuities in the topographic surface. When interpolating elevations from the DEM in the vicinity of a breakline, only points on the same side of the breakline are used. Spot heights may also be included with the breaklines; if a peak lies between elevation posts, interpolation between the posts will not give the true height in that area.

An alternative DEM format is the *triangulated irregular network (TIN)*. Instead of storing elevation values at a regular grid of points, a TIN uses an irregularly-spaced set of points to approximate the terrain surface as a series of triangles. A number of

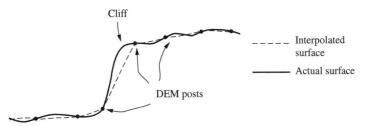

Figure 8-3 DEM with breaklines at a cliff.

methods are used to generate TINs (Chen and Guevara, 1987; Polis and McKeown, 1992; Lee, 1991), in order to obtain the most efficient representation of the terrain with the fewest points. The primary advantage of TIN representations is that they require significantly fewer points to represent the terrain with the same accuracy. The representation accuracy can be increased by adding more points, up to the level of the original DEM (Richbourg and Stone, 1997). Figures 8-4a and 8-4b show two TINs produced from the DEM in Fig. 8-2 with approximately 750 and 7000 points, respectively. The original DEM contains 167,653 points. The TIN in Fig. 8-4a has an RMS error of 7.7 meters and a maximum error of 26.2 meters, compared to the original DEM. The TIN in Fig. 8-4b has an RMS error of 1.4 meters and a maximum error of 4.1 meters. The 1.4-meter RMS error is essentially negligible compared to the 15-meter accuracy specification of the original USGS DEM.

A disadvantage of TIN representations is that determining the elevation at a point is more complicated. With a gridded DEM, the row, column coordinates of the point within the DEM can be determined by subtracting the corner coordinates of the DEM and dividing by the grid spacing. Points that lie between grid nodes are then determined by interpolation. With a TIN, the triangle containing the point of interest must be determined by search, which requires more computation, even though a number of efficient search methods exist.

DEMs are currently produced by both manual and automated methods. In manual production, either the stereoplotter sets the floating mark at the horizontal position of each post and the operator places it on the ground, or the system drives along a profile while the operator keeps the mark on the ground. The DEM may also be interpolated from manually generated contours, although it is now more common for contours to be generated from a DEM. Automated systems (Chapter 6) use computer vision techniques to perform the operator's task of determining the ground surface elevation by matching corresponding portions of two stereo images.

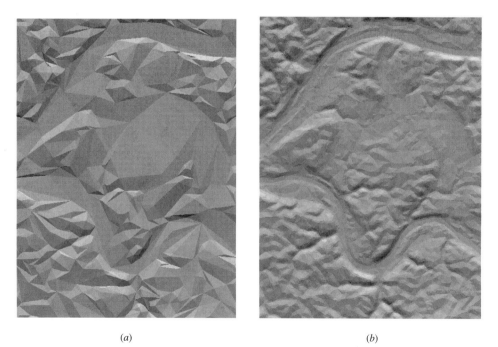

(a)   (b)

Figure 8-4 Triangulated irregular networks (TINs) produced from the DEM in Fig. 8-2, using approximately (a) 750 and (b) 7000 points.

Both production methods have their strengths and weaknesses. Manual methods are typically reliable, but are slow and expensive for large areas. Automated methods can be fast and relatively inexpensive, but fail on complicated scenes, such as urban areas or forests, and in featureless areas. Manual editing of automated results is nearly always required. In some systems, the operator can specify beforehand any areas in which he believes the automated process will not work well. The stereo matcher then skips these areas, leaving them for the operator to do.

Automated systems match on the visible surfaces, such as the tops of buildings or trees (Fig. 8-5), instead of the terrain surface that we want to represent in the DEM. While

Figure 8-5 Automated stereo elevation model results, including elevations on the tops of trees and buildings. Courtesy of Dr. Steven Cochran, Digital Mapping Laboratory, Carnegie Mellon University.

some progress has been made toward automatically recognizing trees and buildings and dropping the elevations to the ground, this process still requires extensive manual editing.

A common problem for both manual and automated methods is banding or *corn rows*, which give the DEM a ridged appearance (Fig. 8-6). This may be caused by systematic offsets between adjacent elevation profiles, or it may be an artifact of interpolating the DEM from contours. Operators often have a tendency to let the floating mark dig into the ground when profiling uphill and to let it float slightly above the ground when profiling downhill. If adjacent profiles are done in opposite directions, to reduce the amount of image movement required, the profile elevations will differ by the amount of this systematic offset. Automated systems may also have systematic errors in their correlation along profiles, leading to the same problems. Even if the overall amplitude of the regular pattern is within accuracy specifications, systematic errors

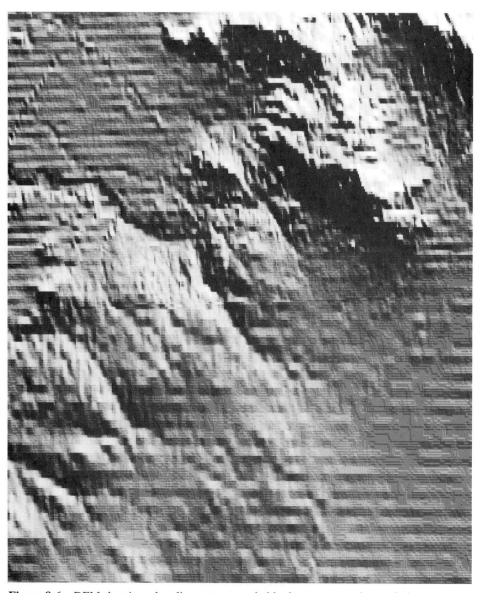

Figure 8-6   DEM showing a banding pattern, probably due to contour interpolation.

have an effect on relative accuracy and also detract from the appearance of products generated from the DEM.

Automated systems may produce spikes or holes due to noise in the data or bad matches. Large anomalies can usually be detected and corrected automatically; the harder problem is errors close to the average terrain variation in magnitude, which must be recognized and corrected manually.

8.2.2 Digital Vector Data

Planimetric features are represented as digital vector data. In its simplest form, digital vector data consists of a set of 2-D or 3-D coordinates that describe the outline of a feature, along with some way of identifying the type of feature. This simple data form may be thought of as the digital equivalent of a hardcopy planimetric map.

Feature types may simply be identified by the file name, as when a file contains only roads or only buildings, or by an *attribute*, which is an identification code included in the data structure. Attribute codes may refer to very broad classes of features (e.g., buildings or roads), or be based on hierarchical classifications (e.g., four-lane divided highways, two-lane highways, unpaved roads, etc.).

Along with positions and object types, we want to represent the *topology* of the data set. Topology is concerned with properties that are not changed by spatial deformation, such as connectivity. In geospatial data sets, we use topological relationships to model networks, such as roads or drainage, and area properties. Having topological information allows us to query the dataset based on connectivity (for path planning or drainage calculations, for example) and adjacency (to determine which parcels adjoin a certain road, for example).

Basic topological elements are of three types: *nodes*, *lines*, and *faces* (Fig. 8-7). Nodes define line endpoints and intersections. Lines are ordered sets of points between nodes. Note that a point is not the same thing as a node—a point specifies a location only, while a node specifies connectivity. There may be several points between nodes on a line. A line cannot cross itself, and connects to or crosses other lines only at nodes. Lines may describe linear features such as roads or may be the boundary of a face. A line has a left side and a right side, looking in the direction of increasing point number, with an identified face on each side. Faces may represent physical regions, such as an area with a defined soil type, or may just be the interior region of a set of

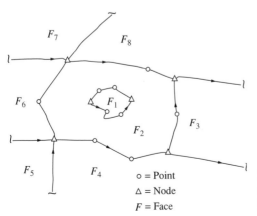

o = Point
△ = Node
F = Face

Figure 8-7 Portion of a topologically structured map, showing points, nodes, lines, and faces.

Figure 8-8 Generalization of a linear feature by point removal.

intersecting lines with no particular physical significance. Every data set has at least two faces, one bounded by the data set and one outside it.

Digital representations are discrete, meaning that we are trying to describe the continuous outline of the object with a finite number of points. We can keep adding points to the description of irregular objects and gain more detail; however, this added detail may not be displayable, useful, or even meaningful.

Generalization (Muller, et al., 1995; McMaster and Shea, 1992) refers to the reduction in displayed detail as map scale decreases. For example, a large-scale map will show bends in a river, both sides of a road, and most buildings; a small-scale map will show only the main path of the river, the centerline of the road, and an outline of the urban area. A similar process of generalization is performed with digital vector data. Digital features are generalized by removing points (Douglas and Peucker, 1973), thereby making the representation less detailed, as shown in Fig. 8-8. Although this may seem like a fairly simple operation, generalization becomes much more complicated when the interactions between adjacent features and different types of features must be taken into account. As the scale is reduced, the relationships between features must be maintained. A road adjacent to a river cannot intersect or cross the river as both are generalized.

Generalization is also determined by the proximity and types of adjacent features. An isolated building is more likely to be represented on large-scale maps than a building of the same size in an urban area. The consistent generalization of data is still a cartographic art. Automated generalization is a problem that has yet to be satisfactorily solved for all feature types (Brassel and Weibel, 1988).

One major difference between digital and hardcopy vector data is that the digital data has no inherent scale. A feature in a digital data file can be shown at any size on a display screen; expressing the scale as the ratio of the displayed size to the feature's real size is therefore meaningless. However, we can use the concept of generalization for an approximate scale description. There are well-established levels of generalization for each standard map scale. For instance, roads are represented to the same level of detail on all 1:24,000 maps and to another level of detail on 1:250,000 maps, although the same road will be represented very differently at the two map scales. The amount of detail captured in the data indicates the type of map for which it would be suitable, and we can use that as a very rough indication of the "scale" of the data. In many cases, the digital data has been derived by digitizing an existing map and, in this case, the scale in terms of generalization is relatively well-defined.

8.2.3 Orthoimagery

An *orthoimage* is an image based on an orthographic projection, rather than the perspective projection of a regular frame photograph (Fig. 8-9). In a frame image the rays of light pass through a single point, the perspective center, before intersecting the image plane. Points at the same horizontal location but at different elevations, such as the top and bottom of a building, will therefore be imaged at different locations in the image. This is called *relief displacement* (Chapter 2). The scale of the frame image varies with the elevation of the terrain, due to the central projection. In an orthographic projection, the projecting rays are perpendicular to a

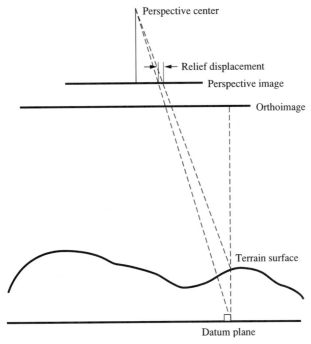

Figure 8-9 Perspective and orthographic image geometry, showing the effect of relief displacement.

horizontal reference plane. Changing the elevation of a point does not affect its projection, so the scale of an orthoimage is constant.

Orthoimages are produced by first obtaining or generating a DEM of the area. This elevation information is then used to remove the elevation effects from the perspective image by reprojection. Orthoimages are often produced from more than one source image to obtain the required coverage for the final product. In addition, selecting only the center portions of images minimizes the relief displacement shown by buildings or elevated objects. Another way to reduce the relief displacement in images intended for orthoimage production is to use lenses with longer focal lengths than the standard 6-inch lens.

There are two basic approaches to generating an orthoimage, forward projection and backward projection (Novak, 1992). In *forward projection* (Fig. 8-10), pixels in the source image are projected onto the DEM and their object space coordinates are determined; the object space points are then projected into the orthoimage. Since the spaces between the points projected into the orthoimage vary due to the terrain variation and perspective effects, the final orthoimage pixels must be determined by interpolating between the projected points.

EXAMPLE 8-1 *Producing an Orthoimage by Forward Projection*

Assume that we are producing a UTM orthoimage with its top left corner at UTM coordinates Easting 588,800 meters, Northing 4,479,000 meters, in Zone 17. The GSD is 0.5 meters. After the region of the source image to be used to produce the orthoimage is selected, the next step is to determine what the spacing of the projected pixels should be. To produce a 1-meter orthoimage, for instance, the pixel spacing in the source image should correspond to no more than a 0.5-meter spacing on the ground. Rapidly changing elevations in the scene will affect the spacing, so a finer sampling may be required.

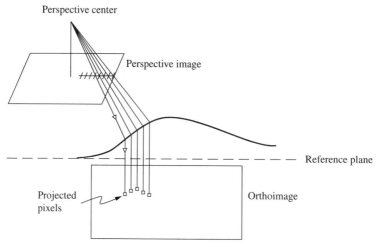

Figure 8-10 Orthoimage production using forward projection. Each pixel in the original perspective image is projected into object space and the resulting object space point is projected into the orthoimage. This results in irregularly spaced points in the orthoimage and requires interpolation to produce a regular array of pixels.

Each pixel in the source image region is projected into the DEM and a set of object space coordinates is calculated. Assume that we project the pixel at row 7320.0, column 6987.0 into object space and obtain the geodetic coordinates N 40° 26' 24.518", W 79° 56' 44.733" and elevation 288.730 meters.

The corresponding UTM coordinates are (589409.38, 4477144.43). To calculate the row, column coordinates in the orthoimage (assuming that row, column coordinates are measured from the top left corner of the image and that row coordinates increase downward and column coordinates increase to the right)

$$\text{row} = \frac{-(Y - Y_0)}{\text{GSD}} = \frac{-(4477144.43 - 4479000.0)}{0.5} = 3711.14$$

$$\text{col} = \frac{(X - X_0)}{\text{GSD}} = \frac{(589409.38 - 588800.0)}{0.5} = 1218.75$$

Notice that we have obtained fractional row, column coordinates in the orthoimage. Once all the pixels in the source image have been projected, the final orthoimage pixel values will have to be obtained by interpolation.

In *backward projection*, shown in Fig. 8-11, the object space X, Y coordinates corresponding to each pixel of the final orthoimage are calculated. The elevation, Z, at that X, Y location is determined from the DEM, and the X, Y, Z object space coordinates are projected into the source image to obtain the gray level or color value for the orthoimage pixel. Since the projected object space coordinates will not fall exactly at pixel centers in the source image, interpolation or resampling must be done in the source image.

EXAMPLE 8-2 *Producing an Orthoimage by Backward Projection*

We can produce the same orthoimage in Example 8-1 using backward projection. For each pixel in the orthoimage, we must calculate the corresponding position in the source image. To do this, we must first calculate its object space coordinates. If we assume that we want to calculate the value at row 3724, column 800, the corresponding UTM coordinates are

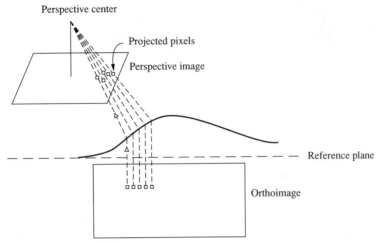

Figure 8-11 Orthoimage production using backward projection. Each pixel in the orthoimage is projected onto the terrain model and the resulting object space point is projected back into the perspective image to determine the corresponding gray value. Interpolation is done in the perspective image.

$$X = X_0 + \text{GSD} \cdot \text{col} = 588800.0 + 0.5(800.0) = 589200.0$$

$$Y = Y_0 - \text{GSD} \cdot \text{row} = 4479000.0 - 0.5(3724.0) = 4477138.0$$

The elevation corresponding to these coordinates is determined from the DEM to be 279.95 meters. These coordinates are then projected into the source image to obtain the image coordinates of row 6334.587, column 6965.817. The gray level for the orthoimage pixel must be interpolated from the adjacent source image pixels.

The depiction of tall objects, such as buildings, that are not included in the terrain model is a problem in orthoimage production. When the orthoprojection is performed on image regions showing objects not in the DEM, the wrong pixels will be extracted from the source image, as shown in Fig. 8-12. This building lean effect can

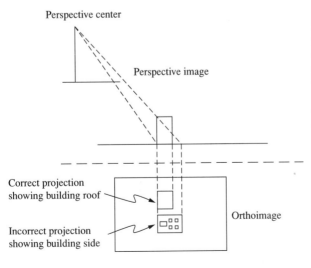

Figure 8-12 The geometry of the building lean problem in orthoimage production. The points at the top and bottom of the building, which have the same horizontal position, should be projected to the same point in the orthoimage. This is possible only if the building height is included in the DEM used to produce the orthoimage; otherwise, the building points will be projected to the terrain surface, resulting in leaning buildings in the orthoimage, as shown in Fig. 8-13.

Figure 8-13   Orthoimage showing the effect of buildings not being included in the DEM.

be particularly noticeable in urban areas, as shown in Fig. 8-13. Some orthoimage production systems use an elevation model containing building outlines, either automatically derived or manually delineated. This problem can be minimized by using long focal length photography and by using only the central part of the photo, since relief displacement increases with shorter focal length and with the radial distance from the image nadir point. In general, points that have multiple elevations, such as a multilevel freeway interchange, cannot be correctly orthorectified unless the photographic view is from directly overhead.

8.2.4   Point Products

Some photogrammetric products consist of a set of point coordinates, possibly with associated covariances. For industrial inspection applications, the coordinates of specified targets on the product or on production fixtures and tooling are determined and compared to the reference values. In aerial work, the control network over a large area may be densified by photogrammetric block triangulation. An initial set of control points are surveyed or located, then additional points to be added to the control network are targeted so that they can be precisely identified and measured on the imagery. The block solution is controlled with the existing known points and simultaneously determines the coordinates of the targeted points. This method can provide a highly accurate and consistent geodetic network over a large area, although it is being replaced by satellite observations.

In some military applications, points determined from a triangulation solution are stored in a database, along with image "chips" that show the point. Future images may then be easily resected using these stored control points.

8.3 GEOGRAPHIC INFORMATION SYSTEMS

One of the primary applications for photogrammetrically-produced data is as the spatial framework for a *geographic information system (GIS)* (Laurini and Thompson, 1992; Burrough and McDonnell, 1998). A GIS consists of four main components:

- a data *input* capability
- a *database* for geographic information storage
- *processing* capabilities for querying and reasoning about the information contained in the database
- a *display* capability

The great advantage of a GIS, and the reason for the rapid growth in its application, is its ability to store large amounts of data in a spatial framework that allows the user to process and redisplay that data in a manner relevant to specific requirements. A paper map is limited to one view of a relatively small subset of available data; a GIS can display any user-selected portion of a large data set, or derive new data sets as required.

The basis of a GIS is its *data model,* which defines how the objects in the GIS are represented. The data model may define objects as discrete entities, such as roads, trees, and houses, each with its own descriptive attributes. Alternatively, continuous field representations may be used for properties that vary continuously, such as elevation, pollution levels, etc.

Another aspect of the data model is the fundamental choice of raster vs. vector. A raster representation, in which each cell of a regular grid is given an attribute or attributes, is better suited for continuous fields. Rasters are simpler to store and to use when performing location queries, spatial filtering, or similar operations. However, large data volumes are produced if the coverage area is large or a small cell size is used. The amount of detail available is limited by the cell size, forcing a trade-off between data volume and system resolution. Vector representations, on the other hand, can be stored efficiently in compact, although complex, data structures. Vectors are especially well-suited for representing networks, since the connectivity information can be encoded using topological relations. The amount of detail on any given entity can be determined as required by the number of points used in its representation. Vector data is therefore better suited for display at different scales than raster data. As a compromise, most current GIS systems support both raster and vector models.

GIS data files may be structured into *layers* and *tiles*. A layer contains data of a particular type; for instance, an urban GIS might have a road layer, a building layer, a political boundary layer, etc. A tile is an arbitrary regular subdivision of the world, analogous to a map sheet. Dividing the database into tiles allows only the relevant portions for a particular problem to be used, instead of having to load the entire database.

The dimensionality of the data stored in the GIS is also an issue. In the majority of cases, data are stored as 2-D objects, with only X, Y coordinates. In some cases, an elevation value may be stored at nodes or posts, but with no elevation information elsewhere. This is often referred to as *2.5-D*. A true 3-D GIS contains three coordinates for every location in the database. This type of representation is important when working with geological strata or complex man-made structures such as multiple levels of utility lines. A 4-D GIS, which contains time-stamped representations of objects, allows the state of the data at any given time to be queried or displayed.

While the ability of a GIS system to display geospatial data is useful, the true power of a GIS is based on its ability to perform spatial queries or selection operations on the data it contains. Elementary types of operations include

- *Quantitative.* Count discrete objects or determine measurements.
- *Query by location.* Retrieve objects at a specified location, within an area, or within a specified distance from some other object.
- *Query by attribute.* Retrieve all objects with a given attribute or whose attributes fall within a given range.
- *Query by topology.* Retrieve adjacent or connected objects.

These basic operations can be combined using *Boolean operators*, often as part of a macro command language, to allow compound queries to be generated and complex spatial reasoning to be performed.

As an example of spatial queries, consider a site location problem. We want to locate a facility at a site that meets the following requirements:

- at least a 20-hectare parcel
- within 10 km of a major highway
- on ground with less than a 10% slope
- located in one of a list of counties with low property taxes

Assuming that all of the necessary data sources exist in the database, a query can be formulated that combines each individual condition to identify any areas that satisfy all the requirements. The layer-by-layer processing will include

- accessing the area attribute of parcels in land parcel records and marking those parcels with area greater than 20 hectares;
- locating major highways, performing a *buffer* operation to define the zone within 10 km of each highway, and marking parcels within that buffer zone;
- accessing the digital elevation model and calculating the maximum slope for each parcel, if it is not stored as an attribute; and
- marking parcels within the political boundaries of the specified counties.

The marked parcels resulting from each separate query are combined and the parcels meeting all the criteria are displayed. If no parcels qualify, the results of each separate query may be examined to identify which requirement is the most difficult to satisfy and to decide whether it can be relaxed. Alternatively, weights can be assigned to the requirements and scores calculated for each parcel to identify those that most closely satisfy the weighted requirements.

Geographic information systems theory is a rapidly growing field, encompassing practical considerations like efficient data storage structures and theoretical concerns such as the semantics of various query and combination operations. The interested student should consult the references for a more in-depth appreciation of GIS implementation and application issues.

8.4 THREE-DIMENSIONAL PRODUCTS

Photogrammetry is often used to generate three-dimensional representations of objects, especially in close-range applications. Three-dimensional representations tend to be designed for specific applications, such as the object representations for computer vision discussed in Chapter 6.

In many terrestrial applications, especially architectural ones, CAD (computer-aided design) formats are used. While older CAD formats simply reproduced drawings in digital form, newer formats contain additional information that allows the computer to reason about object properties such as weight, volume, and thermal conductivity.

Geospatial data is often transformed into 3-D formats to allow more efficient 3-D visualization. One common visualization format is virtual reality modeling language (VRML) (Carey, 1998), which is used for display of 3-D objects on the World Wide Web.

Three-dimensional models of cities currently have a variety of applications (Leberl et al., 1999). Telecommunications planners use urban models to simulate the propagation of radio waves for the determination of optimal transmitter locations. Highly detailed visual models are increasingly used for city planning, tourism, and training purposes. Urban models are valuable tools for architects and planners, because these 3-D models allow them to better understand the relationships between proposed developments and the surrounding areas and to better present the plan to the general public. A realistic interactive model allows the non-architect to better visualize and understand the proposed development. Three-dimensional city models are also used in virtual tourism, which lets potential visitors see an area before they visit. Urban models are used in simulators for training truck drivers and also for military training, to provide experience in urban operations.

When planning the construction of an urban model, the main considerations are the intended use of the model, the amount of interaction desired, and the level of detail to be included. These factors are closely interrelated; for a given set of graphics hardware, the amount of detail present determines the display rate and therefore the level of interaction possible. The minimum acceptable display rate is 20–30 Hz. Up to 60 Hz is required for rapid movements or changes in viewpoint.

Models intended for interactive viewing are usually built from polygons; the frame rate achievable when viewing the model is (roughly) determined by the number of polygons in the scene and the rendering speed of the graphics hardware, usually specified in polygons per second. A more detailed urban model requires a greater number of polygons to represent each building or object. The more polygons that must be rendered, the slower the scene rendering will be. If the model is to be used in a simulator or another interactive mode, the polygon count must be constrained to be within the capabilities of the graphics hardware. A typical trade-off to reduce the number of polygons required is to use simple building models and apply image textures to represent doors, windows, or architectural details, instead of specifically modeling each feature. This produces a reasonably realistic scene, as long as the building features are not viewed closely and do not extend far from the building.

The amount of detail that can be included in a city model depends on the input data available. A complete, detailed 3-D photogrammetric model of a city can be made, but in practice, the construction of such a model is usually too expensive for more than a small area. If 2-D building outlines are included in an existing GIS, these may be extruded to generate 3-D building models using automated matching against aerial imagery (Lammi, 1997) or by assigning typical elevations. In some cases, only highly visible landmark buildings are modeled, and other buildings are represented by generic models. *Street furniture,* such as light poles and fire hydrants, may be placed automatically based on street and intersection positions.

A main determining factor in the amount of detail included in a 3-D model is the amount and type of interaction required. Models for telecommunications simulation have few or no interaction requirements, and can therefore include a large amount of

Figure 8-14 Three-dimensional scene visualization. Image courtesy of TerraSim, Inc.

detail. Models for real-time simulation and training must be capable of being rendered at high frame rates. A common goal in an interactive model is for the user to be able to walk down a street, enter a building, and explore its interior. This is seldom realistic unless very powerful graphics computers are available, due to the number of polygons required to represent building interiors and structure.

Figure 8-14 shows a scene from an urban model of a portion of Pittsburgh. The detailed building models came from photogrammetric modeling and from architectural CAD models; the remainder were extruded from 2-D footprints from a GIS.

8.5 PRODUCT ACCURACY

The accuracy specifications of a product are as important as the data contained in it. Unfortunately, this *metadata* is often not collected or not included in data distributions. Even when accuracy specifications are present, they may be ignored by the users. The ease with which digital data can be transferred between systems, transformed into different representations, and combined with other data from different sources facilitates the loss of accuracy metadata, and may result in data sets with large internal inconsistencies or errors. Legal actions have been brought in cases where maps were used at larger than intended scale, with correspondingly higher accuracy expectations, and damages were incurred. This underscores the importance of preserving accuracy metadata.

A well-defined digital product type must have an associated accuracy specification, set by a government agency, a professional organization, or by agreement between the producer and the user. Each data set of that product type must reference the specification and quantify its compliance with the specification. For example, the Federal Geographic Data Committee (FGDC) specifies how accuracy is defined and tested for data in the National Spatial Data Infrastructure. These standards are then applied by data production agencies, such as the U.S. Geological Survey, to their products.

Geospatial data accuracy must be defined by reference to some independent external standard of higher accuracy. The independent source may be field surveys, such as check profiles for contour and DEM testing, or independent photogrammetric checks using larger-scale imagery. Sufficient check points must be evaluated to obtain a statistically valid result. For example, the FGDC standards require that a minimum of 20 check points be compared between the evaluated and reference data sets.

The accuracy statement is usually based on the root-mean-square (RMS) of the check point differences. If the errors are assumed to be normally distributed, the statistical confidence level can be calculated from the RMS value by using the RMS value as the standard deviation. For one-dimensional errors, such as vertical errors, the 95% confidence region is ±1.96 times the RMS value. For horizontal error, the combined RMS error for both X and Y coordinates is used. Since this is a two-dimensional error distribution, the factor for 95% confidence is 2.447.

While the absolute accuracy is always of interest, another useful statistic often used is the relative accuracy, which describes the internal consistency of the dataset. For a DEM, the relative accuracy specifies the accuracy of the differences in elevation between posts. A DEM might be affected by an overall vertical shift, making its absolute accuracy poor, but still have good relative accuracy.

In some cases, statistics on the size of the largest *blunders,* or gross errors, possibly present in the data are given. By their nature, gross errors are difficult to model statistically, although reliability statistics have been developed for that purpose (Chapter 5). For most products, the blunder limit is a function of the amount of editing and checking performed on the data.

8.6 PHOTOGRAMMETRIC PROJECT PLANNING

Given a desired set of products, photogrammetric project planning is a question of trade-offs between accuracy and cost. The economics of photogrammetric mapping are constantly changing, due to factors such as the transition to digital imaging and production methods, the commercial availability of high-resolution satellite imagery, and basic changes in the marketplace for geospatial information. This section will therefore cover some of the basic considerations in project planning without specific economic information.

All project planning must start by choosing the end product—what exactly does the customer want? Will orthoimagery, elevation data (as contours or DEM), vector planimetry, or land use/land cover information be required? What are the accuracy requirements for the data? Formerly, the determining factor for accuracy was the map scale of the data. However, as discussed in Section 8.2.2, for digital data the more appropriate descriptor is the level of detail included. Are there requirements for when the data must be obtained? In wooded areas, the photography often must be taken after the leaves have fallen from deciduous trees so that the ground is visible.

Once the product is defined, the type of imagery required to produce that product is selected. Choices might include both aerial and satellite imagery, or even LIDAR (Chapter 11) for elevation data. A number of factors must be considered: the spectral characteristics, the resolution of the imagery, its cost, when it can be obtained, and how its use works within the production flow.

Assuming that aerial photography is chosen as the most economical option, the flight characteristics must be determined. We usually want to fly at the highest possible altitude that still allows us to obtain the required accuracy. This covers the area with the fewest images and therefore the fewest stereomodels, which reduces processing and compilation costs. For contouring, the maximum flying altitude is determined by the C-factor (Section 8.1.1); for orthoimage production, it is controlled by the required final resolution of the orthoimage.

The flight line configuration should be arranged so that each flight line is as long as possible, to reduce the amount of time spent turning the aircraft at the end of each line and realigning it at the start of the next line. Turns must be wide and flat, since sharply banked turns can cause the aircraft to block the GPS antenna's view of one or more satellites.

Images along the flight line must overlap to obtain stereo coverage, of course. Sixty percent overlap is the minimum used, although the imagery may be flown with 80% overlap to allow the selection of alternative stereo pairs. Sidelap between flight lines is typically 10 to 20%, to prevent gaps in the coverage. Most aerial cameras are now controlled by GPS, which allows the exposure station to be precisely specified. When this is coupled with stabilized mounts to keep the camera axis vertical, the overlap can be reduced, since the safety margin need not be so large.

Once the flight line layout and block configuration are determined, locations for ground control points must be selected. Current GPS triangulation techniques (Section 5.10.1) have greatly reduced the need for ground control, but some points are still required to establish the datum and to serve as a check against blunders.

PROBLEMS

8.1 With what precision must elevations be determined to map at a 5-meter contour interval, according to U.S. National Map Accuracy Standards?

8.2 A mapping company uses a digital stereoplotter with a C-factor of 2000. What is the greatest altitude at which they can fly to map with 5-foot contours? With 5-meter contours?

8.3 A rectified image is to be used as a map of an area with an elevation variation of ± 50 meters around the reference elevation for the rectification. If the original image was taken with a 152-mm lens from an altitude of 1520 meters above the reference elevation, what is the maximum relief displacement at 100 mm from the principal point? If the rectified photo is enlarged by a factor of 5, what is the scale of the final image? What will the maximum error due to relief displacement be in using it as a map?

8.4 What enlargement factor is required to produce a rectified image map at 1:1000 scale from aerial photography taken at 760 meters above terrain using a 152-mm lens?

8.5 Explain the difference between an orthophoto and a rectified photograph.

8.6 Under what conditions could a contact print be considered an orthophoto?

8.7 What types of features could be extracted from an orthoimage, in conjunction with a DEM? What types of features could not be extracted?

8.8 For which types of features is a raster GIS more appropriate? For which types of features is a vector GIS more appropriate?

REFERENCES

Afek, Y., and Brand, A. 1998. Mosaicking of orthorectified aerial images. *Photogrammetric Engineering and Remote Sensing* 64(2):115–125.

Al-garni, A. M. 1996. Urban photogrammetric data base for multi-purpose cadastral-based information systems: The Riyadh city case. *ISPRS*

Journal of Photogrammetry and Remote Sensing 51(1):28–38.

AL-ROUSAN, N., CHENG, P., PETRIE, G., TOUTIN, T., and VALADANZOEJ, M. J. 1997. Automated DEM extraction and orthoimage generation from SPOT Level 1B imagery. *Photogrammetric Engineering and Remote Sensing* 63(8):965–974.

BALTSAVIAS, E. P. 1996. Digital ortho-images—a powerful tool for the extraction of spatial- and geo-information. *ISPRS Journal of Photogrammetry and Remote Sensing* 51(2):63–77.

BOLSTAD, P., and STOWE, T. 1994. An evaluation of DEM accuracy: Elevation, slope, and aspect. *Photogrammetric Engineering and Remote Sensing* 60(11): 1327–1332.

BRASSEL, K. E., and WEIBEL, R. 1988. A review and conceptual framework of automated map generalization. *International Journal of Geographical Information Systems* 2(3):229–244.

BROWN, D. G., and BARA, T. J. 1994. Recognition and reduction of systematic error in elevation and derivative surfaces from 7.5 minute DEMs. *Photogrammetric Engineering and Remote Sensing* 60(2):189–194.

BURROUGH, P. A., and MCDONNELL, R. A. 1998. *Principles of Geographic Information Systems*. Oxford: Oxford University Press.

CAREY, R. 1998. The virtual reality modeling language explained. *IEEE Multimedia* 5(3):84–93.

CARTER, J. R. 1998. Digital representations of topographic surfaces. *Photogrammetric Engineering and Remote Sensing* 54(11):1577–1580.

CHEN, Z. -T., and GUEVARA, J. A. 1987. Systematic selection of very important points (VIP) from digital terrain model for constructing triangular irregular networks. In CHRISMAN, N. R. ed., *AutoCarto8: International Symposium on Computer Assisted Cartography*. pp. 50–54. Baltimore, MD: American Society for Photogrammetry and Remote Sensing.

CHEN, L. C., and LEE, L. H. 1993. Rigorous generation of digital orthophotos from SPOT images. *Photogrammetric Engineering and Remote Sensing* 59(5):655–661.

DOUGLAS, D. H., and PEUCKER, T. K. 1973. Algorithms for the reduction of the number of points required to represent a digitized line or its character. *The Canadian Cartographer* 10(2):112–123.

FRITSCH, D. 1996. Three-dimensional geographic information systems—status and prospects. In *International Archives of Photogrammetry and Remote Sensing*. Vol. XXXI No. B3 pp. 215–221. Vienna: International Society for Photogrammetry and Remote Sensing.

GILES, P. T., and FRANKLIN, S. E. 1996. Comparison of derivative topographic surfaces of a DEM generated from stereoscopic SPOT images with field measurements. *Photogrammetric Engineering and Remote Sensing* 62(10):1165–1171.

HELMERING, R. J. 1978. *Digital Terrain Models (DTM) Symposium*. Baltimore, MD: American Society of Photogrammetry.

HÖHLE, J. 1996. Experiences with the production of digital orthophotos. *Photogrammetric Engineering and Remote Sensing* 62(10):1189–1194.

HOOD, J., LADNER, L., and CHAMPION, P. 1989. Image processing techniques for digital ortho production. *Photogrammetric Engineering and Remote Sensing* 55(9):1323–1329.

KONECNY, G. 1979. Methods and possibilities for digital differential rectification. *Photogrammetric Engineering and Remote Sensing* 45(6):727–734.

LAMMI, J. 1997. Automatic building extraction using a combination of spatial data and digital photogrammetry. In *Proceedings of the SPIE: Integrating Photogrammetric Techniques with Scene Analysis and Machine Vision III*. Vol. 3072 pp. 223–230. Orlando, FL: SPIE.

LAURINI, R., and THOMPSON, D. 1992. *Fundamentals of Spatial Information Systems*. London: Academic Press.

LEBERL, F. W., WALCHER, W., WILSON, R., and GRUBER, M. 1999. Models of urban areas for line-of-sight analyses. *ISPRS Conference on Automatic Extraction of GIS Objects from Digital Imagery*. Technische Universität München, Germany.

LEE, J. 1991. Comparison of existing methods for building triangular irregular network models of terrain from grid digital elevation models. *International Journal of Geographical Information Systems* 5:267–285.

LI, Z. 1993. Mathematical models of the accuracy of digital terrain model surfaces linearly constructed from least square gridded data. *Photogrammetric Record* 14(82):661–674.

LIGHT, D. L. 1999. C-factor for softcopy photogrammetry. *Photogrammetric Engineering and Remote Sensing* 65(6):667–669.

MCGWIRE, K. C. 1998. Mosaicking airborne scanner data with the multiquadric rectification technique. *Photogrammetric Engineering and Remote Sensing* 64(6):601–606.

MCMASTER, R. B., and SHEA, K. S. 1992. *Generalization in Digital Cartography*. Washington, DC: Association of American Geographers.

MERCHANT, J. W. 1999. Special issue: GIS in state and local government. *Photogrammetric Engineering and Remote Sensing* 65(11).

MULLER, J. C., LAGRANGE, J. P., and WEIBEL, R., eds. 1995. *GIS and Generalization: Methodology and Practice*. London: Taylor and Francis.

NOVAK, K. 1992. Rectification of digital imagery. *Photogrammetric Engineering and Remote Sensing* 58(3):339–344.

PEUQUET, D. J., and MARBLE, D. F., eds. 1990. *Introductory Readings in Geographic Information Systems*. London: Taylor and Francis.

POLIS, M., and MCKEOWN, D. 1992. Iterative TIN generation from digital elevation models. *Proceedings of IEEE Conference on Computer Vision and Pattern Recognition*. pp. 787–790. Urbana, IL.

RICHBOURG, R. F., and STONE, T. 1997. Triangulated irregular network (TIN) representation quality as a function of source data resolution and polygon budget constraints. *Proceedings of the SPIE: Integrating Photogrammetric Techniques with Scene Analysis and Machine Vision III*. Vol. 3072 pp. 199–210.

ROBINSON, G. J. 1994. The accuracy of digital elevation models derived from contour data. *Photogrammetric Record* 14(83):805–814.

SINNING-MEISTER, M., GRUEN, A., and DAN, H. 1996. 3D city models for CAAD-supported analysis and design of urban areas. *ISPRS Journal of Photogrammetry and Remote Sensing* 51(4):196–208.

THAPA, K., and BOSSLER, J. 1992. Accuracy of spatial data used in geographic information systems. *Photogrammetric Engineering and Remote Sensing* 58(6):835–841.

TORLEGÅRD, K., ÖSTMAN, A., and LINDGREN, R. 1986. A comparative test of photogrammetrically sampled digital elevation models. *Photogrammetria* 41(1): 1–16.

Chapter **9**

Close-Range Photogrammetry

The development of analytical photogrammetric techniques has enabled rapid growth in the applications of close-range photogrammetry. The ability to model the camera geometry mathematically, instead of relying on analog restitution, allows the use of a wide variety of nonmetric cameras. Through the use of analytical self-calibration techniques, accuracy has increased to the point where close-range photogrammetry has become a standard technique for precision industrial inspection. The introduction of electronic sensors has also broadened the applicability of close-range photogrammetry. While the resolution of digital sensors, and therefore the accuracy, is typically less than that available with film cameras, the ability to produce real-time measurements and to use sequences of images has led to many new applications.

This chapter discusses the equipment and techniques used in close-range photogrammetry, then gives an overview of some of the current applications of close-range photogrammetry. The chapter concludes with a brief discussion of several photogrammetric techniques using nontraditional types of imagery.

9.1   CAMERAS FOR CLOSE-RANGE PHOTOGRAMMETRY

For years, the only cameras used for close-range photogrammetry were metric cameras such as the Wild P31 (Fig. 9-1), which is no longer in production. The Wild P31 had a 4 by 5 inch format, used glass plates or cut film, and accepted interchangeable lenses to vary the focal length. Metric cameras could be used in conjunction with a theodolite, to allow the photogrammetric survey to be tied into a ground survey.

Figure 9-1 Wild P31 metric camera. Photo courtesy of LH Systems, LLC.

Standard stereometric cameras consist of two metric cameras connected by a rigid base that maintains the camera axes parallel to each other and perpendicular to the base. In some cameras, the base distance is continuously adjustable, or can be varied by substituting bars of different length. Some models allow for adjustment of the common tilt angle of the cameras or for the cameras to be adjusted into a convergent configuration.

More recent metric film cameras include the CRC-1 and CRC-2 (Fig. 9-2) from Geodetic Services, Inc (GSI) (Brown, 1984), designed for the measurement of point targets by analytical methods. The CRC-1 has the same-sized format as an aerial camera (230 by 230 mm) and uses aerial film. These cameras have interchangeable lenses and an internal microprocessor that controls the exposure. The platen flattens the film using a vacuum and also contains a set of 25 reseaux which are projected onto the film from the back during exposure, allowing precise correction of distortions for accurate measurements of point targets. The camera can be easily rotated

Figure 9-2 CRC-2 close-range film camera. Photo courtesy of
Geodetic Services, Inc.

around its axis, to facilitate taking multiple exposures with different orientations from
the same exposure station.

The future of metric cameras is represented by GSI's INCA series of digital cameras
(Fig. 9-3), which are based on solid-state area imaging arrays up to 3072 by 2048 pixels.
Each camera has an embedded computer that controls camera function and also com-
presses the acquired image to reduce image transmission times and storage requirements.
Multiple-camera systems can be used for real-time measurement of targeted points.

Recent years have seen the widespread use of amateur nonmetric film and electronic
cameras for photogrammetric purposes. While the absolute accuracy attainable using
such commercially available cameras is of necessity lower than that for fully calibrated
metric cameras, their low cost and ease of use make them attractive for many applications.

9.2   INSTRUMENTS AND SOFTWARE FOR CLOSE-RANGE
PHOTOGRAMMETRY

Most analog stereoplotters designed for use with aerial imagery were unable to exploit
terrestrial imagery, due to limitations on orientation angles or focal length settings. The

Figure 9-3 INCA-2 close-range digital camera. Photo courtesy of Geodetic Services, Inc.

development of analytical plotters (Chapter 7) and specialized close-range software packages has made the plotting of terrestrial imagery much more practical and efficient.

Monocomparators (Chapter 7) are used for target measurement for high-accuracy point determination with film cameras. The use of standardized targets allows the measurement process to be automated (Brown, 1987).

The development of digital photogrammetric systems has led to the introduction of a number of close-range photogrammetric workstations. A good example of a high-end close-range photogrammetric workstation is FotoG-FMS from Vexcel. The core functionality is a bundle block adjustment with added calibration parameters that supports imagery from frame, x-ray, panoramic, and gamma cameras. A number of automated algorithms are included for feature extraction, image matching, tie point matching, and recognition of targets. Its primary applications are industrial, manufacturing, and facilities management; it therefore supports the import and export of data in standard CAD (computer aided design) formats, so that the photogrammetrically produced measurements and models can be easily used for design or analysis.

A relatively new development is the advent of nonstereo multiple-image softcopy photogrammetric systems running on standard PCs and designed for 3-D modeling of objects for computer graphics or for low- to medium-accuracy measurement applications. Since the system does not support a stereo display, the processing and hardware requirements are greatly reduced. The lack of stereo limits the user to measuring either targets or photo-identifiable points, but for many applications, this is not a problem.

An example of such a system is PhotoModeler, a product of EOS Systems. Photo-Modeler accepts photos from both uncalibrated and calibrated cameras (a camera calibration routine is included). The system defines objects in the scene by combinations of geometric primitives, such as cylinders or planes, and the operator specifies points or edges to be used in the solution. The models produced can be exported in standard CAD or graphics formats.

9.3   MATHEMATICAL MODELS FOR CLOSE-RANGE PHOTOGRAMMETRY

Although the basic projective geometry is the same for close-range and aerial photogrammetry, a number of factors complicate terrestrial photogrammetry applications. The locations and orientations of images are much less regular in close-range setups than in aerial applications, where relatively regular flight lines and standard image spacings are used. In precise close-range work, camera stations typically surround the workpiece in order to obtain convergent image ray geometry at all points. Multiple exposures with varying camera orientations are often taken at each camera station, to strengthen the solution for the detection and removal of systematic errors. These types of block configurations complicate the initial design and specification of the mathematical model and make proper evaluation of the results crucial.

The use of uncalibrated cameras, or the recovery of residual systematic errors for calibrated cameras, further complicates the processing. In aerial photogrammetry, cameras are always set at infinity focus, but in close-range work, each photo may require a different focus setting, thus introducing different principal distances, which must be recovered in the solution. Proper design of the acquisition geometry to allow recovery of the calibration elements is an art; the results must be carefully evaluated to ensure that correlations between parameters or weak imaging geometries have not compromised the accuracy of the results.

In aerial photogrammetry, the reference coordinate system, or datum, is defined by the ground control points. In close-range work, measuring three-dimensional control point coordinates to the same level of accuracy as the photogrammetric observations is difficult and expensive, if not impossible. Therefore, instead of using absolute coordinates in the solution, close-range solutions often use *free net adjustment* (Section 9.6), in which only the minimum information necessary to define scale, and possibly vertical orientation, is supplied and the coordinates are defined only in a relative sense.

9.3.1   Collinearity Representation

The standard collinearity equations (Chapter 4) are the basis of bundle adjustment as applied to close-range photogrammetry. The interior orientation elements are often carried in the solution with *a priori* weights to limit the corrections that can be applied and thereby make the solution more stable. Residual systematic errors may be modeled by added parameters, as described in Chapter 5.

9.3.2   The Direct Linear Transform

The direct linear transform (DLT) (Abdel-Aziz and Karara, 1971; Marzan and Karara, 1975) models the transformation between the comparator or image pixel coordinate system and the object space coordinate system as a linear function. The DLT can be derived from the standard collinearity equations, or alternatively, it can be thought of as an implementation of projective geometry. Systematic error corrections may be included as part of the transformation, although this makes the solution nonlinear.

While the linear solution is more efficient than the full collinearity solution, it can be less stable in many situations. At least six noncoplanar control points, with all three coordinates known, are required for the linear solution.

To derive the DLT equations, we start with the collinearity equations (Chapter 4), but scale the principal distance c differently in the x and y directions to reflect the image aspect ratio:

$$x + \delta x - x_0 = -c_x \frac{m_{11}(X_P - X_C) + m_{12}(Y_P - Y_C) + m_{13}(Z_P - Z_C)}{m_{31}(X_P - X_C) + m_{32}(Y_P - Y_C) + m_{33}(Z_P - Z_C)}$$

$$y + \delta y - y_0 = -c_y \frac{m_{21}(X_P - X_C) + m_{22}(Y_P - Y_C) + m_{23}(Z_P - Z_C)}{m_{31}(X_P - X_C) + m_{32}(Y_P - Y_C) + m_{33}(Z_P - Z_C)} \tag{9-1}$$

$$c_x = c\lambda_x$$
$$c_y = c\lambda_y$$

where x and y are the image coordinates; x_0 and y_0 are the principal point coordinates; c_x and c_y are the principal distance c scaled by different factors λ in the x and y directions; X_P, Y_P, and Z_P are the object space coordinates of the point; X_C, Y_C, and Z_C are the object space coordinates of the perspective center; m_{ij} are the elements of the 3×3 rotation matrix M; and δx and δy are the total lens distortions in the x and y directions, as a function of the K_i coefficients:

$$\delta x = (x - x_0)(K_1 r^2 + K_2 r^4 + \cdots)$$
$$\delta y = (y - y_0)(K_1 r^2 + K_2 r^4 + \cdots) \tag{9-2}$$

As long as the camera is not located at the coordinate origin, the collinearity equations can be transformed into the DLT equations as follows:

$$x + \delta x = \frac{L_1 X_P + L_2 Y_P + L_3 Z_P + L_4}{L_9 X_P + L_{10} Y_P + L_{11} Z_P + 1}$$

$$y + \delta y = \frac{L_5 X_P + L_6 Y_P + L_7 Z_P + L_8}{L_9 X_P + L_{10} Y_P + L_{11} Z_P + 1} \tag{9-3}$$

where

$$L = -\frac{1}{(m_{31}X_C + m_{32}Y_C + m_{33}Z_C)}$$
$$L_1 = L(x_0 m_{31} - c_x m_{11})$$
$$L_2 = L(x_0 m_{32} - c_x m_{12})$$
$$L_3 = L(x_0 m_{33} - c_x m_{13})$$
$$L_4 = x_0 + Lc_x(m_{11}X_C + m_{12}Y_C + m_{13}Z_C)$$
$$L_5 = L(y_0 m_{31} - c_y m_{21}) \tag{9-4}$$
$$L_6 = L(y_0 m_{32} - c_y m_{22})$$
$$L_7 = L(y_0 m_{33} - c_y m_{23})$$
$$L_8 = y_0 + Lc_y(m_{21}X_C + m_{22}Y_C + m_{23}Z_C)$$
$$L_9 = Lm_{31}$$
$$L_{10} = Lm_{32}$$
$$L_{11} = Lm_{33}$$

Note that, in these equations, c_x, c_y, x_0, y_0, x, and y are in the comparator or pixel coordinate systems and units. Therefore, no fiducial marks are required.

The DLT was originally formulated as a linear solution, however, a nonlinear formulation is possible. This makes possible multiple-image solutions with tie points between images, the use of control points without all three coordinates known, and the weighting of individual control points according to their precision.

9.3.3 Conversion from DLT to Physical Parameters

Physical camera parameters can be derived from the DLT parameters using the following relations:

$$L^2 = L_9^2 + L_{10}^2 + L_{11}^2$$

$$x_0 = \frac{L_1 L_9 + L_2 L_{10} + L_3 L_{11}}{L^2}$$

$$y_0 = \frac{L_5 L_9 + L_6 L_{10} + L_7 L_{11}}{L^2}$$

$$c_x^2 = \frac{L_1^2 + L_2^2 + L_3^2}{L^2} - x_0^2$$

$$c_y^2 = \frac{L_5^2 + L_6^2 + L_7^2}{L^2} - y_0^2$$

$$m_{31} = \frac{L_9}{L}$$

$$m_{32} = \frac{L_{10}}{L}$$

$$m_{33} = \frac{L_{11}}{L} \tag{9-5}$$

$$m_{11} = \frac{x_0 m_{31} - \dfrac{L_1}{L}}{c_x}$$

$$m_{12} = \frac{x_0 m_{32} - \dfrac{L_2}{L}}{c_x}$$

$$m_{13} = \frac{x_0 m_{33} - \dfrac{L_3}{L}}{c_x}$$

$$m_{21} = \frac{y_0 m_{31} - \dfrac{L_5}{L}}{c_y}$$

$$m_{22} = \frac{y_0 m_{32} - \dfrac{L_6}{L}}{c_y}$$

$$m_{23} = \frac{y_0 m_{33} - \dfrac{L_7}{L}}{c_y}$$

Orientation angles can be derived from the orientation matrix M, as described in Chapter 4. Note that the derived matrix may need to be orthogonalized, since rounding errors may affect the values of its elements. The camera position is calculated from

$$\begin{bmatrix} X_C \\ Y_C \\ Z_C \end{bmatrix} = - \begin{bmatrix} L_1 & L_2 & L_3 \\ L_5 & L_6 & L_7 \\ L_9 & L_{10} & L_{11} \end{bmatrix}^{-1} \begin{bmatrix} L_4 \\ L_8 \\ 1 \end{bmatrix} \tag{9-6}$$

9.3.4 Singularity Constraint Equations

The DLT, like any uncalibrated camera model, will have stability problems unless a very strong control configuration is used, since it is simultaneously determining the exterior and interior orientation parameters for the camera. Adding to the stability problem is the fact that the DLT uses eleven parameters, while an ideal frame camera has only nine. The two extra parameters model linear scaling of the image in the x and y directions, and can be thought of as the scale factors in the comparator calibration, or as the pixel sizes in the row and column directions for solid-state sensors. These two parameters may be recoverable if the imaging geometry is strong enough, but in general they should not be included. These extra parameters are eliminated by including two nonlinear constraint equations between the parameters (Bopp and Krauss, 1978).

Using the relationships between the rows of the orthogonal matrix M, the *singularity* constraint equations are

$$L^2 = L_9^2 + L_{10}^2 + L_{11}^2$$

$$0 = (L_1^2 + L_2^2 + L_3^2) - (L_5^2 + L_6^2 + L_7^2)$$

$$+ \frac{(L_5 L_9 + L_6 L_{10} + L_7 L_{11})^2 - (L_1 L_9 + L_2 L_{10} + L_3 L_{11})^2}{L^2} \tag{9-7}$$

$$0 = L_1 L_5 + L_2 L_6 + L_3 L_7 - \frac{(L_1 L_9 + L_2 L_{10} + L_3 L_{11})(L_5 L_9 + L_6 L_{10} + L_7 L_{11})}{L^2}$$

Of course, the inclusion of these constraints makes the solution nonlinear. Similar constraints can be written to hold interior orientation elements to known values, using the equations for conversion from DLT parameters to physical parameters.

9.3.5 Homogeneous Coordinates

Another common mathematical model is *homogeneous coordinates*, often used in *projective geometry*. In homogeneous coordinates, only the ratio of the coordinates is of interest; two-dimensional image coordinates x and y are represented as

$$\begin{bmatrix} x \\ y \end{bmatrix} = \begin{bmatrix} x' \\ y' \\ w \end{bmatrix} \tag{9-8}$$

$$x = \frac{x'}{w}, \quad y = \frac{y'}{w}$$

with three-dimensional coordinates similarly represented by four numbers. Homogeneous coordinates allow the consistent representation of points at infinity, by taking the divisor as 0. This facilitates computations involving vanishing points and other projective properties.

Another advantage of homogeneous coordinates is their ability to represent coordinate transformations as matrix multiplications. For instance, a translation by t_x, t_y, t_z is represented as

$$\begin{bmatrix} X' \\ Y' \\ Z' \\ W \end{bmatrix}_{\text{translated}} = \begin{bmatrix} 1 & 0 & 0 & t_x \\ 0 & 1 & 0 & t_y \\ 0 & 0 & 1 & t_z \\ 0 & 0 & 0 & 1 \end{bmatrix} \begin{bmatrix} X' \\ Y' \\ Z' \\ W \end{bmatrix} \tag{9-9}$$

Rotation is expressed in terms of the standard 3×3 orthogonal matrix M, embedded in the 4×4 transformation matrix:

$$\begin{bmatrix} X' \\ Y' \\ Z' \\ W \end{bmatrix}_{\text{rotated}} = \begin{bmatrix} & & & 0 \\ & M & & 0 \\ & 3 \times 3 & & 0 \\ 0 & 0 & 0 & 1 \end{bmatrix} \begin{bmatrix} X' \\ Y' \\ Z' \\ W \end{bmatrix} \tag{9-10}$$

Scaling each axis by s_x, s_y, s_z is written as

$$\begin{bmatrix} X' \\ Y' \\ Z' \\ W \end{bmatrix}_{\text{scaled}} = \begin{bmatrix} s_x & 0 & 0 & 0 \\ 0 & s_y & 0 & 0 \\ 0 & 0 & s_z & 0 \\ 0 & 0 & 0 & 1 \end{bmatrix} \begin{bmatrix} X' \\ Y' \\ Z' \\ W \end{bmatrix} \tag{9-11}$$

A perspective transformation, which reduces the dimensionality from four to three, is expressed as

$$\begin{bmatrix} x' \\ y' \\ w \end{bmatrix} = \begin{bmatrix} 1 & 0 & 0 & 0 \\ 0 & 1 & 0 & 0 \\ 0 & 0 & \frac{1}{f} & 0 \end{bmatrix} \begin{bmatrix} X' \\ Y' \\ Z' \\ W \end{bmatrix} \tag{9-12}$$

To model a camera using homogeneous coordinates, we first write the projective equations (Mundy and Zisserman, 1992):

$$\begin{bmatrix} x' \\ y' \\ w \end{bmatrix} = \underset{3 \times 4}{T} \begin{bmatrix} X' \\ Y' \\ Z' \\ W \end{bmatrix} \tag{9-13}$$

Although the T matrix has 12 elements, only 11 are independent, due to the ratios in the homogeneous coordinates. The abstract parameters in the matrix can be converted to physical parameters in various ways; to relate it to the collinearity formulation, we will express the transformation matrix as the product of two matrices, one containing interior orientation parameters and the other containing exterior orientation parameters.

If the position of the perspective center (in Euclidean coordinates) is (X_C, Y_C, Z_C), the exterior orientation matrix T_E can be written as:

$$T_E = \begin{bmatrix} & & & -T_X \\ & M & & -T_Y \\ & 3 \times 3 & & -T_Z \\ 0 & 0 & 0 & 1 \end{bmatrix} \tag{9-14}$$

where T_X, T_Y, and T_Z are the coordinates of the perspective center rotated into the camera coordinate system:

$$\begin{bmatrix} T_X \\ T_Y \\ T_Z \end{bmatrix} = M \begin{bmatrix} X_C \\ Y_C \\ Z_C \end{bmatrix} \tag{9-15}$$

If we were to use the standard camera model, with only the principal point coordinates x_0, y_0 and principal distance c as parameters, the interior orientation matrix T_I would be

$$T_I = \begin{bmatrix} 1 & 0 & -\dfrac{x_0}{c} & 0 \\ 0 & 1 & -\dfrac{y_0}{c} & 0 \\ 0 & 0 & -\dfrac{1}{c} & 0 \end{bmatrix} \tag{9-16}$$

However, using the interior orientation matrix in this form results in a total of only nine parameters; the original projective transformation matrix had eleven. The two additional parameters describe nonorthogonality of the image coordinate axes, as shown in Fig. 9-4, and a scale difference between the axes.

If the angle between the axes is θ and the scale of the y-axis relative to the x-axis is k_y, the interior orientation matrix becomes

$$T_I = \begin{bmatrix} 1 & -\cot\theta & -\dfrac{x_0}{c} & 0 \\ 0 & \dfrac{k_y}{\sin\theta} & -\dfrac{y_0}{c} & 0 \\ 0 & 0 & -\dfrac{1}{c} & 0 \end{bmatrix} \tag{9-17}$$

The complete perspective transformation is then

$$T = T_I T_E \tag{9-18}$$

The DLT can be thought of as an alternative derivation of the perspective transform. To emphasize this, we can express the DLT in homogeneous coordinates.

$$\begin{bmatrix} x' \\ y' \\ w \end{bmatrix} = \begin{bmatrix} L_1 & L_2 & L_3 & L_4 \\ L_5 & L_6 & L_7 & L_8 \\ L_9 & L_{10} & L_{11} & 1 \end{bmatrix} \begin{bmatrix} X' \\ Y' \\ Z' \\ W \end{bmatrix} \tag{9-19}$$

9.4 IMAGE COORDINATE CORRECTION FOR CLOSE-RANGE PHOTOGRAMMETRY

Although many of the image coordinate corrections are the same for close-range images as for aerial images, there are important differences in how the lens distortion is determined and applied.

As discussed in Chapter 3, lens distortion is usually divided into two components, *radial* and *decentering*. Radial distortion affects the position of image points radially from the principal point (precisely speaking, the principal point of symmetry) of the camera. This is also known as *symmetric* distortion, since it is a function of radial distance only and is the same for any angle around the principal point. Decentering distortion, caused by the improper alignment of the lens components during manufacturing, has both tangential and radial components. Tangential distortion at a point occurs in a direction perpendicular to the radial line from the principal point to the point; its magnitude varies as a function of the radial distance from the principal point and the orientation of the radial line with respect to a reference direction. The *asymmetric* radial component of the decentering distortion is added to the symmetric radial component.

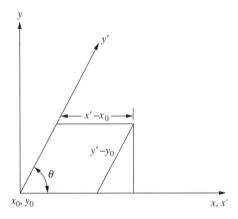

Figure 9-4 Nonorthogonality of coordinate axes.

A complicating factor in close-range photogrammetry is the theoretical variation of lens distortion with object distance. This variation is a function of the distance at which the lens is focused and also of the position of the object of interest relative to the plane of best focus, within the depth of field (Brown, 1971; Fryer and Brown, 1986). There is currently some disagreement over whether lens distortion needs to be corrected for the object distance or for the position of the object within the field.

9.4.1   Radial Lens Distortion Correction

The radial distortion, δr, of a lens focused at infinity can be written as a function of radial distance r:

$$\delta r = K_1 r^3 + K_2 r^5 + K_3 r^7 + \ldots \tag{9-20}$$

in which the K_i are the polynomial coefficients determined by calibration. Only the first one or two coefficients are significant for most lenses. Higher-order coefficients become non-negligible as the lens field of view becomes larger. As discussed in Section 3.2.8, the correction may be applied using the functional form or by a lookup table.

9.4.2   Decentering Distortion Correction

Decentering distortion is more complicated than radial distortion, since it involves both tangential and radial asymmetric components. The radial and tangential components, δr and δt, are (Brown, 1966)

$$\begin{aligned}
\delta r &= 3P(r)\sin(\phi - \phi_0) \\
\delta t &= P(r)\cos(\phi - \phi_0)
\end{aligned} \tag{9-21}$$

where $P(r)$ is the value of the tangential distortion profile at the radial distance of the point, ϕ_0 is the angle between the axis of maximum tangential distortion and the x image axis, and ϕ is the angle between the line from the principal point to the point of interest and the x image axis (Section 3.2.8).

9.4.3   Variation of Lens Distortion with Zoom

Metric cameras are not usually equipped with zoom lenses, but there are many applications where zoom lenses are useful or even necessary. The computer vision community has a particular interest in zoom lenses for use in robotic control (Wiley and Wong, 1995; Willson, 1994; Li and Lavest, 1996).

Zoom lenses work by moving one or more lens elements relative to the rest of the lens, introducing two types of changes in the lens parameters. First, the actual lens design changes with the varying relationships between the lens components. Second, the tolerances within the mechanical components lead to variations in the lens geometry, especially in the lens alignment, which influences the decentering distortion, and in the location of the principal point.

A zoom lens is calibrated by performing a standard camera calibration at a number of different zoom/focus settings. Different researchers have included varying numbers of parameters in zoom lens calibration. Willson (1994) included terms for focal length, lens distortion, principal point location, and the change in front nodal point position as the lens is zoomed. The number of zoom and focus settings required in the calibration is determined by the lens design, since the changes in parameters are determined by the nonlinear motion of the moving elements. Once the calibration parameters are obtained at each setting, they can be used in a lookup table, or polynomials can be fit to model each parameter's behavior as a function of focal length.

9.4.4 Treatment of Refraction in Multiple-Media Photogrammetry

Some close-range applications involve imagery taken through multiple media. A common example is underwater photography, in which the camera is enclosed in a waterproof housing (Rinner, 1969). The rays from the object travel through the water, the glass of the viewing window, and then through air to the camera lens.

The change in *refractive index* (Chapter 3) at the interfaces between different media results in refraction, or bending, of the light rays. Correction for this refraction (Torlegård and Lundaly, 1974) requires modeling the shape of the boundary between the two media, then using Snell's Law to calculate the refraction of the ray at the boundary. Snell's Law is

$$n_i \sin i = n_r \sin r \qquad (9\text{-}22)$$

where n_i and n_r are the refractive indices of the media on the incident and refracted sides of the boundary, respectively, and i and r are the angles between the incident and refracted rays and the surface normal.

Alternatively, if underwater imagery is involved, the principal distance and distortion corrections can be scaled by a factor of 1.3 to correct for the difference in refractive index between water and air. Any remaining refraction distortion can be corrected by self-calibration.

9.5 CALIBRATION PROCEDURES FOR CLOSE-RANGE CAMERAS

A standard laboratory calibration can be performed for metric close-range cameras, similar to those for aerial cameras described in Chapter 3. The most common calibration method for high-accuracy applications is to perform an initial laboratory calibration and then refine the camera parameters as part of a bundle adjustment. The simultaneous determination of interior and exterior orientation parameters was first proposed for aerial cameras (Brown, 1968) and then applied to terrestrial work (Brown, 1971).

As discussed in Chapter 5, self-calibration involves the inclusion of the camera interior orientation parameters in a bundle adjustment solution. Along with the physical camera parameters, such as principal distance and principal point coordinates, added parameters may be used to model residual systematic errors. Although the concept is simple, its application in practice can be problematic, and requires careful evaluation during the data reduction process.

The problem of correlation between interior and exterior orientation parameters must be addressed in any application of self-calibration. As discussed in Chapter 5, one of the consequences of weak geometric configurations is that the effects of small changes in the interior orientation parameters cannot be distinguished from the effects of changes in the exterior orientation parameters. For instance, with narrow-angle lenses, a small change in principal point location produces image displacements nearly identical to those produced by a small tilt or shift of the camera. This phenomenon is known as *projective compensation*. The parameters involved and the amount of compensation vary according to the imaging geometry. This can cause problems when calibration parameters from one configuration are used in another configuration in which parameter correlations and therefore the compensation mechanisms are different.

To combat the effects of the correlation between interior and exterior orientation parameters, strong imaging geometries with multiple images taken in strongly convergent configurations must be used. Many systematic errors are correlated with a particular direction on the image; these errors can be isolated by taking multiple images from each camera station, with the camera rotated on its axis.

Laboratory calibration can also be done using self-calibration software, for a number of different geometric configurations. One common calibration procedure uses a dense array of surveyed control points, which are well-distributed across the camera field of view and in depth, to allow for calibration at different focal distances. If the position of the camera (actually, the position of the front nodal point of the lens) can be precisely established, the known positions of the control points can be used in the calibration instead of requiring multiple images. This procedure is still subject to systematic error in measurement, particularly to error in determining the camera location. Of course, using multiple identifiable points without known locations in conjunction with multiple images is exactly the self-calibration technique, as applied to a standard point determination problem.

Calibration can also be done with purely geometric information instead of known coordinates, although not all interior parameters can be determined. In *plumb-line* calibration (Brown, 1971), an array of straight lines (not necessarily plumb) is imaged several times with the camera rotated on its axis. Since the image of a straight line is a straight line, any curvature of the lines in the image must be due to radial and decentering distortion. By measuring a number of points along each imaged line (an operation well-suited to automated measurement), a least squares solution can be performed to obtain the lens distortion coefficients. This method cannot determine the principal point or principal distance, since no absolute geometric information is involved.

Another purely geometric calibration technique is the use of vanishing point geometry for the recovery of focal length (Wang and Tsai, 1991; Kanatani, 1992). The vanishing points in an image are determined by the tilt of the image and the principal distance of the camera. Given information about the orientation of lines in space (e.g., vertical or horizontal) and the vanishing points, we can determine both tilt and principal distance.

The calibration of electronic cameras, either scanning or solid-state (Chapter 3), is of particular interest to the computer vision and robotics communities. The ultimate calibration goals in these close-range applications are different than those in aerial photogrammetry, in that what would be considered low accuracy for a photogrammetric application (for example, object point accuracies of 1/1000 to 1/2000 of the distance from the camera to the point) are often acceptable. The most important criterion is the speed of the camera calibration process, in both the data acquisition and computation stages. Note that "calibration" in the computer vision community often refers to the determination of both the camera interior parameters and also its position and orientation; in other words, it refers to calibrating the whole camera/workplace/robot setup, instead of just the camera's interior parameters.

A number of calibration procedures have been developed for use in machine vision. The procedure described by Tsai (1987) is representative. The camera images a planar target not parallel to the image plane containing aligned black square targets, at a set of distances from the camera (Fig. 9-5). The corners of the black squares are used as control points in the calibration. Parameters are calculated in steps, proceeding from exterior orientation to interior orientation. This simplifies and optimizes the processing, but may affect the accuracy of the individual parameters. However, the accuracy attained is suitable for most computer vision applications. The calculation steps are

1. The rotation and translation between the object space and camera coordinate systems (i.e., camera position and orientation) are determined using the control point coordinates.

2. The principal distance is determined as part of the projection from 3-D camera coordinates into 2-D image coordinates.

3. Radial lens distortion (one term only) is determined. Two critical assumptions are made to facilitate the computation. First, the principal point is assumed to be located at the center of the image format, which eliminates two of the three geometric elements of interior orientation (x_0 and y_0). Second, only radial distortion is assumed to be present. These two assumptions allow the calculation of radial distortion independent of camera location, orientation, and focal length, since according to these assumptions, image points are only displaced radially from the principal point.

4. The uncertainty scale factor, which models error in the digitization of the image, is determined if a multi-planar calibration is being performed.

When calibrating a film camera, we usually think only in terms of the properties of the lens and its relationship to the imaging plane. For electronic sensors, other

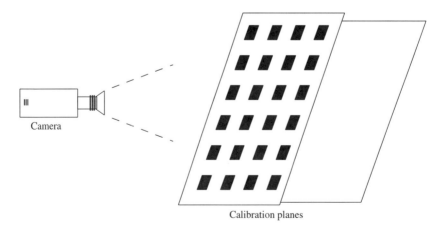

Camera

Calibration planes

Figure 9-5 Tsai camera calibration setup.

factors must also be considered. Scanning sensors, such as older video cameras, may be affected by variation in the beam scanning velocity and geometry. Cameras using solid-state (e.g., CCD) sensors have stable internal geometry, but the image geometry, especially in the line direction, may be affected by timing variations in digitization of the signal from the chip (Luhmann and Wester-Ebbinghaus, 1987; Dähler, 1987). The temperature of the sensor and of the environment may also affect the sensor calibration.

With proper preparation of targets and approximate knowledge of camera positions, the camera setup calibration procedure can be performed nearly automatically. For applications in which the measurements are performed repeatedly, such as industrial applications, the effort required to make the procedure automatic is more than repaid by the increase in productivity.

To automate the measurement procedure, each control target is uniquely coded so that it can be visibly identified. This enables the system to associate the control coordinates with image measurements for each camera. When enough targets are identified, a resection is performed for each camera to establish approximate positions and orientations; then an overall bundle adjustment is performed to establish strong relative geometry. Once the camera positions are established, the images can be relatively oriented and the points to be measured can be determined by epipolar matching among the images. Another bundle adjustment is then performed with the new points, and testing is done to ensure that all matches were done correctly and that the precision of the coordinates is adequate.

9.6   CLOSE-RANGE PHOTOGRAMMETRIC SOLUTIONS: NETWORK DESIGN

Close-range photogrammetric work nearly always uses bundle adjustment (Chapter 5), with the point coordinates and camera stations as unknowns. The solution may include the interior orientation parameters of the cameras as well as added parameters to model the effects of residual systematic error.

For maximum accuracy, the imaging geometry must be carefully designed and the results rigorously examined to ensure that the results meet specifications. The improper use of self-calibration techniques or added parameters can actually decrease the accuracy of the final point coordinates.

Even though the basic mathematics is the same for close-range and aerial photogrammetry, a number of additional factors must be considered in designing an imaging configuration for close-range solution.

The overall precision is a function of how accurately the direction of each ray can be measured and of the geometric strength of the ray intersections, which is determined by both the number of rays determining each point and the angles between them. Given these basic determinants, the design problem is then to specify a configuration of images and points that will attain the specified accuracy, precision, and reliability (Chapter 5). The design should use the minimum number of images possible, to reduce data acquisition and reduction costs. Practical constraints such as limitations on camera placement, point visibility, and lighting must also be satisfied.

Initial network designs are usually selected based on experience with previous similar situations. Multiple convergent images should be used for maximum accuracy, with the camera stations selected so that points are visible in as many images as possible. If the correction of systematic errors is important, multiple exposures at each station, with the cameras rotated on their optical axes, may be used.

The initial layout should be simulated to ensure that the required level of accuracy will be reached. In a simulation, the positions of the target points and the proposed image positions are supplied and the image coordinates of the points are calculated. A preliminary bundle adjustment is then run using the simulated image coordinates. The covariances from the simulated solution give an indication of the results to be expected from the actual project, barring any uncorrected systematic errors or other problems. Most current close-range photogrammetric programs include a simulation capability, many with a graphical interface to allow interactive specification and evaluation.

While distances can be measured with high accuracy in industrial settings, the measurement of accurate three-dimensional coordinates is expensive and difficult. Indeed, the extreme difficulty of generating accurate 3-D coordinates by other means gives close-range photogrammetry its economic advantages. Since photogrammetric observations are of higher accuracy than other methods, we do not want to distort the accuracy of the photogrammetric solution by including lower-accuracy control information. Therefore, we use only the minimum amount of external control information necessary to define a datum or reference coordinate system.

The problem of datum definition is essentially the problem of absolute orientation (Chapter 7). We are performing a block relative orientation, then doing an absolute orientation on the resulting model. To do a complete absolute orientation, the necessary information is the position of the block coordinate origin, the orientation of the coordinate frame, and the scale factor between the photogrammetric model and the world coordinate system. These can be specified in a number of ways; for instance, we could use two points with X, Y, Z coordinates known and another with the Z coordinate specified. Alternatively, we could define one point to be the origin, define the X axis as running from the origin through another point at a known distance, and then specify that another point also has the Z coordinate 0. We may want to supply only part of the information necessary to specify the datum; for instance, we often want to fix the scale, but are not concerned about the orientation or position of the model.

The coordinate system can be arbitrarily specified, but the selection of a coordinate system does have an effect on the calculated precision of the object coordinates (Fraser, 1982). In order to obtain a consistent estimate, we prefer to get the minimum average covariance for the point coordinates. This is done using the technique of *free net adjustment* (Papo and Perelmuter, 1982), which determines the coordinate system configuration that minimizes the trace of the point covariance matrix. (The *trace* is the sum of the diagonal elements of a matrix, as described in Appendix A; for a covariance matrix, the trace is the sum of the point variances.)

9.7 APPLICATIONS OF CLOSE-RANGE PHOTOGRAMMETRY

Close-range photogrammetry has a number of applications with a wide range of object sizes, accuracy requirements, and equipment specifications. This section gives a brief look at a few of the more common applications. Further discussion and descriptions of other applications can be found in Karara (1989) and Atkinson (1996) and in current photogrammetric journals and conferences.

9.7.1 Industrial Inspection

Close-range photogrammetry has several advantages for industrial inspection:

1. *Speed.* Photogrammetric inspection allows rapid data collection, requiring only the time to place measurement targets and take the images, thereby

minimizing the impact on industrial processes. Several current systems use electronic cameras, which allow nearly real-time data collection and coordinate generation. In a hazardous environment, speed can also translate into increased safety of the personnel involved.

2. *Precision.* Photogrammetric techniques have produced precisions in the 1:1,000,000 range (Fraser, 1992), in terms of the ratio between the standard deviation of the point coordinates and the largest dimension of the object measured. This ratio is widely used as a precision measure for industrial inspection, since it is independent of camera configuration considerations.

3. *Reliability.* The redundancy inherent in a multi-ray bundle solution ensures that bad measurements are detected and eliminated, and thus do not invalidate the results of the inspection.

4. *Documentation.* Photogrammetric inspection provides a detailed record of the object at the time of the inspection, allowing later viewing or measurement of other aspects of the object.

Industrial inspection applications may be divided into two main classes (Fraser, 1993): *offline* procedures, in which images (usually on film but possibly digital) are obtained at the workplace and the data reduction performed elsewhere, and *online* applications, in which point coordinates or object measurements are produced in real time for robotic control or dimensional inspection.

Offline photogrammetric procedures are used to inspect a variety of objects, including aircraft tooling fixtures (Schwartz, 1982), antennas (Fraser, 1992), and ships (Siegwarth et al., 1984; Kenefick, 1977). Large-format film cameras are used to meet the highest accuracy requirements, while digital cameras may be used for applications that require only moderate or low accuracy. Points to be dimensioned are typically marked by special targets, to allow accurate and unambiguous identification. Special retroreflective targets may be used in conjunction with strobe flashes placed at the camera positions. The camera aperture is adjusted so that the scene itself is underexposed and only the strobe-illuminated reflective targets show clearly, making the imagery ideal for measurement by automated comparators (Brown, 1987).

In online photogrammetric applications, electronic cameras are mounted in fixed positions around the workplace and their positions and orientations are determined. This is usually done by imaging a calibration object placed within the workplace, or by using external fixed targets. Once the camera positions and orientations are determined, points or objects within the workspace can be positioned by simple intersection. Well-defined targets are used to make the recognition task easier, and may be coded to allow automatic identification by the system. If the orientation of an object such as a robot arm is required, multiple points are observed. Online measurement systems are typically used in assembly line applications, where dimensional inspection information is needed while the part is in the workstation, and for real-time robotic guidance.

For interactive measurement, the operator may use a touch probe (Fig. 9-6). A touch probe contains a measuring point located at a precisely known offset relative to a set of attached targets. To determine a position, the targets are automatically located by the camera system and their positions calculated. The position of the measurement tip is then determined from the targets' positions and the known offset between the targets and probe.

One example of an online measurement system is the Mapvision system, which is used for industrial applications ranging from the inspection of tubing and ship propellers

Figure 9-6 Touch probes for interactive measurements. Photo courtesy of Geodetic Services, Inc.

to the real-time control of an assembly-line robot responsible for sealing seams in car bodies. Specifications for the Mapvision III E system call for accuracy of 1:20,000 and measurement time for a predefined point of 0.05 seconds. Figure 9-7 shows an application for robotic control in an automobile assembly plant, where the Mapvison system determines the position of an automobile body, then guides the robots applying sealant.

9.7.2   Architectural Documentation

Photogrammetry is often used for the documentation, preservation, and restoration of historic buildings, since it allows the rapid capture of detailed and accurate information on building facades and details. Even if no plotting is done, the photographs remain as precise records of the state of the building at that point in time and can be examined or plotted at a later date.

Much of the impetus for the development of architectural photogrammetry in recent years has been UNESCO's (United Nations Educational, Scientific, and Cultural Organization) desire to document and preserve cultural monuments around the world. This is the focus of the International Committee for Architectural Photogrammetry (CIPA) and is also an ongoing topic of the International Society for Photogrammetry and Remote Sensing's (ISPRS) Commission V, which is concerned with close-range photogrammetry.

Architectural photogrammetry may be done for many different purposes. In some cases, only an overall understanding of the building and its design is desired. In other cases, the construction and condition of a building must be precisely documented for preservation or reconstruction. Building decorations, such as attached sculptures or wall paintings, may also be the subject of the work. The photography may be done at different scales to best capture the various levels of the building design. Stereo pairs

Figure 9-7 Mapvision system used for automobile body production. Figure courtesy of Oy Mapvision, Ltd.

covering the whole facade can be used to measure structural integrity and to show the relationships between the architectural details and decorations, while photographs or stereo pairs of isolated details such as stone carvings give precise dimensions and documentation for study or restoration.

Standard products for photogrammetric building documentation include, first, the photographs and stereo pairs themselves. A detailed set of images, with the capability to extract precise measurements when required, is a valuable description of a building. In some cases, the photographs are rectified with respect to a wall or facade to provide a plan view, providing an inexpensive and detailed representation.

Historically, the most common photogrammetric product for building documentation has been the line drawing, since this is the representation most used by architects. Production of a line drawing can be complicated, depending on the amount of detail to be shown and also the definition of that detail. Lines may not be well-defined on older buildings or on buildings with complex shapes.

Most architectural work is now done using computer-aided design (CAD) systems; these representations can be output by most analytical or digital plotters. As an example, the photograph in Fig. 9-8 is one of a set of 16 of University Hall at Purdue University taken with a nonmetric camera. A self-calibrating block adjustment solution was performed to determine point coordinates on the tower, which were used to define the CAD model of the tower rendered in Fig. 9-9.

9.8 OTHER FORMS OF CLOSE-RANGE PHOTOGRAMMETRY

A number of applications of close-range photogrammetric techniques have been developed in recent years that are not necessarily *photographic* as we normally think of

Figure 9-8 Photograph of University Hall at Purdue University, one of a block of 16 taken with a Hasselblad nonmetric camera.

the term. In some of these nontraditional cases, such as x-ray or scanning electron microscope (SEM) imagery, the image is formed by a completely different process than traditional photography. In other cases, such as moiré imaging, one camera is replaced by a projector, thereby making the object geometry more explicit. This section briefly describes some of these techniques and points out the connections between them and standard photogrammetric techniques.

Figure 9-9 CAD model of University Hall at Purdue University derived from close-range photogrammetry.

9.8.1 Hologrammetry

A *hologram* is a three-dimensional image of an object made by recording the interference pattern between two laser beams. A normal photograph records only the intensity of the incident light, discarding the phase information of the light waves. This information loss means that the original light wave cannot be completely reconstructed. A hologram records both the amplitude and phase of the incident light, in the form of an interference pattern that can be used to reconstruct the original light rays from the object.

Holograms are recorded with lasers, since they are a *coherent* light source. Coherent light consists of only one wavelength, with all the waves in phase—in other words, the peaks and troughs of all the light waves are aligned. Coherent light can form an interference pattern. Normal light is incoherent, meaning that it consists of a number of different wavelengths and that the waves at any particular wavelength are out of phase.

Several techniques exist for recording holograms. One of the most common methods is the *Fresnel* hologram. As shown in Fig. 9-10, a laser beam is divided into two parts by a beam splitter. One part illuminates the object and is reflected onto the recording film, while the second part, the *reference beam*, goes directly to the film. The two beams interfere and the resulting interference pattern is recorded on the film. Since the size of the interference fringes is on the order of the wavelength of the light, the recording setup must be very stable to reduce vibrations. Holograms are recorded on high-resolution film, which must be processed carefully to prevent shrinkage or distortion.

A Fresnel hologram is viewed by illuminating it with another laser beam, the *reconstruction beam* (Fig. 9-11). Depending on the relative geometry of the hologram and the reference beam, different types of images may be formed. If the reconstruction beam's geometry relative to the hologram duplicates that of the reference beam, a *virtual image* is formed. This virtual image is geometrically identical to the original object, except for any distortions which may have occurred in the recording process. Looking at the hologram is similar to looking at the original object through a window corresponding to the hologram; as the observer's eye moves, the viewpoint changes.

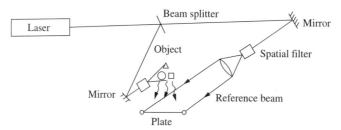

Figure 9-10 Recording a Fresnel hologram.

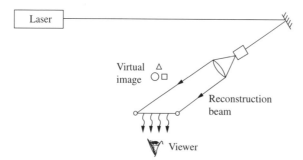

Figure 9-11 Reconstruction of Fresnel virtual image.

Holograms may be used for measuring objects in several different ways (Mikhail, 1974). To measure a small object, an illuminated floating mark can be placed on a three-dimensional coordinate measuring instrument and moved by the operator through the hologram's virtual image.

Another hologrammetric technique uses interference techniques to measure small deformations or strains. Instead of using an unmodified reference beam and another beam reflected by the object, the reference beam may be reflected by the object in its unstressed state and recorded, then the object stressed and the beam recorded again. This will result in an image of the object with interference fringes superimposed. Each fringe corresponds to a deformation of the object by the amount of the wavelength of the laser beam.

9.8.2 Raster Photogrammetry (Structured Light)

Many objects have a relatively featureless surface, making the identification and measurement of points or contours on their surfaces difficult. Human skin, for example, is relatively uniform and featureless; generating accurate shape or contour information for medical studies from standard imaging techniques is therefore difficult.

One way to deal with this is to project a randomly textured pattern onto the surface when the stereo imagery is taken. In this way, the operator or an automated stereo matching algorithm can match on the projected texture.

Another approach is to project a regular pattern onto the object. In stereo photogrammetry, we normally think of rays from an object point going to both images; if we instead replace one of the cameras with a projector and send a ray of light to the object, it will mark a well-defined point and be seen by the remaining camera. The basic geometry remains exactly the same and the data reduction process is analogous.

Image coordinates in the *raster image* (the projected raster grid) are defined by the (row, column) offset from the raster origin. In the camera image, the line or intersection of lines is identified, either by numbers or marks projected along with each line or by a pattern of heavy and light lines. Image coordinates may be generated automatically by scanning the camera image and detecting the lines and intersections. If only lines are projected, instead of a raster grid, the image measurements are defined only in the direction perpendicular to the line.

Control points in the scene are measured in the camera image as in normal photogrammetry. In the raster image, the control points must be measured relative to the raster grid; however, the control points will not necessarily fall on raster grid intersections. In this case, the image coordinates must be estimated by interpolation of the four nearest raster grid points.

Once the control points are measured, a bundle adjustment is performed to determine the camera and point locations. The surface of the object can then be modeled.

A variation on this technique is to scan a laser beam across the object while imaging its intersection with the object surface. The 3-D location of the scanned point is determined using the relative geometry of the scanner and camera. An example of this type of system is the Cyberware Model 15, shown in Fig. 9-12. As an infrared laser is scanned across the object, the position of the laser spot is detected by a video sensor. The object is simultaneously illuminated by white light, acquiring a full color image at the same time as the scanning. Such scanners are used extensively in computer graphics applications, for modeling objects or environments in three dimensions while simultaneously capturing their appearance.

Figure 9-12 Cyberware Model 15 scanner, scanning a bicycle part mounted on a turntable. Photo courtesy of Cyberware, Inc.

9.8.3   Moiré Topography

A *moiré pattern* is an interference pattern resulting from two superimposed gratings. The geometry of the moiré fringes is determined by the spacing and orientation of the grids. If we know the geometry of the grids and can image the moiré pattern formed on a surface, we can determine the topography of the surface.

There are a number of ways to produce moiré patterns and to use them to determine object topography (Karara, 1989). In the *shadow moiré* technique, a reference grid is placed close to the object and illuminated so that it casts a shadow on the object. The shadow of the reference grid is distorted by the object topography. When viewed from another point, the original grid and its shadow form a different moiré pattern. If the light source and viewing point are at the same distance from the reference grid, the moiré fringes represent contours of depth with respect to the reference grid.

In *projection moiré*, a reference grid is projected onto the object. The projection is distorted by the object's topography. This distorted version of the grid is imaged through the reference grid, and a moiré pattern is formed on the image plane.

The basic geometry of moiré pattern imagery is, in some ways, similar to that of a pair of images. While the grids and camera can be arranged such that the fringes represent lines of equal elevation, the depth interval between fringes does not remain constant due to the perspective nature of the imaging. Another difficulty is determining the relative elevation change between adjacent fringes, that is, whether the object surface is closer or farther away.

A common application of moiré imaging is for medical measurements, where its noninvasive nature and ability to determine depths over a large area with no defined features (such as the human body) are valuable. The imagery obtained serves to document the condition of the patient for later analysis or comparison. Moiré techniques are also used for dimensional or shape inspection in industrial applications.

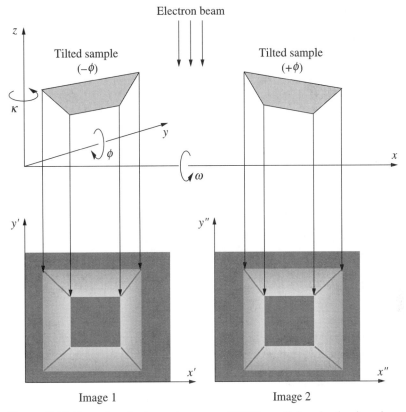

Figure 9-13 Scanning electron microscope (SEM) parallel projection imaging geometry, used to obtain stereo views of an object by tilting the stage. Figure courtesy of Dipl-Ing. Matthias Hemmleb, Technical University of Berlin.

9.8.4 Scanning Electron Microscopy

The scanning electron microscope (SEM), as its name suggests, uses an electronic "lens" to scan a beam of electrons across a small object. The reflected electrons are recorded to form an image. The specimen is mounted on a small stage, which can be tilted to optimize the incident angle of the electron beam on the surface. As a side effect, this tilting capability allows the acquisition of stereo images (Fig. 9-13).

SEM images can be modeled as parallel projections, since the electron beam effectively has a very small field of view (Ghosh, 1975; 1976). The residual perspective effects can be treated as a distortion, along with the beam distortions, which are analogous to lens distortion. SEM images may be exploited using analytical or digital stereoplotters.

9.8.5 X-Ray Photogrammetry

An x-ray image, or *radiograph*, is formed by x-rays generated by the emitter and then passing through the object and onto the film (Fig. 9-14). A radiograph is therefore a shadow, with the image determined by the object's transparency to the x-rays. X-ray measurements are used in medical applications, particularly for skeletal measurements, and for industrial inspection.

The geometry of a radiograph is complicated by the fact that the relationship between the film and the emitter is not fixed and must be recalibrated for each image.

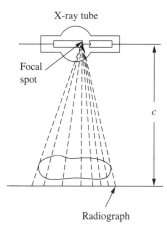

Figure 9-14 X-ray imaging geometry, showing emitter, object, and radiograph.

This requires that the emitter be positioned, a calibration object imaged, and then the object of interest be imaged with the same setup. If the object is small enough, the calibration object can be imaged in the same radiograph. An additional complication is the fact that an x-ray source is not actually a point, but is instead the anode of the x-ray tube. It is modeled as a point, however, allowing a central projection model to be used (Veress et al., 1977).

Calibration consists of determining the location of the principal point of the emitter, defined as the foot of the perpendicular from the film plane to the emitter focal point. This is analogous to the nadir point, or the vertical vanishing point, in aerial photogrammetry (Chapter 2).

Most calibration objects determine the principal point by using the property that the images of vertical lines converge at the vertical vanishing point. (In this case, *vertical* is defined as perpendicular to the film plane.) Calibration objects consist of markers opaque to x-rays, such as steel or lead balls, mounted in a structure transparent to x-rays. The Moffitt calibration fixture (Moffitt, 1972) shown in Fig. 9-15 uses two Lucite sheets parallel to the film plane and separated by four rods. Each rod has radiopaque balls at the top and bottom, thereby defining four vertical lines. When imaged, each vertically-aligned pair of targets defines a line pointing at the principal point. The image coordinates of the targets are measured, and the intersection of the lines is calculated to obtain the principal point.

Radiographs can also be taken in stereo to enable three-dimensional measurement. Stereo can be obtained by using simultaneous images from two emitters, by moving the emitter and taking two exposures, or by moving or rotating the object and film. The principal point is determined for each radiograph using the calibration object. Three-dimensional coordinates can then be determined using parallax calculations.

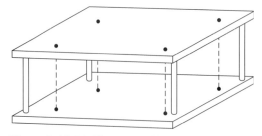

Figure 9-15 Moffitt x-ray calibration fixture.

PROBLEMS

9.1 Given the following physical camera parameters, calculate the corresponding DLT parameters.

$$\text{pixel size} = 0.025 \text{ mm}$$
$$\text{focal length} = 50 \text{ mm}$$
$$x_0 = y_0 = 0.0$$
$$\omega = \phi = \kappa = 0 \, (M = I)$$
$$X_C = 1000.0, \; Y_C = 2000.0, \; Z_C = 200.0$$

9.2 Given the following DLT parameters, calculate the physical camera interior, position, and orientation parameters.

$$L_1 = 10.0$$
$$L_2 = 0.0$$
$$L_3 = 0.0$$
$$L_4 = -10000.0$$
$$L_5 = 0.0$$
$$L_6 = 10.0$$
$$L_7 = 0.0$$
$$L_8 = -20000.0$$
$$L_9 = 0.0$$
$$L_{10} = 0.0$$
$$L_{11} = -0.005$$

9.3 Using homogeneous coordinates in 3-D, express the following operations:
  (a) Translation by 10, 15, and 20 along the X, Y, and Z axes respectively.
  (b) Scaling by 1.0, 1.5, and 1.2 along the X, Y, and Z axes respectively.
  (c) Rotation by 45 degrees around the X-axis.

9.4 What operations are represented by the following homogeneous coordinate transformation matrix?

$$\begin{bmatrix} 1.0 & 0.0 & 0.0 & 10.0 \\ 0.0 & 1.5 & 0.0 & 15.0 \\ 0.0 & 0.0 & 1.2 & 20.0 \\ 0.0 & 0.0 & 0.0 & 1.0 \end{bmatrix}$$

9.5 Express the camera parameters in Problem 9.1 as a homogeneous perspective transform matrix.

9.6 Express the DLT parameters in Problem 9.2 as a homogeneous perspective transform matrix.

9.7 Discuss the approximations in the Tsai camera calibration method and their possible ramifications.

REFERENCES

ABDEL-AZIZ, Y. I., and KARARA, H. M. 1971. Direct linear transform from comparator coordinates into object-space coordinates. In *ASP Symposium on Close-Range Photogrammetry*. Falls Church, VA: pp.1–18. American Society for Photogrammetry.

ATKINSON, K. B., ed. 1996. *Close Range Photogrammetry and Machine Vision*. Caithness, Scotland: Whittles Publishing.

BOPP, H., and KRAUSS, H. 1978. An orientation and calibration method for non-topographic applications.

Photogrammetric Engineering and Remote Sensing 44(9):1191–1196.

BROWN, D. C. 1966. Decentering distortion of lenses. *Photogrammetric Engineering and Remote Sensing* 32(3):444–462.

BROWN, D. C. 1968. *Advanced Methods for the Calibration of Metric Cameras.* Technical Report DA-44-009-AMG-1457. Melbourne, FL: DBA Systems.

BROWN, D. C. 1971. Close-range camera calibration. *Photogrammetric Engineering and Remote Sensing* 37(8): 855–866.

BROWN, D. C. 1972. Calibration of close-range cameras. In *International Archives of Photogrammetry and Remote Sensing.* Vol. XIX Ottawa, Canada: ISPRS.

BROWN, D. C. 1984. A large format, microprocessor controlled film camera optimized for industrial photogrammetry. In *International Archives of Photogrammetry and Remote Sensing.* Vol. XXV No. 5. Rio de Janeiro: ISPRS.

BROWN, D. C. 1987. Autoset, an automated monocomparator optimized for industrial photogrammetry. In *International Conference and Workshop on Analytical Instrumentation.* Phoenix, AZ: ISPRS.

DÄHLER, J. 1987. Problems in digital image acquisition with CCD cameras. In *Proceedings of the ISPRS Intercommission Conference on Fast Processing of Photogrammetric Data.* pp. 48–59. Interlaken, Switzerland: ISPRS.

FRASER, C. S. 1982. Optimization of precision in close-range photogrammetry. *Photogrammetric Engineering and Remote Sensing* 48(4):561–570.

FRASER, C. S. 1992. Photogrammetric measurement to one part in a million. *Photogrammetric Engineering and Remote Sensing* 58(3):305–310.

FRASER, C. S. 1993. A resume of some industrial applications of photogrammetry. *ISPRS Journal of Photogrammetry and Remote Sensing* 48(3):12–23.

FRASER, C. S. 1997. Automation in digital close-range photogrammetry. In *Proceedings of First Trans-Tasman Surveyors Conference.* pp. 1–10. Newcastle: Institution of Surveyors Australia.

FRASER, C. S., and MALLISON, J. A. 1992. Dimensional characterization of a large aircraft structure by photogrammetry. *Photogrammetric Engineering and Remote Sensing* 58(5):539–543.

FRASER, C. S., and SHAO, J. 1998. Scale-space methods for image feature modeling in vision metrology. *Photogrammetric Engineering and Remote Sensing* 64(4):323–328.

FRASER, C. S., and SHORTIS, M. R. 1992. Variation of distortion within the photographic field. *Photogrammetric Engineering and Remote Sensing* 58(6):851–855.

FRYER, J., and BROWN, D. C. 1986. Lens distortion for close-range photogrammetry. *Photogrammetric Engineering and Remote Sensing* 52(1):51–58.

GANAPATHY, S. 1984. Decomposition of transformation matrices for robot vision. *Pattern Recognition Letters* 2:401–412.

GANCI, G., and HANDLEY, H. 1998. Automation in videogrammetry. *ISPRS Comm. V Symposium on Real-Time Imaging and Dynamic Analysis.* Hakodake, Japan, June 2–5, 1998.

GHOSH, S. K. 1975. Photogrammetric calibration of a scanning electron microscope. *Photogrammetria* 31(3): 91–114.

GHOSH, S. K. 1976. Scanning electron micrography and photogrammetry. *Photogrammetric Engineering and Remote Sensing* 42(5):649–657.

HEMMLEB, M., ALBERTZ, J., SCHUBERT, M., GLEICHMANN, A., and KÖHLER, J. 1996. Digital microphotogrammetry with scanning electron microscope. In *International Archives of Photogrammetry and Remote Sensing.* Vol. XXXI No. B5 pp. 225–230. Vienna: International Society for Photogrammetry and Remote Sensing.

HIERHOLZER, E. 1994. Calibration of a video rasterstereographic system. *Photogrammetric Engineering and Remote Sensing* 60(6):745–750.

INDEBETOUW, G., and CZARNEK, R., eds. 1992. *Selected Papers on Optical Moiré and Applications.* Vol. MS64. Bellingham, WA: Society of Photo-Optical Instrumentation Engineers (SPIE).

JOKINEN, O., and HAGGRÉN, H. 1998. Statistical analysis of two 3-D registration and modeling strategies. *ISPRS Journal of Photogrammetry and Remote Sensing* 53(6):320–341.

KANATANI, K. 1992. Statistical analysis of focal-length calibration using vanishing points. *IEEE Journal of Robotics and Automation,* 8:767–775.

KARARA, H. M., ed. 1989. *Non-Topographic Photogrammetry.* Falls Church, VA: American Society for Photogrammetry and Remote Sensing.

KENEFICK, J. F. 1977. Applications of photogrammetry in shipbuilding. *Photogrammetric Engineering and Remote Sensing* 43(9):1169–1175.

KOBAYASHI, K., and MORI, C. 1997. Relations between the coefficients in the projective transformation equations and the orientation elements of a photograph. *Photogrammetric Engineering and Remote Sensing* 63(9):1121–1127.

LEVOY, M. 1999. The digital Michelangelo project. In *Second International Conference on 3D Digital Imaging and Modeling.* Ottawa, Canada: IEEE Computer Society. pp. 2–11.

LI, M. X., and LAVEST, J. M. 1996. Some aspects of zoom lens camera calibration. *IEEE Transactions on Pattern Analysis and Machine Intelligence* 18(11):1110–1114.

LI, R. 1997. Mobile mapping: An emerging technology for spatial data acquisition. *Photogrammetric Engineering and Remote Sensing* 63(9):1085–1092.

LUHMANN, T., and WESTER-EBBINGHAUS, W. 1987. On geometric calibration of digitized video images of CCD arrays. In *Proceedings of the ISPRS Intercommission Conference on Fast Processing of Photogrammetric Data* pp. 35–47. Interlaken, Switzerland: ISPRS.

MARZAN, G. T., and KARARA, H. M. 1975. A computer program for direct linear transform of the collinearity condition and some applications of it. In *Proceedings of the ASP Symposium on Close-Range Photogrammetry.* pp. 420–475. Champaign, IL: American Society of Photogrammetry.

MIKHAIL, E. M. 1974. Hologrammetry: Concepts and applications. *Photogrammetric Engineering* 40(12):1407–1422.

MOFFITT, F. H. 1972. Stereo X-ray photogrammetry applied to orthodontic measurements. In *International Archives of Photogrammetry* Vol. XIX(5). Ottawa, Canada: ISPRS.

MUNDY, J., and ZISSERMAN, A., eds. 1992. *Geometric Invariance in Computer Vision.* Cambridge, MA: MIT Press.

PAPO, H. B., and PERELMUTER, A. 1982. Free net analysis in close-range photogrammetry. *Photogrammetric Engineering and Remote Sensing* 48(4):571–576.

PATORSKI, K., and KUJAWINSKA, M. 1993. *Handbook of the Moiré Fringe Technique.* Amsterdam, the Netherlands: Elsevier Science Ltd.

RINNER, K. 1969. Problems of two-medium photography. *Photogrammetric Engineering* 35(3):275–282.

SCHWARTZ, D. 1982. Close-range photogrammetry for aircraft quality control. In *Proceedings of the American Society of Photogrammetry 48th Annual Meeting.* pp. 353–360. Denver, CO: American Society for Photogrammetry.

SEMPLE, J. G., and KNEEBONE, G. T. 1952. *Algebraic Projective Geometry.* Oxford: Clarendon Press.

SEQUEIRA, V., NG, K., WOLFART, E., GONÇALVES, J. G. M., and HOGG, D. 1999. Automated reconstruction of 3D models from real environments. *ISPRS Journal of Photogrammetry and Remote Sensing* 54(1):1–22.

SIEGWARTH, J. D., LABRECQUE, J. F., and CARROLL, C. L. 1984. Volume uncertainty of a large tank calibrated by photogrammetry. *Photogrammetric Engineering and Remote Sensing* 50(8):1127–1134.

STRAT, T. M. 1984. Recovering the camera parameters from a transformation matrix. In *Proceedings of the DARPA Image Understanding Workshop.* pp. 264–271 New Orleans, LA: Science Applications International Corp.

SUN, L. 1992. A microcomputer-based electron microscope digital image 3D processing system. In *International Archives of Photogrammetry and Remote Sensing* Vol. XXIX No. B5 pp. 475–481. Washington, D.C.: International Society for Photogrammetry and Remote Sensing.

TORLEGÅRD, A. K., and LUNDALY, T. L. 1974. Underwater analytical systems. *Photogrammetric Engineering* 40(3):287–293.

TSAI, R. Y. 1987. A versatile camera calibration technique for high-accuracy 3D machine vision metrology using off-the-shelf TV cameras and lenses. *IEEE Journal of Robotics and Automation* 3(4):323–344.

VERESS, S. A., LIPPERT, F. G., and TAKAMOTO, T. 1977. An analytical approach to X-ray photogrammetry. *Photogrammetric Engineering and Remote Sensing* 43(12):1503–1510.

WANG, L. L., and TSAI, W. H. 1991. Camera calibration by vanishing lines for 3-D computer vision. *IEEE Transactions on Pattern Analysis and Machine Intelligence* 13(4):370–376.

WILEY, A. G., and WONG, K. W. 1995. Geometric calibration of zoom lenses for computer vision metrology. *Photogrammetric Engineering and Remote Sensing* 61(1):69–74.

WILLSON, R. G. 1994. Modeling and calibration of automated zoom lenses. Technical Report CMU-RI-TR-94-03. Pittsburgh, PA: Robotics Institute, Carnegie Mellon University.

Chapter **10**

Analysis of Multispectral and Hyperspectral Image Data

David Landgrebe
School of Electrical and Computer Engineering,
Purdue University, West Lafayette, IN

10.1 INTRODUCTION, BACKGROUND, AND HISTORY

The beginning of the space age is generally defined by the launching of Sputnik in October 1957. Along with many other effects, this event caused people to consider how to use space-based technology for practical purposes. An application of immediate interest was the observation of the Earth from space and, in particular, the observation of the Earth's atmospheric conditions. The first Earth observational satellite, TIROS 1, was launched on April 1, 1960. The newly formed U.S. National Aeronautics and Space Administration then began to launch additional observational satellites at the rate of about two per year.

It was not long before the use of satellites to monitor the land, in addition to the atmosphere, was considered (Landgrebe, 1997). Early in the 1960s, engineers began to develop techniques for monitoring the Earth's land resources, both renewable ones such as those of agriculture and forestry, and nonrenewable ones such as those of geologic interest. The initial questions to be addressed were what kind of land resource information would be most useful and what kind of sensor system and analysis approach should be used to obtain it. Imagery similar to photographs, analyzed using standard photo interpretation technique, was considered. However, it was quickly recognized that photographic cameras were not a good sensor choice because the data (imagery) would have to be transmitted to the Earth electronically. Photo interpretation was also problematical due to the large quantities of such imagery and the cost of manual methods. The advantage that the space vantage point provides is the ability to monitor large areas in a nearly instantaneous fashion, thus potentially leading to very large amounts of imagery.

The resources of the land are characterized by a much more detailed nature, requiring much higher spatial resolution than that needed for atmospheric monitoring. Since data volume increases as the square of spatial resolution, very large quantities of data would indeed need to be analyzed economically in order to obtain useful information on land resources from imagery. This immediately suggested some type of automated or computer-assisted analysis, rather than entirely manual methods. However, straightforward computer image analysis methods did not seem a good choice. To be able to identify a field of corn, for example, image analysis methods would require spatial resolution high enough to identify individual plants as corn by their leaf structure, etc., as we humans do it.

A more fundamental look at the problem suggested that perhaps the required identification could be carried out based on the spectral reflectance properties of the different materials. The idea was to identify materials by measuring the energy emanating from individual pixels as a function of wavelength, then using the emerging computer-implemented pattern recognition technology to label each pixel as to its contents. The advantage of this approach is that it makes very efficient use of whatever spatial resolution is used. The contents of each pixel can be identified individually, rather than having to use a collection of pixels to label an object as the smallest item in the scene to be labeled.

This approach, labeling a given pixel based on the distribution of its electromagnetic energy as a function of wavelength, became known as the *multispectral approach*. Research to test and develop this approach during the 1960s used aircraft data with 12 to 18 spectral bands, located in the visible, reflective infrared, and thermal infrared portions of the electromagnetic spectrum.

The type of sensor needed for this approach is one that measures the energy intensity in each of a number of wavelengths at once. The first satellite to carry such a sensor was Landsat 1, launched in July 1972. Landsat 1 carried an instrument known as MSS (for multispectral scanner). The MSS collected image data with pixels 80 meters across in four spectral bands per pixel with a signal-to-noise ratio sufficient to support a 6-bit data system (i.e., $2^6 = 64$ shades of gray in each band).

For some years, this low number of bands per pixel was the primary factor limiting the performance of multispectral sensors. Sampling the spectrum at only four locations was useful for rather simple problems, but did not provide the detail needed to discriminate between more subtle classes. A second-generation system, known as *Thematic Mapper*, was first launched in 1984 and served as the primary multispectal space sensor through the 1980s and 1990s. It sampled the spectrum at seven locations with pixels 30 meters across and had an 8-bit (256 shades of gray) data system; a significant improvement, but still quite limiting.

Advances in sensor technology have now made much more complete sampling of the spectrum possible. Sensor systems now sample the spectrum in hundreds of locations with signal-to-noise ratios justifying 10 to 12-bit data systems. This much more complex data moves the key problem of information extraction from sensor limitations to the analysis process. Thus, it is very important to have a thorough understanding of such complex data and adequately powerful and sophisticated algorithms for analyzing it.

10.2   STATISTICAL PATTERN RECOGNITION AND CLASSIFICATION

The analysis of a multispectral image data set is based on spectral processing techniques rather than image processing techniques. That is, the advantage of data that has been collected in a number of spectral bands is that it is possible to label pixels individually. This is because it is possible to discriminate between different materials in a scene based on the difference in the spectral response of the various materials, rather than spatial characteristics, such as the physical shapes of scene elements. Being able to label pixels individually makes very efficient use of the spatial resolution of the sensor. Spatial resolution is one of the most expensive parameters of data collection.

Analysis of multispectral data begins with viewing the measurements made on each pixel as an N-dimensional vector, where N is the number of spectral bands sensed. The most common product of such analysis is a thematic map, which is a digital image of the scene in which each pixel in the scene is labeled with one of the desired classes of the data set.

The analysis process may be viewed as a mapping from points in an N-dimensional vector space to a one-dimensional set of M labels specified by the analyst. The way this can best be accomplished is via a method called *pattern recognition* (Swain and Davis, 1978; Richards and Jia, 1999; Fukunaga, 1990). To see how this mapping might be done, consider a case where measurements are made in only two spectral bands for each pixel, and so $N = 2$. Assume there are three classes in the data set, $M = 3$. In two-dimensional space, the data might look as shown in Fig. 10-1. Here, the abscissa value of a pixel indicates the magnitude of the response in band 1 and the ordinate indicates the response of the pixel in band 2. The classification problem then comes down to dividing this two-dimensional space into three nonoverlapping regions so that any outcome is uniquely associated with one of the three classes.

10.2.1 Discriminant Functions

A very powerful way to accomplish this partitioning is based upon so-called *discriminant functions*. Assume that one can determine M functions of the N-dimensional vector X, $\{g_1(X), g_2(X), \ldots, g_M(X)\}$, such that the value of $g_i(X)$ is larger than all the others when X is from class i. Then a decision rule to accomplish the desired mapping from data to classes would be

Let ω_i denote the ith class.

Then, decide measured (vector) value X is in class i ($X \in \omega_i$) if and only if
$g_i(X) \geq g_j(X)$ for all $j = 1, 2, \ldots, M$.

Note that this rule is particularly easy to implement in computer code. Given the discriminant functions, to classify a new pixel, the algorithm has only to calculate the magnitude of the M discriminant functions and select the largest one.

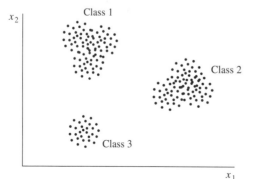

Figure 10-1 Two-dimensional data plotted for three hypothetical classes.

But how does one determine what the discriminant functions should be? In answering this, one must note that the spectral response for any given material is best described as a distribution. Any given material in a scene will

- exist in a number of different states
- be observed through many different columns of atmosphere
- be illuminated and observed from a number of different angles
- have a number of different topographic positions
- be near to any number of other kinds of materials

All of these and a number of other variables have an effect on the spectral response. Some of these effects tend to be diagnostic of the material, but some are not. Thus the spectral response expected from a given material will not be a single point in the N-dimensional feature space referred to above. It will be a distribution in this space.

If this distribution can be accurately described in terms of a probability density function, then the density function itself can be used as a discriminant function for that class. The value of a probability density function at any point is a quantitative measure of how likely that value is to occur. Thus, if one has the density function for all M classes, assignment of a data vector to a particular class is a matter of evaluating the value of all M densities at a given point to see which one indicates the greatest likelihood. This type of analysis scheme is referred to as *maximum likelihood classification*.

This leads to another question, namely how to determine the class probability density functions for a practical circumstance. One usually begins with the data set on the one hand and one or more classes to be identified on the other. The most effective way for the user to specify what is desired of the analysis process is to label some examples of each class in the data set to be analyzed. These samples are called *training samples,* or *design samples*, and this phase of the analysis process is the most critical part of analyzing a data set.

For optimal performance, the list of classes must be

- *Of informational value.* Obviously, the user must specify at some point the classes about which information is desired.
- *Separable.* There is no reason to specify classes that cannot be discriminated based upon the spectral features at hand.
- *Exhaustive.* There must be a logical class to which every pixel in the scene can be assigned.

Note that the user imposes the first of these three requirements, while the latter two are conditions determined by the data. The user's desires and the properties of the data must be brought together in the analysis process. This part of the task is not a trivial one, as these three conditions must be met simultaneously. It is often not trivial to be able to specify, even by direct example in the data set, what the limits of each desired class are intended to be, to specify the limits so that the classes turn out to be adequately separable, and at the same time to anticipate all of the spectral classes that are present in the data set at hand. Often, a good deal of practice and skill are needed, and there are many possible algorithms (Landgrebe, 1999) and tools that can aid in the procedure. Thus we will need to spend a good deal of time studying this process.

10.2.2 Training a Classifier

A wide variety of methods have been devised to train classifiers. Some do not appear to involve estimating the class density function, although they nearly always amount to that. One of the most common schemes is to model each class density function in terms of one or more Gaussian probability density functions. The Gaussian density function in one dimension is given by

$$p(x|\omega_i) = \frac{1}{\sqrt{2\pi}\sigma_i} \exp\left[\frac{-(x-\mu_i)^2}{2\sigma_i^2}\right] \tag{10-1}$$

where x is the variable of the density and corresponds to the measured value of radiance of the pixel, ω_i designates class i, μ_i indicates the mean or average value of x for class i, and σ_i^2 indicates the variance of x around μ_i. In the usual case of more than one dimension ($N > 1$), though ω_i remains a scalar, x and μ_i become vectors of dimension N, and σ_i^2 becomes an N-dimensional matrix.

From probability theory (Cooper and McGillem, 1999), it is shown that

$$p(\omega_i|x) = \frac{p(x|\omega_i)p(\omega_i)}{p(x)} \tag{10-2}$$

The notation used here is as follows:

$p(\omega_i|x)$ is the probability of class ω_i, given the measured value x

$p(x|\omega_i)$ is the probability of measured value x, given class ω_i

$p(\omega_i)$ is the probability that class ω_i occurs in the data

$p(x)$ is the probability that measured value x occurs in the data

The relationship described by Eq. 10-2, known as *Bayes' Theorem*, is very useful in classification. It is the quantity on the left, the likelihood of class ω_i, given the measurement x, that we seek to maximize by picking the correct class i, thus providing the minimum error rate. The term $p(x|\omega_i)$, the probability of x, given class ω_i, is available as a result of the training process. The quantity $p(\omega_i)$ is known as the *prior probability*, or the probability of the given class before the data is evaluated. The same relationship is valid if each of the quantities are density functions rather than discrete probabilities.

Since the denominator quantity, $p(x)$, is the same for all classes, the quantity in the numerator should be used as the discriminant function. Bayes' Theorem then ensures that the error rate will be a minimum. Such a minimum error rate classifier is known as a *Bayes classifier*. For calculation purposes, the discriminant function can be simplified a bit further, as follows.

In the N-dimensional case, the Gaussian probability density function is written in vector notation as

$$p(x|\omega_i) = (2\pi)^{-N/2}|\Sigma_i|^{-1/2}\exp\left[-\frac{1}{2}(x-\mu_i)^T\Sigma_i^{-1}(x-\mu_i)\right] \tag{10-3}$$

The Bayes rule classifier, also called the *minimum a posteriori error rule* would now be: Decide x is in class ω_i if and only if

$$p(x|\omega_i)p(\omega_i) \geq p(x|\omega_j)p(\omega_j) \text{ for all } j = 1, 2, \ldots, M \tag{10-4}$$

Now if $p(x|\omega_i)p(\omega_i) \geq p(x|\omega_j)p(\omega_j)$ for all $j = 1, 2, \ldots, M$, then it is also true that

$$\ln p(x|\omega_i)p(\omega_i) \geq \ln p(x|\omega_j)p(\omega_j) \text{ for all } j = 1, 2, \ldots, M \tag{10-5}$$

Thus, we may take the following as an equivalent discriminant function that requires substantially less computation time. (Note in this expression that we have dropped the factor involving 2π since it would be common to all class discriminant functions and thus does not contribute to the discrimination.)

$$g_i(x) = \ln p(\omega_i) - \frac{1}{2}\ln |\Sigma_i| - \frac{1}{2}(x - \mu_i)^T\Sigma_i^{-1}(x - \mu_i)$$

or

$$2g_i(x) = \ln\left(\frac{p^2(\omega_i)}{|\Sigma_i|}\right) - (x - \mu_i)^T\Sigma_i^{-1}(x - \mu_i) \tag{10-6}$$

For the same reason, we may drop the leading factor of two. Note also that the first term on the right in Eq. 10-6 only must be computed once per class, and only the last term must be computed for each measurement to be classified.

Thus, in this case, the determination of the needed discriminant functions has been reduced to estimating the mean vector μ_i and the covariance matrix Σ_i for each class.

10.2.3   Classes and Subclasses

The derivation of the discrimant functions described above assumes that the probability distribution for a given class is a Gaussian distribution, meaning that the class has a single mode with the familiar e^{-x^2} shape. This is often not a very good model for a class. It may not allow sufficient flexibility to the user. For example, in an agricultural area, the planting season of corn in the spring might have been interrupted by a rainy period, so that some of the corn was planted early and some late. Thus, later in the season the corn canopies might be found in two different states. The user, on the other hand, might not wish to distinguish between these two states, and simply desire one class called corn.

This circumstance can easily be handled by training two spectral classes for the user class "corn," allowing the classification to take place on this basis, then combining the two together after the classification. The two spectral classes would be referred to as subclasses of the class "corn." By using a varying number of subclasses for a desired user class, arbitrarily complex non-Gaussian distributions can be modeled successfully.

10.2.4   Training Samples and Estimation Precision

A disadvantage of this class/subclass scheme is that more spectral classes must be trained, and this requires more training samples. It is characteristic of the remote sensing situation that the number of training samples available is always less than might be desirable. The process of labeling samples within a data set to be analyzed so they can be used for training can often be a complex task. It is usually situation-specific and so each case must be dealt with in a unique fashion.

Thus, on the one hand, one would like to spend as little time on the training sample labeling process as possible. On the other hand, the precision with which the classes are defined, and therefore the accuracy of the resulting classification, is strongly dependent upon the number of samples available by which to estimate the parameters of the class description. So far, we have only described one classifier algorithm, the maximum likelihood Gaussian one. There are many others, some of which we will examine later. This strong dependence on the number of training samples is true regardless of the algorithm used. One cannot expect a good quality output from the classifier

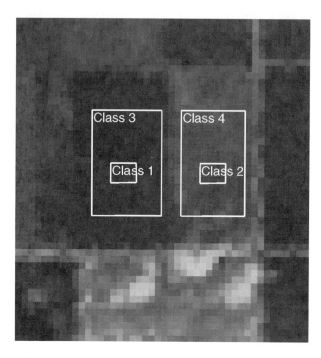

Figure 10-2 A small portion of a Landsat TM scene with test areas for two agricultural fields.

unless there is a good quality input to it, in terms of an accurate and precise quantitative description of the classes desired.

A simple example may help to make the situation clearer. Figure 10-2 shows a small portion of a Landsat TM frame over an agricultural area with test areas for two agricultural fields marked. If the areas marked Class 1 and Class 2, each containing 12 pixels, were to be used as training areas for those classes, and it was proposed to use Thematic Mapper bands 1 and 3 for the classification, the estimated mean values and covariance matrices would be

$$\mu_1 = \begin{bmatrix} 83.4 \\ 25.7 \end{bmatrix} \quad \mu_2 = \begin{bmatrix} 85.2 \\ 29.3 \end{bmatrix}$$

$$\Sigma_1 = \begin{bmatrix} 1.17 & 0.06 \\ 0.06 & 0.24 \end{bmatrix} \quad \Sigma_2 = \begin{bmatrix} 1.66 & 0.73 \\ 0.73 & 2.97 \end{bmatrix}$$

$$\rho_{13_1} = 0.11 \quad \rho_{13_2} = 0.33$$

Figure 10-3 shows a plot of the 12 data points, showing band 1 (abscissa) vs. band 3 (ordinate). In addition, the *area of concentration* is shown for each class for a Gaussian density with the same mean vector and covariance matrix as the training data.

The elements of μ_i are the mean or average values of the 12 training pixels in band 1 (upper element of μ_i) and band 3 (lower element of μ_i). The elements of Σ_i, on the major diagonal of the matrix, are the variances, σ_j^2 of the data in the individual bands. Thus, for band 1 of class 1, $\sigma_1^2 = 1.17$ quantifies how much the data varies about its mean value of 83.4, while $\sigma_3^2 = 0.24$ quantifies the variation in band 3 about its mean value of 25.7. The off-diagonal element of Σ_i is the covariance value between band 1 and band 3, σ_{13}. It relates to how the data in bands 1 and 3 vary with respect to one another.

A normalized form of this covariance element is computed as

$$\rho_{13} = \frac{\sigma_{13}}{\sqrt{\sigma_1^2 \sigma_3^2}}$$

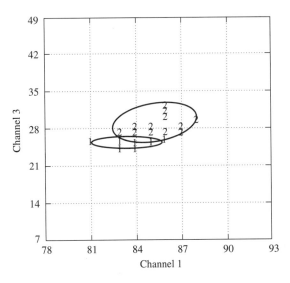

Figure 10-3 A scatter plot of the training samples for Classes 1 and 2. The ovals show the area of concentration for a Gaussian density with the same mean vector and covariance matrix.

This quantity is the correlation coefficient of the data between the two bands. A correlation coefficient can vary over a range $-1 \leq \rho_{jk} \leq +1$, indicating the degree to which data in the two bands tend to vary in the same direction (positive correlation) or in opposite directions (negative correlation). This turns out to be very useful information for a classifier. Two bands that have positive correlation tend to be distributed along a line slanted at 45 degrees upward to the right, assuming equal variances. The higher the correlation, the more closely the data approaches the line. The distribution is similar for negative correlation, but along a line upward to the left. Correlation values near zero imply distributions that tend to be circularly distributed, not having any favored direction.

Correlation between bands may in some instances seem undesirable. For example, it can suggest redundancy. However, in the case of classification, another more positive interpretation is appropriate. The mean value of a class defines where the class distribution is located in the feature space. The covariance matrix provides information about the shape of the distribution. Here, it is seen that the higher the correlation between features, the more concentrated the distribution is about a 45-degree line. Zero correlation means it is circularly distributed, thus occupying a greater area (volume) in the feature space. In this case, there may be a greater likelihood that the distribution will overlap with a neighboring one.

With this in mind, let us return to the consideration of the effect of the size of the training set. Consider defining the same user classes, but with a larger number of training samples. If, instead of the areas marked Class 1 and Class 2 in Fig. 10-2, the areas marked Class 3 and Class 4, each containing 200 points, are used as training samples, the corresponding results would be (Fig. 10-4)

$$\mu_3 = \begin{bmatrix} 83.5 \\ 26.2 \end{bmatrix} \quad \mu_4 = \begin{bmatrix} 86.9 \\ 31.2 \end{bmatrix}$$

$$\Sigma_3 = \begin{bmatrix} 1.86 & 0.13 \\ 0.13 & 1.00 \end{bmatrix} \quad \Sigma_4 = \begin{bmatrix} 3.31 & 2.42 \\ 2.42 & 4.43 \end{bmatrix}$$

$$\rho_{13_3} = 0.09 \quad \rho_{13_4} = 0.63$$

Comparing μ_1 with μ_3 and μ_2 with μ_4 shows that there is relatively little change, indicating that the locations of the two distributions were reasonably well determined by

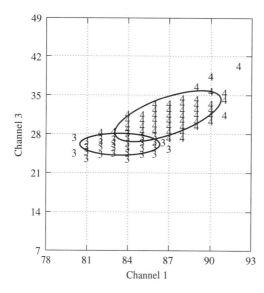

Figure 10-4 A scatter plot of the training samples for Classes 3 and 4 of the areas in Figure 10-2. The ovals show the area of concentration for a Gaussian density with the same mean vector and covariance matrix as the data.

the smaller training sets. However, the change in the corresponding covariance matrices is greater. The implication of this change is that the shape of the distributions was not as well determined by the smaller training sets.

This is a well-known result. The mean vector is known as a *first-order statistic*, because it involves only one variable. The covariance matrix is called a *second-order statistic*, because it involves the relationship between two variables; the correlation shows how two variables relate to one another. Higher-order statistics involve the relationships between more variables. To perfectly describe an arbitrary class density function would require knowing the value of statistics of all orders. However, this would require an infinite number of samples by which to estimate the statistics of all orders.

It is also the case that as the order of the statistic grows, the estimation process using a finite number of samples becomes more problematic. This is why one would in general expect that the mean vector would be reasonably well estimated with a smaller number of samples than would the covariance matrix, the circumstance we observed in the above example.

10.2.5 Training Samples and the Number of Bands

Another factor in the training process that relates to the size of the training set has to do with the number of bands or spectral features that are to be used. Some years ago, Hughes (1968) derived a very general but very useful theoretical result that bears on the problem at hand. The result is shown in Fig. 10-5. This graph, which was derived relative to pattern recognition problems in general rather than specifically to remote sensing data, shows the relationship between expected classification accuracy (averaged over the ensemble of all classifiers) and measurement complexity. Measurement complexity relates to the number of discrete locations in the feature space. In the case of multispectral data, measurement complexity thus relates to the number of spectral bands and the number of discrete values in each spectral band. If the data has N spectral bands and there are R discrete values in each band, then the total number of discrete locations in the feature space is R^N. The parameter m in Fig. 10-5 is the number of training samples used. The graph assumes a two-class problem with the two classes equally likely.

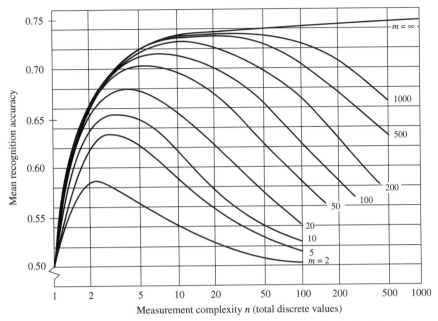

Figure 10-5 Mean recognition accuracy vs. measurement complexity for the finite training case.

If $m \longrightarrow \infty$, implying perfectly precise definition of the class statistics, the expected accuracy increases continuously with increasing measurement complexity, rapidly at first, but then more slowly. However, in a practical circumstance with a finite number of training samples, the curve has a maximum, indicating that there is an optimum measurement complexity. Using too many spectral bands would result in less than optimal performance. Notice also that the peak of accuracy moves upward and to the right as m is increased. This indicates that greater accuracy can be expected in general by increasing the number of spectral bands, but to achieve it, greater numbers of training samples would be required.

This graph is a theoretical result, and since it is quite general, analysis results may not conform to it exactly in any specific case. However, practical analyses do tend to show this general behavior. That is, if one were to analyze a given data set with a fixed (finite) number of training samples, varying the number of spectral bands, the accuracy of the result would tend to increase with increasing number of spectral bands to a point, then decrease. This reinforces the importance of choosing the right number of spectral features to use in any given analysis, using as many training samples as possible. It also makes clear that the number of bands to be used, the number of training samples, and the expected accuracy are variables that are all interrelated. We shall return to this point shortly.

10.2.6 Other Classification Algorithms

In Section 10.2.2, the Gaussian probability density function was described as a suitable discriminant function for many circumstances. The advantage of this model for multivariate data is that it utilizes both first-order and second-order variations of the data in feature space. It thus results in decision boundaries in feature space that are, in general, segments of second-order surfaces. One of the forms in which the Gaussian density was expressed as a discriminate function is

$$g_i(x) = \ln p(\omega_i) - \frac{1}{2}\ln|\Sigma_i| - \frac{1}{2}(x - \mu_i)^T\Sigma_i^{-1}(x - \mu_i) \qquad (10\text{-}7)$$

However, as has just been seen, the size of the training set to be used in estimating the parameters of a class is limited. The imprecise parameter estimates that result from this limitation in the training set size causes the error rate to be higher than necessary. We also noted that the estimation precision tends to affect higher-order statistics most. In this case, that means there may be more problems with the estimated covariance matrix, Σ_i, than with the mean vector, μ_i. As a result, a simpler classifier may outperform this more complex one.

The classifier known as the *Fisher linear discriminant* is one simplification of the Gaussian distribution classifier. In this case, it is assumed that all classes have a common class covariance and thus the training samples from all classes may be used to estimate the one common covariance matrix. Even though the decision boundary in feature space is now restricted to be segments of linear surfaces instead of second-order surfaces, higher accuracy can result from the greater estimation precision. The discriminate function in this case becomes

$$g_i(x) = \ln p(\omega_i) - \frac{1}{2}(x - \mu_i)^T\Sigma^{-1}(x - \mu_i) \qquad (10\text{-}8)$$

Note the differences between Eq. 10-8 and Eq. 10-7. The Σ now has no subscript, since there is only one, and the second term on the right of Eq. 10.7 can be dropped, as it would be the same for all classes.

A further simplification results in the *Minimum distance to means* classifier. In this case, the covariance term is dropped completely, again resulting in linear decision boundaries, but without influence in their orientation based upon the shape of the class distributions in feature space. In this case, the discriminate function becomes

$$g_i(x) = \ln p(\omega_i) - \frac{1}{2}(x - \mu_i)^T(x - \mu_i) \qquad (10\text{-}9)$$

An additional advantage in this case is that eliminating the covariance matrix from the discriminant functions means that significantly less computation is needed in the classification process. There are even simpler classifiers than that defined by Eq. 10-9, which require even less computation. However, one quickly reaches a point of diminishing returns, depending on the complexity of the task, with performance of the classifier falling below acceptable standards.

There are many classifiers that are more complex, or utilize different approaches. Some are based upon different schemes for modeling the class distributions, and some on how the locations of the decision boundaries are located and how the training process is implemented. Popular classifiers of the former type are *K-nearest neighbor* schemes, those based upon the theory of *fuzzy sets*, and *Parzen density estimators*. In the latter category, *neural network* implementations are an example.

Neural networks and Parzen density estimators are said to be *non-parametric* or *distribution-free* schemes, in that the class probability density functions have no initially prescribed forms and thus may be seen as perfectly general. However, in fact, all classifiers necessarily have parameters, and the more general they are, the more parameters they must have to describe that generality quantitatively. The Hughes phenomenon of Fig. 10-5 makes clear that there is a price to be paid for that generality. The greater the complexity of the class description in terms of the number of parameters used, the greater the size of the training set required to adequately quantify the required amount of detail.

10.2.7 Clustering: Unsupervised Classification

As was pointed out earlier, achieving a quantitative definition of the classes to be used is perhaps the most critical step in the classification process. A maximally effective set of classes must be (1) of informational value, (2) separable, and (3) exhaustive. The use of training samples to define the classes is referred to as *supervised classification,* because, via the training samples, the human analyst is supervising the definition of the classifier and thereby ensuring that the resulting classes will be of informational value. However, simply listing training samples for the desired set of classes does not necessarily lead to a set of classes that are separable or exhaustive.

Another frequently useful type of classification is called *clustering* or *unsupervised classification.* Though it is seldom useful for directly achieving a final classification, it can be very helpful in establishing a set of training data that meets the three required conditions stated above. The basic concept is to group the pixels into an appropriate number of clusters in feature space, in a manner that satisfies an appropriate optimality criterion. We will illustrate the concept with a (perhaps oversimplified) example using two-dimensional real data. Figure 10-6 shows in image form a small area from an agricultural region. The dotted box designates an area that appears to be dominated by two informational classes. Without any information about what is contained in the two agricultural fields, suppose we wish to divide the pixels in the dotted box into two groups based upon their spectral similarity in all the bands contained in the data set.

The data from the dotted box area of Fig. 10-6 in two spectral bands is plotted in Fig. 10-7. The desired classification of the pixels into two groups can be accomplished by using a clustering algorithm. There are many different clustering algorithms in the literature. In general, three basic capabilities are needed in a working clustering algorithm:

1. a measure of similarity or distance between points
2. a measure of similarity or distance between point sets
3. a clustering criterion

In essence, one uses the measure of distance between points to decide which points are close to one another in *N*-dimensional space. Points that are close together are

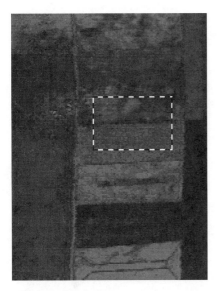

Figure 10-6 An image of a small area in an agricultural region with a test area including parts of two fields marked.

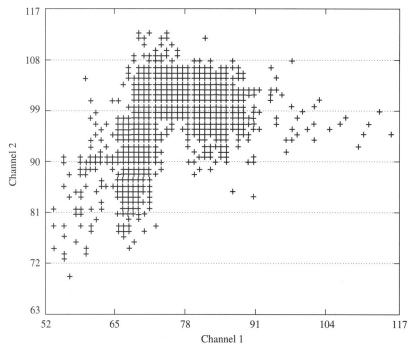

Figure 10-7 A scatter plot of the data in two bands from the region marked in Figure 10-6.

grouped to form point sets, or clusters. The measure of distance between point sets is used to determine whether the clusters are sufficiently distinct from neighboring clusters. The clustering criterion is then used to determine if each cluster is sufficiently compact and also adequately separable from the other clusters. The process is usually an iterative one in which points are compared with tentative cluster centers and then the clusters are tested to see if they meet the cluster criterion.

To begin the process, one must have a means for establishing initial cluster centers as a starting point. A typical sequence of steps for a clustering algorithm is as follows.

1. Estimate or specify the number of clusters needed, and select (often arbitrarily) an initial cluster center for each.

2. Assign each point to the nearest cluster center, using the measure of distance between points.

3. Compute the mean value of the points assigned to each cluster, and compare this with the previous center. If the mean value is not at the cluster center, assign the mean value as the cluster center and return to Step 2 using the new cluster centers.

4. Determine if the clusters have the characteristics required

   a. Are they sufficiently compact? This might be done, for example, by using the squared sum of the distances between the points and their cluster center. If they are not sufficiently compact, subdivide any that are too distributed, choose centers for the new clusters, and go back to Step 2.

   b. Are they sufficiently separated from other clusters, using the measure of distance between point clusters? If not, combine those that are too close together, assume a new cluster center and go back to Step 2. If so, clustering is complete.

Notice that such a procedure has the ability to end up with either more or fewer cluster centers than it started with.

It should be noted that clustering is usually carried out on higher-dimensional data, for which it is not possible to have points plotted to visualize what clusters might be appropriate or even where would be logical points for the initial cluster centers. This is a major reason an algorithm is needed, rather than simply doing the clustering manually. We use this two-dimensional situation here only to show the concept.

Continuing the example of Figs. 10-6 and 10-7, we must select the number of clusters desired and the initial location of the cluster centers. We will seek two clusters in this case. One way to select initial cluster centers is to space them equally along the major diagonal of the data to be clustered. The two points indicated by the round dots in Fig. 10-8 are the result of doing so in this case.

Then, using the algorithm steps listed with simple Euclidean distance as the measure of difference between points and the sum of the squared distances as the clustering criterion, the cluster centers move in the directions shown in Fig. 10-8 in nine iterations to the final points indicated by the diamonds in Fig. 10-8. The final cluster labels of the points are shown in Fig. 10-9.

A clustering algorithm can be very useful in helping to establish training sets for desired classes, because it gives an initial indication of what data might be separable from what. When presented with a seven-band data set for an area containing perhaps ten or so classes, for example, one might have no idea how to set up classes and subclasses initially. By first clustering the data, one could begin to tell which areas of the data set are similar to one another spectrally and which are not. Note, however, that clustering or unsupervised classification cannot normally be expected to indicate informational classes which are separable for you. For example, using the cluster results on the example would give the classification result shown in Fig. 10-10. Clearly, as a

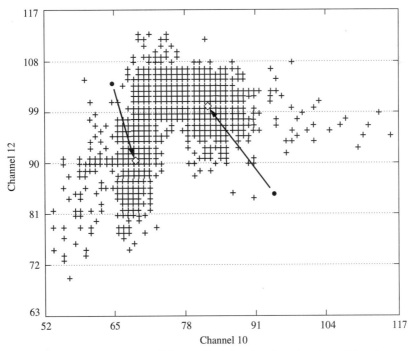

Figure 10-8 The scatter plot of Figure 10-7 showing the migration of the cluster centers during the iterations of the cluster process.

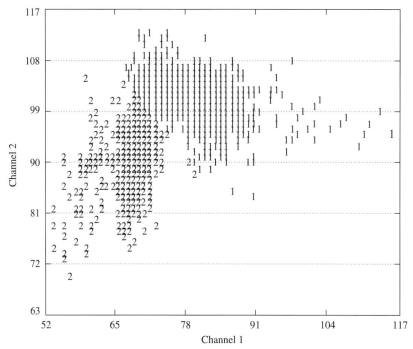

Figure 10-9 The scatter plot of Fig. 10-7 showing the final cluster assignments of the pixels resulting from the clustering.

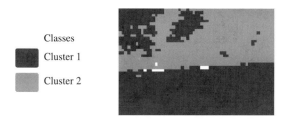

Classes

Cluster 1

Cluster 2

Figure 10-10 A thematic presentation showing how the pixels of Figure 10-6 were assigned to the two classes as a result of the clustering.

final classification, the error rate is higher than desirable and, by itself, poorer than what is possible, even using only two bands.

10.3 FEATURE REDUCTION AND SPECTRAL TRANSFORMATIONS

As was shown in Section 10.2.5, the use of too many features could lead to suboptimal results because of parameter estimation imprecision due to the limited size of training sets. It is not a matter of having sensors with smaller numbers of bands, because a different set of spectral features is optimal for every different problem. Thus, generally, it is desirable to have data gathered in a large number of spectral bands, but to then be able to determine the best subset of spectral features to use for classification in a problem-specific way. A number of feature reduction methods have appeared in the literature. A representative sampling of them is given in this section.

10.3.1 Subset Selection

One of the most straightforward ways to reduce the dimensionality of the data set is to simply select a subset of the available bands. A convenient means for finding an optimal spectral band subset is to use a separability measure to find the best (most separable) M bands of the N available, $M < N$. Given the training samples for a desired set of classes, for example, one might seek to find the best 4 bands out of the 7 available. This method rests on the ability to project classification accuracy from the training data and each possible subset without actually doing the classification for the entire data set.

There are a number of methods for projecting the relative classification accuracy from training statistics. One common and very useful method is to use a statistical distance measure, which indicates the relative separation between two distributions in N-dimensional space. A particularly effective measure for this purpose is the *Bhattacharyya distance*. In terms of the mean vectors μ_i and the covariance matrices Σ_i of two classes, the Bhattacharyya distance is given by

$$B = \frac{1}{8}(\mu_1 - \mu_2)^T \left[\frac{\Sigma_1 + \Sigma_2}{2}\right]^{-1} (\mu_1 - \mu_2) + \frac{1}{2}\ln\left[\frac{\left|\frac{1}{2}(\Sigma_1 + \Sigma_2)\right|}{\sqrt{|\Sigma_1||\Sigma_2|}}\right] \qquad (10\text{-}10)$$

This quantity is known to have a nearly linear relationship with classification accuracy. Notice that the first term on the right measures the separability due to the difference in mean values between the two classes, while the second term on the right measures the separability due to the covariance matrices.

One can use such a metric to examine the various subsets of size M that exist in the N dimensions. To see how this might work, consider the following example. Assume that there are 9 classes defined for a 12-band data set and it is desired to find the best 4 of the 12 bands to use for classification. Then the Bhattacharyya distance between each pair of the 9 classes would be computed for each of the possible subsets of 4 bands out of the 12. Use of the binomial coefficient

$$\binom{12}{4} = \frac{12!}{4!(12-4)!} = 495 \qquad (10\text{-}11)$$

shows that there are 495 4-tuples in 12-dimensional data that must be examined. There are 28 possible class pairs in 9 classes. Thus, one would compute the Bhattacharyya distance between each of the 28 class pairs for each of the 495 4-tuples. Then one might rank order the results based on the average Bhattacharyya distance for each 4-tuple. The result might be as shown in Table 10-1.

Table 10-1 suggests that, based upon the average of the Bhattacharyya distances between class pairs, bands 1, 6, 9, and 12 would be the best 4 bands of the 12 to use, because the average of the interclass distances is 11.5, which is greater than that for any other 4-tuple row.

Another possibility is to choose the 4-tuple with the largest minimum interclass distance. For example, in Table 10-1, the pair of classes 1 and 5 appears to be more difficult to separate. Looking down the 15 column, one sees that bands 1, 6, 9, and 10 have a Bhattacharyya distance of 1.84, compared with only 1.69 for 1, 6, 9, and 12. Thus, even though the average for 1, 6, 9, and 10 is slightly smaller than that for 1, 6, 9, and 12, the 4-tuple of 1, 6, 9, and 10 may do a better job on the difficult problem of separating classes 1 and 5.

This method of finding an optimum subset is very effective in many cases. It works especially well when the number of bands available is not too large (< 10 or so). However, it begins to present problems as the dimensionality increases. For example, to find

Table 10-1   Bhattarcharyya Interclass Distances

	Class pairs	1 2	1 3	1 4	1 5	1 6	...	1 9	2 3	...
Rank Bands	Average									
1 1 6 9 12	11.50	20.6	3.75	5.16	1.69	12.5	...	30.3	16.5	...
2 1 6 9 11	11.23	20.2	3.37	4.81	1.65	12.2		30.0	16.3	
3 1 6 9 10	10.87	19.8	4.73	4.26	1.84	11.7		29.8	16.4	
4 2 6 9 12	10.65	19.2	3.40	5.44	1.22	12.1		30.3	16.4	
5 1 6 8 9	10.59	19.8	1.61	4.74	1.53	14.2	...	30.5	16.6	...
6 1 7 9 12	10.53	18.5	3.82	4.51	1.34	12.6		23.7	14.4	
7 1 6 8 12	10.43	15.9	3.90	5.34	1.63	14.2		23.7	11.4	
8 1 6 10 12	10.30	15.1	6.12	4.41	1.58	10.5		23.7	11.6	
9 1 8 9 12	10.26	16.8	3.87	4.44	1.22	14.2		21.0	13.8	
10 6 9 10 12	10.23	19.3	5.96	4.82	1.25	10.1	...	27.6	17.1	...
.		.				.			.	
.		.				.			.	
(495 rows)		.				.			.	

the best 10 of 50 bands would require a table as in Table 10-1 with 10,272,278,170 rows. For the best 10 of 100, the number of 10-tuples becomes 1.73×10^{13}.

Another limitation of this approach, especially in high-dimensional cases, stems from its exclusive nature. For example, in a case of 100 bands available, picking only 10 completely excludes any separability characteristics of the other 90. Though the contribution of the rejected bands to separability may be somewhat smaller than the 10 picked, it may be significant in aggregate. Is there some way this problem can be reduced, and still achieve the desired dimensionality reduction? Fortunately, there is. It comes by allowing linear combinations of the original bands to form new features.

10.3.2 Principal Components Transformation

Suppose one has data distributed over a two-dimensional region as shown in Fig. 10-11. Since the region is diagonally oriented, data from the two axes are correlated. Define a new set of axes such that the data are horizontally or vertically oriented, so that in this new space, the data will not be correlated (Fig. 10-12).

A transformation to accomplish this is known as a *principal components transformation*. The name derives from the fact that, after deriving the new coordinate values, in

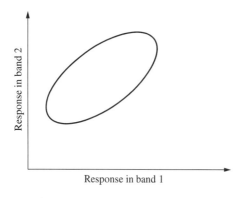

Figure 10-11 A hypothetical data distribution in two-dimensional feature space, for purposes of studying a transformation concept.

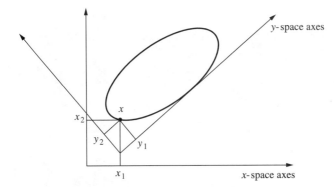

Figure 10-12 Principal component transformation of the data shown in Fig. 10-11. The *x*-axes are the original components and the *y*-axes are the principal components.

the form of $y_1 = a_{11}x_1 + a_{12}x_2$ and $y_2 = a_{21}x_1 + a_{22}x_2$, the one with the largest variance is chosen as the first one, and the second one, orthogonal to the first, is the one with the next highest variance, and so on. Data in the new coordinate system will be uncorrelated from feature to feature. Since data in the new coordinate system are now composed of a combination of bands, the term *feature* will be used to refer to them instead of *band*.

Following is a brief example illustrating the effect of a principal component transformation. Figure 10-13 shows images of the first four Thematic Mapper bands for a representative agricultural area. There is good contrast in all four bands, indicating a moderate dynamic range and thus a normal variance for the data in each band. The actual variances for the four bands are 50.41, 24.01, 75.69, and 408.04, respectively.

The result of a principal component calculation carried out on this data is shown in Table 10-2. The eigenvalues for the new features are the variances of the data in the new coordinates. It is seen that the first principal component now has a variance very much larger than the others, and only the first two have a variance of significant size. The coefficients (the a_{ij}'s) for forming the new features are given in Table 10-3. Images made with the new components are shown in Fig. 10-14. The first two have significant contrast, but the dynamic range of the last two is so small as to show mostly noise.

Principal components analysis has some substantial advantages. It is common and widely known, and conceptually, it is easy to understand. Since it is not case-specific and the individual class definitions are not used, the entire data set can be used to determine the required transformation. This means that the parameter estimation problem is minimized. It performs well in lower-dimensional situations, such as for MSS or Thematic Mapper data.

However, it, too, has some significant shortcomings, especially in higher-dimensional cases. Because the transformation does not use the individual class information, it cannot be optimal with respect to class discrimination. It really optimizes the *representation* of the whole data set, rather than the *discrimination between classes.*

In high-dimensional cases, a principal components transformation can actually be detrimental. Suppose, for example, in a 100-band data set that one of the classes to be identified has a spectral feature that is completely and clearly diagnostic of the class that occurs in only one of the 100 bands. Such narrow diagnostic features occur, for example, in geologic mapping problems, where a specific mineral may have a molecular absorption feature in a very narrow region. Since the variation in only one of the 100 bands would be a very small portion of the total variation, this narrow-band feature would be manifest in only the high-numbered principal components, and would therefore probably not be present in the components selected for use. In short, principal component analysis will often de-emphasize any narrow-band feature.

Another possible problem with this transformation results from the fact that the data are represented digitally and thus with a finite range. A well-designed remote sensor data system

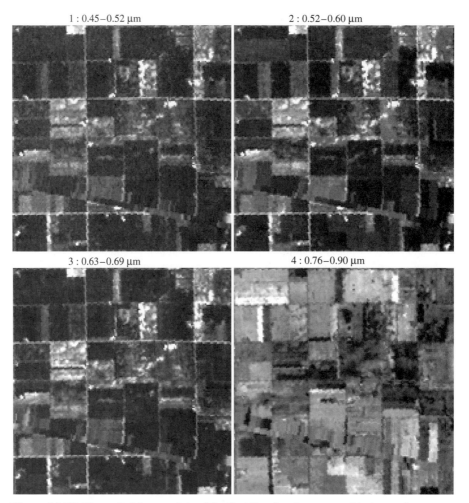

Figure 10-13 Thematic Mapper data of the first four bands expressed in image form.

Table 10-2 Principal Component Transformation of Data from Fig. 10-13

Component	Eigenvalue	Percent	Cumulative percent
1	460.4417	82.3284	82.3284
2	93.7505	16.7629	99.0913
3	3.1477	0.5628	99.6541
4	1.9346	0.3459	100.0000

Table 10-3 Coefficients for Principal Component Transformation of Data from Fig. 10-13

Component	Band 1	Band 2	Band 3	Band 4
1	0.21241	0.12417	0.28065	−0.92774
2	0.54514	0.40520	0.63343	0.37067
3	−0.81072	0.28293	0.51247	0.00727
4	0.02079	−0.86043	0.50732	0.04307

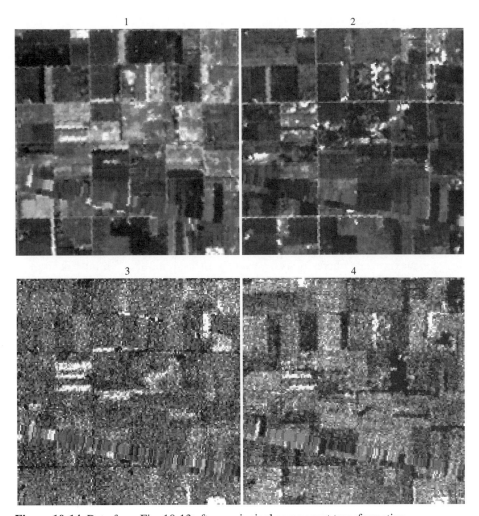

Figure 10-14 Data from Fig. 10-13 after a principal component transformation.

utilizes a significant portion of the available dynamic range over the ensemble of data to be collected. For example, if a sensor system has a 10-bit data system, this means that the digital data would have $2^{10} = 1024$ possible digital values. Then, based upon the expected brightness of the targets to be sensed and the sensitivity of the sensors, the amplifier gains preceding the conversion to digital form would be set so that a good portion of these 1024 values would be used. A principal component transformation significantly alters the data ranges of the various features, making some very much larger and some very much smaller. The features with smaller ranges would be dropped to accomplish the desired dimensionality reduction. On the other hand, it is possible and indeed quite likely that the larger dynamic ranges would substantially exceed the 1024 dynamic range, and some corrective action would need to be taken, reducing its effectiveness. However and worse yet, if this saturation effect went unnoticed, there would be significant distortion in the transformed data.

10.3.3 Discriminant Analysis

The principal components transformation is based upon the global covariance of the entire data set and is thus not explicitly sensitive to inter-class structure. It often works as a feature reduction tool because, in remote sensing data, classes are frequently distributed in the direction of maximum data scatter. Discriminant analysis is a method that is optimized based on class separability. Consider the hypothetical situation depicted in Fig. 10-15. In Fig. 10-15a, the two classes are not separable with one feature axis in the original space, nor with regard to the principal component axis. In Fig. 10-15b, the classes would be separable with one feature if the dotted axis could be found. The problem is to find this axis.

Inspection of Fig.10-15b reveals that the primary axis of this new transformation should be oriented such that the classes have the maximum separation between their means on this axis, while at the same time they should appear as small as possible in their individual spreads. If the former is characterized by σ_B in the figure and the latter by σ_{W1} and σ_{W2}, then it is desired to find the new axis such that

$$\frac{\sigma_B^2}{\sigma_W^2} = \frac{\text{between-class variance}}{\text{average within-classes variance}}$$

is maximized, where σ_W^2 is the average of σ_{W1}^2 and σ_{W2}^2. In matrix form, the within-class scatter matrix Σ_W and the between-class scatter matrix Σ_B may be defined (Fukunaga, 1990) as

$$\Sigma_W = \sum_i p(\omega_i)\Sigma_i$$

$$\Sigma_B = \sum_i p(\omega_i)(\mu_i - \mu_o)(\mu_i - \mu_o)^T$$

$$\mu_o = \sum_i p(\omega_i)\mu_i$$

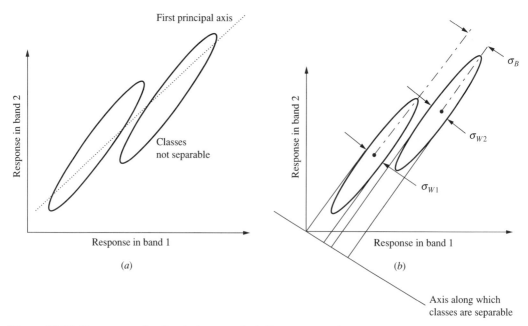

Figure 10-15 The concept for discriminant analysis feature extraction.

where μ_i, Σ_i, and $p(\omega_i)$ are the mean vector, the covariance matrix, and the prior probability of class ω_i, respectively. The criterion for optimization is defined as

$$J_1 = \text{tr}(\Sigma_W^{-1}\Sigma_B) \qquad (10\text{-}12)$$

New feature vectors are selected to maximize this criterion. This method, referred to as *discriminant analysis feature extraction* (DAFE), like principal components analysis, results in new features that are linear combinations of the original bands. However, DAFE maximizes class separability rather than overall data variation. It is usually a very effective feature extraction method, being a rather short computation.

However, this method also has some shortcomings. Since discriminant analysis mainly utilizes class mean differences, the feature vectors defined by discriminant analysis are not reliable if mean vectors are near to one another, a circumstance that is not uncommon, especially in high-dimensional cases. Also, by using the lumped covariance in the criterion, discriminant analysis may lose information contained in class covariance differences. Another problem with the criteria functions using scatter matrices is that the criteria generally do not have a direct relationship to the error probability. Further, the features defined are only reliable up to one less than the number of classes. Nevertheless, it is an effective and practical means for deriving effective features in many circumstances.

10.3.4 Decision Boundary Feature Extraction

Another approach to feature extraction has been devised that does not have the limitations of the preceding ones. It is based directly upon the decision boundary in feature space and the training samples that define it (Lee and Landgrebe, 1993). Discriminately informative features have a component that is normal to the decision boundary at least at one point, while discriminately redundant features are orthogonal to a vector normal to the decision boundary at every point on the boundary. Based upon this distinction, a decision boundary feature matrix (DBFM) may be defined to extract discriminately informative and discriminately redundant features from the decision boundary. The rank of the DBFM is the smallest dimension where the same classification accuracy can be obtained as from the original feature space, and the eigenfunctions of the DBFM corresponding to nonzero eigenvalues are the features necessary to achieve the same accuracy as in the original feature space. The calculation process uses the training samples themselves, rather than statistics from them, to determine the location of the decision boundary, and then from that, the DBFM. The details are contained in the referenced work.

This method of feature extraction does not have any of the limitations imposed by the previous methods. It works well whether the classes have similar means or not, and it is not limited in any way by the number of classes. However, it is a much longer calculation than either of the previous two and it does not perform well when the training sample sets are small.

Thus there are a variety of methods available by which to reduce the dimensionality of the analysis process, each with its strengths and limitations. It is thus important to understand these strengths and weaknesses in terms of the details of the analysis situation being dealt with.

10.3.5 Ad Hoc Transforms

There have been many other *ad hoc* transforms introduced over the years. They are primarily of value for image enhancement purposes when the data are to be analyzed by image interpretation techniques (Richardson and Wiegand, 1977). One of the more common ones occurring in the literature is the normalized difference vegetative index (NDVI). Depending on the particular sensor, it is calculated as

$$\text{NDVI} = \frac{\text{IR} - \text{VIS}}{\text{IR} + \text{VIS}} \qquad (10\text{-}13)$$

where IR is the digital count of an IR band for a pixel, and VIS is the digital count of a visible band. The idea is that the greater the value of the IR return over the visible, the larger the vegetative index. The sum in the denominator is used to normalize the quantity, thus the name. Note that this transformation is sensor-dependent, since the value of NDVI would depend not only upon the net sensor system gain in the channels used, but also upon the specific bandwidth and location as well. It thus provides a relative indication of the vegetation, rather than an absolute measure, and even in this application, it is dependent upon the degree to which the relative size of the IR band response to that in the visible implies vegetation and only vegetation.

10.4 A PROCEDURE FOR ANALYZING MULTISPECTRAL DATA

It should be apparent from the preceeding discussion that the key to a successful analysis of a multispectral data set of any dimensionality is in the definition of the desired classes. As previously noted, for a completely successful analysis, the classes must be exhaustive, separable, and of informational value. Each class must also be defined with adequate precision. Thus the primary characteristic of any procedure for analyzing multispectral data should be to define a set of classes satisfying these conditions as precisely as possible.

The analysis of a multispectral or hyperspectral image data set may follow any of a number of approaches and processing steps. However, a typical generic list of steps might be to proceed in an interactive mode through the following steps, after appropriate offline preprocessing.

1. *Data review.* To gain general familiarity with the data set, its quality, and its general characteristics, a qualitative review is first carried out. This is usually done, at least in part, by viewing the data in multiband B/W and color IR image form. Thus some type of image display is needed.

2. *Class definition.* Quantitative definition of the set of classes meeting the above three conditions is usually accomplished by the analyst labeling, within the data itself, a set of pixels large enough to be adequately representative of each class.

3. *Feature determination.* The specific features to be used in the analysis must be identified or calculated. If the dimensionality is low (about 10 or less), this may simply be a process of selecting an optimal subset of the available spectral bands. For higher-dimensional situations, feature extraction procedures such as those described in Section 10.3 should be used in order to obtain optimal performance without the need for an excessive number of training samples per class.

4. *Analysis.* The specific analysis algorithm is applied to the data set to carry out the desired identification or discrimination.

5. *Evaluation of results.* Both quantitative and qualitative means are used to determine the quality and characteristics of the results obtained. Based on this assessment, it may be desirable to return to one of the previous steps and repeat a portion of the process.

Given that the labeling of the training samples is so key to the success of the analysis, the analyst must decide how many training samples are needed and what information to incorporate to do the labeling.

The optimum number of training samples cannot be precisely determined. If a maximum likelihood classifier involving both first-order and second-order statistics is to be used, there must be at least one sample more than the dimensionality of the data to be classified, in order that the covariance matrix not be singular. However, usually this is not nearly enough. A number of samples many times the number of features is required to achieve good performance. This is why feature selection or feature extraction procedures become so useful in the analysis process.

The question of where the information comes from to label the training samples for each class is also difficult to state precisely. The difficulty arises because of the wide variety of situations possible in the collection of the data, the classes in the ground scene, and the level of prior knowledge the analyst may have about the scene. Though it may at first seem desirable for the process to be automatic, this rarely turns out to be the case. The Earth's surface is a very dynamic and variable place, with the spectral responses of classes of interest in many cases changing on a week-to-week and even a minute-to-minute basis and also on a kilometer-to-kilometer basis. Thus, except for very simple classes, it is not likely that one can successfully apply prior measurements or measurements made at another location to a data set without operator involvement in the process. Further, for most classes of ground cover that are likely to be of interest, the limits of the definition of what is desired for each class can only be established by the analyst during the labeling process.

For example, in an urban area, the specification of a class called "roof" must define the types of roof that are to be included, including what materials are to be included, whether the size or height of buildings will be restricted, whether a parking garage that has cars parked on the "roof" will be included, etc. For an agricultural situation, the definition of a class called "corn" must specfiy how complete the canopy of a corn field must be before the label for the field changes from "bare soil" to "corn" and how much weed content is to be permitted in a pixel. If an area has become "lodged" (laid over) due to storm effects or animal activities, it must still be included in some class definition. All of these kinds of considerations ultimately must be specified in the definition of a class by labeling pixels that are to be considered descriptive and representative of the class the user desires. In all but very special cases, the labeling of training samples will be an interactive process and desirably so. It is also clear that a class cannot be adequately defined by a single spectrum, as is often implied by the term *spectral signature*.

So where does the information come from that the analyst uses to label training samples? Here are some examples.

- *Observations from the ground.* In some circumstances, it may be possible for an observer to be present on the ground at or near the time that the image data is being collected. Observations could be recorded on a perhaps crude map of a small portion of the area, for example.

- *Observations of the ground.* It may not be necessary to actually be on the ground at the time of data collection. Rather, one may be able to identify examples of desired classes from images either of the data itself or of higher-resolution images from another sensor. For example, one might have photographs taken from a low-altitude aircraft (MacDonald and Hall, 1980; Swain and Davis, 1978). The example from Washington, DC, mall data cited below is another illustration where this means was used.

- *Deterministic features from the pixel spectra.* In some cases, it may be possible to label individual pixels based upon features that can be seen in a spectral plot of a pixel. An example where this may be possible is in the use of hyperspectral data for the mapping of minerals based upon specific narrow-band absorption features shown by the minerals. The name *imaging spectroscopy* is sometimes used for this, because of the similarity with that which the chemical spectroscopist does in the laboratory. However, in the case at hand (Hoffbeck and Landgrebe, 1996), instead of attempting to classify the pixels directly, a more robust and accurate performance is obtained by labeling a significant number of pixels by manually examining the spectrum of each, then using the sets of labeled pixels as training for a classifier.

There are certainly other possibilities than these few examples. Clearly, the method by which the needed class definition information is obtained is very case-dependent.

REFERENCES

COOPER, G. R., and McGILLEM, C. D. 1999. *Probability Methods of Signal and System Analysis.* 3rd edition. Oxford: Oxford University Press.

FUKUNAGA, K. 1990. *Introduction to Statistical Pattern Recognition.* San Diego, CA: Academic Press.

HOFFBECK, J. P., and LANDGREBE, D. A. 1996. Classification of remote sensing images having high spectral resolution. *Remote Sensing of Environment* 57(3):119–126.

HUGHES, G. F. 1968. On the mean accuracy of statistical pattern recognizers. *IEEE Transactions on Information Theory* 14(1):55–63.

KETTIG, R. L., and LANDGREBE D. A. 1976. Computer classification of remotely sensed multispectral image data by extraction and classification of homogeneous objects. *IEEE Transactions on Geoscience Electronics* 14(1):19–26.

LANDGREBE, D. A. 1980. The development of a spectral-spatial classifier for earth observational data. *Pattern Recognition* 12(3):165–175.

LANDGREBE D. 1997. The evolution of landsat data analysis. Special Issue commemorating the 25th anniversary of the launch of Landsat 1, July 1972. *Photogrammetric Engineering and Remote Sensing* 63(7):859–867.

LANDGREBE, D. 1999. Information extraction principles and methods for multispectral and hyperspectral image data. Chapter 1 in *Information Processing for Remote Sensing.* Chen, C. H., ed. River Edge, NJ: World Scientific Publishing.

LEE, C., and LANDGREBE, D. A. 1993. Feature extraction based on decision boundaries. *IEEE Transactions on Pattern Analysis and Machine Intelligence* 15(4): 388–400.

MacDONALD, R. B., and HALL, F. G. 1980. Global crop forecasting. *Science* 208:670–679.

RICHARDS, J. A., and JIA, X. 1999. *Remote Sensing Digital Image Analysis: An Introduction.* 3rd edition, Berlin: Springer-Verlag.

RICHARDSON, A. J., and WIEGAND, C. L. 1977. Distinguishing vegetation from soil background information. *Photogrammetric Engineering and Remote Sensing* 43(12):1541–1552.

SWAIN, P. H., and DAVIS, S. M., eds. 1978. *Remote Sensing: The Quantitative Approach.* McGRAW-HILL.

SWAIN and DAVIS, eds. 1978. Large Area Land-Use Inventory. In *Remote Sensing: The Quantitative Approach.* McGRAW-HILL pp. 309–314.

Chapter 11

Active Sensing Systems

Thomas P. Ager
National Imagery and Mapping Agency

11.1   AN OVERVIEW OF IMAGING RADAR

Radar is an *active* imaging technology, in which the sensor provides its own source of illumination. As such, radar imaging is independent of sunlight and can operate both day and night. Radar sensors use the microwave region of the electromagnetic spectrum, which penetrates cloud cover and other atmospheric particles due to its relatively long wavelengths, giving radar a compelling advantage over optical sensors. Unique among all imaging technologies, radar is a cloud-penetrating, day-night remote sensing system.

The term *radar* is an acronym for *ra*dio *d*etection *a*nd *r*anging, which is an apt description of how radar works. In the case of radar imaging, an antenna transmits

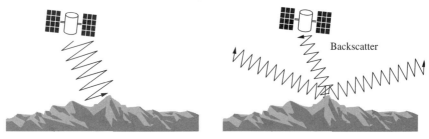

Figure 11-1 Radar backscatter.

microwave energy to the ground as a series of pulses. When a pulse strikes an object, it scatters in all directions. As shown in Fig. 11-1, a small portion of the signal, called the *backscatter*, is returned to the radar and received by the antenna. The strength of the backscatter signal and the transit time from transmission to receipt are recorded by the radar. The backscatter amplitudes define pixel brightness values, and the time delays and the known speed of light are used to derive ranges to ground objects.

A simple method of radar imaging, called *side-looking airborne radar* (SLAR), is shown in Fig. 11-2. In a SLAR system, the antenna transmits a fan-shaped beam in a direction orthogonal to the direction of flight. The backscatter signals arrive sequentially from objects within the radar beam as a function of their range to the antenna. The inset in Fig. 11-2 shows how the backscatter amplitudes vary across the scene from near- to far-range.

As the radar sensor moves along its flight path, a different section of ground is illuminated with each transmitted pulse. Since the motion of the radar is continuous, the illumination of the ground forms a series of overlapping scans. A radar image can be formed by combining scan swaths.

11.2   RADAR WAVELENGTHS AND FREQUENCIES

Imaging radar sensors are built to respond to a narrow spectral range within the microwave region of the spectrum. When radar was developed in secrecy during World War II,

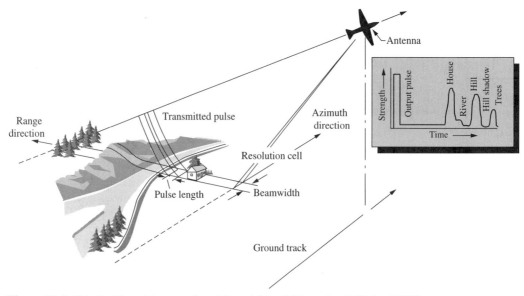

Figure 11-2 Side-looking airborne radar. Adapted from Lillesand and Kiefer (1999).

Table 11-1   Radar Wavelength Bands

Band	Wavelength (cm)	Frequency (GHz)
K	0.83–2.75	36–10.9
X	2.75–5.21	10.9–5.75
C	5.21–7.69	5.75–3.9
S	7.69–19.4	3.9–1.55
L	19.4–76.9	1.55–0.39
P	76.9–133	0.39–0.225

the individual wavelength bands were given obscure designations, such as X-band and C-band, that are still in use. As with all spectral bands, these divisions are entirely arbitrary. The wavelength / frequency ranges assigned to each radar band are given in Table 11-1. The shortest radar wavelengths are designated K-band. These wavlengths would theoretically provide the best radar resolution, but they are partially blocked by water vapor and their cloud-penetration capability is limited. In fact, K-band radars are used by ground-based weather systems to track heavy cloud-cover and storms. For this reason, X-band is typically the shortest wavelength range used for imaging radars.

It is possible for radar sensors to operate in a multi-spectral mode and make use of several different bands. For example, NASA's SIR-C space shuttle mission used a large radar antenna that transmitted X-band, C-band, and L-band radiation. Each of these produced different scattering patterns, which were used by NASA to generate color-composite images.

11.3   RADAR IMAGING FUNDAMENTALS

11.3.1   Radar Imaging Angles

Some of the important angles involved in radar imaging are shown in Fig. 11-3.

Depression angle: The angle between the local horizontal at the antenna and the *radar line of sight*. On a radar image, the depression angle decreases as the range increases, especially in the case of an aircraft sensor.

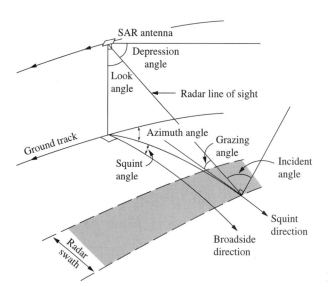

Figure 11-3 Radar angles.

Look angle: The angle between the vertical from the antenna to the ground and the radar line of sight. The look angle, which is sometimes called the *elevation angle*, is the complement of the depression angle.

Incident angle: The angle between the radar line of sight and the local vertical to the geoid at the target location.

Grazing angle: The angle between the radar line of sight and a plane tangent to the geoid at the target location. The grazing angle is the complement of the incident angle.

Azimuth angle: The angle between the direction of flight and the radar line of sight. For many radar sensors the azimuth angle is 90 degrees.

Squint angle: The angle between the cross-track direction, which is orthogonal to the flight direction, and the radar line of sight. The squint angle is the complement of the azimuth angle, and in many cases it is zero.

11.3.2   Factors Affecting Reflectivity

11.3.2.1   Collection Geometry and Topography

The incident angle is the major factor affecting the strength of a radar return. The reflectivity decreases as the incident angle increases because most of the radar energy is reflected away from the sensor. The local incident angle accounts for the influence of topography; it is formed by the normal to the terrain surface and the radar beam, as shown in Fig. 11-4. An increase in slope increases the strength of the return. This effect is greatest when the normal to the slope is coincident to the line of sight of the radar.

11.3.2.2   Surface Roughness

The radar backscatter also increases as surface roughness increases. Surface roughness is largely determined by the ground cover, and minor variations in roughness measured at the centimeter level are important. The effect of variations in surface roughness changes for different radar wavelengths. For example, X-band is much more sensitive to small fluctuations than the very long L-band waves.

A surface is smooth relative to radar energy if its height variation is less than one-eighth of the radar wavelength. These specular surfaces reflect radar waves away from the antenna, and their backscatter signals are very weak. For this reason, smooth features such as water appear dark or black on radar images.

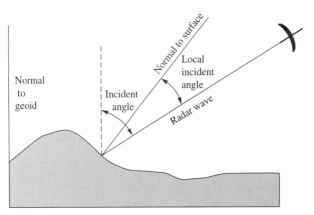

Figure 11-4 Local incident angle and terrain slope.

Diffuse reflectors disperse energy at all angles, and a moderate portion is returned to the sensor. This occurs when the surface roughness ranges from one-eighth to one-half of the radar wavelength. A rough surface, which has height variations greater than one-half the wavelength, has a strong backscatter signal.

11.3.2.3   Dielectric Constant

The dielectric constant is a measure of the reaction of a material to the presence of an electric field. Materials with high dielectric constants are very good reflectors of radar energy. Water has a dielectric constant near 80, while the value for dry land surface ranges from 3 to 8. Since, when its surface is calm, water is a specular reflector with a high dielectric constant, it strongly reflects energy away from the antenna and it appears black on radar images. However, if the water surface is disturbed in a storm, the wave crests provide strong returns to the sensor and they are white.

Radar is useful for making soil moisture maps. The combination of the high dielectric constant of water and the surface roughness provided by soil or vegetation creates bright return areas for moist soil areas that stand out from adjacent dry regions.

11.3.3   Elevation Effects on Image Geometry

11.3.3.1   Foreshortening

Foreshortening refers to a compression of terrain slopes that face the incoming radar energy. This occurs due to the nature of radar as a ranging system. As shown in Fig. 11-5, the points on the facing slope of a hill have similar ranges and thus are near each other on the output radar image. The facing slopes appear to lean in a direction orthogonal to the sensor velocity vector. On the radar image, the steepness of a foreshortened slope has an exaggerated appearance.

When the depression angle is shallow, foreshortening is minimized. This condition is opposite to the geometric effect of relief displacement on an optical image, which is minimized for very steep angles. Since foreshortening is associated with facing slopes where the local incident angle is small, the compressed slopes are normally very bright on the image.

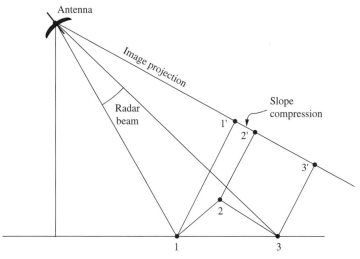

Figure 11-5 Foreshortening.

11.3.3.2   Layover

Layover is an extreme case of slope foreshortening. When the depression and slope angles are both large, the local incident angle can become negative. In such a case, the top of an object is displaced so much that its position on the image extends beyond the pixels associated with the bottom of the object. The range to the top of the object is actually less than the range to the bottom of the object, as shown in Fig. 11-6.

Layover can produce unfamiliar images of buildings in which the base of the building is located on the image to the inside of the building's roof. On steep radar images, such as those taken from the 67-degree depression angle of ERS-2, hills take a very oblique position and the peaks cover the land surface at the hill's base.

11.3.3.3   Radar Shadows

Radar shadows occur when the far side of an object is not illuminated because the radar energy is blocked by the front of the object, as shown in Fig. 11-7. Unlike their counterparts on optical imagery, radar shadows are completely black. Shadows are dark gray on optical images because sunlight is scattered by the atmosphere. Even though the sun's rays may not shine directly on a portion of a hill, some photons strike the shadowed area via reflection from atmospheric particles. The long wavelengths of radar energy are not attenuated in this way, so radar shadows are actually no-return areas. Radar shadows occur on backslopes that are steeper than the depression angle. The length of the shadow increases as the depression angle decreases.

Although radar shadows can obscure features that may be needed for map extraction or some other use, they can also be helpful in identifying image features. Sometimes an object can be identified because its shape is clearly outlined by its shadow. In addition, shadows help in terrain visualization by providing a natural relief-shading effect on the image.

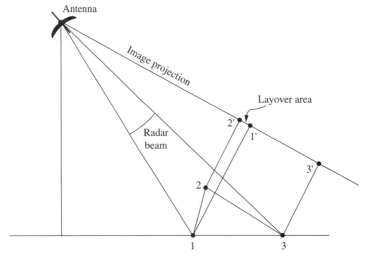

Figure 11-6 Layover.

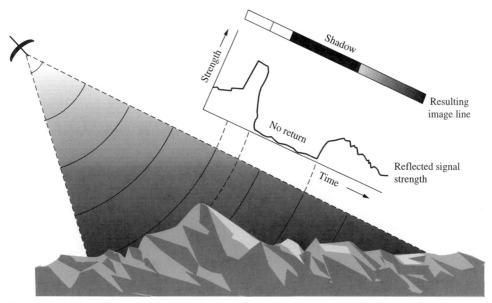

Figure 11-7 Radar shadow.

11.3.4   Polarization

Because synthetic-aperture radar (SAR) consists of coherent waves, the sensor controls the orientation in space, or polarization, of the electric field of the transmitted waves. The transmitted microwaves can vibrate horizontally (H) or vertically (V), but most radar sensors transmit horizontally polarized waves. When a radar signal reflects from a surface object, its polarization is sometimes modified to include both horizontal and vertical components. The portion of the return that has been modified to a new polarization is much weaker than the portion that is returned in the original form.

Some radar sensors have multi-polarization capability. That is, they can transmit and receive both H and V waves. This is useful because different polarizations produce different scattering patterns. SAR data that include horizontally and vertically polarized backscatters have increased information content compared with single-polarity sensors. The possible polarization combinations of transmit and receive are HH, VV, HV, and VH.

11.4   THE RADAR EQUATION

Before reviewing the details of radar imaging, it is useful to consider how the interplay of parameters such as transmitted power, wavelength, antenna size, and platform altitude affect the amount of power received by a radar antenna. The following derivation is described in greater detail in Henderson and Lewis (1998).

A radar antenna can be considered a point source that emits energy in all directions as a series of concentric spheres. The power per unit area of these spheres is equal to the transmitted power (P_T) divided by the surface area of a sphere of radius R:

$$\text{power density} = \frac{P_T}{4\pi R^2}$$

(11-1)

Side-looking imaging radars concentrate power in one direction. This is referred to as the *gain* of the antenna (*G*). In this case, there is a modified power density equation that quantifies the amount of energy arriving at the imaged area:

$$\text{power arriving at target} = \frac{P_T G}{4\pi R^2} \tag{11-2}$$

The amount of energy reflected towards the radar is a function of the target's reflectivity and surface area. The target acts like a transmitter with a reflecting pattern determined by the geometry of the object. This is called the *radar cross section* (σ), and when combined with the power striking the target, we have

$$\text{power reflected toward radar} = \sigma\frac{P_T G}{4\pi R^2} \tag{11-3}$$

This energy travels from the object to the antenna, and it degrades in power as the surface area of the reflected energy widens with distance:

$$\text{power arriving at radar} = \sigma\frac{P_T G}{4\pi R^2}\frac{1}{4\pi R^2} \tag{11-4}$$

The energy actually recorded by the radar is the product of the power arriving at the antenna and the effective aperture of the antenna, *A*, which is given by

$$A = \frac{\lambda^2 G}{4\pi} \tag{11-5}$$

where λ is the wavelength of the radar signal. Combining the terms in Eqs. 11-4 and 11-5 yields the common form of the radar equation, which quantifies the power recorded by the radar, P_R:

$$P_R = P_T\sigma\frac{G^2\lambda^2}{(4\pi)^3 R^4} \tag{11-6}$$

When expressed in terms of the antenna size, the equation for the received power becomes

$$P_R = P_T\sigma\frac{l_h^2 l_v^2}{4\pi\lambda^2 R^4} \tag{11-7}$$

where l_h and l_v refer to the horizontal and vertical dimensions of the antenna, respectively.

Naturally, the power received increases with increased transmitted power, and it is also directly related to the target's radar cross section. It is crucial to note that in both forms of the equation the received energy decreases as a function of the fourth power of the range. This condition has a severe impact on spacecraft radar sensors, which require much more power than airborne radars. For example, the RADARSAT sensor has an average transmit power of 300 watts, antenna dimensions of 15 m by 1.5 m, and emits C-band radiation with a 5.6-cm wavelength. Since it orbits at an altitude near 800 km, ranges as large as 1250 km can be expected. For an object at that range with a radar cross section of 10 m^2, the received power at the antenna is only 1.6×10^{-17} watts! This slight amount of incoming energy must be amplified before it can be processed.

11.5   REAL-APERTURE RADAR

The simple scheme of SLAR imaging described earlier is a *real-aperture* system. This technique uses a long antenna that is mounted to the side of an aircraft. In real-aperture radar, each pulse creates a single image line. Since the pulse repetition frequency

(PRF) is typically about 2000 pulses per second, the aircraft moves forward only a small distance between each image line.

Individual pixels on a SLAR image are described by range and azimuth coordinates and their brightness values are determined by the strength of the return pulse. The range coordinate is simply the distance between the sensor and the object in the direction orthogonal to the flight path, and the azimuth coordinate is equivalent to a scan line number or a time associated with a scan (Fig. 11-8).

11.5.1   Slant-Range Resolution

The straight-line distance between a ground object and the antenna is called the *slant range*. The ability of the radar to discriminate two closely spaced objects along the slant range is a function of the pulse width. The fan-shaped radar beam has a thickness determined by how long a transmitter is turned on. For example, a transmitter may be active for only a portion of a microsecond and emit a pulse that is more than 100 meters long. The pulse propagates at light speed and the pulse length is given by

$$\text{pulse length} = c\tau \tag{11-8}$$

where τ is the pulse duration and c is the speed of light.

Resolution along the slant range is one-half the pulse length:

$$\text{slant range resolution} = \frac{c\tau}{2} \tag{11-9}$$

This becomes evident by considering two points downrange of an antenna along the same broadside direction. Imagine a 100-meter pulse length and two objects that are 60

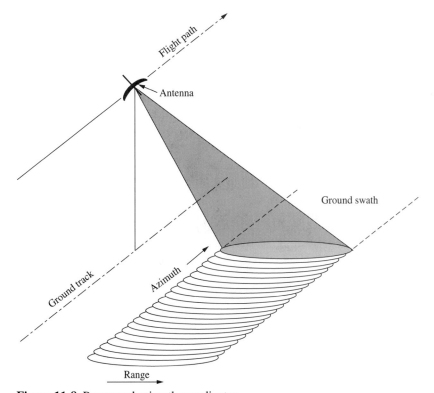

Figure 11-8 Range and azimuth coordinates.

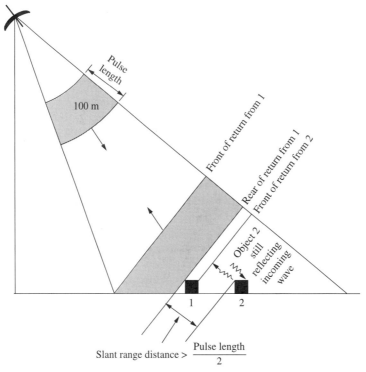

Figure 11-9 Slant-range resolution.

meters apart in the slant-range direction. As shown in Fig. 11-9, when the pulse encounters the first object, a portion of it is reflected to the antenna while the remainder of the wave continues downrange toward the second object. By the time the pulse begins to reflect off the second object, more than half of it has already reflected from the first object. The entire pulse completes the reflection from the first object before the front of the pulse returned from the second object arrives at the first object. In this example, where the two objects are separated by more than one-half the pulse length, the two objects return two distinct echoes and will have two distinct range positions on the radar image. In the case where the slant-range distance between two objects is less than one-half the pulse length, the echoes are mixed together and the objects cannot be separated by the radar.

11.5.2   Ground-Range Resolution

The *ground range* is the distance between a radar scatterer and the nadir position below the antenna. Resolution in the ground range is important because it describes the ability of the radar to distinguish objects along the ground surface. Given a flat ground surface, the relationship between slant range and ground range is simple (Fig. 11-10):

$$\text{ground range resolution} = \frac{c\tau}{2\cos\delta} \tag{11-10}$$

where δ is the depression angle. As the depression angle increases, the ground range resolution degrades. For very steep depression angles, ground-range resolution becomes immense. Radar imaging does not work for steep angles.

Range resolution, whether in ground- or slant-range, is dependent on the pulse duration. Even with a pulse duration as short as one-third of a microsecond, range

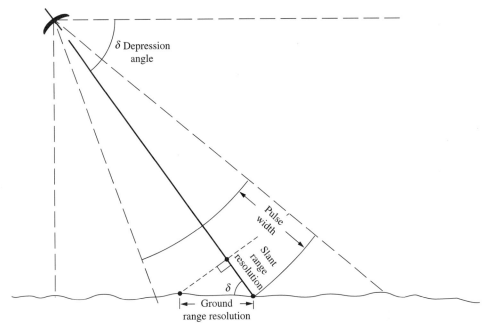

Figure 11-10 Ground-range resolution.

resolution is modest. For this reason, sophisticated methods of improving range reso-lution have been developed. The technique of transmitting frequency modulated pulses will be discussed in the section on synthetic-aperture radar.

11.5.3   Azimuth Resolution

In real-aperture radar, the beam is wide in the range direction and narrow in the azi-muth direction. It is important to keep the azimuth beamwidth very narrow because azimuth resolution is directly dependent on this angle. The beamwidth is small for long antennas and is approximated by (Carrara et al., 1995):

$$\beta \approx \frac{0.89\lambda}{D} \tag{11-11}$$

where β is the angular width of the beam, λ is the radar wavelength, and D is the length of the antenna in the azimuth direction.

Azimuth resolution is simply the product of the slant range and the beamwidth in radians, see Fig. 11-11:

$$\rho_{AZ} = \text{slant range} \times \beta \tag{11-12}$$

Because the beam widens with distance, the resolution in the azimuth direction de-grades from near- to far-range.

The limitations of azimuth resolution for real-aperture radar are obvious and se-vere. Real-aperture systems require long antennas. Resolution degrades across an image from near to far-range, and also as sensor altitude increases because of an in-crease in range. In fact, real-aperture radar imaging is not even possible for spacecraft. Even given a narrow beamwidth of 1.5 milliradians and a low Earth orbit of 300 km, a real-aperture radar would have an azimuth resolution in excess of 600 meters.

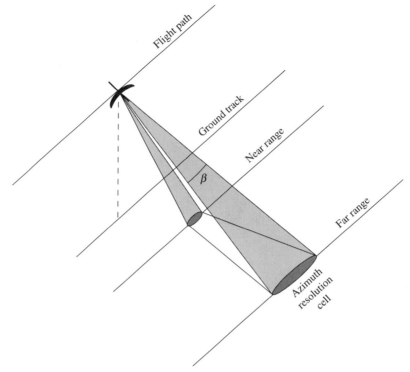

Figure 11-11 Azimuth resolution.

These problems have directed radar imaging technology away from the simple brute force technique of real-aperture processing. It has largely been replaced by a technique called *synthetic-aperture radar* (SAR), which synthesizes a very large aperture from a series of radar pulses transmitted by a small antenna. This is done using coherent radiation and accounting for the Doppler phase shifts of the returned pulses.

11.6 SAR PROCESSING

Radar imaging based on real-aperture processing has limited utility because its azimuth resolution is so constrained. For this reason, radar engineers developed the SAR technique to focus the radar beam in the azimuth dimension. SAR sensors do not attempt to improve azimuth resolution by transmitting narrow beams. Rather, the SAR technique makes use of very wide beams and long exposure times. In contrast to real-aperture radar, the wide beam actually improves resolution, since the SAR technique uses the wider beam to increase the length of the effective aperture by accounting for the Doppler phase shifts of the radar returns.

The forward motion of the sensor during imaging results in a Doppler shift in the frequency of the returned energy (Fig. 11-12). That is, there is an increase in frequency when an object is in the forward part of the beam and a decrease in frequency as it passes into the trailing portion of the beam. A SAR system transmits a coherent reference wave, determines the Doppler *phase history* of the radar return signals, and focuses the shifted returns to obtain detailed azimuth resolution.

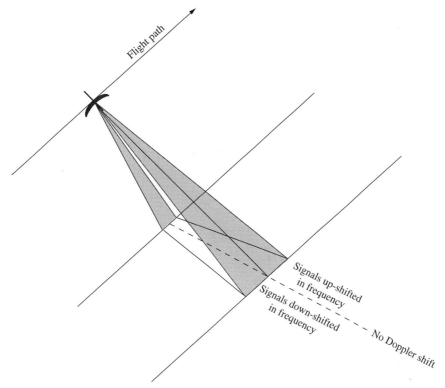

Figure 11-12 Doppler effect.

Recording the changing phase data or phase history during the integration time as the sensor moves forward is equivalent to imaging an object with a very long antenna and a narrow beam. While the actual range and Doppler measurements are made by a small antenna, the long collection period associated with a wide beam creates a synthetic aperture of length L equal to the distance the sensor travels during the illumination of an object (Fig. 11-13).

11.6.1   SAR Azimuth Resolution

In SAR imaging, a wide beam generates a Doppler phase history with a large bandwidth and a long synthetic aperture. The result is a narrow synthetic beamwidth that is a function of the radar wavelength (λ), the length of the synthetic aperture (L), and the Doppler angle (α_D) to the target measured from the velocity vector (Fig. 11-14):

$$\beta_S \approx \frac{\lambda K}{2L \sin \alpha_D} \tag{11-13}$$

where K is a mainlobe broadening factor of 0.89, as described in Carrara et al. (1995).

SAR azimuth resolution is merely the product of the synthetic beamwidth, β_S, and range at the scene center, R_C:

$$\rho_{AZ} = R_C \times \beta_S \tag{11-14}$$

$$\rho_{AZ} \approx \frac{R_C \lambda K}{2L \sin \alpha_D} \tag{11-15}$$

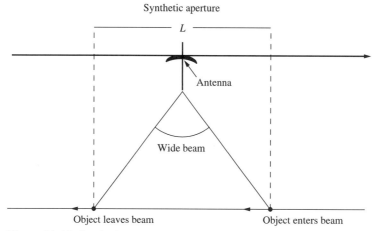

Figure 11-13 Synthetic aperture.

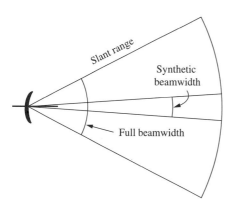

Figure 11-14 Full beam vs. synthetic beam.

11.6.2   Strip-Map SAR

A *strip-map SAR* transmits a fan-shaped beam at a fixed angle and illuminates the ground area with a series of overlapping scans. The illumination direction for a strip-map sensor is typically orthogonal, or broadside, to the flight direction. In this case we have a synthetic aperture, L, that is equal to the product of the physical beamwidth and the range at the scene center (Fig. 11-15):

$$L = R_C \times \beta \tag{11-16}$$

Substituting Eq. 11-16 into Eq. 11-15, we have

$$\rho_{AZ} \approx \frac{R_C \lambda K}{2R_C \beta \sin \alpha_D} \tag{11-17}$$

$$\rho_{AZ} \approx \frac{\lambda K}{2\beta \sin \alpha_D} \tag{11-18}$$

Now substituting for K and the physical beamwidth, we have

$$\rho_{AZ} \approx \frac{0.89\lambda}{2(0.89\lambda/D)\sin \alpha_D} \tag{11-19}$$

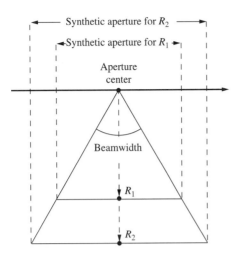

Figure 11-15 Synthetic aperture for strip-map SAR.

and azimuth resolution for strip-map SAR is given by

$$\rho_{AZ} \approx \frac{D}{2 \sin \alpha_D} \qquad (11\text{-}20)$$

A SAR sensor that has a squint capability to point its antenna away from broadside has increased agility and can acquire a greater selection of targets than a simple broadside sensor. However, the resolution of a squint-mode radar degrades as the squint angle $(90 - \alpha_D)$ gets larger.

Finally, with broadside collection, the Doppler angle is 90 degrees, or the squint angle is zero, and azimuth resolution becomes

$$\rho_{AZ} \approx \frac{D}{2} \qquad (11\text{-}21)$$

Strip-map SAR azimuth resolution is dependent only on the size of the antenna, and it therefore has the remarkable characteristic of being independent of the distance between the sensor and the ground. RADARSAT is a strip-map sensor that orbits at 800 km, and its antenna is 15 m by 1.5 m. The theoretical maximum azimuth resolution of 7.5 meters closely matches the 8-meter published value for RADARSAT's highest-resolution mode.

SAR azimuth resolution does not degrade with distance because the length of the synthetic aperture naturally increases for objects at far ranges. These objects are illuminated longer and thus they are imaged by longer synthetic apertures. The resolution of all other remote sensing systems is dependent on range to the imaged area.

11.6.3 Spotlight SAR Imaging

Strip-map SAR azimuth resolution improves as the antenna gets shorter. However, other design considerations prevent the antenna from being extremely small. The antenna gain is proportional to the square of the antenna length, and we have seen that the received power is proportional to the product of the squares of both antenna dimensions. A radar sensor loses power and the signal-to-noise ratio decreases as the antenna gets smaller. For airborne systems, it is possible to increase power to a level that provides good strip-map resolution, but there are practical limits on the resolution available to strip-map sensors aboard radar spacecraft.

A more sophisticated model of SAR azimuth resolution uses the increased bandwidth available to a steered beam. The discussion thus far has assumed a collection with a constant beam angle, usually broadside to the velocity vector. In a technique called *spotlight imaging* (Fig. 11-16), the SAR radiation is steered to illuminate a fixed ground area during the entire imaging time. In this way, the integration time increases, the synthetic aperture is elongated, and the bandwidth of the Doppler phase shifts increases. The beam is steered either through a physical motion of an antenna fixed to a gimbal or via a phased-array method in which the antenna does not move but the direction of radar illumination is manipulated electronically.

As before, the azimuth resolution is approximated as the product of the range at the scene center and the parameters of the synthetic beamwidth:

$$\rho_{AZ} \approx \frac{R_C \lambda K}{2L \sin \alpha_D} \tag{11-22}$$

In this case, the resolution is determined mostly by the synthetic aperture length. While the effect of range is not directly removed as in strip-map SAR, a spotlight radar can be designed to provide excellent azimuth resolution because the ratio λ/L can become so small. A large range value can be offset by an extended synthetic aperture generated through long integration times. If the radar is designed to maintain beam coherence during a long collection period, then azimuth resolution can become extremely small.

Given a spacecraft radar with sufficient power, signal stability, and a steered beam, integration times of several seconds in spotlight mode are conceivable. At orbital speeds of approximately seven kilometers per second, the synthetic aperture would be several kilometers long and an azimuth resolution as small as one meter is possible.

Naturally, there are trade-offs between resolution and ground coverage. Strip-map SAR provides modest resolution and wide coverage, while spotlight SAR illuminates a single ground area and provides maximum resolution but limited coverage.

11.6.4 Improving Range Resolution

The standard slant range resolution developed earlier is one-half the pulse length, and so very short pulse durations are needed to increase resolution. However, there are

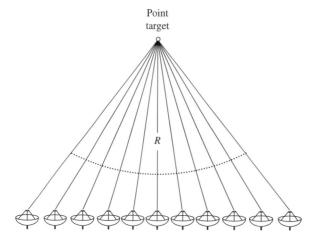

Figure 11-16 Spotlight SAR imaging.

limitations on how short the pulse duration can be due to restrictions on peak power of the radar. To improve range resolution, the radar signals are transmitted with modulated frequencies to provide extensive bandwidth in the range direction.

We have seen that the SAR process improves azimuth resolution by providing signal bandwidth in azimuth. While frequency modulation in range is not directly part of the SAR process, it makes use of similar signal processing theory and provides fine range resolution to match the impressive azimuth resolution available via SAR processing.

Sophisticated radar systems transmit a pulse with a frequency that has been modulated in a linear manner. This is called a chirped pulse because sound waves processed in this way sound like a bird's chirp. The return signal of a chirped pulse is demodulated by removing the frequency of the transmitted pulse. Range processing involves the application of Fourier transforms to these difference signals. This matching of the return and the original pulse, known as *matched filtering*, creates an effective pulse that is extremely short. In fact, a pulse of bandwidth B can be compressed to an effective time duration of $1/B$ (Carrara et al., 1995). Thus range resolution now becomes

$$\rho_R \approx \frac{c}{2B} \tag{11-23}$$

The ERS-2 sensor has a pulse duration of 37 μs, which creates a pulse width of 5.5 km, but the range bandwidth of this radar is 15.6 MHz. Thus, ERS-2 has a raw slant-range resolution of about 10 m. Frequency-modulated pulses can have a bandwidth of hundreds of megahertz. While the current SAR satellites do not have such large range bandwidths, it is possible to build a system with a range resolution as small as 1 meter.

Modern SAR sensors can now produce images of very high quality. Because of the extensive signal bandwidth available in both the azimuth and range directions, it is now possible to build extremely high-resolution SAR sensors. Since distance does not degrade range resolution and its effect on azimuth resolution can be overcome by spotlight sensors, high-resolution SAR imaging is possible even from orbiting spacecraft.

11.6.5   Multi-Look Processing

Despite the high range and azimuth resolution of SAR images, image quality can be corrupted by a random noise effect called *speckle*. Radar speckle is the appearance of random variation in the gray tones throughout a SAR image. Speckle is caused by interference patterns characteristic of the coherent radar energy.

Multi-look is an image processing technique that reduces speckle. During image formation, the radar data are divided into sub-beams. That is, the full aperture is arranged into a series of sub-apertures, each of which can be used to generate a SAR image of lower resolution than the full-aperture image. Each of these sub-apertures provides an independent *look*, and these looks are combined via averaging. In multi-look processing, the synthetic aperture, L, has been shortened to L/N, where N is the number of sub-apertures. There is, however, a matching loss of resolution. For example, the ERS-2 antenna has dimensions of 10 m by 1 m, and its maximum azimuth resolution is 5 m. However, all of the finished image products for this sensor are three-look images, and the output azimuth resolution is closer to 15 m.

Sometimes sub-apertures are arranged so that they overlap, and they are not independent looks. The effective number of looks for such correlated sub-apertures is less than the actual number of looks. RADARSAT processing uses sub-apertures that overlap by 39%. The RADARSAT product specifications document states that a four-look image actually has 3.1 effective looks.

The smoothing process to reduce speckle can also be applied to a full resolution pixel image. In this case, the full aperture is generated by the SAR processor, and afterward the pixels of the full resolution image are averaged. For example, a 4-by-4 group of two-meter pixels might be averaged to become eight-meter pixels.

A multi-look image has lower resolution than a full-aperture image, but it has improved image quality because of reduced speckle. A SAR image generated at maximum resolution may have impressive resolution statistics, but random noise may reduce image utility.

11.6.6   Dynamic Range

The complex radar phase information has a tremendous dynamic range, and radar sensors record an enormous span of intensity levels. This range is so large that it is measured using a logarithmic scale called *decibels* (dB):

$$dB = 10 \log \frac{\text{maximum intensity}}{\text{minimum intensity}}$$

Most SAR sensors have a dynamic range of 40–50 dB. In other words, the maximum intensity level of the raw radar data may be 10,000 to 100,000 times the value of the minimum intensity value. In comparison, optical sensors are based on electronic detector reactions to incoming photons and they produce 8-bit (256 gray levels) or 11-bit (2048 gray levels) data.

The dynamic range of SAR data does not directly provide a benefit in viewing an image because the human visual system can only discriminate about 40 intensity levels. However, these data are available to image processing algorithms, such as automated feature extraction tools, that may benefit from the large dynamic range inherent to the SAR process.

11.7   SAR GEOMETRY MODEL

Now that an overview of imaging radar has been presented, we will turn our attention to the geometric modeling of SAR images, and the relationship between SAR image coordinates and their corresponding ground coordinates. In the SAR image formation processor, the two measurements that are made for a given scatterer are its range and Doppler frequency shift. The range is determined by the time it takes the radar pulse to make its round trip from the sensor to the scatterer and back to the sensor. The Doppler frequency shift, which arises because of relative velocity between the SAR sensor and the scatterer, is given by

$$f_D = \frac{2}{\lambda} |\vec{S}| \cos \alpha_D \tag{11-24}$$

where $\vec{S}$ is the sensor velocity vector, $|\vec{S}|$ is the vector's length, λ is the radar wavelength, and α_D is the Doppler angle.

11.7.1   Spotlight SAR Modeling

On a spotlight SAR image, each scatterer has an associated range to the antenna phase center and a Doppler angle measured relative to the vehicle velocity vector. The image is created from a huge volume of range and Doppler phase data collected over a long synthetic aperture. However, the output of the SAR process is a set of data in which all range and Doppler values are referenced to the midpoint of the synthetic aperture

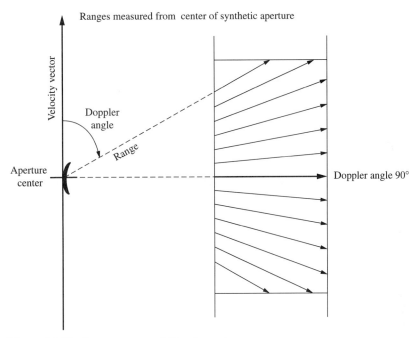

Figure 11-17 Output range and Doppler values.

(Fig. 11-17). A single scatterer generates range and phase data for all positions of the antenna during imaging, but on the final SAR product it is described by a single range and Doppler coordinate pair.

11.7.2 Strip-Map SAR

As with spotlight SAR, strip-map images are built from the range and phase data collected over the synthetic aperture. In this case, the imaging is performed at a fixed angle relative to the velocity vector, and all points on the output image have identical Doppler angles. On strip-map images, each line is associated with a specific sensor position. Thus, a strip-map image can be considered a range and time model (Fig. 11-18). If a time-tagged ephemeris is provided with the image, a unique sensor position can be determined for each line.

11.7.3 The Range Condition

The range measurement constrains the possible location of the scatterer to be on a sphere of radius equal to the observed range and centered at the SAR sensor, as shown in Fig. 11-19. The range condition is modeled via a simple equation for a sphere:

$$R = \sqrt{(X - X_S)^2 + (Y - Y_S)^2 + (Z - Z_S)^2} \qquad (11\text{-}25)$$

in which (X, Y, Z) and (X_S, Y_S, Z_S) are the coordinates of the ground point and sensor, respectively.

For spotlight images, the sensor position is the center of the synthetic aperture and it is provided directly as part of the image support data. Sensor positions for strip-map images are computed from time coordinates and sensor ephemeris data.

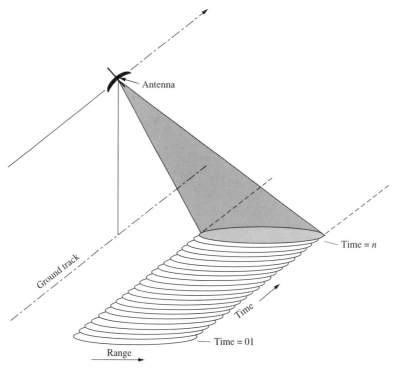

Figure 11-18 Range and time coordinates.

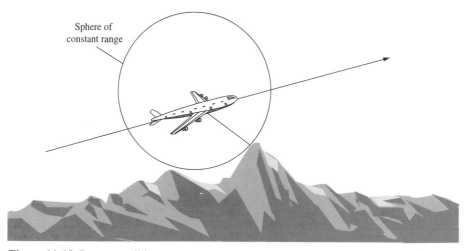

Figure 11-19 Range condition.

11.7.4 The Doppler Cone Condition

The Doppler angle, α_D, is measured between the sensor velocity vector, $\vec{S}$, and the range vector between the sensor and the ground location (Fig. 11-20). In three-dimensional space, this constrains the scatterer to be on a cone with an apex at the sensor position and an axis coincident with the SAR velocity vector. The vector expression for the Doppler cone is given by the dot product of two vectors

$$\vec{S} \cdot \vec{R} = \cos\alpha_D |\vec{S}||\vec{R}| \qquad (11\text{-}26)$$

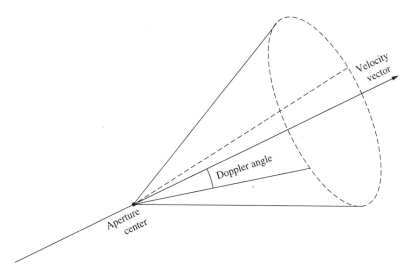

Figure 11-20 Doppler cone condition.

in which $\vec{S}$ is the velocity vector and $\vec{R}$ is the range vector. The quantity $\left| \vec{R} \right|$ represents the length of the range vector, and it is equal to the radius, R, of the range sphere. Expanding this equation into component form yields

$$\dot{X}_S(X - X_S) + \dot{Y}_S(Y - Y_S) + \dot{Z}_S(Z - Z_S) =$$

$$\cos \alpha_D \sqrt{\dot{X}_S^2 + \dot{Y}_S^2 + \dot{Z}_S^2} \sqrt{(X - X_S)^2 + (Y - Y_S)^2 + (Z - Z_S)^2} \tag{11-27}$$

where $\dot{X}_S$, $\dot{Y}_S$, and $\dot{Z}_S$ are the sensor velocity vector's components in the X, Y, and Z directions.

Broadside strip-map images have a Doppler angle of 90 degrees, and the entire right side of Eq. 11-27 is zero. In effect, the Doppler cone becomes a plane that is orthogonal to the velocity vector.

The combined sphere and cone conditions yield a sphere-cone intersection. The ground point associated with image location (R, α_D) is somewhere on the circle of intersection of these two figures (Fig. 11-21). Since the resulting circle of intersection is orthogonal to the velocity vector, all elevated objects layover in that orthogonal direction (Fig. 11-22). Due to fluctuations in surface elevation, it is possible for two distinct points to have identical range and Doppler values. These points are placed at the same pixel location on the image and are not resolvable.

Contrary to the optical model expressed by the collinearity condition, SAR image coordinates are not tied directly to the sensor and the ground. Rather, SAR coordinates are used to relate the sensor to the ground via a range and Doppler formulation. We cannot sketch a picture of the SAR image in the sphere/cone intersection graphic. A SAR image is not directly produced during imaging and it is used only indirectly to determine ranges and angles from the sensor to ground locations.

11.7.5   Sensor Orientation

The SAR coordinate system consists of measures of range and Doppler values. Since the Doppler angle is based on the vehicle velocity vector, the angular motion of the

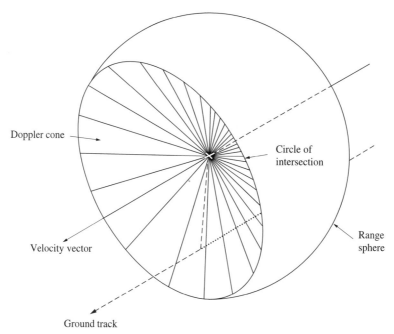

Figure 11-21 Intersection of range sphere and Doppler cone. Adapted from Henderson and Lewis (1998).

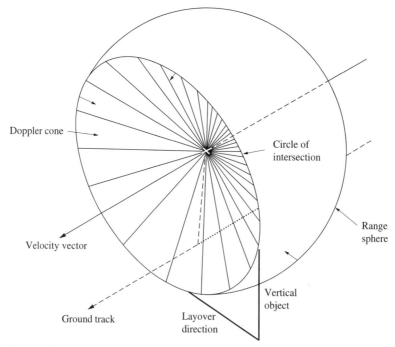

Figure 11-22 Orthogonal displacement of layover. Adapted from Henderson and Lewis (1998).

actual antenna does not affect the SAR coordinates. The SAR image-to-ground mathematical model has no sensor attitude data. The orientation of the antenna is important only in ensuring that the antenna illuminates the proper ground area. Estimates of sensor attitude based on star trackers (for satellite SAR) or gyroscopes (for airborne SAR) are not needed for computing ground locations. Since sensor attitude can be difficult to estimate accurately, this condition is a significant benefit to SAR modeling as compared to optical image modeling.

11.8   SAR IMAGE COORDINATES

11.8.1   SAR Processing

In the days when digital computational power was limited, it was very difficult and slow to process SAR data with a computer. Instead, range, phase, and amplitude data were recorded on a medium called SAR *signal film*, which was focused in a follow-on process to create a SAR image. On modern computer hardware, SAR processing can now occur very quickly and signal film processing has been replaced by digital methods. These involve computational phase history processing (PHP) that is normally implemented via Fourier transformations. The SAR process generates range, Doppler, and brightness values for all of the objects illuminated by the radar. It does not directly produce an image.

The phase and amplitude data for individual scatterers are captured in two parameters of a complex number that has an in-phase channel (I) and a quadrature component (Q). The gray value of a SAR pixel is computed in a detection process that combines the I and Q values:

$$\text{pixel magnitude} = \sqrt{I^2 + Q^2} \tag{11-28}$$

The pixel magnitude data are mapped to image locations as 8-bit or 11-bit gray values. However, unlike optical sensors, where the image surface is defined by the physical film or CCD detector array, the SAR image surface is defined mathematically. SAR processors may define the image surface to correspond to the slant plane, a ground tangent plane, the Earth ellipsoid surface, or it may be defined in some other fashion such as a map projection. Regardless of the choice of image surface, a sampling process associates the SAR range and Doppler values with pixel locations on the output image surface.

11.8.2   Slant Plane Images

The slant plane is a fundamental surface in all SAR imaging systems. During image collection, there are actually multiple slant planes. The wide swath of the fan-shaped radar beam includes a large range of angles from near- to far-range. However, the range and Doppler data must be projected onto a single image surface. This is often chosen to be a slant plane determined by the vehicle velocity vector, an *aperture center point* (ACP), S, that specifies a single sensor position within the synthetic aperture, and a *ground reference point* (GRP), G_0, in the center of the illuminated area. The basic slant plane geometry for a radar flying in a straight and level path is shown in Fig. 11-23.

Figure 11-23 displays a three-dimensional Cartesian coordinate system that specifies the slant plane and which can be established via a set of unit vectors centered on G_0. The range unit vector, $\hat{U}_R$, connects G_0 and S, the cross-range unit vector, $\hat{U}_C$, is

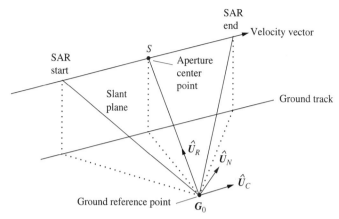

Figure 11-23 Slant plane geometry.

orthogonal to the range vector and is positive in the direction of flight; and the normal unit vector, $\hat{U}_N$, completes a right-handed coordinate system. Note that in broadside collection $\hat{U}_R$ is normal to $\vec{V}$, and $\hat{U}_C$ is parallel to $\vec{V}$.

The slant plane's relationship to image pixel coordinates is defined by specifying the image coordinates of the slant plane origin, the orientation of the pixel coordinate system with respect to the slant plane axes, and the pixel size. Typically, the image y-axis is aligned with $\hat{U}_R$ and the x-axis with $\hat{U}_C$.

11.8.3 Mapping Range, Doppler, and Magnitude to Image Space

The raw signal collected by the SAR sensor contains range and Doppler information, which must be processed to form the SAR image. There are numerous image formation algorithms. The choice of algorithm depends on the collection mode, the desired resolution, the required accuracy, and the computational resources available. Fine-resolution spotlight SAR data are often processed using a polar format algorithm (Carrara et al., 1995) that produces the range and Doppler coordinates. Frequently, additional processing modifies the range and Doppler data to fit an orthogonal image space for ease of exploitation.

Given an image coordinate system aligned to the slant plane, the processed SAR data must be mapped to their respective image space locations. This involves interpolating the range and Doppler data. In this sampling process, image pixel dimensions are chosen based on the inherent resolution of the SAR data and theoretical Nyquist sampling considerations. Normally, the output pixel dimensions are smaller than the corresponding SAR ground resolution.

For example, RADARSAT offers a georeferenced fine resolution product that has 8-meter resolution in both range and azimuth and pixel dimensions of 6.25 meters in each axis. However, this product is undersampled relative to the optimal Nyquist rate, and there is a potential loss of information in such an image. For this reason, RADARSAT also offers the same data in an extra-fine resolution image. In this case, the same 8-meter SAR data are output with 3.125-meter pixels. These pixels are more than double the radar resolution throughout the image, so there is no loss of data due to inadequate sampling.

SAR image pixel sizes are not determined by detector/ground-area ratios; they are chosen by considering aspects of resolution, Nyquist sampling theory, and image file size. Once this decision is made, the range, Doppler, and magnitude data are interpolated into the image space coordinate system arranged on the slant plane or some other surface.

11.8.4   Ground Plane Images

Although slant plane images are the most natural output of the SAR processor, they are not ideal for exploitation. Just as with their oblique optical image counterparts, SAR slant plane images typically suffer from an oblique appearance and a distortion of angles. As a result, many SAR vendors offer other output image representations. One common form is called a *ground plane* image. In this case, the SAR magnitude values are projected onto a plane tangent to the Earth ellipsoid at the center of the imaged area (Fig. 11-24). A ground plane image is similar to a vertical or rectified tilted optical image. Ground plane and other image representations, such as standard map projections, may be generated by the SAR processor directly or they may be produced by resampling the slant plane SAR image.

In order to resample a slant plane image to create a ground plane image, each slant plane pixel must be placed at the ground plane location that has the identical range and Doppler values of the original image. This is equivalent to intersecting the projection circles of the range and Doppler coordinates with the ground plane for all slant plane pixels.

For ground plane images, the image support data specify the origin of the ground plane, the orientation of its X, Y, and Z axes (a common choice is X = east, Y = north, Z = up), the image coordinates that correspond to the origin, the orientation of the pixel coordinate system relative to the ground plane X-Y axes, and the pixel spacing. Each pixel in this ground plane corresponds to a particular range and Doppler angle.

11.8.5   Latitude and Longitude Coordinates

It is possible to geocode SAR images with latitude and longitude coordinates for all of the range and Doppler values. These are generated by intersecting the range sphere and Doppler cone with the Earth ellipsoid or the ground plane, as described in Section 11.9. In this way, it is possible to associate range and Doppler data with latitude and longitude for all pixels in the imaged area. Of course, these latitude and longitude coordinates are not precise Earth surface locations because they are computed via an intersection with a zero elevation surface or a flat surface with an elevation set to the tasked aim point elevation.

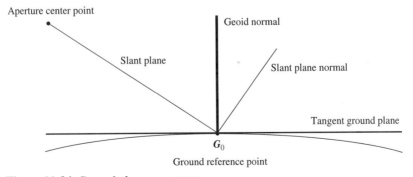

Figure 11-24 Ground plane geometry.

Any pixel in a SAR image may be mapped to an ellipsoid or ground plane latitude and longitude position. The accuracy of the mapping is determined by several factors. Sensor position and velocity measurement errors are typically the dominant error sources. Since the sensor position and velocity can be measured quite accurately with GPS, modern SAR systems are capable of very accurate geopositioning. Other error sources include a variety of range measurement errors, such as atmospheric and radio frequency timing errors. The quantitative effect of these errors is analyzed in Section 11.12. In particular, note that the positioning accuracy of the SAR is not affected by small sensor attitude measurement errors, which are the dominant error source for optical sensors.

Given the calculated latitude and longitude values, a SAR image can be output as a standard map projection, but the planimetric locations will retain the displacements caused by the projection to an arbitrary elevation surface and feature layover. If a digital elevation model is available, the image can be orthorectified and geocoded to remove these effects. This requires an intersection of the original range and Doppler values with the topographic surface to calculate correct horizontal positions.

11.8.6 Transformations from Image Pixel to Range and Doppler

A SAR image is created via transformations of (range, Doppler) to (line, sample) or (latitude, longitude). For accurate positioning from a SAR image, the user needs to know the mapping of (line, sample) or (latitude, longitude) back to (range, Doppler). The pixel latitude and longitude values on images that have not been orthorectified cannot be used for accurate positioning. Only the original range and Doppler values can be used to compute accurate Earth surface locations. This is done by projection to an Earth elevation model or by intersection from multiple images.

For digital images, there are no standard transformations to convert (line, sample) coordinates to (range, Doppler) or time coordinates. However, as long as the SAR vendor defines the velocity vector, the output image surface, the ACP, the image and ground space coordinates of the GRP, and the pixel dimensions, then the transformation from line and sample coordinates to range and Doppler data can be computed.

For example, given a spotlight image referenced to the slant plane with a defined velocity vector, ACP, and GRP, it is possible to compute range and Doppler coordinates for any pixel using the following method:

1. *Compute the slant plane unit vectors.* The range unit vector, cross-range unit vector, and slant plane normal unit vector, illustrated in Fig. 11-23, can be computed using the following equations:

$$\hat{U}_R = \frac{\vec{S} - \vec{G}_0}{|\vec{S} - \vec{G}_0|} \tag{11-29}$$

$$\hat{U}_N = \frac{\hat{U}_R \times \dot{\vec{S}}}{|\hat{U}_R \times \dot{\vec{S}}|} \times K_L \tag{11-30}$$

$$\hat{U}_C = \hat{U}_N \times \hat{U}_R \tag{11-31}$$

where $\vec{S}$ is the ACP position vector, $\vec{G}_0$ is the GRP position vector, $\dot{\vec{S}}$ is the vehicle velocity vector, and the K_L term, which has a value of either $+1$ or

−1, orients the slant plane normal vector for right-looking or left-looking illumination directions, respectively.

2. *For the pixel of interest, p, calculate the image space offsets from the GRP.*

$$\Delta \text{line} = \text{line}_p - \text{line}_{\text{GRP}} \tag{11-32}$$

$$\Delta \text{sample} = \text{sample}_p - \text{sample}_{\text{GRP}} \tag{11-33}$$

3. *Calculate the local space ground distance for this offset.*

$$X' = \Delta \text{line} \times \text{line dimension of pixel in meters} \tag{11-34}$$

$$Y' = \Delta \text{sample} \times \text{sample dimension of pixel in meters} \tag{11-35}$$

4. *Calculate the position vector,* $\vec{g}$*, of the measured point in Earth-centered-fixed (ECF) coordinates.* This is done by multiplying the local space ground distance by the unit vectors and adding the result to the GRP position vector. Note that $\vec{g}$ represents the location of the point in the slant plane.

$$\vec{g} = \begin{bmatrix} U_{C_X} & U_{R_X} & U_{N_X} \\ U_{C_Y} & U_{R_Y} & U_{N_Y} \\ U_{C_Z} & U_{R_Z} & U_{N_Z} \end{bmatrix} \begin{bmatrix} X' \\ Y' \\ 0 \end{bmatrix} + \begin{bmatrix} X_{G_0} \\ Y_{G_0} \\ Z_{G_0} \end{bmatrix} \tag{11-36}$$

where the elements in the 3×3 matrix are components of the unit vectors $\hat{U}_C$, $\hat{U}_R$, and $\hat{U}_N$.

5. *Determine the range to the sensor.*

$$R = |\vec{g} - \vec{S}| \tag{11-37}$$

6. *Determine the Doppler angle.* This is derived from the dot product of the velocity vector and the range vector, or

$$\alpha_D = \cos^{-1}\left(\frac{\vec{S}}{|\vec{S}|} \cdot \frac{\vec{g} - \vec{S}}{R} \right) \tag{11-38}$$

A strip-map image would be processed in a similar manner, except that there is no central ACP. Instead, each line is associated with a specific sensor position. The time of an individual line, and hence the sensor position, can be computed from support file information specifying the time of the first and last lines. Given the time of a line, the sensor position is computed from a time-tagged ephemeris, and the range vector can then be calculated. The Doppler angle is always 90 degrees for broadside images; it is a fixed known angle for squinted images.

To simplify the exploitation of SAR images on softcopy workstations, it is useful to derive a polynomial model to associate (line, sample) with (range, Doppler). This process involves generating a grid of image points and applying the rigorous technique just described to derive the range and Doppler values of all grid points. A least squares fit can then be used to derive polynomial coefficients that relate (line, sample) to (range, Doppler). The polynomial is then used to quickly derive range and Doppler values for any measured image location. Once a pixel location is converted to its corresponding (range, Doppler) coordinates, the radar projection equations are used to determine ground coordinates. While this is a replacement model to the rigorous equations, when the coefficients are estimated on the basis of a dense grid, and the fit errors are small, exploitation accuracy will be quite adequate.

11.8.7 Hardcopy Images

Hardcopy strip-map SAR film can be supplied with range and time reference marks or simply with time reference marks. In the latter case, range is derived by accounting for image scale and the range offset to the first pixel location:

$$R = y \times \text{scale} + \text{range offset} \tag{11-39}$$

For a detailed discussion of hardcopy image geometry see Leberl (1990).

11.9 CALCULATION OF GROUND COORDINATES

11.9.1 Monoscopic Intersection

The following equations intersect the projection circle with the Earth ellipsoid or a surface at constant height above the ellipsoid, see Fig. 11-25. The vector $\vec{G}$ represents the location of the point on the Earth ellipsoid or at height h above it. ECF coordinates are assumed. The range condition is

$$R = |\vec{G} - \vec{S}| \tag{11-40}$$

where R is calculated from the pixel values by Eq. 11-37. The Doppler cone angle condition is

$$\cos \alpha_D = \frac{\vec{S}}{|\vec{S}|} \cdot \frac{\vec{G} - \vec{S}}{|\vec{G} - \vec{S}|} \tag{11-41}$$

The equation for an ellipsoid is

$$\frac{G_X^2 + G_Y^2}{(a + h)^2} + \frac{G_Z^2}{(b + h)^2} = 1 \tag{11-42}$$

where G_X, G_Y, and G_Z are the three components of $\vec{G}$, a and b are the semimajor and semiminor axes of the reference Earth ellipsoid, and h is the elevation above the ellipsoid surface.

Equations 11-40–11-42 form a system of three nonlinear equations in three unknowns, G_X, G_Y, and G_Z, so they must be solved via an iterative numerical process

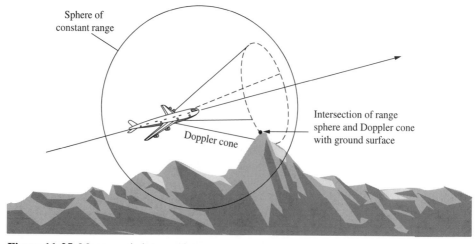

Sphere of constant range

Doppler cone

Intersection of range sphere and Doppler cone with ground surface

Figure 11-25 Monoscopic intersection.

that linearizes the equations in each iteration. See Mikhail (1976) for appropriate solution techniques to nonlinear systems of equations.

11.9.2   Stereoscopic Intersection

For stereoscopic SAR images, the following four equations intersect two projection circles to determine the three-dimensional ground coordinates, see Fig. 11-26. The two range conditions are

$$R_1 = \left| \vec{G} - \vec{S}_1 \right| \qquad R_2 = \left| \vec{G} - \vec{S}_2 \right| \tag{11-43}$$

The Doppler cone angle conditions are

$$\cos \alpha_{D1} = \frac{\vec{S}_1}{\left| \vec{S}_1 \right|} \cdot \frac{\vec{G} - \vec{S}_1}{\left| \vec{G} - \vec{S}_1 \right|} \qquad \cos \alpha_{D2} = \frac{\vec{S}_2}{\left| \vec{S}_2 \right|} \cdot \frac{\vec{G} - \vec{S}_2}{\left| \vec{G} - \vec{S}_2 \right|} \tag{11-44}$$

This is an overdetermined system of four nonlinear equations in three unknowns, and as in the monoscopic case, they must be solved via an iterative numerical least squares process that linearizes the equations in each iteration. For more than two images, the same approach can be used, and this is referred to as a *multi-ray intersection*. Again, see Mikhail (1976) for appropriate solution techniques to nonlinear least squares problems.

11.10   DETERMINING HEIGHTS FROM LAYOVER AND SHADOW MEASUREMENTS

Just as it is possible to derive the height of a vertical object on an optical image by measuring its relief displacement or shadow length, similar computations can be made

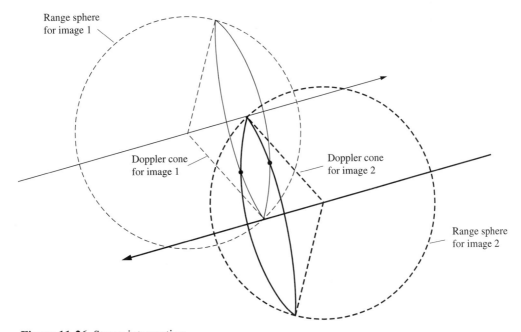

Figure 11-26 Stereo intersection.

with SAR images. The accuracy of the layover process increases as layover gets larger, and determining heights from layover works best at steep depression angles. Conversely, the accuracy of the shadow process increases when shadows are long, and accuracy is maximized at shallow depression angles. Thus, the effects of layover and shadow measurements in deriving heights are opposite to one another. When possible, it is best to use both procedures.

11.10.1   Heights from Layover

Layover geometry is shown in Fig. 11-27. The following algorithm may be used to determine heights of vertical objects:

1. Using the image coordinates of the object's base and the pixel to range-Doppler mapping discussed in Section 11.8.6, compute range-Doppler coordinates of the base.

2. Using these range-Doppler coordinates, perform a monoscopic image-to-ground projection (Eqs. 11-40–11-42) and record the resulting horizontal ground coordinates, $\vec{G}_B$. This intersection should use the best available estimate for the terrain surface elevation (h).

3. Repeat the image-to-ground projection, but this time use ($h + 1$) for the assumed elevation. Record the associated horizontal ground coordinates, $\vec{G}_C$, and then convert vectors $\vec{G}_B$ and $\vec{G}_C$ to local tangent plane coordinates ($\vec{G}_{TB}$, $\vec{G}_{TC}$) using standard coordinate transformation equations.

4. Compute the surface layover distance, D_{LU}, corresponding to a vertical object of unit height.

$$D_{LU} = \sqrt{(G_{TB_X} - G_{TC_X})^2 + (G_{TB_Y} - G_{TC_Y})^2} \qquad (11\text{-}45)$$

where the X and Y subscripts denote the X and Y coordinates in the tangent plane system.

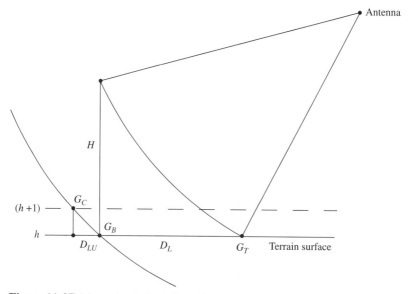

Figure 11-27 Measuring heights from layover.

5. Using the image coordinates of the object's top, perform a monoscopic image-to-ground projection to the surface elevation, h. Record the horizontal ground coordinates that result, $\vec{G}_T$, and as before convert the vector to the local tangent plane, $\vec{G}_{TT}$.

6. Compute the horizontal distance between this position and the location of the object's base. This is the layover distance, D_L, for the vertical object of interest.

$$D_L = \sqrt{(G_{TT_X} - G_{TB_X})^2 + (G_{TT_Y} - G_{TB_Y})^2} \qquad (11\text{-}46)$$

7. Divide this layover distance by the layover distance for an object of unit height to yield the height, H, of the vertical object.

$$H = \frac{D_L}{D_{LU}} \qquad (11\text{-}47)$$

11.10.2   Heights from Shadow

Shadow geometry is shown in Fig. 11-28. The following algorithm may be used to determine heights of vertical objects:

1. As was done in steps one and two of the preceding section, compute the range-Doppler coordinates of the object's base and then perform a monoscopic image-to-ground projection. Record the resulting horizontal ground coordinates, $\vec{G}_B$. This intersection should use the best available estimate for the terrain surface elevation (h).

2. Compute the range vector, $\vec{R}_B$, and unit vector, $\vec{U}_B$, from the sensor position, $\vec{S}$, to the ground location of the object's base.

$$\vec{R}_B = \vec{G}_B - \vec{S} \qquad (11\text{-}48)$$

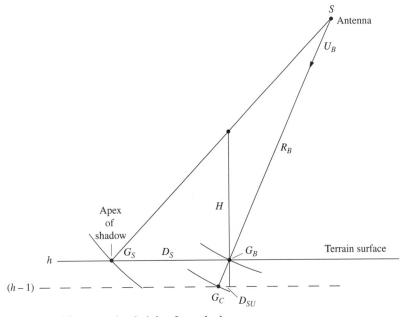

Figure 11-28 Measuring heights from shadows.

$$\vec{U}_B = \frac{\vec{R}_B}{|\vec{R}_B|} \qquad (11\text{-}49)$$

The unit vector is the vector describing the radar illumination direction near the feature of interest.

3. Convert vectors $\vec{G}_B$ and $\vec{U}_B$ to local tangent plane coordinates $(\vec{G}_{TB}, \vec{U}_{TB})$ using standard transformation equations.

4. Using the position of the object's base as a starting location, project the unit vector to a surface whose height is one unit below the surface elevation $(h - 1)$. Record the corresponding horizontal ground coordinates, G_{TC_X}, and G_{TC_Y}.

$$G_{TC_X} = G_{TB_X} + U_{TB_X}/U_{TB_Z} \qquad (11\text{-}50)$$

$$G_{TC_Y} = G_{TB_Y} + U_{TB_Y}/U_{TB_Z} \qquad (11\text{-}51)$$

where the X, Y, Z subscripts denote the X, Y, Z coordinates in the tangent plane system.

5. Determine the shadow distance for an object of unit height by computing the horizontal distance between this location and the object's base.

$$D_{SU} = \sqrt{(G_{TB_X} - G_{TC_X})^2 + (G_{TB_Y} - G_{TC_Y})^2} \qquad (11\text{-}52)$$

6. Using the image coordinates of the object's shadow tip, perform a monoscopic image-to-ground projection to the surface elevation, h. Record the horizontal ground coordinates that result, $\vec{G}_S$, and convert to the local tangent system, $\vec{G}_{TS}$.

7. Compute the horizontal distance, D_S, between the object's base and the ground position of the shadow's tip. This provides the shadow distance corresponding to the vertical object of interest.

$$D_S = \sqrt{(G_{TS_X} - G_{TB_X})^2 + (G_{TS_Y} - G_{TB_Y})^2} \qquad (11\text{-}53)$$

8. Divide this shadow distance by the shadow distance for an object of unit height to yield the height of the vertical object.

$$H = \frac{D_S}{D_{SU}} \qquad (11\text{-}54)$$

11.11 SAR STEREO GEOMETRY CONSIDERATIONS

In order to derive three-dimensional ground coordinates, it is necessary to have two or more images of the target area collected with different projection geometries. The relative geometry of the stereo images strongly influences geometric accuracy and visual fusion. However, since SAR is an active sensor using coherent illumination to produce an image, the preferred relative collection geometries for SAR stereo are considerably different than for optical systems.

This section discusses the general categories of SAR stereo collection and compares them to geometries that are typical for optical sensors. As with optical imagery, the best SAR stereo collections are obtained when the scene illumination conditions are similar, but when relief is displaced differently on the two images.

11.11.1 Comparison to Optical Stereo

For optical imagery, relief is displaced along the line of sight from the sensor to the ground target. To obtain good stereo, two images are collected so that line-of-sight vectors have a moderate convergence angle between them. Another desirable condition is to collect the two images with similar sun angles in order to improve both visual fusion and automatic matching methods.

As explained in the discussion about the Doppler cone condition (Section 11.7.4), relief on a SAR image is displaced along the projection circle defined by the range sphere and the Doppler cone, not along the line of sight from the sensor to the ground target. This projection circle is normal to the slant plane, and the SAR slant plane normal vectors are analogous to the line-of-sight vectors for optical images. For SAR stereo, we refer to the angle between the slant plane normal vectors as the convergence angle, and as with optical imagery, a moderate angle is needed for good stereo fusion.

However, in the case of SAR, a large stereo base does not necessarily result in large parallax differences. The base-to-height ratio used for optical stereo imagery is not very meaningful for SAR images; the angle between the slant plane normal vectors is the significant measure of SAR stereo. In fact, the geometric condition obtained from SAR images having slant plane normal vectors N_1 and N_2 are exactly the same as what is obtained from optical images that have line-of-sight vectors N_1 and N_2, as shown in Fig. 11-29. For the sun angle condition useful for optical images, the analogous condition for SAR is to have similar radar illumination angles.

11.11.2 Single Flight Path

Single flight path geometry is illustrated in Fig. 11-30. In this scenario, a single vehicle is used to image the target area on a flight line where one image is squinted forward and the other is squinted backward. Although this is the desired stereo geometry configuration for optical sensors, this does not provide good stereo geometry for radar systems. For straight-line flight paths, the velocity vector and range vectors of conjugate points lie in the same slant plane, and the angle between the associated slant plane normal vectors is zero. The sphere/cone projection circles are coincident for conjugate points, and there is no stereo effect at all.

For satellite SAR platforms, the curved orbit results in slight changes in slant plane geometry between the two images. As shown in Fig. 11-30, this is not very

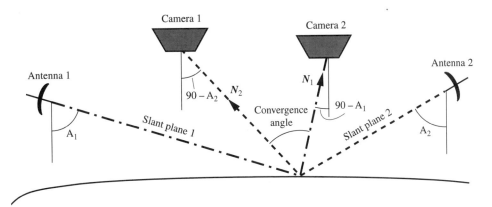

Figure 11-29 SAR and optical convergence angles.

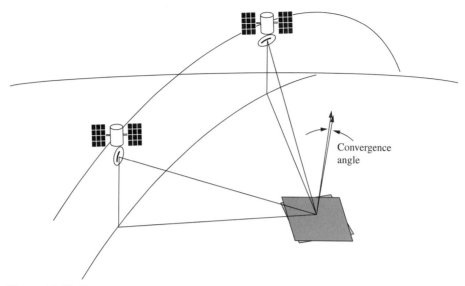

Figure 11-30 Single flight path stereo.

significant. In this case, there are very small changes in relief displacement and minimal or no vertical exaggeration. In addition, the large delta in viewing azimuths causes the illumination conditions of the two images to be different, making stereo fusion difficult.

11.11.3 Parallel Flight Paths

Same-side parallel flight path geometry is illustrated in Fig. 11-31. In this scenario, the flight paths are parallel, but they are shifted so that the images are collected with different depression angles and similar azimuth angles. Since layover direction is determined by the direction of the vehicle velocity vectors, relief is displaced in the

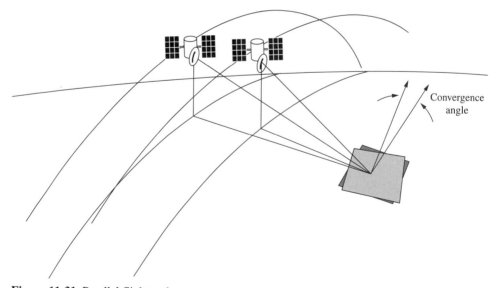

Figure 11-31 Parallel flight paths stereo.

same direction in both images, but by different amounts due to the different depression angles. The slant planes of the two images are clearly different but the angle between the slant plane normal vectors is small; it is typically in the range of 10 degrees or less. This provides some amount of stereo effect, but considerably less than the 40 degrees or more available with optical stereo geometry.

Large differences in the depression angles create a greater stereo effect, but they may cause difficulty in image fusion because of the different illumination conditions. Working with images where the difference in grazing angles is more than 10 degrees can be difficult because significant differences in shadow lengths can degrade automated stereo matching, especially in rugged terrain. Another concern regarding stereo pairs from multiple flight paths is that there may be a significant temporal difference between the two images. This is of particular concern for satellite images, which may be collected weeks apart from one another.

11.11.4 Crossing Flight Paths

In this scenario, the two images of the stereo pair are collected using crossing flight paths as shown in Fig. 11-32. For satellite platforms, this would require that one image be taken on an ascending orbit pass, and the other on a descending orbit pass, or by imaging from two satellites with different inclinations. Because relief displacement in SAR imagery is directed perpendicular to the vehicle velocity vector, these two images would have layover oriented in different directions, thus providing the desired stereo effect.

The ideal geometry for crossed flight paths occurs when the images are acquired from nearly identical locations in space, which occurs near the flight path crossing. In this case, both images have the same grazing and azimuth angle characteristics, which means the illumination conditions are very similar, and the images look very much alike. These two images, however, will have relief displaced in different directions, making this the ideal geometry for stereo viewing and for automated elevation extraction algorithms that rely on automatic matching of conjugate pixel data.

Unfortunately, the crossing paths geometry requires multiple vehicles in order to provide collection opportunities over wide areas. In addition, the ascending-descending crossing path requires the SAR sensor to image on different sides of the vehicle. Such

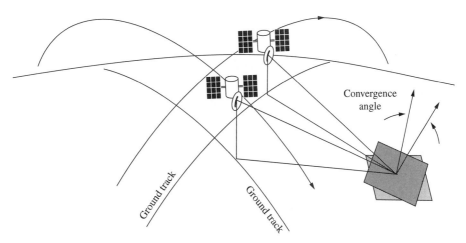

Figure 11-32 Crossing flight paths stereo.

collection is not possible for a SAR sensor that images to only one side, such as the right-looking RADARSAT sensor.

11.11.5 Opposite-Side Flight Paths

The geometry for opposite-side flight paths is shown in Fig. 11-33. In this scenario, the flight paths are on opposite sides of the ground target, causing the layover and shadows configurations to be in opposite directions on the two images. While this geometry provides large convergence angles and significant vertical exaggeration for accurate three-dimensional measuring, the large differences in illumination conditions may make stereo matching nearly impossible. With opposite-side stereo imagery, matching is especially difficult in rugged areas. However, the accuracy available from opposite-side stereo pairs may be beneficial for applications where a few discrete point measurements are needed. Such points may be manually measured via a split-screen stereo arrangement.

11.11.6 Summary of SAR Stereo Geometry

The requirements for SAR stereo geometry are identical to optical stereo collection. That is, a strong stereo pair is characterized by a convergence angle of moderate magnitude and similar illumination conditions for the two images. The crossing flight paths scheme meets these needs and it is the best stereo arrangement for SAR, but this method requires multiple sensors to support wide-area coverage. Unfortunately, crossing flight paths collection is not yet a practical collection alternative.

It is possible to collect useful stereo images from parallel passes of a SAR sensor, but collection constraints need to be considered carefully. In parallel-pass collection, the SAR range-Doppler geometry model requires differing illumination conditions to generate divergent slant planes, but the active nature of SAR imaging means that stereo images collected under different conditions will have differing shadow configurations that make stereo matching very challenging. In addition, the rather small magnitude of parallel-pass convergence angles is a serious accuracy limitation. Finally, the temporal nature of SAR stereo collection adds the further complication of ground cover changes over time.

There is great potential for SAR to generate detailed and accurate terrain data from a type of SAR stereo processing that makes use of the coherent nature of SAR radiation. *Radar interferometry,* which is discussed in the following section, has been used on multiple airborne and spaceborne SAR systems.

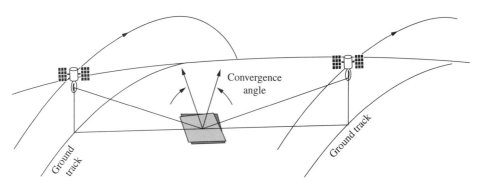

Figure 11-33 Opposite-side flight paths stereo.

11.12   INTRODUCTION TO IFSAR

A single SAR image allows the positions of scatterers in the scene to be determined in two dimensions. SAR imaging resolves a scene into range and Doppler frequency (cross-range) components. We have seen that the SAR image coordinates for an arbitrary scatterer can be related to their corresponding range and Doppler frequency observations. These range and Doppler observations constrain the position of the scatterer to be along the arc satisfying the range sphere and Doppler cone angle conditions.

SAR interferometry provides a way to determine the third dimension of the scatterer's position; that is, it provides a means to determine the scatterer's location along the range-Doppler arc. In SAR interferometry, two SAR images are collected with a very small difference in their collection geometries. A single sensor may collect the two images on two different imaging passes, or they may be collected on a single pass over the target if the SAR sensor has two displaced receiving antennas.

The phase of a complex SAR image pixel includes a component related to the range from the sensor to the scatterer. This pixel phase ϕ is given by

$$\phi = \frac{4\pi R}{\lambda} + \cdots \tag{11-55}$$

where λ is the wavelength corresponding to the radar center frequency f_c ($\lambda = c/f_c$) and R is the scalar range from the sensor to the scatterer. The ellipsis (. . .) represents other terms affecting the pixel phase which will disappear when the interferometric phase difference is computed.

By subtracting the phase between corresponding pixels from the two interferometric SAR images, we obtain an interferometer phase

$$\Delta\phi = \frac{4\pi(R_2 - R_1)}{\lambda} \tag{11-56}$$

which is related to the difference in range $(R_2 - R_1)$ from the scatterer to the SAR phase centers for the respective images. Since the two images being differenced have very similar collection geometry and identical wavelength, the additional phase terms, represented by the ellipsis in the pixel phase Eq. 11-55, are identical and cancel out in the subtraction.

Given a particular observed value of $\Delta\phi$, the locus of points satisfying the interferometer phase Eq. 11-56 form a hyperboloid of revolution with an axis of revolution given by the vector between the two SAR phase centers. At large distances from the sensor, this equi-phase surface is approximately a cone.

The combination of the interferometric phase observation with the range and Doppler frequency observations described earlier is illustrated in Fig. 11-34. The interferometer phase provides an additional constraint which enables the scatterer's three-dimensional position relative to the SAR sensor to be determined.

11.12.1   Geolocation Using the IFSAR Observations

Interferometric SAR utilizes three types of observations to geolocate points in the SAR image. The range, Doppler, and interferometer phase observation equations can be summarized as follows:

$$|\vec{R}| - R = 0 \tag{11-57}$$

Range sphere   Doppler cone (Disk)   Interferometric phase cone

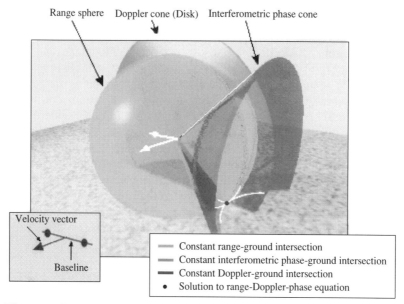

Velocity vector

Baseline

——— Constant range-ground intersection
······· Constant interferometric phase-ground intersection
——— Constant Doppler-ground intersection
•   Solution to range-Doppler-phase equation

Figure 11-34 Interferometric SAR imaging.

$$\frac{2}{\lambda}\vec{S}\cdot\vec{U}_R - f_D = 0 \tag{11-58}$$

$$\frac{2n\pi}{\lambda}\vec{B}\cdot\vec{U}_R - \phi = 0 \tag{11-59}$$

where $\vec{R}$ is the range vector from the sensor to the scatterer, R is the scalar range, $\vec{S}$ is the sensor velocity vector, $\vec{U}_R$ is the line-of-sight unit vector, $\vec{B}$ is the interferometer baseline vector, and ϕ is the observed interferometric phase. The factor $n = 2$ indicates collections where the two phase centers each transmit and receive independently (monostatic mode), and $n = 1$ indicates when one phase center transmits and both phase centers receive (the receive-only channel operates in bistatic mode).

Given knowledge of the SAR sensor's position and baseline attitude with respect to some coordinate frame of interest, for example, WGS-84, the location of points in the SAR image can be determined in the desired coordinate frame. To locate the scatterer position relative to the SAR sensor, we must solve the three observation equations for the three unknown components of $\vec{R}$. Once $\vec{R}$ is known, the scatterer position, $\vec{G}$, can be determined with respect to a desired coordinate frame using the known position of the SAR sensor, $\vec{S}$, in that coordinate frame as follows:

$$\vec{G} = \vec{S} + \vec{R} \tag{11-60}$$

Noting that the magnitude of the range vector is directly observable, we can rewrite the previous equation as follows:

$$\vec{G} = \vec{S} + R\vec{U}_R \tag{11-61}$$

The range observation makes the system of equations nonlinear. Normal approaches for linearizing and iteratively solving the equations are possible, but a simpler approach was developed independently by Madsen et al. (1993) and Abshier

(1992). This approach uses a clever choice of basis vectors along with the Doppler equation (Eq. 11-58), the interferometer phase equation (Eq. 11-59), and a unit magnitude constraint to solve for $\vec{U}_R$. With this result and the directly observable scalar range, R, the scatterer's position is determined using Eq. 11-61.

11.12.2   Interferometric SAR Processing

A block diagram of the basic SAR interferometry process is shown in Fig. 11-35. The inputs to the process are two channels of SAR phase history data, along with ancillary information, such as radar parameter and motion measurement data. Image formation generates a complex SAR image from each channel of the phase history data. Though not shown on the diagram, the two images are registered, if necessary, at this point in the process. Image registration may be required if the images are collected on two separate collection passes, or if the data are collected with a large baseline. The registered complex imagery is conjugate multiplied pixel-by-pixel, and the complex product is filtered to smooth the interferogram and reduce the effects of speckle noise. Phase detection yields the phase difference represented by Eq. 11-56 modulo 2π. The resulting wrapped fringe map must be unwrapped to determine the actual $\Delta\phi$.

After phase unwrapping, the absolute phase difference is still ambiguous by an integer multiple of 2π. Several techniques exist to resolve this remaining phase ambiguity, including the use of external control points (Adams et al., 1996), comparison with overlapping data, stereo parallax, and split-spectrum interferometry. Once the absolute phase is recovered, the data can be geocoded using the range, Doppler, and interferometer phase observations. The geocoded data are then resampled into the desired output map projection. The associated SAR magnitude imagery can be orthorectified once the digital terrain elevation (DTE) data is complete.

In addition to the DTE and orthorectified SAR imagery, a correlation output consisting of the correlation magnitude value at each output post is often produced.

11.12.3   Error Sources (Performance-Limiting Factors)

The accuracy of an interferometric SAR measurement is affected by a number of error sources. The most significant error sources are

- Sensor position error
- Sensor velocity error
- Baseline attitude error
- Baseline length error
- Interferometer phase error
- Range measurement errors

The impact of these errors on the position measurements is most easily understood by looking at a simplified, broadside-looking collection geometry as illustrated in Fig. 11-36. The sensor is located at point $P = (P_x, P_y, P_z)$. The broadside-looking SAR is illuminating a swath that includes some arbitrary point $A = (A_x, A_y, A_z)$. The upper right portion of Fig. 11-36 shows a detail of the interferometer baseline and illustrates several important angles. The interferometer measures a phase that depends on the wavelength and the difference in range, $\Delta R = R_2 - R_1$, from the point to each of the phase centers shown in Eq. 11-55. The range difference ΔR is determined by the length of the baseline, and the interferometer incidence angle, θ.

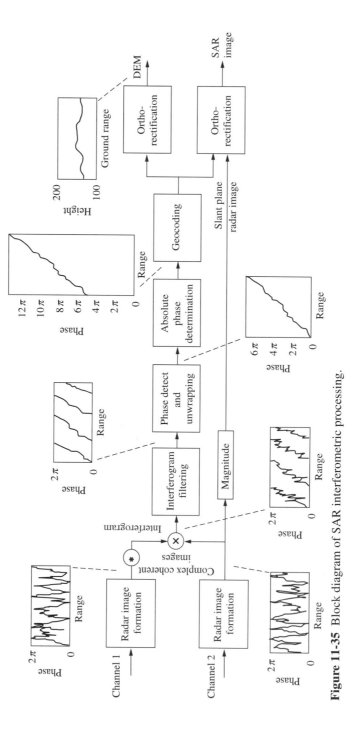

Figure 11-35 Block diagram of SAR interferometric processing.

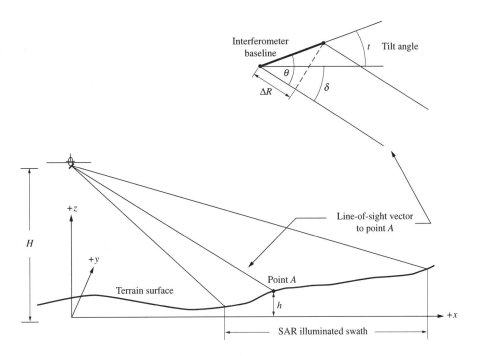

Figure 11-36 Geometry of broadside-looking interferometric SAR collection geometry.

The interferometer incident angle can be considered the sum of two angles, the baseline tilt angle, t, and the depression angle of the line of sight from the horizontal reference, δ.

Point A's position can be readily computed given the sensor position, P, the baseline length, B, the baseline attitude, t, the range R of the scatterer, and the depression angle δ as follows:

$$A_x = P_x + R\cos\delta \tag{11-62}$$

$$A_y = P_y \tag{11-63}$$

$$A_z = P_z + R\sin\delta \tag{11-64}$$

The depression angle is not directly observable, but can be computed from the interferometer phase, the baseline attitude, baseline length, and the wavelength

$$\delta = \theta - t = \cos^{-1}\left(\frac{\lambda\phi}{2n\pi B}\right) - t \tag{11-65}$$

Position error sensitivity can be determined by looking at the partial derivatives of the equations used to calculate A_x, A_y, and A_z (Eqs. 11-62–11-64) with respect to errors in the various parameters. The sensitivity equations for each of the significant error sources are given in Table 11-2, along with a numerical example including a typical value for the parameter error and the resulting position error. It is important to note that the typical values vary significantly from system to system and depend heavily on specific design characteristics. The nominal parameters for the examples in the table are slant range $R = 14$ km, wavelength $\lambda = 3.1$ cm (X-band), sensor velocity $V = 200$ m/s, a 1-m-long baseline oriented horizontally ($t = 0°$), and a depression angle $\delta = 45°$.

Table 11-2 Interferometric SAR Error Sources and Sensitivities

Error	Sensitivity	Typical value	Resulting error
Position	$\Delta A_x = \Delta P_x$	$\Delta P_x = 1$ m	$\Delta A_x = 1$ m
	$\Delta A_y = \Delta P_y$	$\Delta P_y = 1$ m	$\Delta A_y = 1$ m
	$\Delta A_z = \Delta P_z$	$\Delta P_z = 1$ m	$\Delta A_z = 1$ m
Velocity	$\Delta A_y = \dfrac{\Delta V_{\text{los}}}{V} \cdot R$	$\Delta V_{\text{los}} = 1$ cm/s	$\Delta A_y = 0.7$ m
Baseline attitude	$\Delta A_x = -R \sin \delta \cdot \Delta t$	$\Delta t = 200 \, \mu\text{rad}$	$\Delta A_x = 2$ m
	$\Delta A_z = -R \cos \delta \cdot \Delta t$	$\Delta t = 200 \, \mu\text{rad}$	$\Delta A_z = 2$ m
Baseline length	$\Delta A_x = -R \sin \delta \cdot \cot \theta \dfrac{\Delta B}{B}$	$\Delta B = 0.1$ mm	$\Delta A_x = 1$ m
	$\Delta A_z = -R \cos \delta \cdot \cot \theta \dfrac{\Delta B}{B}$	$\Delta B = 0.1$ mm	$\Delta A_z = 1$ m
Interferometer phase	$\Delta A_x = -R \sin \delta \cdot \dfrac{\lambda \Delta \phi}{2 n \pi B \sin \theta}$	$\Delta \phi = 2°$	$\Delta A_x = 1.2$ m
	$\Delta A_z = -R \cos \delta \cdot \dfrac{\lambda \Delta \phi}{2 n \pi B \sin \theta}$	$\Delta \phi = 2°$	$\Delta A_z = 1.2$ m
Range	$\Delta A_x = \cos \delta \cdot \Delta R$	$\Delta R = 10$ cm	$\Delta A_x = 0.07$ m
	$\Delta A_z = -\sin \delta \cdot \Delta R$	$\Delta R = 10$ cm	$\Delta A_z = 0.07$ m

11.12.4 Example IFSAR Systems and Products

The ERIM IFSARE system is a dual-channel X-band SAR installed in a Learjet 36. This system rapidly produces geocoded SAR imagery and digital terrain elevation data. The IFSARE is a single-pass interferometric SAR system that can collect day or night, through cloud cover in reasonable weather. System height accuracy and swath widths are a function of collection altitude, as shown in Table 11-3. Accuracies are

Table 11-3 Performance Parameters of IFSARE

	Altitude	
	6,100 m	**12,200 m**
Ground swath width (km)	6	10
Collection rate (km²/min)	60–72	100–120
DEM post spacing (m)	7.5	15
Pixel spacing (m)	2.5, 5.0, or 10.0	2.5, 5.0, or 10.0
Absolute elevation accuracy (m, one sigma)	1.5	3.0
Absolute horizontal position accuracy (m, one sigma)	2	3
Local elevation variation (m)	0.4–1.0	0.8–1.4

valid over rugged or featureless terrain with −20 dB reflectivity. The data collection rate at high altitude is greater than 100 square kilometers per minute. No ground control points are required to achieve absolute accuracy other than a single differential ground GPS position reference nominally within 200 kilometers of the collection area. Since the system is dual channel, interferometric data is collected in a single pass, eliminating pass-to-pass temporal decorrelation as well as the expense of a second pass. Figures 11-37 and 11-38 show sample products from the IFSARE system.

11.13 THE EMERGENCE OF SAR IMAGING

Beyond the obvious benefit of assured access independent of cloud-cover patterns and sunlight, radar sensors can be designed to provide images with a combination of useful properties, including: very high resolution, expansive dynamic range, multiple spectral bands, cross-polarizations, and good geometric accuracy. The image displayed in Fig. 11-39 shows the dramatic landforms of the Phang Hoei Range of north central Thailand. This image is a black-and-white presentation of multi-frequency, multi-polarization data acquired by NASA's SIR-C imaging radar in October 1994. The SIR-C/X-SAR radar includes L-band, C-band, and X-band sensors in different combinations of HH and HV polarizations.

In addition to image formation, special techniques that exploit the coherent nature of radar energy, such as interferometry, can be used to generate nonstandard products like high-density terrain models. While radar imagery will never match the

Figure 11-37 Sarajevo IFSARE radar magnitude image.

Figure 11-38 Sarajevo IFSARE shaded relief image.

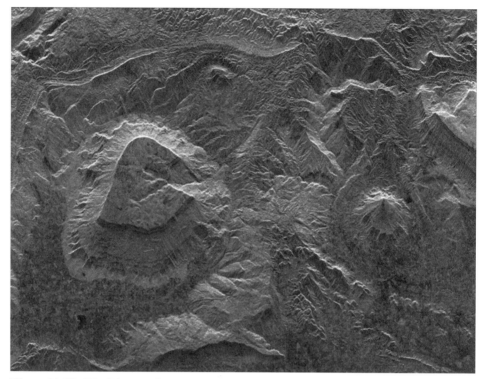

Figure 11-39 SIR-C image of north central Thailand.

visual detail and clarity of optical images, radar imaging has now emerged as a significant remote sensing technology.

11.14 LIDAR

The original airborne application of laser distance measurement was laser profilers, which generated elevations of a single line directly below the aircraft. Applications of laser profilers were limited, due to the single line of elevations produced and also to the limited positioning precision of available aircraft navigation sensors. In recent years, applications of LIDAR (LIght Detection And Ranging) sensors have grown rapidly, due to two technological developments. First, electro-mechanical technology has improved scanners and lasers to the point that they can be accurately and reliably controlled in a moving aircraft. Second, aircraft positioning using Global Positioning System, or GPS, and Inertial Navigation Systems, or INS, technology has advanced to the point that aircraft position and orientation can be continuously determined to high levels of precision. This section discusses the basic design and operating principles of laser ranging sensors and covers some of their recent applications.

11.14.1 Laser Ranging Technology

A schematic of laser ranging scanning is shown in Fig. 11-40. The main components of a LIDAR system are the laser, the scanning mechanism and projection optics, the receiver optics, and the platform navigation sensors.

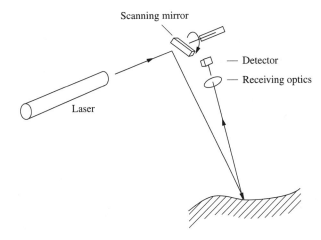

Figure 11-40 Schematic of a laser scanner.

Lasers used for topographic laser ranging typically operate in the infrared region of the spectrum with wavelengths from around 1064 nm up to 1540 nm in some cases. Blue-green lasers, at around 532 nm, are used for bathymetric work due to their water penetration capabilities.

Distances are measured by sending out pulses of light, about 10 nanoseconds long, and recording the time required for each pulse to return from the terrain surface. Current systems operate at pulse rates from a few hundred pulses per second up to 10,000 pulses per second. A higher pulse rate allows for a wider coverage swath, a higher aircraft speed, or closer spacing of elevation points. However, using a higher pulse rate increases the laser power requirements, since as the laser's power is spread over more pulses, each return is weaker and may be harder to reliably recognize and time.

Each pulse covers a finite area determined by the instantaneous field of view (IFOV) of the scanner. The returned pulse is therefore a combination of the elevations within the field of view. A single pulse may have multiple returns, as when it is partially reflected from tree canopies, undergrowth, or the ground. Some scanner systems record these multiple reflections to aid in removing vegetation reflections from the terrain model. The intensity of the reflected laser beam can also be recorded to generate an intensity image of the area of coverage.

The scanning system may be based on rotating optical elements, such as mirrors or prisms, or deflecting elements, such as galvanometers. The scanner must accurately point the laser beam at a high rate. Scanners may operate at up to 300 scans per second.

Scanning airborne laser ranging systems may be mounted on either fixed-wing or rotary-wing aircraft and may be internally mounted in a camera port or contained in an attached pod. The most important element of the platform is its navigation system. High-accuracy differential GPS is essential, using an aircraft receiver in conjunction with one or more fixed ground receivers. An inertial navigation system is operated in conjunction with the GPS, to provide aircraft orientation information and also position information between GPS updates.

Laser scanning technology continues to improve rapidly. Current attainable accuracies are about 15 cm in absolute elevation and about 5 cm in relative elevation (between adjacent points). Digital elevation models with grid spacings of 1 meter can be generated.

11.14.2 LIDAR Processing

Laser scanning produces large volumes of data, which requires several processing steps to produce the final DEM.

The first step is to calculate the position of each returned pulse. This is essentially a vector problem; to calculate the position of the reflecting surface, we need the starting point of the vector (the scanner position), the direction of the vector, and the length of the vector. The aircraft navigation information, along with the offset between the GPS antenna and the scanner aperture, gives the position of the scanner at the time the pulse was emitted. The direction of the laser ray is calculated from the aircraft orientation, obtained from the INS, combined with any angular offsets between the scanner chassis and the INS, and also the angular position of the scanner for the pulse. A stabilized mount may be used for the scanner, removing the influence of aircraft orientation. The time of flight of the pulse gives the length of the vector from the scanner to the reflecting surface.

Multiple returns for a single pulse are filtered next, to remove the vegetation canopy. If only the terrain is of interest, elevated objects such as buildings or trees must also be recognized and removed. Scanning areas with elevated objects will result in "shadows," or missing data, behind the objects. The terrain in the shadow must be interpolated from adjacent points. To minimize shadowing, the total scan angle is kept narrow; however, this reduces the area covered, increasing the number of flights required.

The final elevation points are irregularly spaced horizontally due to the changes in surface elevation and also due to the scanning geometry and aircraft motion effects. To form a standard DEM the data must be interpolated to a regular grid.

Figure 11.41 and 11.42 show two examples of LIDAR data over two entirely different sites.

Figure 11-41 LIDAR image of a rural site. Courtesy of TerraPoint, LLC.

Figure 11-42 LIDAR image of an urban site. Courtesy of TerraPoint, LLC.

11.14.3 Applications of LIDAR

Laser scanning has a great many advantages over conventional surveying methods. It provides high-accuracy data over large areas within extremely short time periods. Many service firms advertise that they can deliver a finished DEM within 24 hours after data acquisition. Since the sensor provides its own illumination, it can also operate at night.

Laser scanning does have its disadvantages, however. Each flight line covers a relatively narrow swath, necessitating merging data from multiple flights to map large areas. Shadowing can be a problem in areas with large terrain relief or in urban areas with tall buildings.

Laser scanning has a number of current applications. The most common one is DEM generation, whether for general mapping purposes or for more specialized uses such as flood plain or coastal mapping. When forested regions are scanned, information on tree height and density can be obtained along with the underlying terrain elevations. A specialized application for which laser scanning is well-suited is right-of-way mapping for transportation or power lines. When mapping power lines, the actual shape of the power line can be determined and checked for ground clearance.

PROBLEMS

11.1 An aircraft carrying a radar system flies 3000 m above the ground. Show the layover effect on a building located 1800 m from the broadside of the aircraft with the sizes as given in the figure below (compute R_Q and R_P).

11.2 If you are the designer of the airborne radar system in Problem 11.1, what can you do to overcome this

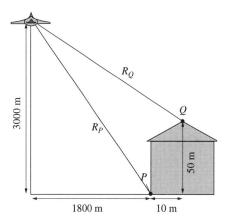

layover problem? Prove this numerically. What other effect will you still see on the image of the building?

11.3   Below are the system specifications of RADARSAT:
- Altitude: 798 km
- Wavelength: 5.6 cm (C-band)
- Transmit pulse length: 42.0 μs
- Antenna size (azimuth × range dimension): 15.0 × 1.5 m

Assuming that the system transmits a simple pulse (not a chirp), what is the resolution along the slant-range direction? What does this mean if we translate it into the (flat) ground resolution, assuming $45°$ as the depression angle?

11.4   The slant-range resolution can be improved if a chirp is used instead of a simple pulse. One of the chirps used in RADARSAT has a bandwidth of 17.3 MHz. Using the specifications in Problem 11.3, what is the slant-range resolution with the chirped pulse?

11.5   Using the specifications in Problem 11.3 and a depression angle of $45°$:
(a) What is the azimuth resolution that can be achieved by this system if the real-aperture technique is used?
(b) If the synthetic-aperture technique is applied, what is the azimuth resolution? Remember that RADARSAT is a strip-map sensor.

11.6   A simple SAR interferogram is given below (the value of each gray value is given in the legend). If the phase at point A is 0, compute the phase at point B. Use a simple integration method (i.e., starting from A, compute the phase difference between two consecutive fringes, and add this difference to the previous fringe to get the value of the next fringe; assuming that the difference between two consecutive fringes lies in the interval $[-\pi, \pi]$).

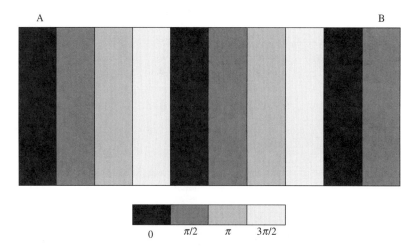

REFERENCES

ABSHIER, J. O., 1992. IFSARE Geometric Model. ERIM Technical Memo. Ann Arbor, MI: Veridian ERIM International, Inc.

ADAMS, G. F., AUSHERMAN, D. A., CRIPPEN, S. L., SOS, G. T., and WILLIAMS, B. P. 1996. The ERIM Interferometric SAR: IFSARE, *IEEE National Radar Conference* pp. 249–254. Ann Arbor, MI: Veridian ERIM International, Inc.

CARRARA, W. G., GOODMAN, R. S., and MAJEWSKI, R. M. 1995. *Spotlight Synthetic Aperture Radar: Signal Processing Algorithms.* Norwood, MA: Artec House. Chapters 1, 2, 3, 9.

CRAIG, R. D. 1999. SAR Radargrammetric Algorithms. BAE Systems Technical Memo, San Diego, CA: BAE Systems.

CUCCI, A. 1992. The TOPSAR interferometric radar topographic mapping instrument. *IEEE Transactions on Geoscience and Remote Sensing* 30(5):933–940.

EUROPEAN SPACE AGENCY. 1995. *SAR Products CCT Specifications,* Rev. 2.1.

FLOOD, M., and GUTELIUS, B. 1997. Commercial implications of topographic terrain mapping using scanning airborne laser radar. *Photogrammetric Engineering and Remote Sensing* 63(4):327–366.

HENDERSON, F. M., and LEWIS, A.J., eds. 1998. Principles and applications of imaging radar. In *Manual of Remote Sensing,* 3rd edition. Vol. 2. Chapters 1, 2, 3, 4.

JAKOWATZ, C. V., WAHL, D. E., EICHEL, P. H., GHIGLIA, D. C., and THOMPSON, P. A., 1996. *Spotlight-Mode Synthetic Aperture Radar: A Signal Processing Approach.* Kluwer Academic Publishers: Boston, MA. Interferometry pp. 273–348.

LEBERL, F. 1990. *Radargrammetric Image Processing.* Norwood, MA: Artec House.

LILLESAND, T. M., and KIEFER, R. W. 1999. *Remote Sensing and Image Interpretation.* 4th edition. New York: John Wiley and Sons.

MADSEN, S., ZEBKER, H., and MARTIN, J. M. 1993. Topographic mapping using radar interferometry: Processing techniques. *IEEE Transactions of Geoscience and Remote Sensing* 31(1):246–255.

MEASURES, R. 1991. *Laser Remote Sensing: Fundamentals and Applications.* Malabar, FL: Kreiger.

MIKHAIL, E. M. 1976. *Observations and Least Squares.* New York: Dun-Donnelley.

RADARSAT INTERNATIONAL. 1997. *RADARSAT Data Products Specification,* Issue 2.

RODRIGUEZ, E., and MARTIN, J. M. 1992. Theory and design of interferometric synthetic aperture radars. *IEEE Proceedings-F* 139(2):147–159.

Appendix A

Mathematics for Photogrammetry

A.1   COORDINATE SYSTEMS

A.1.1   Two-Dimensional Coordinate Systems

A reference system of coordinates in a plane is defined by three elements, two associated with a point origin and one with a direction. This is the minimum requirement. The units of measurement, which represent a scale, may also be included, thus leading to a total of four elements (see the four-parameter transformation in Section A.4).

   Any point can be located within the reference system by two coordinates. Figure A-1 depicts two commonly used two-dimensional coordinate systems: *polar* (r, θ) and

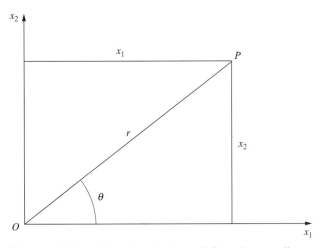

Figure A-1 Two-dimensional polar and Cartesian coordinate systems.

Cartesian or *rectangular* (x_1, x_2). A point P can be located by the angle θ, measured from the reference direction x_1, and the range r, measured from the reference point O. Alternatively, the position of the point P may be determined by its two distances from two perpendicular axes, x_1 and x_2. These distances are called the Cartesian coordinates (x_1, x_2) of any point P. The Cartesian coordinates can be obtained from the polar coordinates (r, θ) by

$$x_1 = r \cos \theta$$
$$x_2 = r \sin \theta$$

(A-1)

Inversely, (r, θ) may be derived from (x_1, x_2) using

$$r = \sqrt{x_1^2 + x_2^2}$$
$$\theta = \tan^{-1} \frac{x_2}{x_1}$$

(A-2)

A.1.2 Three-Dimensional Coordinate Systems

A coordinate system in three-dimensional space requires six elements for its definition, three associated with a reference point and three with orientation. If the linear unit of measurement is also to be fixed, the minimum number of required elements becomes seven (see also the seven-parameter transformation in Section A.4).

Figure A-2 shows two systems of three-dimensional coordinates: *spherical* (r, α, β) and *Cartesian* or *rectangular* (x_1, x_2, x_3). In the spherical system, any point P is located by its distance r from the origin, the angle α between the x_1-axis and r' (the projection or r onto the x_1-x_2 plane), and the angle β between r and r'. The Cartesian coordinate system is composed of three mutually perpendicular axes x_1, x_2, and x_3. The system depicted in Fig. A-2 is *right-handed* since a right-threaded screw rotated by an angle less than 90° from $+x_1$ to $+x_2$ would advance in the direction of $+x_3$. In general, any point may be determined in space by three coordinates (x_1, x_2, x_3). The relations between the spherical and Cartesian coordinate systems are as follows ($r' = r \cos \beta$):

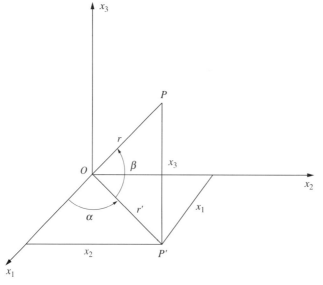

Figure A-2 Three-dimensional spherical and right-handed Cartesian coordinate systems.

$$x_1 = r' \cos\alpha = r \cos\alpha \cos\beta$$
$$x_2 = r' \sin\alpha = r \sin\alpha \cos\beta \qquad \text{(A-3)}$$
$$x_3 = r \sin\beta$$

$$r = \sqrt{x_1^2 + x_2^2 + x_3^2}$$
$$\alpha = \tan^{-1}\frac{x_2}{x_1} \qquad \text{(A-4)}$$
$$\beta = \sin^{-1}\frac{x_3}{r}$$

A.2   VECTOR ALGEBRA

A *vector* is an entity which has a magnitude and a direction. In two- and three-dimensional spaces, it is a directed line segment from one point to another. This will later be generalized to multi-dimensional cases. A vector may be represented by a single lower-case boldface letter, for example, a, or by $\overrightarrow{PQ}$ to represent the vector from point P to point Q. An example is shown in Fig. A-3. The projections of the vector a on the x_1 and x_2 axes are a_1 and a_2, which are called the vector *components*. We represent the vector components in a column matrix,

$$a = \begin{bmatrix} a_1 \\ a_2 \end{bmatrix}$$

It is clear from Fig. A-3 that $a_1 = x_{1Q} - x_{1P}$ and $a_2 = x_{2Q} - x_{2P}$. Therefore, the components of a vector can be obtained by subtracting the coordinates of its initial point P from the coordinates of its terminal point Q. Since the vectors p and q in Fig. A-3 begin at the origin,

$$p = \begin{bmatrix} x_{1P} \\ x_{2P} \end{bmatrix} \quad \text{and} \quad q = \begin{bmatrix} x_{1Q} \\ x_{2Q} \end{bmatrix}$$

and the vector a is given by

$$\begin{bmatrix} a_1 \\ a_2 \end{bmatrix} = \begin{bmatrix} x_{1Q} \\ x_{2Q} \end{bmatrix} - \begin{bmatrix} x_{1P} \\ x_{2P} \end{bmatrix}$$

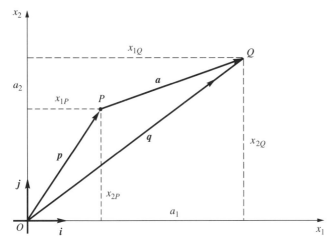

Figure A-3 A vector in a plane.

or

$$a = q - p \tag{A-5}$$

Expanding to three dimensions, we write

$$a = \begin{bmatrix} a_1 \\ a_2 \\ a_3 \end{bmatrix}$$

The *length* of the vector is designated by $|a|$ and is given by

$$|a| = \sqrt{a_1^2 + a_2^2 + a_3^2} \tag{A-6}$$

A vector's *direction* is given either by the angles it makes with the axes, α, β, and γ, or by their cosines. The latter are called *direction cosines* and are given by

$$\cos\alpha = \frac{a_1}{|a|} \quad \cos\beta = \frac{a_2}{|a|} \quad \cos\gamma = \frac{a_3}{|a|} \tag{A-7}$$

Since a direction in space is completely determined by only two angles (in Fig. A-2, $\overrightarrow{OP}$ is fixed by α and β), only two of the direction cosines are independent. Consequently, the direction cosines are related by the equation

$$\cos^2\alpha + \cos^2\beta + \cos^2\gamma = 1 \tag{A-8}$$

which can be readily proven from Eqs. A-6 and A-7. We may now generalize a vector to n dimensions, or

$$a = \begin{bmatrix} a_1 \\ a_2 \\ \vdots \\ a_n \end{bmatrix}$$

A.2.1 Vector Operations

Two vectors are *equal*, $a = b$, when all of the corresponding components of the vectors are equal, $a_1 = b_1, a_2 = b_2, ..., a_n = b_n$.

Vectors are added or subtracted by adding or subtracting each individual component. Therefore, $c = a \pm b$ means that $c_1 = a_1 \pm b_1, c_2 = a_2 \pm b_2, ..., c_n = a_n \pm b_n$. Addition and subtraction of vectors are *commutative*, $a + b = b + a$; *associative*, $(a + b) + c = a + (b + c)$; and *transitive*, $a = c$ and $a = b$ if $b = c$. Only vectors with the same number of components may be added or subtracted.

A *scalar* is a quantity which has only magnitude but no direction, such as mass, temperature, time, etc., and will be designated here by a lower-case Greek letter. Vectors of any number of components can be multiplied by a scalar by multiplying each component by the scalar:

$$\lambda a = \begin{bmatrix} \lambda a_1 \\ \lambda a_2 \\ \vdots \\ \lambda a_n \end{bmatrix}$$

Multiplication of vectors by scalars also has the following properties:

$$\begin{aligned} \lambda(\mu a) &= (\lambda\mu)a = \mu(\lambda a) \\ (\lambda + \mu)a &= \lambda a + \mu a \\ \lambda(a + b) &= \lambda a = \lambda b \\ |\lambda a| &= \lambda|a| \end{aligned} \tag{A-9}$$

Any vector a is reduced to a *unit vector* a^0 when dividing its components by its length, which is a scalar:

$$a^0 = \frac{a}{|a|}$$

The components of a^0 are the direction cosines of a. Unit vectors along the coordinate axes are called *base vectors* and are given by

$$i = \begin{bmatrix} 1 \\ 0 \\ 0 \end{bmatrix} \qquad j = \begin{bmatrix} 0 \\ 1 \\ 0 \end{bmatrix} \qquad k = \begin{bmatrix} 0 \\ 0 \\ 1 \end{bmatrix} \tag{A-10}$$

(see Fig. A-3 for i and j). Any vector in three-dimensional space is uniquely expressed as

$$a = a_1 i + a_2 j + a_3 k \tag{A-11}$$

The right-handed system introduced in Section A.1.2 can be generalized for three vectors a, b, and c. If they are not coplanar, and have the same initial point, then they are said to form a *right-handed* system if a right-threaded screw rotated through an angle *less than 180°* from a to b would advance in the direction of c.

A.2.2   Vector Products

Dot (or Scalar) Product

The *dot product* or *scalar product* of two vectors is

$$a \cdot b = \sum_{p=1}^{n} a_p b_p = a_1 b_1 + a_2 b_2 + \ldots + a_n b_n \tag{A-12}$$

This is also called *inner product*. It is a scalar and has the following properties:

$$a \cdot b = b \cdot a$$
$$a \cdot (b + c) = a \cdot b + a \cdot c$$
$$\lambda(a \cdot b) = (\lambda a) \cdot b = a \cdot (\lambda b) = (a \cdot b)\lambda \tag{A-13}$$
$$i \cdot i = j \cdot j = k \cdot k = 1$$
$$i \cdot j = j \cdot k = k \cdot i = 0$$

The dot product of a vector with itself is equal to the square of its length, or

$$a \cdot a = a_1^2 + a_2^2 + \ldots + a_n^2 = |a|^2 \tag{A-14}$$

If θ is the angle between two vectors a and b (in two- or three-dimensional space), it can be shown that

$$a \cdot b = |a||b| \cos \theta \tag{A-15}$$

It follows that if a is perpendicular to b, then $a \cdot b = 0$.

Cross (or Vector) Product

The *cross product* or *vector product* of two vectors, $a \times b$ (read "a cross b"), is another vector c which is perpendicular to both a and b and in a direction such that a, b, and c (in this order) form a right-handed system. The length of c is given by

$$|c| = |a \times b| = |a||b| \sin \theta \tag{A-16}$$

where θ is the angle between a and b. This quantity is the area of the parallelogram determined by a and b. If $a = a_1 i + a_2 j + a_3 k$ and $b = b_1 i + b_2 j + b_3 k$, then c is given by the determinant (see Section A.3.4 for the definition of *determinant*)

$$c = a \times b = \begin{vmatrix} i & j & k \\ a_1 & a_2 & a_3 \\ b_1 & b_2 & b_3 \end{vmatrix} = \begin{bmatrix} a_2 b_3 - b_2 a_3 \\ b_1 a_3 - a_1 b_3 \\ a_1 b_2 - b_1 a_2 \end{bmatrix} \tag{A-17}$$

The cross product has the following properties:

$$a \times b = -(b \times a)$$
$$a \times (b + c) = a \times b + a \times c$$
$$a \cdot (a \times c) = 0$$
$$|a \times b|^2 = |a|^2 |b|^2 - (a \cdot b)^2 \tag{A-18}$$
$$i \times i = j \times j = k \times k = 0$$
$$i \times j = k, \quad j \times k = i, \quad k \times i = j$$

For two nonzero vectors, if $a \times b = 0$, then a and b are parallel.

Scalar Triple Product

The *scalar triple product* of three vectors a, b, and c is a scalar given by the determinant

$$a \times b \cdot c = \begin{vmatrix} a_1 & a_2 & a_3 \\ b_1 & b_2 & b_3 \\ c_1 & c_2 & c_3 \end{vmatrix} \tag{A-19}$$

which is equal to the volume of the parallelepiped determined by a, b, and c. If the scalar triple product is zero, then the three vectors are coplanar. It has the following properties:

$$a \times b \cdot c = b \times c \cdot a = c \times a \cdot b$$
$$a \times b \cdot c = a \cdot b \times c$$

Planes and Lines

If p_0 is a vector from the origin to a given point in a plane, n is a nonzero vector normal to the plane, and p is a vector from the origin to any other point in the plane, then the equation of the plane takes the form

$$(p - p_0) \cdot n = 0 \quad \text{or} \quad p \cdot n - p_0 \cdot n = 0 \tag{A-20}$$

Let $n = Ai + Bj + Ck$, $p_0 = X_0 i + Y_0 j + Z_0 k$, and $p = Xi + Yj + Zk$. Then Eq. A-20 becomes

$$A(X - X_0) + B(Y - Y_0) + C(Z - Z_0) = 0$$

or

$$AX + BY + CZ + D = 0 \tag{A-21}$$

where $D = -(AX_0 + BY_0 + CZ_0)$. Two planes are parallel when they have a common normal vector n, and are perpendicular when their normals are perpendicular, or $n_1 \cdot n_2 = 0$.

If p_0 represents a given point on a line, p is any other point on the line, and v is a given nonzero vector parallel to the line, then

$$p = p_0 + \lambda v \tag{A-22}$$

is an equation of the line. In component form, it yields three scalar equations describing the parametric form (λ is the running parameter)

$$X = X_0 + \lambda v_x$$
$$Y = Y_0 + \lambda v_y \qquad \text{(A-23)}$$
$$Z = Z_0 + \lambda v_z$$

If λ is eliminated, one gets the usual two-equation form of a straight line in space.

A.3 MATRIX ALGEBRA

A.3.1 Definitions

A *matrix* is a group of numbers or scalar functions collected in a two-dimensional (rectangular) array. The following are examples of matrices:

$$\begin{bmatrix} 1 & -2 & 0 \\ -6 & 4 & 3 \end{bmatrix}, \quad \begin{bmatrix} 1 \\ 5 \end{bmatrix}, \quad \begin{bmatrix} 7 & 9 & 3 \end{bmatrix}, \quad \begin{bmatrix} a & b \\ c & d \end{bmatrix}$$
$$\quad (a) \qquad\qquad (b) \qquad (c) \qquad\quad (d)$$

Every matrix has a specified number of rows and columns. Thus the example matrix (a) has 2 rows and 3 columns and is said to be a 2×3 matrix (read "two by three"). Similarly, (b) is a 2×1 matrix, (c) is a 1×3 matrix, and (d) is a 2×2 matrix. The two numbers representing the rows and columns are referred to as the *matrix dimensions*.

A matrix is designated by a boldface capital letter. Thus, an $m \times n$ matrix can be symbolically written as

$$\underset{m \times n}{\boldsymbol{A}} = \begin{bmatrix} a_{11} & a_{12} & \dots & a_{1n} \\ a_{21} & a_{22} & \dots & a_{2n} \\ \dots & \dots & \dots & \dots \\ a_{m1} & a_{m2} & \dots & a_{mn} \end{bmatrix}$$

A lower-case letter with a double subscript designates an element in a matrix. Thus a_{ij} represents a typical element of the matrix $\boldsymbol{A}$. The first subscript, i, refers to the number of the row containing a_{ij}, starting with 1 at the top and proceeding down to m at the bottom. The second subscript, j, refers to the number of the column containing a_{ij}, starting with 1 at the left and proceeding to n at the right. Thus a_{ij} lies at the intersection of the ith row and jth column. For example, a_{23} in the previous example matrix (a) is 3, while a_{12} in matrix (c) is 9. The smallest matrix dimension is 1×1.

A.3.2 Types of Matrices

A *square matrix* is a matrix in which the number of rows equals the number of columns. If $\boldsymbol{A}$ is a square matrix with m rows and m columns, $\boldsymbol{A}$ is of *order m*. The *principal diagonal* (or *main diagonal*) of a square matrix is composed of all elements a_{ij} for which $i = j$. The following are examples of square matrices:

$$\boldsymbol{A} = \begin{bmatrix} 1 & 2 \\ 3 & 4 \end{bmatrix}, \qquad \boldsymbol{B} = \begin{bmatrix} a & b & c \\ d & e & f \\ g & h & k \end{bmatrix}$$

The main diagonal of $\boldsymbol{A}$ is composed of the elements 1 and 4; the main diagonal of $\boldsymbol{B}$ contains the elements a, e, and k.

A *row matrix* is a matrix composed of only one row. It is designated by a lower-case boldface letter. The following are examples of row matrices:

$$\underset{1 \times n}{\boldsymbol{a}} = \begin{bmatrix} a_1 & a_2 & \dots & a_n \end{bmatrix}, \qquad \underset{1 \times 3}{\boldsymbol{d}} = \begin{bmatrix} 1 & 2 & 4 \end{bmatrix}$$

A *column matrix*, or *column vector*, is a matrix composed of only one column. For example,

$$\underset{m \times 1}{\boldsymbol{b}} = \begin{bmatrix} b_1 \\ b_2 \\ \vdots \\ b_m \end{bmatrix}, \qquad \underset{2 \times 1}{\boldsymbol{c}} = \begin{bmatrix} -1 \\ 3 \end{bmatrix}$$

This is the same definition of a vector introduced in Section A.2.

A *diagonal matrix* is a square matrix in which all elements not on the main diagonal are zero. For example,

$$\boldsymbol{D} = \begin{bmatrix} d_{11} & 0 & \dots & 0 \\ 0 & d_{22} & \dots & 0 \\ \dots & \dots & \dots & \dots \\ 0 & 0 & \dots & d_{mm} \end{bmatrix}$$

where

$$d_{ij} = 0 \text{ for all } i \neq j$$
$$d_{ij} \neq 0 \text{ for some or all } i = j$$

The following are examples of diagonal matrices:

$$\boldsymbol{A} = \begin{bmatrix} 1 & 0 & 0 \\ 0 & 0 & 0 \\ 0 & 0 & -3 \end{bmatrix}, \qquad \boldsymbol{B} = \begin{bmatrix} p & 0 & 0 \\ 0 & q & 0 \\ 0 & 0 & r \end{bmatrix}$$

A *scalar matrix* is a diagonal matrix whose main diagonal elements are *all* equal to the same scalar. For example,

$$\boldsymbol{A} = \begin{bmatrix} a & 0 & \dots & 0 \\ 0 & a & \dots & 0 \\ \dots & \dots & \dots & \dots \\ 0 & 0 & \dots & a \end{bmatrix}$$

where $a_{ij} = 0$ for all $i \neq j$ and $a_{ij} = a$ for all $i = j$ and

$$\boldsymbol{B} = \begin{bmatrix} 2 & 0 & 0 \\ 0 & 2 & 0 \\ 0 & 0 & 2 \end{bmatrix}$$

are scalar matrices.

A *unit* or *identity matrix* is a diagonal matrix whose main diagonal elements are all equal to 1. A unit matrix will always be referred to by $\boldsymbol{I}$. Thus,

$$\boldsymbol{I} = \begin{bmatrix} 1 & 0 & \dots & 0 \\ 0 & 1 & \dots & 0 \\ \dots & \dots & \dots & \dots \\ 0 & 0 & \dots & 1 \end{bmatrix}$$

in which $a_{ij} = 0$ for all $i \neq j$ and $a_{ij} = 1$ for all $i = j$. A *null* or *zero matrix* is a matrix whose elements are all zero. It is denoted by a boldface zero, $\boldsymbol{0}$.

A *triangular matrix* is a square matrix whose elements above (or below), but not including, the main diagonal are all zero. An *upper* triangular matrix takes the form

$$A = \begin{bmatrix} a_{11} & a_{12} & \cdots & a_{1m} \\ 0 & a_{22} & \cdots & a_{2m} \\ \cdots & \cdots & \cdots & \cdots \\ 0 & 0 & \cdots & a_{mm} \end{bmatrix}$$

with $a_{ij} = 0$ for $i > j$. The matrix

$$A = \begin{bmatrix} -1 & 2 & 4 \\ 0 & 1 & 0 \\ 0 & 0 & 3 \end{bmatrix}$$

is an example of an upper triangular matrix of order 3. A *lower* triangular matrix is of the form

$$A = \begin{bmatrix} a_{11} & 0 & \cdots & 0 \\ a_{21} & a_{22} & \cdots & 0 \\ \cdots & \cdots & \cdots & \cdots \\ a_{m1} & a_{m2} & \cdots & a_{mm} \end{bmatrix}$$

where $a_{ij} = 0$ for $i < j$. The matrix

$$B = \begin{bmatrix} 18 & 0 \\ 2 & -11 \end{bmatrix}$$

is a lower triangular matrix of order 2.

A.3.3 Basic Matrix Operations

Several matrix operations are similar to their equivalents for scalars: equality, addition, subtraction, and multiplication. Division does not exist in matrix algebra; instead another operation, *inversion* replaces it. Additional operations are specific to matrices without scalar equivalents: transpose, multiplication by a scalar, and trace.

Two matrices A and B are *equal* if they are of the same dimensions and each element $a_{ij} = b_{ij}$ for all i and j. Matrices of different dimensions cannot be equal. The sum of two matrices A and B is possible only if they are of equal dimensions, and the elements of the resulting matrix C are $c_{ij} = a_{ij} + b_{ij}$ for all i and j. The following relations apply to addition (and subtraction) of matrices:

$$A + B = B + A$$
$$A + (B + C) = (A + B) + C = A + B + C \quad \text{(A-24)}$$
$$A + (-A) = 0$$

where 0 is the zero or null matrix and $-A$ is the matrix composed of $-a_{ij}$ as elements. For example, if

$$A = \begin{bmatrix} 1 & -2 & 0 \\ 0 & 4 & 6 \end{bmatrix}, \qquad B = \begin{bmatrix} 1 & -3 & 2 \\ 0 & 2 & 6 \end{bmatrix}, \qquad C = \begin{bmatrix} x & y & z \\ u & v & w \end{bmatrix}$$

and $C = B - A$, to compute the values of the six elements x, y, z, u, v, and w of C, first compute $B - A$:

$$\begin{bmatrix} 1 & -3 & 2 \\ 0 & 2 & 6 \end{bmatrix} - \begin{bmatrix} 1 & -2 & 0 \\ 0 & 4 & 6 \end{bmatrix} = \begin{bmatrix} 0 & -1 & 2 \\ 0 & -2 & 0 \end{bmatrix}$$

then form

$$C = \begin{bmatrix} x & y & z \\ u & v & w \end{bmatrix} = \begin{bmatrix} 0 & -1 & 2 \\ 0 & -2 & 0 \end{bmatrix}$$

Thus, $x = 0$, $y = -1$, $z = 2$, $u = 0$, $v = -2$, and $w = 0$.

Multiplication of a matrix by a scalar α results in another matrix $B = \alpha A$ whose elements are $b_{ij} = \alpha a_{ij}$ for all i and j. The following relations hold for scalar multiplication (λ and μ are scalars):

$$\begin{aligned} \lambda(A + B) &= \lambda A + \lambda B \\ (\lambda + \mu)A &= \lambda A + \mu A \\ \lambda(AB) &= (\lambda A)B = A(\lambda B) \\ \lambda(\mu A) &= (\lambda \mu)A \end{aligned} \tag{A-25}$$

The product of two matrices is another matrix. The two matrices must be *conformable for multiplication*, which means that the number of columns of the first matrix must equal the number of rows of the second matrix. Thus, if A is an $m \times q$ matrix and B is a $q \times n$ matrix, the product AB, *in that order*, is another matrix C with m rows (as in A) and n columns (as in B). Each element c_{ij} in C is obtained by multiplying each one of the q elements in the ith row of A by the corresponding element in the jth column of B and adding. Algebraically, this is written as

$$c_{ij} = a_{i1}b_{1j} + a_{i2}b_{2j} + \dots + a_{iq}b_{qj} = \sum_{k=1}^{q} a_{ik}b_{kj} \tag{A-26}$$

To illustrate matrix multiplication:

$$\underset{2 \times 1}{C} = \underset{2 \times 3}{A} \ \underset{3 \times 1}{B} = \begin{bmatrix} 1 & 0 & 2 \\ 2 & 1 & 0 \end{bmatrix} \begin{bmatrix} 1 \\ 5 \\ 3 \end{bmatrix} = \begin{bmatrix} (1 \times 1) + (0 \times 5) + (2 \times 3) \\ (2 \times 1) + (1 \times 5) + (0 \times 3) \end{bmatrix} = \begin{bmatrix} 7 \\ 7 \end{bmatrix}$$

Matrix multiplication is *not* commutative; that is, in general $FG \neq GF$ even if the dimensions of the matrices allow multiplication in both directions (e.g., $m \times n$ and $n \times m$, or square matrices of the same order). For example:

$$FG = \begin{bmatrix} 1 & 2 \\ 5 & 0 \end{bmatrix} \begin{bmatrix} 3 & 4 \\ 0 & 2 \end{bmatrix} = \begin{bmatrix} 3 & 8 \\ 15 & 20 \end{bmatrix}$$

$$GF = \begin{bmatrix} 3 & 4 \\ 0 & 2 \end{bmatrix} \begin{bmatrix} 1 & 2 \\ 5 & 0 \end{bmatrix} = \begin{bmatrix} 23 & 6 \\ 10 & 0 \end{bmatrix}$$

with the obvious result that $FG \neq GF$.

The following relationships regarding matrix multiplication hold:

$$\begin{aligned} &AI = IA = A \\ &AB \neq BA \\ &A(BC) = (AB)C = ABC \quad (associative\ law) \\ &A(B + C) = AB + AC \quad (distributive\ law) \\ &(A + B)C = AC + BC \quad (distributive\ law) \end{aligned} \tag{A-27}$$

One important property of matrix multiplication which distinguishes it from scalar multiplication is that the product of two matrices can be the null or zero matrix without either matrix being the zero matrix, or $AB = 0$ with $A \neq 0$, $B \neq 0$, as for example:

$$AB = \begin{bmatrix} 1 & 1 \\ 0 & 0 \end{bmatrix} \begin{bmatrix} 2 & 3 \\ -2 & -3 \end{bmatrix} = \begin{bmatrix} 0 & 0 \\ 0 & 0 \end{bmatrix}$$

Also, $AB = AC$ does *not* imply $B = C$.

The *transpose* of the $m \times n$ matrix A is an $n \times m$ matrix formed from A by interchanging rows and columns such that the ith row of A becomes the ith column of the transposed matrix. We denote the transpose of A by A^T. If $B = A^T$, it follows that $b_{ij} = a_{ji}$ for all i and j. For example, if

$$B = \begin{bmatrix} -1 & 6 \\ 0 & 3 \\ -5 & 0 \end{bmatrix} \text{ then } B^T = \begin{bmatrix} -1 & 0 & -5 \\ 6 & 3 & 0 \end{bmatrix}$$

The following relationships apply to the transpose of a matrix:

$$\begin{aligned} (A + B)^T &= A^T + B^T \\ (\alpha A)^T &= \alpha(A^T) \\ (A^T)^T &= A \\ (AB)^T &= B^T A^T \end{aligned} \tag{A-28}$$

Note the reverse order in the matrix multiplication (last) relationship.

A square matrix is *symmetric* if it is equal to its transpose; A is symmetric if $A^T = A$. Since transposing a matrix does not change the elements of the main diagonal, the elements above the main diagonal of a symmetric matrix are mirror images of those below the diagonal. For example,

$$\begin{bmatrix} 3 & 2 & -1 \\ 2 & 0 & 6 \\ -1 & 6 & 4 \end{bmatrix} \text{ and } \begin{bmatrix} a & b \\ b & c \end{bmatrix} \text{ are symmetric matrices.}$$

Diagonal, scalar, and identity matrices are symmetric, since each is equal to its transpose. For any matrix A (not necessarily square), both AA^T and A^TA are symmetric. If B is a symmetric matrix of suitable dimensions, then for any matrix A, both ABA^T and A^TBA are also symmetric.

If a is a column matrix (or vector), then a^Ta is a positive scalar which is equal to the sum of the squares of its elements, for example

$$a^Ta = \begin{bmatrix} a_1 & a_2 & a_3 \end{bmatrix} \begin{bmatrix} a_1 \\ a_2 \\ a_3 \end{bmatrix} = a_1^2 + a_2^2 + a_3^2$$

This is the same as the dot or inner product of the vector a with itself, which is equal to the square of its length.

A square matrix is called *skew-symmetric* if it is equal to the negative of its transpose, or $A^T = -A$ and $a_{ij} = -a_{ji}$ for all i and j. It follows that the diagonal elements of a skew-symmetric matrix are all zero, and the only matrix that is both symmetric and skew-symmetric is the null or zero matrix, $\mathbf{0}$. An example of a skew-symmetric matrix is

$$\begin{bmatrix} 0 & 1 & -3 \\ -1 & 0 & 6 \\ 3 & -6 & 0 \end{bmatrix}$$

For any square matrix A, the matrix $(A + A^T)$ is symmetric and $(A - A^T)$ is skew-symmetric.

The *trace* of a *square* matrix is the scalar which is equal to the sum of its main-diagonal elements. It is denoted by tr(A). For example, the trace of

$$A = \begin{bmatrix} 1 & 4 & 7 \\ 2 & 5 & 8 \\ 3 & 6 & 9 \end{bmatrix}$$

is

$$\text{tr}(A) = 1 + 5 + 9 = 15$$

The following are properties of the trace:

$$\begin{aligned} \text{tr}(A) &= \text{tr}(A^T) \\ \text{tr}(\lambda A) &= \lambda \text{tr}(A) \\ \text{tr}(A + B) &= \text{tr}(A) + \text{tr}(B) \\ \text{tr}(AB) &= \text{tr}(BA) \\ \text{tr}(FAF^{-1}) &= \text{tr}(A) \end{aligned} \qquad \text{(A-29)}$$

where F is a nonsingular matrix.

A.3.4 Matrix Inverse

As mentioned earlier, *division* of matrices is not defined, and $AB = AC$ does not imply $B = C$. In place of division, the concept of *matrix inversion* is used. The *inverse* of a *square* matrix A, *if it exists*, is the unique matrix A^{-1} with the following property:

$$AA^{-1} = A^{-1}A = I \qquad \text{(A-30)}$$

where I is the identity matrix. Thus, for

$$A = \begin{bmatrix} 3 & 1 \\ 2 & 1 \end{bmatrix}$$

the matrix

$$A^{-1} = \begin{bmatrix} 1 & -1 \\ -2 & 3 \end{bmatrix}$$

is its inverse because

$$\begin{bmatrix} 1 & -1 \\ -2 & 3 \end{bmatrix} \begin{bmatrix} 3 & 1 \\ 2 & 1 \end{bmatrix} = \begin{bmatrix} 1 & 0 \\ 0 & 1 \end{bmatrix}$$

and

$$\begin{bmatrix} 3 & 1 \\ 2 & 1 \end{bmatrix} \begin{bmatrix} 1 & -1 \\ -2 & 3 \end{bmatrix} = \begin{bmatrix} 1 & 0 \\ 0 & 1 \end{bmatrix}$$

The properties of the inverse are

$$\begin{aligned} (AB)^{-1} &= B^{-1}A^{-1} \\ (A^{-1})^{-1} &= A \\ (A^T)^{-1} &= (A^{-1})^T \\ (\lambda A)^{-1} &= \frac{1}{\lambda}A^{-1} \end{aligned} \qquad \text{(A-31)}$$

Note the reversal of the order for distribution over a product, just as for the transpose. A square matrix that has an inverse is called *nonsingular*. A matrix that does not have an inverse is called *singular*.

It was shown previously that AB can equal 0 without either $A = 0$ or $B = 0$. If, however, either A or B is nonsingular, then the other matrix must be a null matrix. Hence, the product of two nonsingular matrices cannot be a null or zero matrix.

In order to present a method for computing a matrix inverse, the concept of determinants is first introduced. Associated with each *square* matrix A is a unique scalar value called the *determinant* of A. It is denoted either by $\det(A)$ or by $|A|$. Thus, for

$$A = \begin{bmatrix} 3 & 1 \\ 1 & 2 \end{bmatrix}$$

the determinant is expressed as

$$|A| = \begin{vmatrix} 3 & 1 \\ 1 & 2 \end{vmatrix}$$

The student should be careful to differentiate between the square brackets used for the matrix and the vertical lines used for the determinant. The computation of the value of the determinant, which is a scalar, will now be defined.

The determinant of order n (for an $n \times n$ square matrix) can be defined recursively in terms of determinants of order $n - 1$. In order to apply this procedure, the determinant of a 1×1 matrix must be defined. Accordingly, for a matrix consisting of a single element, the determinant is defined as the value of the element, that is, for $A = \begin{bmatrix} a_{11} \end{bmatrix}$, $|A| = \det(A) = a_{11}$.

If A is an $n \times n$ matrix, and one row and one column of A are deleted, the resulting matrix is an $(n - 1) \times (n - 1)$ *submatrix* of A. The determinant of such a submatrix is called a *minor* of A, and it is designated by m_{ij}, where i and j correspond to the deleted row and column, respectively. More specifically, m_{ij} is known as the *minor of the element a_{ij} in A*. For example, consider

$$A = \begin{bmatrix} a_{11} & a_{12} & a_{13} \\ a_{21} & a_{22} & a_{23} \\ a_{31} & a_{32} & a_{33} \end{bmatrix}$$

Each element of A has a minor. The minor of a_{11}, for example, is obtained by deleting the first row and first column from A and taking the determinant of the 2×2 submatrix that remains,

$$m_{11} = \begin{vmatrix} a_{22} & a_{23} \\ a_{32} & a_{33} \end{vmatrix}$$

The *cofactor* c_{ij} of an element a_{ij} is defined as

$$c_{ij} = (-1)^{i+j} m_{ij} \tag{A-32}$$

Obviously, when the sum of the row number i and column number j is even, $c_{ij} = m_{ij}$, and when $i + j$ is odd, $c_{ij} = -m_{ij}$.

The determinant of an $n \times n$ matrix A can now be defined as

$$|A| = a_{11}c_{11} + a_{12}c_{12} + \ldots + a_{1n}c_{1n} \tag{A-33}$$

which states that the determinant of A is the sum of the products of the elements of the first row of A and their corresponding cofactors. (It is equally possible to define $|A|$

in terms of any other row or column, but for simplicity, we used the first row.) On the basis of the definition, the 2×2 matrix

$$A = \begin{bmatrix} a_{11} & a_{12} \\ a_{21} & a_{22} \end{bmatrix}$$

has cofactors $c_{11} = |a_{22}| = a_{22}$, and $c_{12} = -|a_{21}| = -a_{21}$, the determinant of A is

$$|A| = a_{11}c_{11} + a_{12}c_{12} = a_{11}a_{22} - a_{12}a_{21}$$

Thus, for example,

$$\begin{vmatrix} 3 & -1 \\ 2 & 4 \end{vmatrix} = (3)(4) + (-1)(-2) = 14$$

The *cofactor matrix* C of a matrix A is the square matrix of the same order as A in which each element a_{ij} is replaced by its cofactor c_{ij}. For example, the cofactor matrix of

$$A = \begin{bmatrix} 1 & 2 \\ -3 & 4 \end{bmatrix}$$

is

$$C = \begin{bmatrix} 4 & 3 \\ -2 & 1 \end{bmatrix}$$

The *adjoint matrix* of A, denoted by adj (A), is the transpose of its cofactor matrix,

$$\text{adj}(A) = C^T \tag{A-34}$$

It can be shown that

$$A \cdot \text{adj}(A) = \text{adj}(A) \cdot A = |A|I \tag{A-35}$$

Comparison of Eqs. A-30 and A-35 leads directly to a procedure for evaluating the inverse from the adjoint matrix, namely,

$$A^{-1} = \frac{\text{adj}(A)}{|A|} \tag{A-36}$$

For example

$$A = \begin{bmatrix} 3 & 1 \\ 2 & 1 \end{bmatrix}, |A| = 1, C = \begin{bmatrix} 1 & -2 \\ -1 & 3 \end{bmatrix}, \text{adj}(A) = \begin{bmatrix} 1 & -1 \\ -2 & 3 \end{bmatrix}$$

and the inverse of A is

$$A^{-1} = \frac{\text{adj}(A)}{|A|} = \begin{bmatrix} 1 & -1 \\ -2 & 3 \end{bmatrix}$$

Note that for a 2×2 matrix, the adjoint matrix is simply

$$\begin{bmatrix} a_{22} & -a_{12} \\ -a_{21} & a_{11} \end{bmatrix}$$

A square matrix is called *orthogonal* if its inverse is equal to its transpose, or $A^{-1} = A^T$. Thus a matrix M is orthogonal when

$$M^TM = MM^T = I \tag{A-37}$$

The columns of an orthogonal matrix are mutually orthogonal vectors of unit length. Also, for an orthogonal matrix,

$$|M| = \pm 1 \tag{A-38}$$

When $|M| = +1$, then M is called *proper orthogonal;* otherwise it is termed *improper orthogonal.* The product of two orthogonal matrices is also an orthogonal matrix.

A.3.5 Matrix Inverse by Partitioning

Let A be an $n \times n$ square nonsingular matrix whose inverse is to be evaluated. We partition A in the form

$$A = \begin{bmatrix} \overset{s}{A_{11}} & \overset{m}{A_{12}} \\ A_{21} & A_{22} \end{bmatrix} \begin{matrix} s \\ m \end{matrix}$$

where A_{11} is $s \times s$, A_{12} is $s \times m$, A_{21} is $m \times s$, A_{22} is $m \times m$, and $m + s = n$. The inverse A^{-1} exists (since A is nonsingular) and we shall denote it, in the correspondingly partitioned form, by

$$A^{-1} = B = \begin{bmatrix} B_{11} & B_{12} \\ B_{21} & B_{22} \end{bmatrix}$$

From the basic definition of an inverse we have $AA^{-1} = AB = I$, or in the partitioned form

$$\begin{bmatrix} A_{11} & A_{12} \\ A_{21} & A_{22} \end{bmatrix} \begin{bmatrix} B_{11} & B_{12} \\ B_{21} & B_{22} \end{bmatrix} = \begin{bmatrix} I_s & 0 \\ 0 & I_m \end{bmatrix}$$

where I_s and I_m are identity matrices of orders s and m, respectively. This leads to

$$\begin{aligned} A_{11}B_{11} + A_{12}B_{21} &= I_s \\ A_{11}B_{12} + A_{12}B_{22} &= 0 \\ A_{21}B_{11} + A_{22}B_{21} &= 0 \\ A_{21}B_{12} + A_{22}B_{22} &= I_m \end{aligned} \tag{A-39}$$

The solution of Eqs. A-39 when A_{11}^{-1} exists is given by

$$\begin{aligned} B_{11} &= A_{11}^{-1} - A_{11}^{-1} A_{12}B_{21} \\ B_{12} &= -A_{11}^{-1} A_{12}B_{22} \\ B_{21} &= -B_{22} A_{21}A_{11}^{-1} \\ B_{22} &= (A_{22} - A_{21}A_{11}^{-1} A_{12})^{-1} \end{aligned} \tag{A-40}$$

Alternatively, when A_{22}^{-1} exists, the solution is

$$\begin{aligned} B_{11} &= (A_{11} - A_{12} A_{22}^{-1} A_{21})^{-1} \\ B_{12} &= -B_{11}A_{12} A_{22}^{-1} \\ B_{21} &= -A_{22}^{-1} A_{21} B_{11} \\ B_{22} &= A_{22}^{-1} - A_{22}^{-1} A_{21} B_{12} \end{aligned} \tag{A-41}$$

If A is originally a symmetric matrix, then $A_{21} = A_{12}^T$ and, correspondingly, $B_{21} = B_{12}^T$.

In inverting by partitioning we end up computing directly the inverse of matrices of a smaller order than the original matrix. Inversion by partitioning can be performed in more than one step. It is usually used when one of the submatrices of A has a structure that can be easily inverted, such as a diagonal or block-diagonal structure.

The *rank* of a matrix is the order of the largest nonzero determinant that can be formed from the elements of the matrix by appropriate deletion of rows or columns (or both). Thus a matrix is said to be of *rank m* if and only if it has *at least one nonsingular submatrix of order m*, but has *no* nonsingular submatrix of order more than *m*. A nonsingular matrix of order *n* has a rank *n*. A matrix with zero rank has elements that must all be zero.

The inverse A^{-1} is defined for only square matrices, and exists when the rank of A is equal to its order. A more general inverse may be defined for rectangular matrices with arbitrary rank. It is called the *generalized inverse*, denoted by A^-, and satisfies the relation

$$AA^-A = A \tag{A-42}$$

This condition is not sufficient to define a unique A^-. Additional conditions are imposed on A^- such as

$$
\begin{aligned}
A^-AA^- &= A^- \\
(AA^-)^T &= AA^- \\
(A^-A)^T &= A^-A
\end{aligned}
\tag{A-43}
$$

If we impose all four conditions in Eqs. A-42 and A-43, the inverse is called a *pseudo inverse* and is denoted by A^+.

A.3.6 The Eigenvalue Problem

For a square matrix A of order n, we seek a nonzero vector x and a scalar λ such that

$$Ax = \lambda x \tag{A-44}$$

This is called the *eigenvalue problem*. A solution λ_0 and x_0 to this problem is called an *eigenvalue* (or proper value or characteristic value) and the corresponding *eigenvector* (or proper vector or characteristic vector) of the matrix A. An eigenvector, if one exists, can be determined only to a scalar multiple, for if λ_0, x_0 satisfy Eq. A-44, then λ_0, αx_0, where α is an arbitrary scalar, will also.

Equation A-44 can be rewritten as

$$(A - \lambda I)x = 0 \tag{A-45}$$

which represents a set of homogenous linear equations. For a nontrivial solution to this set of equations, the following condition must be satisfied:

$$|A - \lambda I| = 0 \tag{A-46}$$

Equation A-46 represents a real polynomial equation of degree n:

$$b_n(-\lambda)^n + b_{n-1}(-\lambda)^{n-1} + \ldots + b_0 = 0 \tag{A-47}$$

where

$$b_n = 1$$

$$b_{n-1} = a_{11} + a_{22} + \ldots + a_{nn} = \sum_{i=1}^{n} a_{ii} = \text{tr}(A) = \text{ trace of } A$$

$$\vdots \tag{A-48}$$

$$b_{n-r} = \text{ sum of all principal minors of order } r \text{ of } A$$

$$\vdots$$

$$b_0 = |A| = \text{ determinant of } A$$

Equation A-47 is called the *characteristic equation* of A, or the *eigenvalue equation*. The matrix $(A - \lambda I)$ is called the *characteristic matrix*. There are n roots for Eq. A-47, counting multiplicity. These are the n eigenvalues of A, $\lambda_1, \lambda_2, \ldots, \lambda_n$. For an eigenvalue λ_i, we solve the set of (homogeneous) linear equations $(A - \lambda_i I)x = 0$ to determine the components of the corresponding eigenvector x_i. In general, λ_i and x_i are either real or complex numbers and vectors, respectively.

If the matrix A is *symmetric*, then

1. The eigenvalues are real.
2. The eigenvectors are all mutually orthogonal, that is, $x_i^T \cdot x_j = x_j^T \cdot x_i = 0$.

As an example, the characteristic polynomial for the matrix

$$A = \begin{bmatrix} 1 & 2 \\ 2 & 1 \end{bmatrix}$$

is

$$\begin{bmatrix} (1 - \lambda) & 2 \\ 2 & (1 - \lambda) \end{bmatrix} = \lambda^2 - 2\lambda - 3 = 0$$

from which

$$\lambda_1 = 3 \text{ and } \lambda_2 = -1$$

are the eigenvalues. Note that both eigenvalues are real. For $\lambda_1 = 3$, we have

$$\begin{bmatrix} 1 & 2 \\ 2 & 1 \end{bmatrix} \begin{bmatrix} x_1 \\ x_2 \end{bmatrix} = \begin{bmatrix} 3x_1 \\ 3x_2 \end{bmatrix}$$

or $x_1 = (1, 1)$ is an eigenvector. For $\lambda_2 = -1$, we have

$$\begin{bmatrix} 1 & 2 \\ 2 & 1 \end{bmatrix} \begin{bmatrix} x_1 \\ x_2 \end{bmatrix} = \begin{bmatrix} -x_1 \\ -x_2 \end{bmatrix}$$

or $x_2 = (1, -1)$ is an eigenvector. These two vectors are orthogonal since

$$x_1^T x_2 = 1 - 1 = 0$$

A.3.7 Bilinear and Quadratic Forms

If A is a square matrix of order n and x and y are two arbitrary n-dimensional vectors, then the scalar

$$u = x^T A y \qquad \text{(A-49)}$$

is called a *bilinear form*. If, however, the matrix A is also *symmetric*, then

$$v = x^T A x \qquad \text{(A-50)}$$

is called a *quadratic form* with the *kernel A*.

The matrix A is called *positive definite* if $v > 0$ for all $x \neq 0$, and we write $A > 0$. If $v \geq 0$ for all x and there exists a nonzero vector x for which equality holds, we say A is *positive semidefinite* (or *nonnegative definite*) and write $A \geq 0$. There are corresponding definitions for *negative definite* and *nonpositive definite*. If there exist vectors x_1 and x_2 such that $x_1^T A x_1 > 0$ and $x_2^T A x_2 < 0$, we say A is *indefinite*.

For a positive definite matrix A it is necessary and sufficient that

$$a_{11} > 0, \begin{vmatrix} a_{11} & a_{12} \\ a_{21} & a_{22} \end{vmatrix} > 0, \ldots, |A| > 0 \tag{A-51}$$

Thus the matrix

$$B = \begin{bmatrix} 3 & -2 & 1 \\ -2 & 3 & 1 \\ 1 & 1 & 4 \end{bmatrix}$$

is positive definite, because $3 > 0$ and

$$\begin{vmatrix} 3 & -2 \\ -2 & 3 \end{vmatrix} = 9 - 4 = 5 > 0$$

and

$$|B| = 3(11) + 2(-9) + 1(-5) = 10 > 0$$

A quadratic form represents, in general, a conic section of some kind. Considering the two-dimensional case for simplicity, we write $x^T A x$ with A symmetric, or

$$a_{11}x_1^2 + 2a_{12}x_1 x_2 + a_{22}x_2^2 = b$$

which is the equation of an ellipse.

A.4 LINEAR TRANSFORMATIONS

A general *linear transformation* of a vector x to another vector y takes the form

$$y = Mx + t \tag{A-52}$$

Each element of the y vector is a linear combination of the elements of x plus a translation or shift represented by an element of the t vector. The matrix M is called the *transformation matrix*, which is in general rectangular, and t is called the translation vector. For our use we restrict M to being square and nonsingular, thus the inverse relation exists, or

$$x = M^{-1}(y - t) \tag{A-53}$$

in which case it is called *affine transformation*. Although both Eqs. A-52 and A-53 apply to higher-dimensional vectors, we will limit our discussions, without loss of generality, to the more practical two- and three-dimensional spaces where the elements of the transformations can be depicted geometrically.

A.4.1 Two-Dimensional Linear Transformations

There are six *elementary* transformations, each representing a single effect, which are geometrically represented in Figs. A-4a-f. Initially, four vectors representing the corners of a square are referenced to the (x_1, x_2) coordinate system (solid lines in Figs. A-4a-f). Each of the six elementary transformations operates on the square and the resulting y_1, y_2 coordinates are plotted to show the effect of each transformation on the location, orientation, size, and shape of the square (dotted lines in Figs. A-4a-f). The effects of the transformations can be shown either by displaying the new figure in the same coordinate system, or by changing the coordinate system. It is easier for the student to visualize these transformations if the new figure is drawn in the same coordinate system as the original figure, which we did in Figs. A-4*a-f*. However, as we discuss each elementary transformation, we will comment on the second interpretation when appropriate.

1. *Translation*

$$y = x + t \qquad M = I \tag{A-54}$$

The square is shifted in the x_1 direction and in the x_2 direction, as shown in Fig. A-4*a*. Alternatively, the solid square remains and the coordinate axes are shifted in the opposite direction, as shown by the dotted axes in Fig. A-4*a*.

2. *Uniform Scale*

$$y = Mx \qquad M = U = \begin{bmatrix} u & 0 \\ 0 & u \end{bmatrix} = uI \tag{A-55}$$

The square is enlarged by the uniform scale u (=1.5 in Fig. A-4*b*), which results from all four point coordinate pairs being multiplied by u. Alternatively, the solid square is referenced to a scaled coordinate system in the same position, with the units along the axes $1/u$ of the original units.

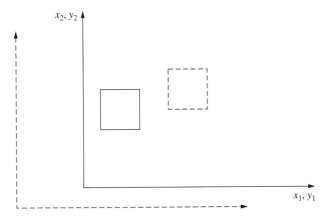

Figure A-4*a*   Translation.

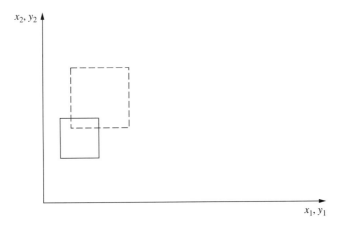

Figure A-4*b*   Uniform scale.

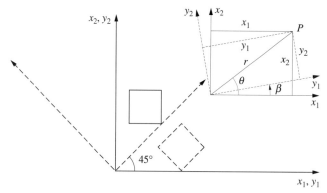

Figure A-4c   Rotation.

3. Rotation

$$y = Mx \qquad M = R = \begin{bmatrix} \cos\beta & \sin\beta \\ -\sin\beta & \cos\beta \end{bmatrix} \tag{A-56}$$

The square retains its shape, but is rotated through θ about the origin of the co-ordinate system. In Fig. A-4c, the coordinate system is also rotated by $-\theta$ to coincide with the original axes. The dotted axes in Fig. A-4c show the trans-formed axes which refer to the original solid square. The elements of R are de-rived from the inset in Fig. A-4c as follows:

$$y_1 = r\cos(\theta - \beta) = r\cos\theta\cos\beta + r\sin\theta\sin\beta$$
$$y_2 = r\sin(\theta - \beta) = r\sin\theta\cos\beta - r\sin\theta\sin\beta$$

or

$$y_1 = x_1\cos\beta + x_2\sin\beta$$
$$y_2 = -x_1\sin\beta + x_2\cos\beta$$

or

$$\begin{bmatrix} y_1 \\ y_2 \end{bmatrix} = \begin{bmatrix} \cos\beta & \sin\beta \\ -\sin\beta & \cos\beta \end{bmatrix} \begin{bmatrix} x_1 \\ x_2 \end{bmatrix}$$

The matrix R is proper orthogonal, $R^{-1} = R^T$ and $|R| = +1$. Rotation matrices do not change the length of the vector, so $|x| = |y|$. Considering the square of the vector length

$$y^T y = (Mx)^T Mx = x^T M^T Mx = x^T x$$

or

$$x^T(M^T M - I)x = 0$$

which for a nontrivial solution means that $M^T M = I$, thus showing that M is an orthogonal matrix.

4. Reflection

$$y = Mx \qquad M = F = \begin{bmatrix} -1 & 0 \\ 0 & 1 \end{bmatrix} \tag{A-57}$$

Fig. A-4d shows reflection of the x_1-axis (i.e., about the x_2-axis). F is improper orthogonal, $F^{-1} = F$ and $|F| = -1$.

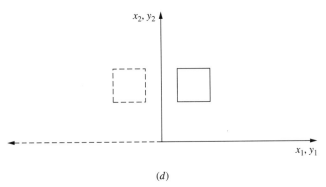

(*d*)

Figure A-4*d*   Reflection.

5. *Stretch (Nonuniform Scale Factors)*

$$y = Mx \qquad M = S = \begin{bmatrix} s_1 & 0 \\ 0 & s_2 \end{bmatrix} \tag{A-58}$$

The square is transformed into a rectangle as shown in Fig. A-4*e*.

6. *Skew (Shear)*

$$y = Mx \qquad M = K = \begin{bmatrix} 1 & k \\ 0 & 1 \end{bmatrix} \tag{A-59}$$

The square is transformed into a parallelogram, as shown in Fig. A-4*f*.

From these elementary transformations, several affine transformations may be constructed using various sequences. The following are two of the commonly used two-dimensional transformations in photogrammetry.

Four-Parameter Transformation

$$\begin{bmatrix} y_1 \\ y_2 \end{bmatrix} = \begin{bmatrix} u & 0 \\ 0 & u \end{bmatrix} \begin{bmatrix} \cos\beta & \sin\beta \\ -\sin\beta & \cos\beta \end{bmatrix} \begin{bmatrix} x_1 \\ x_2 \end{bmatrix} + \begin{bmatrix} t_1 \\ t_2 \end{bmatrix} \tag{A-60a}$$

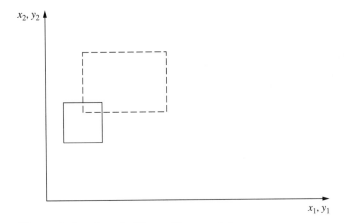

Figure A-4*e*   Stretch (Non-uniform scale).

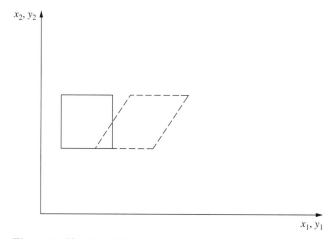

Figure A-4f   Skew (Non-perpendicularity of axes).

or

$$y_1 = ux_1 \cos\beta + ux_2 \sin\beta + t_1$$
$$y_2 = -ux_1 \sin\beta + ux_2 \cos\beta + t_2$$

(A-60b)

or

$$y_1 = ax_1 + bx_2 + c$$
$$y_2 = -bx_1 + ax_2 + d$$

(A-60c)

or

$$\begin{bmatrix} y_1 \\ y_2 \end{bmatrix} = \begin{bmatrix} a & b \\ -b & a \end{bmatrix}\begin{bmatrix} x_1 \\ x_2 \end{bmatrix} + \begin{bmatrix} c \\ d \end{bmatrix}$$

(A-60d)

The inverse transformation is given by

$$\begin{bmatrix} x_1 \\ x_2 \end{bmatrix} = \frac{1}{u}\begin{bmatrix} \cos\beta & -\sin\beta \\ \sin\beta & \cos\beta \end{bmatrix}\begin{bmatrix} y_1 - t_1 \\ y_2 - t_2 \end{bmatrix}$$

(A-60e)

or

$$\begin{bmatrix} x_1 \\ x_2 \end{bmatrix} = \frac{1}{a^2 + b^2}\begin{bmatrix} a & -b \\ b & a \end{bmatrix}\begin{bmatrix} y_1 - c \\ y_2 - d \end{bmatrix}$$

(A-60f)

This transformation has four parameters: a uniform scale, a rotation, and two translations.

Six-Parameter Transformation

$$\begin{bmatrix} y_1 \\ y_2 \end{bmatrix} = \begin{bmatrix} s_1 & 0 \\ 0 & s_2 \end{bmatrix}\begin{bmatrix} 1 & k \\ 0 & 1 \end{bmatrix}\begin{bmatrix} \cos\beta & \sin\beta \\ -\sin\beta & \cos\beta \end{bmatrix}\begin{bmatrix} x_1 \\ x_2 \end{bmatrix} + \begin{bmatrix} t_1 \\ t_2 \end{bmatrix}$$

(A-61a)

or

$$\begin{bmatrix} y_1 \\ y_2 \end{bmatrix} = \begin{bmatrix} a & b \\ d & e \end{bmatrix}\begin{bmatrix} x_1 \\ x_2 \end{bmatrix} + \begin{bmatrix} c \\ f \end{bmatrix}$$

(A-61b)

The six parameters of this transformation are: two scales, one skew factor (lack of perpendicularity of the axes), one rotation, and two shifts. The inverse transformation is given by

$$\begin{bmatrix} x_1 \\ x_2 \end{bmatrix} = \frac{1}{ae - bd}\begin{bmatrix} e & -b \\ -d & a \end{bmatrix}\begin{bmatrix} y_1 - c \\ y_2 - f \end{bmatrix}$$

(A-61c)

A.4.2   Three-Dimensional Linear Transformations

Similar to the two-dimensional case, affine transformation in three dimensions can be factored into several elementary transformations: translation, uniform scale, rotation, reflection, nonuniform scale, etc. We will limit consideration, however, to the seven-parameter transformation, which is used extensively in photogrammetry. It is composed of a uniform scale change, three translations, and three rotations. We consider first rotations in three-dimensional space.

Rotations of a Three-Dimensional Coordinate System

There are three elementary rotations, one about each of the three axes. They are frequently performed in sequence, one after the other. A set of three of these sequential rotations is shown in Fig. A-5, where x is the original system, x' is once rotated, and x'' is twice rotated. The convention is as follows:

1. β_1 about x_1-axis, positive rotation advances $+x_2$ to $+x_3$
2. β_2 about x_2'-axis, positive rotation advances $+x_3'$ to $+x_1'$
3. β_3 about x_3''-axis, positive rotation advances $+x_1''$ to $+x_2''$

Each of the three elementary rotations is represented in matrix form as follows:

$$\begin{bmatrix} x_1' \\ x_2' \\ x_3' \end{bmatrix} = \begin{bmatrix} 1 & 0 & 0 \\ 0 & \cos\beta_1 & \sin\beta_1 \\ 0 & -\sin\beta_1 & \cos\beta_1 \end{bmatrix} \begin{bmatrix} x_1 \\ x_2 \\ x_3 \end{bmatrix} = M_{\beta_1} \begin{bmatrix} x_1 \\ x_2 \\ x_3 \end{bmatrix} \tag{A-62a}$$

where x_1, x_2, and x_3 are the coordinates before rotation and x_1', x_2', and x_3' are the coordinates after rotation. Similarly, a rotation of $+\beta_2$ about the x_2'-axis and $+\beta_3$ about the x_3''-axis are given by

$$\begin{bmatrix} x_1'' \\ x_2'' \\ x_3'' \end{bmatrix} = \begin{bmatrix} \cos\beta_2 & 0 & -\sin\beta_2 \\ 0 & 1 & 0 \\ \sin\beta_2 & 0 & \cos\beta_2 \end{bmatrix} \begin{bmatrix} x_1' \\ x_2' \\ x_3' \end{bmatrix} = M_{\beta_2} \begin{bmatrix} x_1' \\ x_2' \\ x_3' \end{bmatrix} \tag{A-62b}$$

$$\begin{bmatrix} y_1 \\ y_2 \\ y_3 \end{bmatrix} = \begin{bmatrix} x_1''' \\ x_2''' \\ x_3''' \end{bmatrix} = \begin{bmatrix} \cos\beta_3 & \sin\beta_3 & 0 \\ -\sin\beta_3 & \cos\beta_3 & 0 \\ 0 & 0 & 1 \end{bmatrix} \begin{bmatrix} x_1'' \\ x_2'' \\ x_3'' \end{bmatrix} = M_{\beta_3} \begin{bmatrix} x_1'' \\ x_2'' \\ x_3'' \end{bmatrix} \tag{A-62c}$$

The three rotations in Eqs. (A-62) are often referred to as *elementary* rotations, since they may be used to construct any required set of sequential rotations. By successive substitution, the total rotation matrix is obtained:

$$y = x''' = M_{\beta_3} M_{\beta_2} M_{\beta_1} x = Mx \tag{A-63}$$

in which M is now a function of the three rotation angles β_1, β_2, and β_3. The most commonly used set of sequential rotations is given the symbols ω, ϕ, and κ where $\omega \equiv \beta_1$, $\phi \equiv \beta_2$, $\kappa \equiv \beta_3$. In this case, the matrix M, which rotates the object coordinate system (X, Y, Z) parallel to the photo coordinate system (x, y, z) is given by

$$M = \begin{bmatrix} \cos\phi\cos\kappa & \cos\omega\sin\kappa + \sin\omega\sin\phi\cos\kappa & \sin\omega\sin\kappa - \cos\omega\sin\phi\cos\kappa \\ -\cos\phi\sin\kappa & \cos\omega\cos\kappa - \sin\omega\sin\phi\sin\kappa & \sin\omega\cos\kappa + \cos\omega\sin\phi\sin\kappa \\ \sin\phi & -\sin\omega\cos\phi & \cos\omega\cos\phi \end{bmatrix}$$

$$\tag{A-64}$$

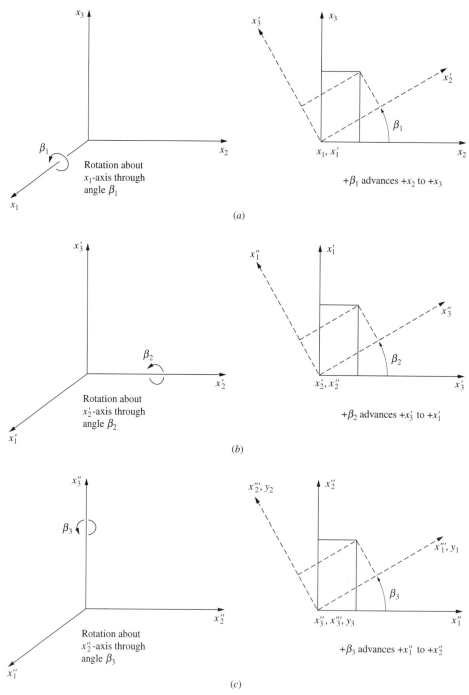

Figure A-5 Rotating a Cartesian system about each axis.

in which ω is about the X-axis, ϕ about the once-rotated Y-axis, and κ is about the twice-rotated Z-axis. The matrix M is orthogonal, since M_ω, M_ϕ, M_κ are each orthogonal.

Another set of sequential rotations consists of azimuth α, tilt t, and swing s, in this order. Azimuth is the clockwise angle from the ground $+Y$-axis (north) to the principal plane, tilt is the angle between the optical axis and local vertical (plumb line), and swing is the clockwise angle between the photo +y-axis and the photo principal line containing the nadir point. The three individual rotation matrices are

$$\begin{bmatrix} X' \\ Y' \\ Z' \end{bmatrix} = \begin{bmatrix} \cos\alpha & -\sin\alpha & 0 \\ \sin\alpha & \cos\alpha & 0 \\ 0 & 0 & 1 \end{bmatrix} \begin{bmatrix} X \\ Y \\ Z \end{bmatrix} = M_\alpha \begin{bmatrix} X \\ Y \\ Z \end{bmatrix} \tag{A-65a}$$

$$\begin{bmatrix} X'' \\ Y'' \\ Z'' \end{bmatrix} = \begin{bmatrix} 1 & 0 & 0 \\ 0 & \cos t & \sin t \\ 0 & -\sin t & \cos t \end{bmatrix} \begin{bmatrix} X' \\ Y' \\ Z' \end{bmatrix} = M_t \begin{bmatrix} X' \\ Y' \\ Z' \end{bmatrix} \tag{A-65b}$$

$$\begin{bmatrix} x \\ y \\ z \end{bmatrix} = \begin{bmatrix} X''' \\ Y''' \\ Z''' \end{bmatrix} = \begin{bmatrix} -\cos s & -\sin s & 0 \\ \sin s & -\cos s & 0 \\ 0 & 0 & 1 \end{bmatrix} \begin{bmatrix} X'' \\ Y'' \\ Z'' \end{bmatrix} = M_s \begin{bmatrix} X'' \\ Y'' \\ Z'' \end{bmatrix} \tag{A-65c}$$

Then, the total matrix is

$$M = M_s M_t M_\alpha \tag{A-65d}$$

or

$$\begin{bmatrix} x \\ y \\ z \end{bmatrix} = \begin{bmatrix} -\cos s\cos\alpha - \sin s\cos t\sin\alpha & \cos s\sin\alpha - \sin s\cos t\cos\alpha & -\sin s\sin t \\ \sin s\cos\alpha - \cos s\cos t\sin\alpha & -\sin s\sin\alpha - \cos s\cos t\cos\alpha & -\cos s\sin t \\ -\sin t\sin\alpha & -\sin t\cos\alpha & \cos t \end{bmatrix} \begin{bmatrix} X \\ Y \\ Z \end{bmatrix} \tag{A-66}$$

All four matrices, M_α, M_t, M_s, and M, are orthogonal.

Seven-Parameter Transformation

This transformation contains seven parameters: a uniform scale change, three rotations, β_1, β_2, and β_3, and three translations, t_1, t_2, t_3. It takes the general form

$$y = \mu Mx + t \tag{A-67}$$

The orthogonal matrix M is a function of only three independent parameters, in this case the angles β_1, β_2, and β_3. This transformation is useful for different applications such as absolute orientation (Section 5.7), model connection, etc. The orthogonal matrix M may be constructed by other methods besides sequential rotations. Two such methods follow.

Constructing M by One Rotation about a Line

This is also often referred to as the *solid body rotation*. Given a three-dimensional object in two different orientations, there exists a line in space about which the object may be rotated by a finite angle to change it from one orientation to the other. If the said line has direction cosines α, β, and γ and angle of rotation is designated by θ, the rotation matrix is given by

$M =$

$$\begin{bmatrix} \alpha^2(1 - \cos\theta) + \cos\theta & \alpha\beta(1 - \cos\theta) - \gamma\sin\theta & \alpha\beta(1 - \cos\theta) + \beta\sin\theta \\ \alpha\beta(1 - \cos\theta) + \gamma\sin\theta & \beta^2(1 - \cos\theta) + \cos\theta & \beta\gamma(1 - \cos\theta) - \alpha\sin\theta \\ a\gamma(1 - \cos\theta) - \beta\sin\theta & \beta\gamma(1 - \cos\theta) + \alpha\sin\theta & \gamma^2(1 - \cos\theta) + \cos\theta \end{bmatrix}$$

$$\tag{A-68}$$

A Purely Algebraic Derivation of M

The preceding methods of deriving M involve trigonometric functions of angles which may be cumbersome to linearize. The following method avoids this disadvantage and allows for the construction of M by computing its elements as rational functions of three independent parameters. The following skew-symmetric matrix contains only three parameters, a, b, and c:

$$S = \begin{bmatrix} 0 & -c & b \\ c & 0 & -a \\ -b & a & 0 \end{bmatrix} \tag{A-69a}$$

An orthogonal matrix M can be obtained from S using

$$M = (I + S)(I - S)^{-1} = (I - S)^{-1}(I + S) \tag{A-69b}$$

in which I is the identity matrix. The student can prove that M is orthogonal, by simply showing that $M^T M = I$. Using the adjoint method, $(I - S)^{-1}$ is evaluated as

$$(I - S)^{-1} = \frac{1}{1 + a^2 + b^2 + c^2} \begin{bmatrix} (1 + a^2) & (ab - c) & (ac + b) \\ (ab + c) & (1 + b^2) & (bc - a) \\ (ac - b) & (a + bc) & (1 + c^2) \end{bmatrix}$$

and then

$$M = (I - S)^{-1}(I + S) =$$

$$\frac{1}{1 + a^2 + b^2 + c^2} \begin{bmatrix} 1 + a^2 - b^2 - c^2 & 2ab - 2c & 2ac + 2b \\ 2ab + 2c & 1 - a^2 + b^2 - c^2 & 2bc - 2a \\ 2ac - 2b & 2bc + 2a & 1 - a^2 - b^2 + c^2 \end{bmatrix}$$

$$\tag{A-70}$$

A.5   NONLINEAR TRANSFORMATIONS

In addition to the linear transformations discussed so far, we use nonlinear transformations in photogrammetry, in both two and three dimensions. In two dimensions, the eight-parameter transformation and polynomial transformations are used.

Eight-Parameter Transformation

$$y_1 = \frac{a_1 x_1 + b_1 x_2 + c_1}{a_0 x_1 + b_0 x_2 + 1}$$

$$y_2 = \frac{a_2 x_1 + b_2 x_2 + c_2}{a_0 x_1 + b_0 x_2 + 1} \tag{A-71a}$$

This projective transformation from the x to the y coordinate systems has the eight transformation parameters a_0, b_0, a_1, ..., c_2. Its inverse is given by

$$x_1 = \frac{(c_1 - y_1)(b_0 y_2 - b_2) - (c_2 - y_2)(b_0 y_1 - b_1)}{(a_0 y_1 - a_1)(b_0 y_2 - b_2) - (a_2 y_2 - a_2)(b_0 y_1 - b_1)}$$

$$x_2 = \frac{(a_0 y_1 - a_1)(c_2 - y_2) - (a_0 y_2 - a_2)(c_1 - y_1)}{(a_0 y_1 - a_1)(b_0 y_2 - b_2) - (a_0 y_2 - a_2)(b_0 y_1 - b_1)} \tag{A-71b}$$

These equations describe the central projectivity between two planes as given in Section 4.1.4.

Two-Dimensional General Polynomials

$$y_1 = a_0 + a_1x_1 + a_2x_2 + a_3x_1x_2 + a_4x_1^2 + a_5x_2^2 + \ldots$$

$$y_2 = b_0 + b_1x_1 + b_2x_2 + b_3x_1x_3 + b_4x_1^2 + b_5x_2^2 + \ldots$$

(A-72a)

These polynomials can obviously be extended to higher powers in x_1 and x_2. A special case of these is the conformal form given in the following section.

Two-Dimensional Conformal Polynomials

The conformal property preserves the angles between intersecting lines after the transformation. If we impose the two conditions

$$\frac{\partial y_1}{\partial x_1} = \frac{\partial y_2}{\partial x_2} \quad \text{and} \quad \frac{\partial y_1}{\partial x_2} = -\frac{\partial y_2}{\partial x_1}$$

(A-72b)

on the general polynomials in Eqs. A-72a, we get

$$y_1 = A_0 + A_1x_1 + A_2x_2 + A_3(x_1^2 - x_2^2) + A_4(2x_1x_2) + \ldots$$

$$y_2 = B_0 - A_2x_1 + A_1x_2 - A_4(x_1^2 - x_2^2) + A_3(2x_1x_2) + \ldots$$

(A-72c)

Note that the first three terms after the equal signs are the same as the four-parameter transformation given in Eq. A-60c. Eqs. A-72c can also be derived using complex numbers by writing

$$(y_1 + y_2i) = (a_0 + b_0i) + (a_1 + b_1i)(x_1 + x_2i) + (a_2 + b_2i)(x_1 + x_2i)^2 + \ldots$$

in which $i = \sqrt{-1}$. Expanding and equating y_1 to the real part and y_2 to the imaginary part (multiplier of i) on the right-hand side leads to Eq. A.72c.

Three-Dimensional General Polynomials

$$y_1 = a_0 + a_1x_1 + a_2x_2 + a_3x_3 + a_4x_1^2 + a_5x_2^2 + a_6x_1x_2 + a_7x_2x_3 + a_8x_1x_3 + \ldots$$

$$y_2 = b_0 + b_1x_1 + b_2x_2 + b_3x_3 + b_4x_1^2 + b_5x_2^2 + b_6x_1x_2 + b_7x_2x_3 + b_8x_1x_3 + \ldots$$

$$y_3 = c_0 + c_1x_1 + c_2x_2 + c_3x_3 + c_4x_1^2 + c_5x_2^2 + c_6x_1x_2 + c_7x_2x_3 + c_8x_1x_3 + \ldots$$

(A-73a)

We can extend these polynomials to higher order. Unlike the two-dimensional case, conformal transformation does *not* exist in three dimensions beyond the first-order (or linear) case given by the seven-parameter transformation, Eq. A-67. A close approximation, which exists only for second-degree terms, is derived by imposing conditions similar to those in Eq. A-72b on every pair of coordinates in Eq. A-73a. This makes the projections of the three-space onto each of the three planes conformal. Imposing

$$\frac{\partial y_1}{\partial x_1} = \frac{\partial y_2}{\partial x_2} = \frac{\partial y_3}{\partial x_3}$$

$$\frac{\partial y_1}{\partial x_2} = -\frac{\partial y_2}{\partial x_1}, \frac{\partial y_2}{\partial x_3} = -\frac{\partial y_3}{\partial x_2}, \frac{\partial y_1}{\partial x_3} = -\frac{\partial y_3}{\partial x_1}$$

(A-73b)

on the general polynomials in Eq. A-73a leads to

$$y_1 = A_0 + Ax_1 + Bx_2 - Cx_3 + E(x_1^2 - x_2^2 - x_3^2) + 0 + 2Gx_3x_1 + 2Fx_1x_2 + \ldots$$

$$y_2 = B_0 - Bx_1 + Ax_2 + Dx_3 + F(-x_1^2 + x_2^2 - x_3^2) + 2Gx_2x_3 + 0 + 2Ex_1x_2 + \ldots$$

$$y_3 = C_0 + Cx_1 - Dx_2 + Ax_3 + G(-x_1^2 - x_2^2 + x_3^2) + 2Fx_2x_3 + 2Ex_3x_1 + 0 + \ldots$$

(A-73c)

A.6 LINEARIZATION OF NONLINEAR FUNCTIONS

The equations expressing the geometric and physical conditions of a photogrammetry problem are frequently nonlinear, which makes their direct solution difficult and uneconomical. We linearize these equations using series expansion, usually Taylor's series, which in general is given by the following:

$$y = f(x) = f(x^0) + \frac{df}{dx}\bigg|_{x^0} \Delta x + \frac{1}{2!}\frac{d^2y}{dx^2}\bigg|_{x^0}(\Delta x)^2 + \ldots + \frac{1}{n!}\frac{d^ny}{dx^n}\bigg|_{x^0} + \ldots \qquad \text{(A-74)}$$

This gives the value of y at $(x^0 + \Delta x)$ given the value of the function $f(x^0)$ at x^0. Eq. A-74 includes still higher order terms, and therefore we usually drop the second and higher order terms and use the approximation with obvious correspondence in terms.

$$y \cong f(x^0) + \frac{dy}{dx}\bigg|_{x^0} \Delta x = y^0 + j\Delta x \qquad \text{(A-75)}$$

The technique of linearization is demonstrated in Fig. A-6. The curve represents the original nonlinear function $f(x)$, whereas the straight line represents the linearized form given by Eq. A-72. That line is tangent to the curve at the given point a, (x^0, y^0). When Δx is given (or evaluated), the value of the function would be approximated by point b, whose ordinate is $(y^0 + j\Delta x)$, and the exact value from the nonlinear function is point c, with ordinate $f(x^0 + \Delta x)$. The error arising from using the linear form is the line segment bc.

A.6.1 One Function of Two Variables

The Taylor series expansion of a function of two variables is

$$y = f(x_1, x_2)$$

$$= f(x_1^0, x_2^0) + \frac{\partial y}{\partial x_1}\bigg|_{x_1^0, x_2^0} \Delta x_1 + \frac{\partial y}{\partial x_2}\bigg|_{x_1^0, x_2^0} \Delta x_2$$

$$+ \frac{1}{2!}\frac{\partial^2 y}{\partial x_1^2}\bigg|_{x_1^0, x_2^0}(\Delta x_1)^2 + \frac{1}{2!}\frac{\partial^2 y}{\partial x_2^2}\bigg|_{x_1^0, x_2^0}(\Delta x_2)^2 \qquad \text{(A-76)}$$

$$+ \frac{\partial y}{\partial x_1}\bigg|_{x_1^0, x_2^0}\frac{\partial y}{\partial x_2}\bigg|_{x_1^0, x_2^0}(\Delta x_1)(\Delta x_2) + \ldots$$

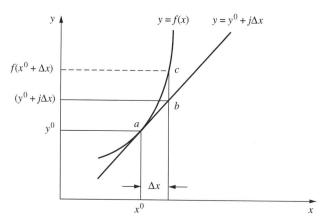

Figure A-6 Linearization.

For linearized form, Eq. A-76 is truncated to

$$y = y^0 + j_1 \Delta x_1 + j_2 \Delta x_2 \tag{A-77}$$

where

$$y^0 = f(x_1^0, x_2^0)$$

$$j_1 = \left. \frac{\partial y}{\partial x_1} \right|_{x_1^0, x_2^0}$$

$$j_2 = \left. \frac{\partial y}{\partial x_2} \right|_{x_1^0, x_2^0}$$

Equation A-77 can be rewritten in matrix form as

$$y = y^0 + [j_1 \; j_2] \begin{bmatrix} \Delta x_1 \\ \Delta x_2 \end{bmatrix}$$

or

$$y = y^0 + \boldsymbol{J}_{yx} \Delta \boldsymbol{x} \tag{A-78}$$

where

$$\boldsymbol{J}_{yx} = \frac{\partial y}{\partial \boldsymbol{x}} = \begin{bmatrix} \dfrac{\partial y}{\partial x_1} & \dfrac{\partial y}{\partial x_2} \end{bmatrix}$$

is the Jacobian of y with respect to $\boldsymbol{x}$.

A.6.2   Two Functions of One Variable

The first-order Taylor series approximations of two functions of x are given by

$$\begin{aligned} y_1 &= f_1(x) \cong y_1^0 + j_1 \Delta x \\ y_2 &= f_2(x) \cong y_2^0 + j_2 \Delta x \end{aligned} \tag{A-79}$$

or

$$\boldsymbol{y} = \boldsymbol{y}^0 + \boldsymbol{J}_{yx} \Delta x$$

where

$$y_1^0 = f_1(x^0)$$
$$y_2^0 = f_2(x^0)$$

$$\boldsymbol{J}_{yx} = \begin{bmatrix} j_1 \\ j_2 \end{bmatrix} = \begin{bmatrix} \left. \dfrac{dy_1}{dx} \right|_{x^0} \\[2ex] \left. \dfrac{dy_2}{dx} \right|_{x^0} \end{bmatrix}$$

A.6.3   Two Functions of Two Variables Each

Two functions of two variables can be linearized by

$$\begin{aligned} y_1 &= f_1(x_1, x_2) \cong y_1^0 + j_{11}\Delta x_1 + j_{12}\Delta x_2 \\ y_2 &= f_2(x_1, x_2) \cong y_2^0 + j_{21}\Delta x_1 + j_{22}\Delta x_2 \end{aligned} \tag{A-80a}$$

or

$$\begin{bmatrix} y_1 \\ y_2 \end{bmatrix} \cong \begin{bmatrix} y_1^0 \\ y_2^0 \end{bmatrix} + \begin{bmatrix} j_{11} & j_{12} \\ j_{21} & j_{22} \end{bmatrix} \begin{bmatrix} \Delta x_1 \\ \Delta x_2 \end{bmatrix} \qquad (A\text{-}80b)$$

or

$$\boldsymbol{y} = \boldsymbol{y}^0 + \boldsymbol{J}_{yx} \Delta x \qquad (A\text{-}80c)$$

where

$$\boldsymbol{y}^0 = \begin{bmatrix} y_1^0 \\ y_2^0 \end{bmatrix} = \begin{bmatrix} f_1(x_1^0, x_2^0) \\ f_2(x_1^0, x_2^0) \end{bmatrix}$$

$$\boldsymbol{J}_{yx} = \frac{\partial \boldsymbol{y}}{\partial \boldsymbol{x}} = \begin{bmatrix} \dfrac{\partial y_1}{\partial x_1}\bigg|_{x_1^0, x_2^0} & \dfrac{\partial y_1}{\partial x_2}\bigg|_{x_1^0, x_2^0} \\[2ex] \dfrac{\partial y_2}{\partial x_1}\bigg|_{x_1^0, x_2^0} & \dfrac{\partial y_2}{\partial x_2}\bigg|_{x_1^0, x_2^0} \end{bmatrix}$$

A.6.4 General Case of *m* Functions of *n* Variables

The previous linearizations can be generalized to *m* functions of *n* variables by

$$\begin{aligned} y_1 &= f_1(x_1, x_2, \ldots, x_n) \\ y_2 &= f_2(x_1, x_2, \ldots, x_n) \\ &\vdots \\ y_m &= f_m(x_1, x_2, \ldots, x_n) \end{aligned} \qquad (A\text{-}81a)$$

The linearized form of Eq. A-81a becomes

$$\boldsymbol{y} \cong \boldsymbol{y}^0 + \boldsymbol{J}_{yx} \Delta x \qquad (A\text{-}81b)$$

where

$$\boldsymbol{y}^0 = \begin{bmatrix} y_1^0 \\ y_2^2 \\ \vdots \\ y_m^0 \end{bmatrix} = \begin{bmatrix} f_1(x_1^0, x_2^0, \ldots, x_n^0) \\ f_2(x_1^0, x_2^0, \ldots, x_n^0) \\ \vdots \\ f_m(x_1^0, x_2^0, \ldots, x_n^0) \end{bmatrix}$$

$$\boldsymbol{J}_{yx} = \frac{\partial \boldsymbol{y}}{\partial \boldsymbol{x}} = \begin{bmatrix} \dfrac{\partial y_1}{\partial x_1}\bigg|_{x^0} & \dfrac{\partial y_1}{\partial x_2}\bigg|_{x^0} & \cdots & \dfrac{\partial y_1}{\partial x_n}\bigg|_{x^0} \\[2ex] \vdots & \vdots & \vdots & \vdots \\[2ex] \dfrac{\partial y_m}{\partial x_1}\bigg|_{x^0} & \dfrac{\partial y_m}{\partial x_2}\bigg|_{x^0} & \cdots & \dfrac{\partial y_m}{\partial x_n}\bigg|_{x^0} \end{bmatrix}$$

$$\Delta x = \begin{bmatrix} \Delta x_1 \\ \Delta x_2 \\ \vdots \\ \Delta x_n \end{bmatrix}$$

Equation A-81b represents the general form with $\boldsymbol{y}$, $\boldsymbol{y}^0$ being $m \times 1$ vectors, $\boldsymbol{J}$ an $m \times n$ Jacobian matrix, and Δx an $n \times 1$ vector. Equations A-75, A-78, and A-80c are special cases of Eq. A-81b.

A.6.5 Differentiation of a Determinant

Some photogrammetric conditions are either in determinant form, or contain determinants. The partial derivative of a $p \times p$ determinant with respect to a scalar is composed of the sum of p determinants, each having the elements of only one row or one column replaced by their derivatives. Thus, given the determinant $d = \begin{vmatrix} D_1 & D_2 & \ldots & D_p \end{vmatrix}$ in which D_i, $i = 1, 2, \ldots, p$, represents its p columns, then

$$\frac{\partial d}{\partial x} = \begin{vmatrix} \frac{\partial D_1}{\partial x} & D_2 & \ldots & D_p \end{vmatrix} + \begin{vmatrix} D_1 & \frac{\partial D_2}{\partial x} & \ldots & D_p \end{vmatrix} + \ldots + \begin{vmatrix} D_1 & D_2 & \ldots & \frac{\partial D_p}{\partial x} \end{vmatrix} \quad \text{(A-82)}$$

An expression similar to Eq. A-82 can be written in which the rows instead of the columns of d are partially differentiated.

A.6.6 Differentiation of a Quotient

A quotient of functions such as $g = U/W$ appears in many photogrammetric condition equations. The partial derivative of g with respect to a variable x is given by

$$\frac{\partial g}{\partial x} = \frac{1}{W} \left[\frac{\partial U}{\partial x} - \frac{U}{W} \frac{\partial W}{\partial x} \right] \quad \text{(A-83)}$$

Both U and W can be general functions, including determinants, of several variables.

A.7 MAP PROJECTIONS

Map projection is concerned with the theory and techniques of proper representation of the curved Earth surface on the plane of a map. When the map is of such large scale as to represent a very limited area, the curvature of the Earth is insignificant and field survey measurements can be directly represented on the map. On the other hand, as the area represented by the map gets larger, this curvature becomes significant and must be dealt with. The Earth is an ellipsoid, which is sometimes approximated by a sphere. Neither of these surfaces can accurately be developed into a plane. Therefore, all map projection methods must contain some distortion. Various methods are selected to best fit the shape of the specific area to be mapped and to minimize the effects of particular distortions.

Locations on the Earth are represented by meridians of longitude, λ, and parallels of latitude, ϕ. On the map these are represented by scaled linear distances, X and Y, using the dimensions of the Earth ellipsoid and selected criteria that the specific map projection must satisfy. These are obtained from transformation equations taking the general functional form of

$$\begin{aligned} X &= f_x(\lambda, \phi) \\ Y &= f_y(\lambda, \phi) \end{aligned} \quad \text{(A-84)}$$

Although all modern map projections are performed by computer programs, several are based on geometric projection of the Earth onto one of three surfaces: a plane, a cylinder, or a cone. It is clear that the cylinder and cone are chosen because they can be developed into a plane—that of the map. When projecting onto a plane, it is tangent to the Earth surface at a point and the projection center is either the center of the Earth, as in the gnomonic projection shown in Fig. A-7a, or the point diametrically opposite to the tangent point, as in the stereographic projection shown in Fig. A-7b. If the projection lines are perpendicular to the plane, we have an orthographic projection, Fig. A-7c.

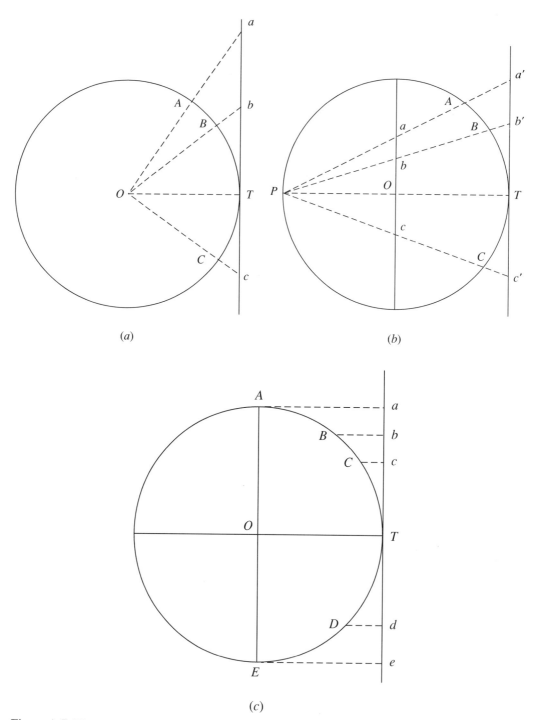

Figure A-7 Three types of map projection: (*a*) gnomonic, (*b*) stereographic, and (*c*) orthographic.

A cone is usually selected with its axis coincident with the Earth's polar axis. It may be tangent to the Earth at one small circle, called *standard parallel*, or intersect the Earth surface in two standard parallels. When developed, the scale will be true (i.e., without any distortion or error) at the standard parallels, as shown in Fig. A-8. Polyconic projections use a series of frustums of cones, each from a separate cone.

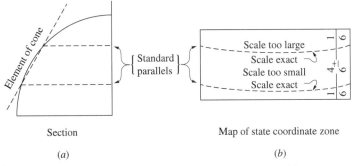

Section

Map of state coordinate zone

(a)

(b)

Figure A-8 Lambert conformal conic projection. Adapted from
Anderson and Mikhail, Surveying: *Theory and Practice,* 7th ed., p.637,
McGraw-Hill.

Similar to a cone, a cylinder may be selected to be tangent to the Earth, or to intersect it. It may be regular (with its axis being the polar axis), transverse (with one tangent meridian), intersecting (with two meridians), or oblique cylindrical, as shown in Fig. A-9.

Equation A-84 will produce a *perfect* map without any distortions if it satisfies *all* of the following conditions:

a. All distances and areas have correct relative magnitudes

b. All azimuths and angles are correctly represented on the map

c. All great circles are shown on the map as straight lines

d. Geodetic longitudes and latitudes are correctly shown on the map

No one map projection can satisfy *all* these conditions. However, each class satisfies some selected conditions. The following are four classes:

1. *Conformal* or *orthomorphic* projection results in a map showing the correct angle between any pair of short intersecting lines, thus making small areas appear correct in *shape*. Since the scale varies from point to point, the shapes of larger areas are incorrect.

2. An *equal-area* projection results in a map showing all areas in proper relative *size*, although these areas may be shown as incorrect shapes and the map may have other defects.

3. In an *equidistant* projection, distances are correctly represented from one central point to other points on the map.

4. In an *azimuthal* projection, the map shows the correct *direction* or azimuth of any point relative to one central point.

For conformal mapping, a new latitude, ψ, called the *isometric latitude* is used in place of ϕ, where

$$\psi = \ln\left[\tan\left(\frac{\pi}{4} + \frac{\phi}{2} \right) \left[\frac{1 - e\sin\phi}{1 + e\sin\phi} \right]^{e/2} \right] \qquad (A\text{-}85)$$

in which $e^2 = (a^2 - b^2)/a^2$, with a, b being the semi-major and semi-minor axes of the Earth ellipsoid, respectively. Then, Eq. A-84 is replaced by

$$X = f_1(\lambda, \psi)$$
$$Y = f_2(\lambda, \psi) \qquad (A\text{-}86)$$

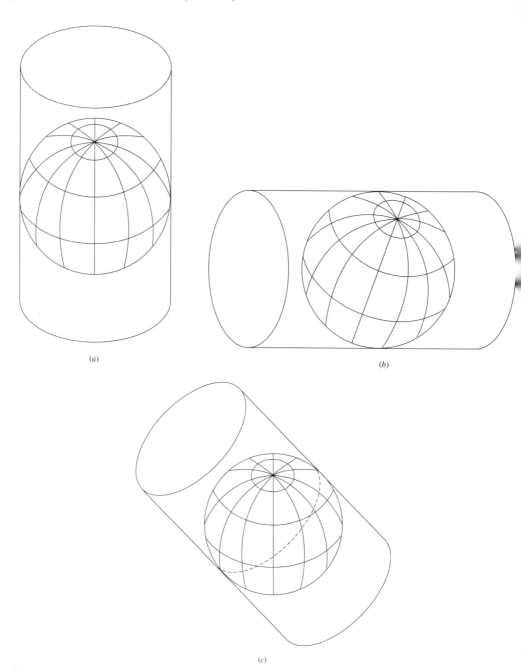

Figure A-9 Projection onto a cylinder: (*a*) regular or vertical, (*b*) transverse or horizontal, and (*c*) oblique.

In order for the mapping in Eq. A-86 to be conformal, the following Cauchy-Riemann conditions must be satisfied:

$$\frac{\partial X}{\partial \lambda} = \frac{\partial Y}{\partial \psi} \quad \text{and} \quad \frac{\partial X}{\partial \psi} = -\frac{\partial Y}{\partial \lambda} \tag{A-87}$$

Two commonly used conformal projections are the Lambert conformal conic projection and the transverse Mercator projection. The former is shown in Fig. A-8,

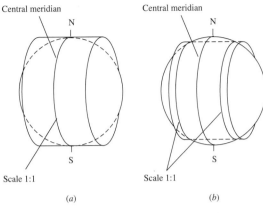

Figure A-10 Transverse Mercator projection. (*a*) Cylinder with one standard line. (*b*) Cylinder with two standard lines.

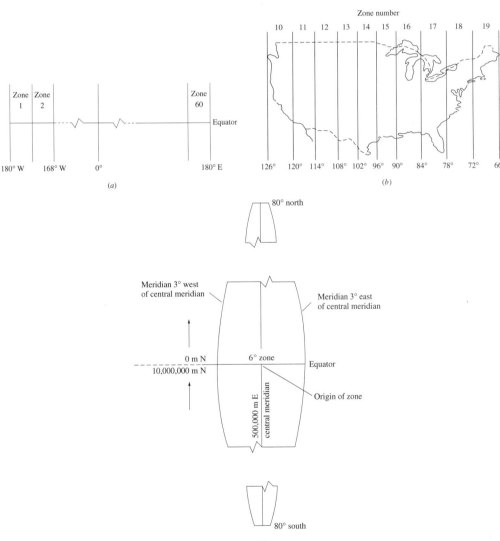

Figure A-11 (*a*) Universal transverse Mercator zones. (*b*) UTM zones in the United States. (*c*) *X* and *Y* coordinates of the origin of a UTM grid zone, from U.S. Army Field Manual *Map Reading*, FM 21–26.

the cylinder is tangent to the ellipsoid, the scale at the central meridian is 1:1. When it is not tangent, the scale at the central meridian is less than 1:1, as shown in Fig. A-10*b*. The central meridian is the origin of the map *X* coordinate, while the origin of the map *Y* coordinate is the equator. This projection is used as a state plane coordinate system for states with greater extent north-south than east-west.

An extensively used map projection system is the *Universal Transverse Mercator*, or UTM. It is shown schematically in Fig. A-11. It is in six-degree-wide zones, with the scale at each central meridian of a zone being 0.9996. The false easting for each central meridian is 500,000 m. A transverse Mercator projection with three-degree-wide zones is possible, where the scale at the central meridian is improved to 0.9999.

Appendix B

Least Squares Adjustment

B.1 INTRODUCTION

Many of the quantities involved in photogrammetry are *random variables* which would assume different values if the measurements used to obtain them were repeated. An elementary example of random variables are the coordinates of an image point on a photograph. If an instrument is used to measure these coordinates several times, it is more than likely that slightly different values would be obtained for each measurement. Each measured value represents an *estimate* of the random variable representing an image coordinate. If the coordinates are measured only once, the measurements are unique and there is then no way of knowing how good these values are. In fact, any errors in those measurements, no matter how large, cannot be detected. It follows, then, that additional measurements offer some advantages. When more measurements are made than the minimum needed, *redundancy* is said to exist among the measurements or observations. If the total number of measurements is n, and it takes a minimum of n_0 measurements to uniquely determine the *model* underlying the problem, then the redundancy (or number of degrees of freedom) is given by

$$r = n - n_0 \qquad (B-1)$$

When redundancy exists, subsets of n_0 measurements from the given n measurements will yield a solution. The solution obtained using one subset is generally different from the solutions obtained using any other subset. This means that the total observations are *inconsistent* with respect to the model and that an *adjustment* must be performed in order to eliminate the inconsistency. After the adjustment, no matter which

387

subset of the measurements is used, the solution will always be the same. This is possible only if the original measurements l are replaced by another set of estimates $\hat{l}$, often called *adjusted observations*, by adding a set of *residuals* or *corrections* v to the measurements. That is,

$$\hat{l} = l + v \tag{B-2}$$

Thus if an image distance x is measured four times with values l_1, l_2, l_3, and l_4, then after the adjustment the four estimated or adjusted observations $\hat{l}_1 = l_1 + v_1$, $\hat{l}_2 = l_2 + v_2$, $\hat{l}_3 = l_3 + v_3$, and $\hat{l}_4 = l_4 + v_4$ will all be equal. Consequently, regardless of which element of $\hat{l}$ is used to evaluate x, we always get the same answer. The role of adjustment is to derive values for the residuals v_i such that the above conditions are satisfied. Clearly, there is a large number of possibilities for a set of four residuals that would make all four $\hat{l}_i$ equal.

Suppose that the four measured image distances are $l_1 = 102.50$ mm, $l_2 = 102.55$ mm, $l_3 = 102.54$ mm, and $l_4 = 102.56$ mm. Three of the large number of possible sets of residuals that will make all four $\hat{l}_i$ in each set equal, are as follows:

Set 1	Set 2	Set 3
v_1 = 0.00 mm	+0.05 mm	+0.03 mm
v_2 = −0.05 mm	0.00 mm	−0.02 mm
v_3 = −0.04 mm	+0.01 mm	−0.01 mm
v_4 = −0.06 mm	−0.01 mm	−0.03 mm

Then $\hat{l}$ is 102.50 mm for set 1, 102.55 mm for set 2, and 102.53 mm for set 3. Thus, an additional criterion must be imposed on the residuals so that their selection is not arbitrary. This criterion is that of *least squares*, which is used extensively in photogrammetric data adjustment.

B.2   MATHEMATICAL MODEL FOR ADJUSTMENT

In Section B.1 we introduced the notion of a *model* underlying the problem. It is referred to as the *mathematical model*, which describes the geometric or physical situation involved in the adjustment problem to be solved. It is important that the reader appreciates the importance of the concept of the mathematical model as well as its adequacy for the photogrammetric problem. As an example, suppose that we are interested in the direction of the line from a camera station to a point target, T. Let that direction be expressed as a horizontal angle α and a vertical angle β, as shown in Fig. B-1. The camera is set up such that its optical axis is horizontal. It follows, then, that $\tan \alpha = x/p$ and $\tan \beta = (y \cos \alpha)/p$, in which p is the camera principal distance and x and y are the measured coordinates of the image of the target. Given these three quantities, α and β can be easily computed. This sounds simple enough, until we begin taking a closer look at the problem and analyzing the factors involved in its solution.

There are several elements involved in the underlying mathematical model that need to be examined carefully. First, how close is the camera optical axis to being horizontal? Next, it is assumed that the optical axis is perpendicular to the image plane and pierces it at the origin of the coordinate system. Furthermore, how accurate is the value of p, which depends on focusing the camera at the time of imaging point T? These elements relate to the camera parameters. Other factors are involved in photographing the target.

The line from C to T in Fig. B-1 is assumed to be a straight line that pierces the plane of the photograph at image point t. Since the light ray from T travels through the

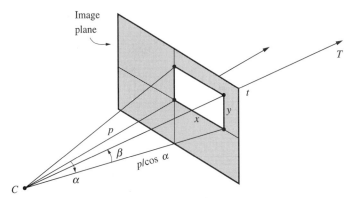

Figure B-1 Mathematical model for imaging a point target.

atmosphere before entering the camera lens, it will certainly bend due to refraction. When it enters the lens, it will be deflected further due to lens distortion. The latent image on the emulsion (for film photography) will undergo additional shifts due to dimensional deformations and lack of flatness. Thus, what began as a relatively simple geometric problem involving a pair of trigonometric functions has progressed to a substantially more involved proposition. The intent, of course, is to point out to the student that whenever measurements (such as x, y) are used to derive other quantities of interest (α and β), the underlying mathematical model must be carefully considered. All possible systematic errors must be corrected, so that only random errors remain in the measurements.

The mathematical model is composed of two parts: a functional model and a stochastic model. The *functional model* is the more obvious part, since it usually describes the geometric or physical characteristics of the photogrammetric problem. Thus the functional model for our example involves the determination of the direction of a straight line in three-dimensional space through the one-time measurement of two image coordinates on a given photograph. If these coordinates are measured more than once, redundant measurements will be present with respect to the functional model. This does not say anything about the properties of the measured coordinates. For example, each coordinate pair may be measured by the same observer, using the same instrument, applying the same measuring technique, and performing the measurements under very similar environmental conditions. In such a case, the measured pairs are said to be equally *precise*. There are certainly other possible cases in which the resulting measurements are not of equal quality. Such statistical variations in the observations are important and must be taken into consideration when using photogrammetric measurements to derive the required information. The *stochastic model* is the part of the mathematical model that deals with the statistical properties of all the elements involved in the functional model. For example, in the case of the line direction, we need to specify the *observables* (x and y but not p), provide information on how well they are measured, and specify if there is reason to believe that there is statistical *correlation* (or interaction) among the measurements and if so, how much. The determination of unique estimates $\hat{\alpha}$ and $\hat{\beta}$ from the redundant measurements depends both on knowing the given equation pairs (the functional model) and knowing the statistical properties of all the measured quantities (the stochastic model).

The statistical properties of the observations are usually expressed by a *covariance matrix*, Σ ; its diagonal elements are *variances*, σ_{ii}^2, and its off-diagonal elements are *covariances*, σ_{ij}. The positive square root of the variance, σ, is called the *standard deviation*. The value of each covariance reflects the degree of correlation between the corresponding observations. When all covariances are zero, the observations are said to be *uncorrelated*. If a scalar, σ_0^2, which is called the *reference variance*, is factored out, the resulting matrix is called the *relative covariance matrix*, or more frequently, the *cofactor matrix Q*. Consequently,

$$\Sigma = \sigma_0^2 \, Q \tag{B-3a}$$

The inverse of the cofactor matrix is called the *weight matrix*, or

$$W = Q^{-1} \tag{B-3b}$$

Often, the reference variance may be chosen to equal unity, or $\sigma_0^2 = 1$; in this case

$$W = \Sigma^{-1} \tag{B-3c}$$

When the random variables involved are uncorrelated, the covariance, cofactor, and weight matrices reduce to the diagonal form. If, further, the observations have equal variances, then both the cofactor and weight matrices become the identity matrix, I.

B.3 THE LEAST SQUARES CRITERION

In the elementary treatment of adjustment by the method of least squares, given in Section B.1, it is assumed that all measurements are independently made and that no *correlation* exists between them; that is, one measurement does not influence another measurement. Under this assumption, one of two conditions could exist: (1) All of the measurements have equal quality; or (2) each measurement can have a different quality. The quality of a measurement is given in terms of its *weight, w.*

Under the condition of equal weights, the least squares criterion states that the sum of the squares of the n residuals or corrections to a set of n measurements must be a minimum, that is

$$\phi = v_1^2 + v_2^2 + \ldots + v_n^2 = \text{a minimum} \tag{B-4a}$$

or in matrix notation

$$\phi = v^T v = \text{a minimum} \tag{B-4b}$$

If each observation or measurement has a different weight, the least squares criterion becomes

$$\phi = w_1 v_1^2 + w_2 v_2^2 + \ldots + w_n v_n^2 = \text{a minimum} \tag{B-4c}$$

or in matrix notation

$$\phi = v^T W v = \text{a minimum} \tag{B-4d}$$

in which

$$W = \begin{bmatrix} w_1 & 0 & \cdots & 0 & 0 \\ 0 & w_2 & \cdots & 0 & 0 \\ \vdots & \vdots & \vdots & \vdots & \vdots \\ 0 & 0 & \cdots & 0 & w_n \end{bmatrix}$$

is the *weight matrix*. If the weights are all equal, then $W = I$, the unit matrix.

B.4 THE TECHNIQUES OF LEAST SQUARES

Although there is only one least squares criterion, there are several techniques by which least squares may be applied. Regardless of which technique is applied, the results of an adjustment of a given set of measurements *must* be the same. The choice of a technique, therefore, is mostly a matter of convenience and/or computational economy as will be explained later.

The first step in a least squares adjustment is to identify the model underlying the problem, as explained in Section B.2, along with the minimum number of observations, n_0, necessary to determine the model. Given $n > n_0$ observations, the redundancy $r = n - n_0$ is determined. For each element in the redundancy (that is, for each degree of freedom), one equation, called a *condition equation* or simply a *condition*, is written that relates the model variables to one another.

As an example, suppose all three angles *A, B,* and *C* of a triangle are measured. The model is $A + B + C = 180°$. Only two of the three angles are needed to obtain the third angle, so $n_0 = 2$. Since $n = 3$, $r = 1$. In the case of an image distance x, $n_0 = 1$, since it takes just one measurement to determine *x*. If, for example, four measurements are made, then $n = 4$ and $r = 3$. Then three *independent* condition equations can be formed. These can be written: $\hat{l}_1 - \hat{l}_2 = 0$, $\hat{l}_2 - \hat{l}_3 = 0$, and $\hat{l}_3 - \hat{l}_4 = 0$. The student should realize that any other equation, such as $\hat{l}_1 - \hat{l}_4 = 0$, will simply be a *dependent* equation that can be derived from the three independent equations, in this case by simply adding them together. When the condition equations are written in terms of only observations and constants, as in this case, the technique is called *adjustment of observations only*. In general, the set of *r* conditions in terms of *n* unknown residuals takes the form

$$\begin{aligned}
a_{11}v_1 + a_{12}v_2 + \ldots + a_{1n}v_n &= f_1 \\
a_{21}v_1 + a_{22}v_2 + \ldots + a_{2n}v_n &= f_2 \\
&\;\;\vdots \\
a_{r1}v_1 + a_{r2}v_2 + \ldots + a_{rn}v_n &= f_r
\end{aligned} \tag{B-5}$$

which in matrix notation becomes

$$\begin{bmatrix} a_{11} & a_{12} & \ldots & a_{1n} \\ a_{21} & a_{22} & \ldots & a_{2n} \\ \vdots & \vdots & \vdots & \vdots \\ a_{r1} & a_{r2} & \ldots & a_{rn} \end{bmatrix} \begin{bmatrix} v_1 \\ v_2 \\ \vdots \\ v_n \end{bmatrix} = \begin{bmatrix} f_1 \\ f_2 \\ \vdots \\ f_r \end{bmatrix} \tag{B-6}$$

or

$$\underset{r \times n}{A} \;\; \underset{n \times 1}{v} = \underset{r \times 1}{f} \tag{B-7}$$

For linear adjustment problems the vector *f* is given by

$$f = d - Al \tag{B-8}$$

in which *d* is a vector of numerical constants and *l* is the vector of numerical values of the measurements. As an example, the three independent conditions for a distance measured four times are

$$\begin{aligned}
\hat{l}_1 &= \hat{l}_2 \quad \text{or} \quad l_1 + v_1 = l_2 + v_2 \\
\hat{l}_2 &= \hat{l}_3 \quad \text{or} \quad l_2 + v_2 = l_3 + v_3 \\
\hat{l}_3 &= \hat{l}_4 \quad \text{or} \quad l_3 + v_3 = l_4 + v_4
\end{aligned}$$

leading to

$$\begin{aligned}
v_1 - v_2 &= -l_1 + l_2 \\
v_2 - v_3 &= -l_2 + l_3 \\
v_3 - v_4 &= -l_3 + l_4
\end{aligned}$$

which can be expressed in matrix form as

$$
\begin{bmatrix} 1 & -1 & 0 & 0 \\ 0 & 1 & -1 & 0 \\ 0 & 0 & 1 & -1 \end{bmatrix}
\begin{bmatrix} v_1 \\ v_2 \\ v_3 \\ v_4 \end{bmatrix}
= -
\begin{bmatrix} 1 & -1 & 0 & 0 \\ 0 & 1 & -1 & 0 \\ 0 & 0 & 1 & -1 \end{bmatrix}
\begin{bmatrix} l_1 \\ l_2 \\ l_3 \\ l_4 \end{bmatrix}
$$

or

$$
\underset{3 \times 4}{A} \; \underset{4 \times 1}{v} = - \underset{3 \times 4}{A} \; \underset{4 \times 1}{l}
$$

where the numerical constant vector d is in this case 0.

In the technique of adjustment of observations only, the number of condition equations is equal to the redundancy or the number of degrees of freedom, and the residuals are the only unknowns. In another technique, an additional set of variables, called *parameters*, are treated as *unknowns* in the condition equations. Frequently, these parameters are actually the quantities of most immediate interest to the photogrammetrist and whose estimates or best values are required. For *each* of the unknown parameters, an additional condition equation must be written to account for its computation. As an example, consider again the distance x measured four times. The best value of the distance could be carried as an unknown parameter, and because $r = 3$, four conditions are written as follows:

$$
\begin{aligned}
\hat{l}_1 - x = 0 \quad &\text{or} \quad l_1 + v_1 - x = 0 \\
\hat{l}_2 - x = 0 \quad &\text{or} \quad l_2 + v_2 - x = 0 \\
\hat{l}_3 - x = 0 \quad &\text{or} \quad l_3 + v_3 - x = 0 \\
\hat{l}_4 - x = 0 \quad &\text{or} \quad l_4 + v_4 - x = 0
\end{aligned}
$$

These equations are characterized by having *only one observation* appear in each equation. This technique is called *adjustment of indirect observations*, sometimes referred to classically as *adjustment by observation* equations. In general, for n observations with a redundancy r and u parameters, the number of condition equations c is

$$
c = r + u \tag{B-9}
$$

In the case of the indirect observations technique, each condition contains one observation. Therefore, the total number of conditions is n and they take the following general functional form:

$$
\begin{aligned}
v_1 + b_{11}\delta_1 + b_{12}\delta_2 + \ldots + b_{1u}\delta_u &= f_1 \\
v_2 + b_{21}\delta_1 + b_{22}\delta_2 + \ldots + b_{2u}\delta_u &= f_2 \\
&\vdots \\
v_n + b_{n1}\delta_1 + b_{n2}\delta_2 + \ldots + b_{nu}\delta_u &= f_n
\end{aligned} \tag{B-10}
$$

in which v_i are the measurement residuals, the b_{ij} are numerical coefficients of the unknown parameters δ_j to be determined by the least squares adjustment, and f_i are the constant terms. In matrix form, Eq. B-10 is expressed as

$$
\begin{bmatrix} v_1 \\ v_2 \\ \vdots \\ v_n \end{bmatrix}
+
\begin{bmatrix} b_{11} & b_{12} & \cdots & b_{1u} \\ b_{21} & b_{22} & \cdots & b_{2u} \\ \vdots & \vdots & \vdots & \vdots \\ b_{n1} & b_{n2} & \cdots & b_{nu} \end{bmatrix}
\begin{bmatrix} \delta_1 \\ \delta_2 \\ \vdots \\ \delta_u \end{bmatrix}
=
\begin{bmatrix} f_1 \\ f_2 \\ \vdots \\ f_n \end{bmatrix}
$$

or

$$
\underset{n \times 1}{v} + \underset{n \times u}{B} \; \underset{u \times 1}{\Delta} = \underset{n \times 1}{f} \tag{B-11}
$$

The vector $\Delta = \begin{bmatrix} \delta_1 & \delta_2 & \dots & \delta_u \end{bmatrix}^T$ contains the u unknown parameters, while the vector f is, for linear problems, given by

$$\underset{n \times 1}{f} = \underset{n \times 1}{d} - \underset{n \times 1}{l} \tag{B-12}$$

in which d is a vector of numerical constants and l are the numerical values of the measurements. Thus the four conditions for the measured distance x are

$$v_1 - x = -l_1$$
$$v_2 - x = -l_2$$
$$v_3 - x = -l_3$$
$$v_4 - x = -l_4$$

or

$$\begin{bmatrix} v_1 \\ v_2 \\ v_3 \\ v_4 \end{bmatrix} + \begin{bmatrix} -1 \\ -1 \\ -1 \\ -1 \end{bmatrix} [x] = -\begin{bmatrix} l_1 \\ l_2 \\ l_3 \\ l_4 \end{bmatrix}$$

or

$$\underset{4 \times 1}{v} + \underset{4 \times 1}{B} \; \underset{1 \times 1}{\Delta} = \underset{4 \times 1}{-l}$$

where in this case d is 0.

A more general technique of least squares applies when the condition equations contain parameters, but some or all the equations contain more than one observation. As an example, we consider fitting a straight line of known slope, a, through observed point coordinates in a plane. The equation is given by $y - ax - b = 0$, in which x, y are observables, a is constant, and b is the unknown parameter. If three points are available, the condition equations are

$$y_1 + v_{y1} - a\,(x_1 + v_{x1}) - b = 0$$
$$y_2 + v_{y2} - a\,(x_2 + v_{x2}) - b = 0$$
$$y_3 + v_{y3} - a\,(x_3 + v_{x3}) - b = 0$$

which in matrix form become

$$\begin{bmatrix} -a & 1 & 0 & 0 & 0 & 0 \\ 0 & 0 & -a & 1 & 0 & 0 \\ 0 & 0 & 0 & 0 & -a & 1 \end{bmatrix} \begin{bmatrix} v_{x1} \\ v_{y1} \\ v_{x2} \\ v_{y2} \\ v_{x3} \\ v_{y3} \end{bmatrix} + \begin{bmatrix} -1 \\ -1 \\ -1 \end{bmatrix} [b] = \begin{bmatrix} ax_1 - y_1 \\ ax_2 - y_2 \\ ax_3 - y_3 \end{bmatrix}$$

The general form of these condition equations is

$$\underset{c \times n}{A} \; \underset{n \times 1}{v} + \underset{c \times u}{B} \; \underset{u \times 1}{\Delta} = \underset{c \times 1}{f} \tag{B-13}$$

The number of condition equations ranges from r to n, or $r \leq c \leq n$. The number of parameters is between zero and n_0, or $0 \leq u \leq n_0$. For linear problems, the vector of constant terms, f, is given by Eq. B-8. In the current example, $d = 0$, since

$$f = \begin{bmatrix} ax_1 - y_1 \\ ax_2 - y_2 \\ ax_3 - y_3 \end{bmatrix} = -\begin{bmatrix} -a & 1 & 0 & 0 & 0 & 0 \\ 0 & 0 & -a & 1 & 0 & 0 \\ 0 & 0 & 0 & 0 & -a & 1 \end{bmatrix} \begin{bmatrix} x_1 \\ y_1 \\ x_2 \\ y_2 \\ x_3 \\ y_3 \end{bmatrix} = 0 - Al$$

B.5   ADJUSTMENT OF INDIRECT OBSERVATIONS

In order to develop the technique of adjustment of indirect observations, an introductory example of fitting a straight line in a plane to a set of three points will be used as shown in Fig. B-2. It is assumed that the ordinates, or y coordinates, are *observables* or *measurements*, while the x coordinates are constants, usually taken at equal intervals. Since it takes $n_0 = 2$ observations or points to uniquely determine the line, and there are $n = 3$ observations, the redundancy is $r = 3 - 2 = 1$. We will treat the line slope a and y-intercept b as two parameters; thus $u = 2$ and the number of conditions is $c = r + u = 1 + 2 = 3 = n$. The three condition equations are the equation of the line written for each point, or

$$y_1 + v_1 - ax_1 - b = 0$$
$$y_2 + v_2 - ax_2 - b = 0$$
$$y_3 + v_3 - ax_3 - b = 0$$

or

$$\begin{bmatrix} v_1 \\ v_2 \\ v_3 \end{bmatrix} + \begin{bmatrix} -x_1 & -1 \\ -x_2 & -1 \\ -x_3 & -1 \end{bmatrix} \begin{bmatrix} a \\ b \end{bmatrix} = \begin{bmatrix} -y_1 \\ -y_2 \\ -y_3 \end{bmatrix} = \begin{bmatrix} f_1 \\ f_2 \\ f_3 \end{bmatrix} \qquad \text{(B-14)}$$

which is of the form

$$\underset{3 \times 1}{v} + \underset{3 \times 2}{B}\ \underset{2 \times 1}{\Delta} = \underset{3 \times 1}{f}$$

For purposes of generalization, let w_1, w_2, and w_3 represent the weights of the three observations y_1, y_2, and y_3, which are assumed to be uncorrelated. The quantity to be minimized in this case is, according to Eq. B-4,

$$\phi = w_1 v_1^2 + w_2 v_2^2 + w_3 v_3^2$$

or

$$\phi = w_1(f_1 + ax_1 + b)^2 + w_2(f_2 + ax_2 + b)^2 + w_3(f_3 + ax_3 + b)^2$$

In order for ϕ to be a minimum, its partial derivative with respect to each unknown, a and b, must be zero. Hence

$$\frac{\partial \phi}{\partial a} = 2w_1 x_1(f_1 + ax_1 + b) + 2w_2 x_2(f_2 + ax_2 + b)$$
$$+ 2w_3 x_3(f_3 + ax_3 + b) = 0$$

$$\frac{\partial \phi}{\partial b} = 2w_1(f_1 + ax_1 + b) + 2w_2(f_2 + ax_2 + b) + 2w_3(f_3 + ax_3 + b)$$
$$= 0$$

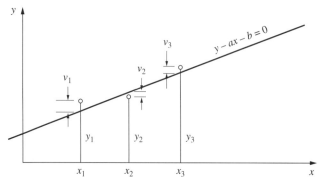

Figure B-2 Fitting a straight line to three data points.

Rearranging gives

$$(w_1x_1^2 + w_2x_2^2 + w_3x_3^2)a + (w_1x_1 + w_2x_2 + w_3x_3)b$$
$$= -(w_1x_1f_1 + w_2x_2f_2 + w_3x_3f_3)$$
$$(w_1x_1 + w_2x_2 + w_3x_3)a + (w_1 + w_2 + w_3)b$$
$$= -(w_1f_1 + w_2f_2 + w_3f_3)$$

(B-15)

Equations B-15 are called *normal equations.* Since the weight matrix is a diagonal matrix,

$$W = \begin{bmatrix} w_1 & 0 & 0 \\ 0 & w_2 & 0 \\ 0 & 0 & w_3 \end{bmatrix}$$

the matrices B and f given in Eq. B-14 are used to obtain the normal equations written in matrix form:

$$\begin{bmatrix} -x_1 & -x_2 & -x_3 \\ -1 & -1 & -1 \end{bmatrix} \begin{bmatrix} w_1 & 0 & 0 \\ 0 & w_2 & 0 \\ 0 & 0 & w_3 \end{bmatrix} \begin{bmatrix} -x_1 & -1 \\ -x_2 & -1 \\ -x_3 & -1 \end{bmatrix} \begin{bmatrix} a \\ b \end{bmatrix}$$

$$= \begin{bmatrix} -x_1 & -x_2 & -x_3 \\ -1 & -1 & -1 \end{bmatrix} \begin{bmatrix} w_1 & 0 & 0 \\ 0 & w_2 & 0 \\ 0 & 0 & w_3 \end{bmatrix} \begin{bmatrix} f_1 \\ f_2 \\ f_3 \end{bmatrix}$$

(B-16)

The student should multiply out the matrices in Eq. B-16 to be convinced that it results in Eq. B-15. The concise matrix form of the normal equations is

$$\underset{2 \times 3}{(B^T} \quad \underset{3 \times 3}{W} \quad \underset{3 \times 2}{B}) \underset{2 \times 1}{\Delta} = \underset{2 \times 3}{B^T} \quad \underset{3 \times 3}{W} \quad \underset{3 \times 1}{f}$$

(B-17)

The relation in Eq. B-17 is general in as much as it can be applied to any size problem, without any restriction on the structure of the weight matrix W. Using the auxiliaries

$$\underset{u \times u}{N} = B^T W B$$
$$\underset{u \times 1}{t} = B^T W f$$

(B-18)

Eq. B-17 becomes

$$N\Delta = t$$

(B-19)

the solution of which is

$$\Delta = N^{-1}t$$

(B-20)

The set of equations in Eq. B-17 (or Eq. B-19) are usually called the *normal equations in the parameters.* The matrices N and t in Eq. B-18 are called the *normal equations coefficient matrix* and the *normal equations constant term vector,* respectively.

EXAMPLE B-1

Consider the following data for the fitting of a straight line through three points as discussed at the beginning of this section:

Point	x	y
1	2.0	2.15
2	4.0	2.90
3	6.0	4.10

Assuming that the y coordinates are uncorrelated and of equal precision ($W = I$), compute the least squares estimates $\hat{a}$ and $\hat{b}$ of the line slope and y-intercept.

SOLUTION   According to the previous analysis, from Eq. B-11, the condition equations are

$$\begin{bmatrix} v_1 \\ v_2 \\ v_3 \end{bmatrix} + \begin{bmatrix} -2 & -1 \\ -4 & -1 \\ -6 & -1 \end{bmatrix} \begin{bmatrix} a \\ b \end{bmatrix} = \begin{bmatrix} -2.15 \\ -2.90 \\ -4.10 \end{bmatrix}$$

or

$$\underset{3 \times 1}{v} + \underset{3 \times 2}{B} \ \underset{2 \times 1}{\Delta} = \underset{3 \times 1}{f}$$

With $W = I$, the normal equations of Eqs. B-17 become

$$(B^T B)\, \Delta = B^T f$$

or

$$\begin{bmatrix} 56 & 12 \\ 12 & 3 \end{bmatrix} \begin{bmatrix} a \\ b \end{bmatrix} = \begin{bmatrix} 40.50 \\ 9.15 \end{bmatrix}$$

$$\begin{bmatrix} \hat{a} \\ \hat{b} \end{bmatrix} = \begin{bmatrix} 56 & 12 \\ 12 & 3 \end{bmatrix}^{-1} \begin{bmatrix} 40.50 \\ 9.15 \end{bmatrix} = \begin{bmatrix} 0.125 & -0.500 \\ -0.500 & 2.333 \end{bmatrix} \begin{bmatrix} 40.50 \\ 9.15 \end{bmatrix} = \begin{bmatrix} 0.488 \\ 1.100 \end{bmatrix}$$

B.6   ADJUSTMENT OF OBSERVATIONS AND PARAMETERS

Although the technique of least squares adjustment of indirect observations is frequently used in photogrammetry (for example, when applying the collinearity equations), other cases (such as coordinate transformations and the coplanarity condition) require the more general technique. Eq. B-13 is a representation of the form of the conditions when observations and parameters are combined together. Under these conditions, the normal equations are of the following general form:

$$[B^T(A W^{-1} A^T)^{-1} B]\Delta = B^T(A W^{-1} A^T)^{-1} f \tag{B-21}$$

The auxiliaries of Eq. B-21 are

$$\begin{aligned} N &= B^T(A W^{-1} A^T)^{-1} B \\ t &= B^T(A W^{-1} A^T)^{-1} f \end{aligned} \tag{B-22}$$

The technique of adjustment of observations only, mentioned in Section B.4, which employs the conditions in the form given by Eq. B-7, is rarely used in photogrammetry. Therefore, its derivation is omitted here and the student should consult Mikhail (1976) for further study. In the discussion in this Appendix, the conditions have all been linear. In practice, however, most photogrammetric problems yield nonlinear conditions. Direct nonlinear least squares adjustment is so complex that it is used very rarely if at all. Consequently, the original nonlinear conditions are linearized using Taylor's series, as presented in detail in Section A.6. Least squares is then applied to the linearized form, and the solution is iterated in order to minimize or eliminate the effect of the higher-order terms neglected in the linearization. This is demonstrated by the following example, which employs the general technique.

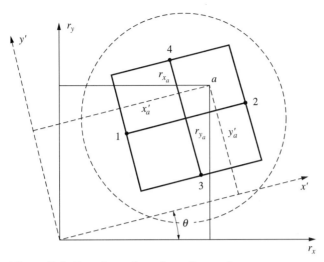

Figure B-3 Transformation of coordinates from comparator readings to photographic fiducial axes.

EXAMPLE B-2

A photographic plate contains four side fiducial marks numbered as shown in Fig. B-3. The calibrated coordinates of the fiducial marks are as follows:

Mark	Calibrated x (mm)	Calibrated y (mm)
1	−117.478	0
2	+117.472	0
3	+0.015	−117.410
4	−0.014	+117.451
IPP	0	0

These four fiducial marks and three image points, r, s, and t, were measured in a comparator, with the following results:

Point	r_x (mm)	r_y (mm)
1	17.856	144.794
2	252.637	154.448
3	140.089	32.326
4	130.400	267.027
r	150.020	66.820
s	210.082	150.010
t	82.822	160.353

Compute the photographic coordinates of r, s, and t with respect to the origin at the indicated principal point using (a) a linear conformal transformation, (b) a six-parameter affine transformation, and (c) an eight-parameter projective transformation.

SOLUTION

(a) Four-Parameter Transformation

First, the parameters a, b, c, and d of Eq. A-60c must be evaluated by the method of least squares. Letting the left-hand side of Eq. A-60c represent the comparator measurements r_x

and r_y for the four fiducial marks, the following eight observation equations can be formed:

$$17.856 + v_1 = -117.478a \qquad +c$$
$$144.794 + v_2 = \qquad +117.478b \qquad +d$$
$$252.637 + v_3 = \quad 117.472a \qquad +c$$
$$154.448 + v_4 = \qquad -117.472b \qquad +d$$
$$140.089 + v_5 = \quad 0.015a \quad -117.410b \quad +c$$
$$32.326 + v_6 = -117.410a \quad -0.015b \qquad +d$$
$$130.400 + v_7 = \quad -0.014a \quad +117.451b \quad +c$$
$$267.027 + v_8 = \quad 117.451a \quad +0.014b \qquad +d$$

In matrix notation these equations become

$$
\begin{bmatrix} v_1 \\ v_2 \\ v_3 \\ v_4 \\ v_5 \\ v_6 \\ v_7 \\ v_8 \end{bmatrix}
+
\begin{bmatrix}
117.478 & 0 & -1 & 0 \\
0 & -117.478 & 0 & -1 \\
-117.472 & 0 & -1 & 0 \\
0 & 117.472 & 0 & -1 \\
-0.015 & 117.410 & -1 & 0 \\
117.410 & 0.015 & 0 & -1 \\
0.014 & -117.451 & -1 & 0 \\
-117.451 & -0.014 & 0 & -1
\end{bmatrix}
\begin{bmatrix} a \\ b \\ c \\ d \end{bmatrix}
=
\begin{bmatrix}
-17.856 \\ -144.794 \\ -252.637 \\ -154.448 \\ -140.089 \\ -32.326 \\ -130.400 \\ -267.027
\end{bmatrix}
$$

which is of the general form of the equations $v + B\Delta = f$. If the observations are assumed uncorrelated and of equal precision, then the weight matrix is $W = I$ and the normal equations become

$$(B^TB)\Delta = (B^Tf)$$

or

$$\underset{4 \times 4}{N} \; \underset{4 \times 1}{\Delta} = \underset{4 \times 1}{t}$$

Using the numerical values of B and f above, the resulting four normal equations (which should be verified by the student) are

$$55{,}180.572a \qquad\qquad -0.0050c \quad +0.0410d \quad -55{,}147.5548 = 0$$
$$55{,}180.5972b \quad +0.0410c \quad 0.0050d \quad +2262.1915 = 0$$
$$-0.0050a \quad +0.0410b \quad +4.0000c \qquad\qquad -540.9820 = 0$$
$$+0.0410a \quad +0.0050b \qquad\qquad 4.0000d \quad -598.5950 = 0$$

The solution for the transformation parameters is obtained either by the method of elimination or, if a computer program is available, by inverting N and premultiplying t with N^{-1} to give $\Delta = N^{-1}t$. From either procedure the values of the four parameters are

$$a = +0.99930227 \qquad c = +135.247 \text{ mm}$$
$$b = -0.04111019 \qquad d = +149.639 \text{ mm}$$
$$a^2 + b^2 = 1.00029508$$

This least squares solution assumes that all of the comparator measurements are of equal precision and that no correlation exists between the points.

The inverse of Eq. A-60c, given in matrix form by Eq. A-60f is then used to compute the transformed coordinates of points r, s, and t in the fiducial coordinate system. For point r, for example

$$x_r = \frac{0.9930227}{1.00029508}(150.020 - 135.247)$$

$$- \frac{-0.04111019}{1.00029508}(66.820 - 149.639) = +11.355 \text{ mm}$$

$$y_r = \frac{-0.04111019}{1.00029508}(150.020 - 135.247)$$

$$+ \frac{0.99930227}{1.00029508}(66.820 - 149.639) = -83.343 \text{ mm}$$

Also

$$x_s = +74.776 \text{ mm}, \quad y_s = -2.705 \text{ mm}$$
$$x_t = -51.933 \text{ mm}, \quad y_t = 12.858 \text{ mm}$$

(b) Six-Parameter Transformation

The parameters a, b, c, d, e, and f of Eq. A-61b must first be evaluated by the method of least squares, assuming equal weights and no correlation. Letting the left-hand side of Eq. A-61b represent the comparator measurements r_x and r_y for the four fiducial marks, the following two sets of observation equations are formed:

$$17.856 + v_1 = -117.478a \qquad\qquad +c$$
$$252.637 + v_3 = +117.472a \qquad\qquad +c$$
$$140.089 + v_5 = \quad +0.015a \quad -117.410b \quad +c$$
$$130.400 + v_7 = \quad -0.014a \quad +117.451b \quad +c$$

to determine a, b, and c and

$$144.794 + v_2 = -117.478d \qquad\qquad +f$$
$$154.448 + v_4 = +117.472d \qquad\qquad +f$$
$$32.326 + v_6 = \quad +0.015\,d \quad -117.410e \quad +f$$
$$267.027 + v_8 = \quad -0.014d \quad +117.451e \quad +f$$

to determine d, e, and f.

It should be obvious that these two sets of four equations have identical numerical coefficients for the unknown parameters. This is because in the six-parameter affine transformation there is no inter-relation between the coefficients of the first and second equations. In matrix notation the two sets of condition equations may be written as

$$v_1 + B\Delta_1 = f_1$$
$$v_2 + B\Delta_2 = f_2$$

where

$$v_1 = [v_1 v_3 ... v_7]^T \qquad v_2 = [v_2 v_4 ... v_8]^T$$
$$\Delta_1 = [a\ b\ c]^T \qquad \Delta_2 = [d\ e\ f]^T$$

$$B = \begin{bmatrix} 117.478 & 0 & -1 \\ -117.472 & 0 & -1 \\ -0.015 & 117.410 & -1 \\ 0.014 & -117.451 & -1 \end{bmatrix}$$

$$f_1 = \begin{bmatrix} -17.856 \\ -252.637 \\ -140.089 \\ -130.400 \end{bmatrix}$$

$$f_2 = \begin{bmatrix} -144.794 \\ -154.448 \\ -32.326 \\ -267.027 \end{bmatrix}$$

Again assuming uncorrelated observations and equal precision, the normal equations become

$$N\Delta_1 = (B^T B)\Delta_1 = B^T f_1 = t_1$$
$$N\Delta_2 = (B^T B)\Delta_2 = B^T f_2 = t_2$$

which may be combined as

$$N[\Delta_1 \vdots \Delta_2] = [t_1 \vdots t_2]$$

or

$$\begin{bmatrix} 27{,}600.7517 & -3.4055 & -0.0050 \\ -3.4055 & 27{,}579.8455 & 0.0041 \\ -0.0050 & 0.0041 & 4.0000 \end{bmatrix} \begin{bmatrix} a & \vdots & d \\ b & \vdots & e \\ c & \vdots & f \end{bmatrix}$$

$$= \begin{bmatrix} 27{,}580.3622 & \vdots & 1129.9524 \\ -1132.2391 & \vdots & 27{,}567.1925 \\ 540.9820 & \vdots & 598.5950 \end{bmatrix}$$

Premultiplying the right-hand side by the inverse of the coefficient matrix will give the values of the six parameters as two adjacent vertical columns, each a 3×1 matrix. The solution results are

$$a = +0.99928070 \qquad b = -0.04113080 \qquad c = +135.247 \text{ mm}$$

$$d = +0.04108960 \qquad e = +0.99932385 \qquad f = +149.639 \text{ mm}$$

$$ae - db = 1.00029509$$

The inverse of Eq. A-61b, given in matrix form in Eq. A-61c, is then used to compute the transformed coordinates of points r, s, and t in the fiducial coordinate system. For point r, for example,

$$x_r = \frac{+0.9932285}{1.00029509}(150.020 - 135.247)$$

$$+ \frac{0.04113080}{1.00029509}(66.820 - 149.639) = +11.353 \text{ mm}$$

$$y_r = \frac{-0.04108960}{1.00029509}(150.020 - 135.247)$$

$$+ \frac{0.99928070}{1.00029509}(66.820 - 149.639) = -83.341 \text{ mm}$$

Also

$$x_s = +74.777 \text{ mm}, \quad y_s = -2.703 \text{ mm}$$
$$x_t = -51.934 \text{ mm}, \quad y_t = +12.857 \text{ mm}$$

(c) Eight-Parameter Transformation

The eight-parameter transformation coefficients a_0, b_0, a_1, b_1, c_1, a_2, b_2, and c_2 of Eq. A-71a are solved uniquely using the comparator coordinates and the calibrated coordinates of the fiducial marks. In this solution, the calibrated coordinates will be placed on the left-hand side of the equation for convenience of solution. Thus

$$\left.\begin{aligned} -117.478 &= \frac{17.856a_1 + 144.794b_1 + c_1}{17.856a_0 + 144{,}794b_0 + 1} \\[2mm] 0 &= \frac{17.856a_2 + 144.794b_2 + c_2}{17.856a_0 + 144.794b_0 + 1} \end{aligned}\right\} \text{point 1}$$

$$\left.\begin{aligned} +117.472 &= \frac{252.637a_1 + 154.448b_1 + c_1}{252.637a_0 + 154.448b_0 + 1} \\[2mm] 0 &= \frac{252.637a_2 + 154.448b_2 + c_2}{252.637a_0 + 154.448b_0 + 1} \end{aligned}\right\} \text{point 2}$$

$$\left.\begin{aligned} +0.015 &= \frac{140.089a_1 + 32.326b_1 + c_1}{140.089a_0 + 32.326b_0 + 1} \\[2mm] -117.410 &= \frac{140.089a_2 + 32.326b_2 + c_2}{140.089a_0 + 32.326b_0 + 1} \end{aligned}\right\} \text{point 3}$$

$$\left.\begin{aligned} -0.014 &= \frac{130.400a_1 + 267.027b_1 + c_1}{130.400a_0 + 267.027b_0 + 1} \\[2mm] +117.451 &= \frac{130.400a_2 + 267.027b_2 + c_2}{130.400a_0 + 267.027b_0 + 1} \end{aligned}\right\} \text{point 4}$$

The solution for the parameters in these eight equations gives

$$a_0 = +0.00000013 \qquad b_0 = +0.00000255$$
$$a_1 = +0.99942724 \text{ mm} \qquad a_2 = -0.04109385$$
$$b_1 = +0.04113506 \qquad b_2 = +0.99938431$$
$$c_1 = -141.323 \text{ mm} \qquad c_2 = -143.971 \text{ mm}$$

Using these values and substituting the comparator coordinates of r, s, and t in the right-hand side of Eq. A-71a gives the transformed coordinates of these points in the fiducial coordinate system. The results are

$$x_r = +11.357 \text{ mm}, \quad y_r = -83.341 \text{ mm}$$
$$x_s = +74.778 \text{ mm}, \quad y_s = -2.685 \text{ mm}$$
$$x_t = -51.931 \text{ mm}, \quad y_t = +12.874 \text{ mm}$$

Overall nonlinear deformations can be eliminated to a certain extent with four fiducial marks by using the conformal polynomial transformation given by Eq. A-72c. The second-degree form given by Eq. A-72c will give two degrees of freedom. If the transformation is extended to include the cubic terms, a unique solution will be obtained with four fiducial marks.

If the plate contains eight calibrated fiducial marks, which are all measured in the comparator, the four-, six-, and eight-parameter transformations to the fiducial system all provide a considerable amount of redundancy.

If the aerial camera contains a calibrated reseau, then all linear and nonlinear film and platen distortions can be essentially eliminated within the limits of measuring accuracy. During the measuring program, the comparator coordinates of the reseau grid intersection nearest each point are measured. All comparator coordinates are then transformed by rotation into a coordinate system parallel to the calibrated reseau axes using Eq. A-56. If u and v are the rotated reseau coordinates closest to the image point, u_c and v_c are their calibrated coordinates, and x and y are the coordinates of the measured image point, then the corrected coordinates x_c and y_c are given by

$$x_c = u_c + (x - u)$$
$$y_c = v_c + (y - v)$$

This computation ignores any deformation between the reseau and the image points. It is of course possible to measure all four grid intersections surrounding a point and to then treat this as a local four-fiducial-mark, nonlinear transformation. However, this computational refinement requires considerably more measurements.

EXAMPLE B-3

Refer to the data for Example B-2, in which the calibrated coordinates x and y of four fiducial marks are given, together with the comparator measurements (observations) of the four fiducials and three additional points, r, s, and t. Compute the coordinates of r, s, and t in the calibration system of coordinates using (a) the four-parameter similarity transformation and (b) the six-parameter affine transformation.

SOLUTION

(a) Four-Parameter Transformation
In Example B-2, the transformation was made *from* the calibrated system *to* the comparator system, so that the observations appear separately on the left-hand side and the technique of adjustment of indirect observations may be applied. After the parameters are computed, the *inverse* of the transformation was applied to the comparator coordinates of the three additional points. To demonstrate the uniqueness of the least squares method, we use the transformation Eq. A-60c with the calibrated coordinates x and y on the left side, and the measured values r_x and r_y, so for any point

$$x = r_x a + r_y b + c$$
$$y = r_y a - r_x b + d$$

or

$$f_x = x - r_x a - r_y b - c = 0$$
$$f_y = y - r_y a + r_x b - d = 0$$

Since r_x and r_y are the observations and a, b, c, and d are the unknown parameters, these equations are nonlinear and must be linearized. Denoting the approximations for the parameters by a^o, b^o, c^o, and d^o, the linearized form is

$$\begin{bmatrix} -a^o & -b^o \\ b^o & -a^o \end{bmatrix} \begin{bmatrix} v_x \\ v_y \end{bmatrix} + \begin{bmatrix} -r_x & -r_y & -1 & 0 \\ -r_y & r_x & 0 & -1 \end{bmatrix} \begin{bmatrix} \delta_a \\ \delta_b \\ \delta_c \\ \delta_d \end{bmatrix}$$

$$= \begin{bmatrix} r_x a^o + r_y b^o + c^o - x \\ r_y a^o - r_x b^o + d^o - y \end{bmatrix}$$

which is of the form

$$Av + B\Delta = f$$

This equation is written for one point; it needs to be repeated four times, once for each fiducial mark. In this case, the total conditions will have matrices with the following dimensions:

$$\underset{8 \times 8}{A} \quad \underset{8 \times 1}{v} \quad + \quad \underset{8 \times 4}{B} \quad \underset{4 \times 1}{\Delta} \quad = \quad \underset{8 \times 1}{f}$$

Under the assumption that the observations are uncorrelated and of equal precision (weight), a computer program was written to implement Eq. B-21, giving the following numerical results.

Let the first approximations be $a^o = 1$, $b^o = 0$, $c^o = d^o = -150$. The matrices in the eight condition equations are therefore:

$$A = -I$$

$$B = \begin{bmatrix} -17.856 & -144.794 & -1 & 0 \\ -144.794 & 17.856 & 0 & -1 \\ -252.637 & -154.448 & -1 & 0 \\ -154.448 & 252.637 & 0 & -1 \\ -140.089 & -32.326 & -1 & 0 \\ -32.326 & 140.089 & 0 & -1 \\ -130.400 & -267.027 & -1 & 0 \\ -267.027 & 130.400 & 0 & -1 \end{bmatrix}$$

$$f = \begin{bmatrix} -14.666 \\ -5.206 \\ -14.835 \\ 4.448 \\ -9.926 \\ -0.264 \\ -19.586 \\ -0.424 \end{bmatrix}$$

The parameter correction vector after the first iteration is

$$\Delta_1 = [-0.0018 \quad 0.0410 \quad 8.7372 \quad 6.0684]^T$$

and the updated approximations become

$$a_1^o = 0.998163901 \quad b_1^o = 0.041045715$$

$$c_1^o = -141.262788 \quad d_1^o = -143.931641$$

After three iterations, the final estimates for the four parameters are

$$\hat{a} = 0.999007492 \quad \hat{b} = 0.041098064$$

$$\hat{c} = -141.262791 \quad \hat{d} = -143.931644$$

The calibrated positions of image points r, s, and t are then directly computed:

$$x_r = (150.020)(0.999007492) + (66.820)(0.041098064) - 141.263 = 11.355 \text{ mm}$$

$$y_r = (66.820)(0.999007492) - (150.020)(0.041098064) - 143.932 = -83.343 \text{ mm}$$

$$x_s = 74.776 \text{ mm} \qquad x_t = -51.933 \text{ mm}$$

$$y_s = -2.705 \text{ mm} \qquad y_t = 12.858 \text{ mm}$$

These values are identical to solution (a) of Example B-2.

(b) Six-Parameter Transformation

$$x = r_x a + r_y b + c$$
$$y = r_x d + r_y e + f$$

or

$$f_x = x - r_x a - r_y b - c = 0$$
$$f_y = y - r_x d - r_y e - f = 0$$

These two equations are nonlinear, and so are linearized to the form

$$\begin{bmatrix} -a^o & -b^o \\ -d^o & -e^o \end{bmatrix} \begin{bmatrix} v_x \\ v_y \end{bmatrix} + \begin{bmatrix} -r_x & -r_y & -1 & 0 & 0 & 0 \\ 0 & 0 & 0 & -r_x & -r_y & -1 \end{bmatrix} \begin{bmatrix} \delta_a \\ \delta_b \\ \delta_c \\ \delta_d \\ \delta_e \\ \delta_f \end{bmatrix}$$

$$= \begin{bmatrix} r_x a^o + r_y b^o + c^o - x \\ r_x d^o + r_y e^o + f^o - y \end{bmatrix}$$

This pair of equations is written once for each of the four fiducial marks, thus leading to a total of eight condition equations of the form

$$\underset{8 \times 8}{A} \ \underset{8 \times 1}{v} + \underset{8 \times 6}{B} \ \underset{6 \times 1}{\Delta} = \underset{8 \times 1}{f}$$

The normal equations after the fourth iteration are given by the following matrices: $N = B^T (AA^T)^{-1} B$; and $t = B^T (AA^T)^{-1} f$. The values of N and t are

$N =$

$$\begin{bmatrix} 101483.943437 & 81524.371973 & 544.796528 & -8339.753821 & -6699.514915 & -44.770323 \\ & 117994.040698 & 602.815764 & -6699.514915 & -9696.521622 & -49.538231 \\ & & 4.028205 & -44.770323 & -49.538231 & -0.331030 \\ (Symmetric) & & & 101483.943437 & 81524.371973 & 544.796528 \\ & & & & 117994.040698 & 602.815764 \\ & & & & & 4.028205 \end{bmatrix}$$

and $t \cong 0$ to seven decimal places; thus $\Delta \cong 0$ to the same number of decimal places. The final estimates for the parameters are

$$\hat{a} = 0.999029054 \quad \hat{b} = 0.041118675 \quad \hat{c} = -141.268792$$
$$\hat{d} = -0.041077471 \quad \hat{e} = 0.998985875 \quad \hat{f} = -143.931194$$

The coordinates of points r, s, and t are computed as follows:

$$x_r = (150.020)(0.999029054) + (66.820)(0.041118675) - 141.269 = 11.353 \text{ mm}$$

$$y_r = (150.020)(-0.041077471) + (66.820)(0.998985875) - 143.931 = -83.341 \text{ mm}$$

$$x_s = 74.777 \text{ mm} \qquad x_t = -51.934 \text{ mm}$$

$$y_s = -2.703 \text{ mm} \qquad y_t = 12.857 \text{ mm}$$

These values are identical to those obtained in part (b) of the solution of Example B-2.

B.7   ADJUSTMENT WITH FUNCTIONAL CONSTRAINTS

Functional constraints occur in practice when some or all of the parameters in the adjustment must conform to some relationships arising from either geometric or physical characteristics of the model. For example, in photogrammetry, points on a lakeshore can be constrained to have the same elevation, which may or may not be known; points chosen on a straight segment of a railroad or highway are constrained to lie on a straight line; two points in the photogrammetric model may be constrained to a known distance; camera stations from an orbiting vehicle may be constrained to an orbit; and so on. The *constraint equation* is therefore defined as an equation that relates only parameters to each other. Its presence in the functional model implies that the parameters are functionally dependent. The functional dependence of the parameters leads to having as many dependent parameters as there are constraint equations. This can be explained in the following manner.

First, from n observations and n_0 minimum number of model variables, the redundancy r is determined, for which r conditions may be formulated among the observations. If we now wish to include u' *functionally independent* parameters, the number of conditions increases accordingly to $(r + u')$, with a maximum of n. Progressing further, we would like to accept the possibility of *functionally dependent* parameters in the model. Thus if $(u' + 1)$ parameters are considered, then one of them will be dependent on the u' independent parameters. For this one parameter, one *constraint equation* must be written to reflect that dependency. In general, if there are u parameters $(u > u')$, some of which are functionally dependent, then s constraints must be written to account for the number of dependent parameters, or $s = u - u'$. Evidently s must always be less than u, or the constraints become an inconsistent set of equations.

$$s < u \tag{B-23}$$

Since the number of constraints s is equal to the number of dependent parameters, it follows that the number of conditions plus constraints must be equal to the redundancy plus the total number of parameters.

$$c + s = r + u \tag{B-24}$$

The two sets of equations, conditions and constraints, are

$$\underset{c \times n}{A}\;\underset{n \times 1}{v} + \underset{c \times u}{B}\;\underset{u \times 1}{\Delta} = \underset{c \times 1}{f} \tag{B-25a}$$

$$\underset{s \times u}{C}\;\underset{u \times 1}{\Delta} = \underset{s \times 1}{g} \tag{B-25b}$$

with

$$\text{rank}(A) = c \qquad \text{rank}(B) = u \qquad \text{rank}(C) = s \tag{B-25c}$$

In the presence of constraints it is possible, in general, that $u > c$, leading to the rank of B being c. However, this is not a very common case and is therefore considered beyond the scope of this Appendix.

Since there are two separate sets of equations, then two sets of vectors of Lagrange multipliers,

$$\underset{c \times 1}{k} \;\;\text{(for the conditions) and}\;\; \underset{s \times 1}{k_c} \;\;\text{(for the constraints)}$$

are needed in order to use the *constrained minima* technique. The total set of normal equations resulting from applying the least-squares criterion is given by

$$
\begin{bmatrix}
-W & A^T & 0 & 0 \\
A & 0 & B & 0 \\
0 & B^T & 0 & C^T \\
0 & 0 & C & 0
\end{bmatrix}
\begin{bmatrix}
v \\
k \\
\Delta \\
k_c
\end{bmatrix}
=
\begin{bmatrix}
0 \\
f \\
0 \\
g
\end{bmatrix}
\tag{B-26}
$$

If the first two equations in Eq. B-26 are used to eliminate v and k, a reduced set of normal equations results:

$$
\begin{bmatrix}
-N & C^T \\
C & 0
\end{bmatrix}
\begin{bmatrix}
\Delta \\
k_c
\end{bmatrix}
=
\begin{bmatrix}
-t \\
g
\end{bmatrix}
\tag{B-27}
$$

which may be solved directly. When N^{-1} exists [rank(B) = u], then:

$$
\Delta = N^{-1}(t + C^T k_c) = N^{-1}t + N^{-1}C^T k_c = \Delta^o + \delta\Delta \tag{B-28}
$$

$$
k_c = (CN^{-1}C^T)^{-1}(g - CN^{-1}t) = M^{-1}(g - C\Delta^o) \tag{B-29}
$$

where

$$
N = B^T(AQA^T)^{-1}B, \quad t = B^T(AQA^T)^{-1}f \tag{B-30}
$$

$$
M = (CN^{-1}C^T), \quad \Delta^o = N^{-1}t, \quad \delta\Delta = N^{-1}C^T k_c \tag{B-31}
$$

The inverse of **M** exists because its rank is equal to its order, s.

The vector Δ^o represents the solution for the parameter vector δ in the absence of constraints. The final parameter vector is therefore

$$
\begin{aligned}
\Delta &= \Delta^o + N^{-1}C^T(CN^{-1}C^T)^{-1}(g - C\Delta^o) \\
&= \Delta^o + N^{-1}C^T M^{-1}(g - C\Delta^o)
\end{aligned}
\tag{B-32}
$$

The cofactor matrix of the estimated parameters is

$$
Q_{\Delta\Delta} = N^{-1}(I - C^T M^{-1} CN^{-1}) \tag{B-33a}
$$

or

$$
Q_{\Delta\Delta} = N^{-1}[I - C^T(CN^{-1}C^T)^{-1}CN^{-1}] \tag{B-33b}
$$

B.8 INNER CONSTRAINTS

A different type of constraint arises when the reference space and its coordinate system are not fully defined. The following subsections address this topic and the way least squares adjustment is modified to handle problems arising from it.

Datum Definition and Minimal Constraints

In the adjustment of geodetic or photogrammetric networks, the relationships between the observations and the point coordinate parameters are expressed by condition equations; for example, the angle or the distance condition equations in geodetic applications and the collinearity equation in photogrammetric applications. These equations all contribute *relative* information rather than *absolute* information about the point positions. The usual procedure to introduce absolute information is by constraining (i.e. fixing) certain point coordinate components. These points are referred to as *control points*. Without such control points, or other constraints, the system of normal equations would be *rank deficient* and hence not uniquely solvable. The rank deficiency is equal to the minimum number of constraints that would be needed to bring the system to *full rank*. In the case of a horizontal network with only angle observations, the rank deficiency is four. In the case of a horizontal network with at least one distance observation, the rank deficiency is three. In the case of a photogrammetric network, the rank deficiency is seven. These rank deficiencies are referred to as *datum defects*, since the presence of the necessary control points would define the datum. Of course, there may be other causes of rank deficiency, such as insufficient observations to define a point. These are another matter altogether, and are referred to as *configuration defects*. These should not occur with careful network planning, and they will not be discussed further here.

If one introduces just enough constraint equations to satisfy the datum defect, then these are known as *minimal constraints*. Different sets of minimal constraints have the interesting property that, although the point coordinate estimates may vary, the observation residuals are invariant. Thus, for residual analysis only, any minimal set of control points is as good as another. However, in practice the choice of control points is very important. The new network must be consistent with existing networks at shared points. More constraints than the minimal requirement will often be used in practice to counteract weak network geometry, or to satisfy particular requirements to match existing point coordinates. If fewer than the minimal number of constraints are used, then the system of normal equations remains rank deficient, and we say that the point coordinates are not *estimable*. Even if the individual point coordinates are not estimable, functions of them may be, for example angles or distances.

For the purpose of illustration, consider the geodetic horizontal triangle network in Fig. B-4 as an example. If only the three angle observations shown are made and no control points are introduced, then the resulting normal equations, of size six by six, have rank two and a rank deficiency of four. The fixing of two control points, or four point coordinates, would resolve this datum defect. These constraints can be implemented very simply by elimination, in effect just replacing the unknowns with numerical constants. However, for generality we assume that the constraints are implemented by the general method of bordering the normal equations, in a manner similar to Eq. B-27, given by

$$\begin{bmatrix} 0 & C \\ C^T & N \end{bmatrix} \begin{bmatrix} \lambda \\ \Delta \end{bmatrix} = \begin{bmatrix} g \\ t \end{bmatrix} \tag{B-34}$$

in which N and t pertain to the normal equations and are given by Eq. B-30. The constraint matrix corresponding to fixing the first two points is

$$C = \begin{bmatrix} 1 & 0 & 0 & 0 & 0 & 0 \\ 0 & 1 & 0 & 0 & 0 & 0 \\ 0 & 0 & 1 & 0 & 0 & 0 \\ 0 & 0 & 0 & 1 & 0 & 0 \end{bmatrix} \tag{B-35}$$

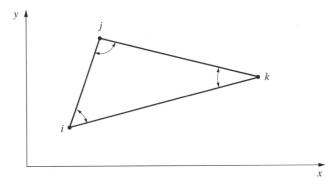

Figure B-4 Triangle network with only angle observations.

g is a vector corresponding to the four known coordinates of the two points, and λ is a 4×1 vector of unknown Lagrange multipliers corresponding to the four constraints.

Inner Constraints or Free Net Adjustment

If a square, symmetric matrix, such as *N* in the previous example, is of full rank, then all of its eigenvalues are nonzero, and its eigenvectors form an orthogonal basis for the row space. If it is not of full rank, it has order *u*, rank $h < u$, and therefore defect $u - h$. Such a matrix will have $u - h$ zero eigenvalues. The *h* eigenvectors associated with the nonzero eigenvalues will form an orthogonal basis for the row space, and the $u - h$ eigenvectors associated with the zero eigenvalues will form an orthogonal basis for the null space. In Fig. B-5, the locus of solutions to the rank-deficient equations and the intersection point of the solution space and the row space are shown schematically. Because of the favorable geometry, this intersection point will have both minimum magnitude and minimum variance compared to other restrictions or constraints. To achieve this solution, we may use the eigenvectors associated with the zero eigenvalues as coefficients in the constraint equations. This strategy will have the following characteristics:

1. It will resolve the rank deficiency from the datum defect and it will therefore permit a unique solution to the system of equations, and

2. Of all the possible solutions to the rank-deficient system, it will select the one with minimum magnitude and minimum variance.

This solution is known in the geodetic and photogrammetric literature as the *inner constraint* solution, or sometimes as the *free net solution*.

This presentation is useful for understanding the geometry of the problem, but there are easier ways to construct the needed constraint matrix. The eigenvectors just described provide a basis of the null space of the rank-deficient matrix. But, as with any vector space, there are many (an infinite number of) such bases. One particular

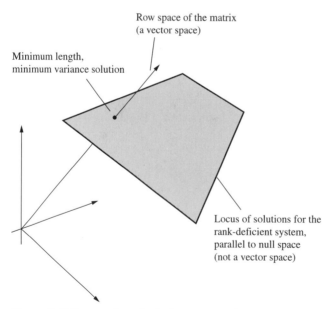

Figure B-5 Intersection of solution space and row space for a rank-deficient system.

basis can be written directly and is therefore the one that is most often used. We will describe this basis, show that it has the required properties, and then give a geometric derivation that provides much insight into the meaning of inner constraints. For the horizontal 2-D network in Fig. B-4, with no distance observations, the following constraint matrix will have the same effect as the (harder to compute) eigenvectors:

$$
C = \begin{bmatrix}
1 & 0 & 1 & 0 & 1 & 0 \\
0 & 1 & 0 & 1 & 0 & 1 \\
Y_i^0 & -X_i^0 & Y_j^0 & -X_j^0 & Y_k^0 & -X_k^0 \\
X_i^0 & Y_i^0 & X_j^0 & Y_j^0 & X_k^0 & Y_k^0
\end{bmatrix}
\tag{B-36}
$$

in which approximate coordinates X^0, Y^0 of the three points are used. If distance observations were present then the datum defect would be one less, and the fourth row of the matrix in Eq. B-36 would not be needed.

For 3-D networks, as usually found in photogrammetry, the datum defect resulting from no control point information is seven. Therefore one needs seven constraint equations to resolve the defect. Using this approach the required inner constraint matrix can be written directly, as

$$
C = \begin{bmatrix}
1 & 0 & 0 & 1 & 0 & 0 & \cdots \\
0 & 1 & 0 & 0 & 1 & 0 & \cdots \\
0 & 0 & 1 & 0 & 0 & 1 & \cdots \\
0 & Z_i^0 & -Y_i^0 & 0 & Z_j^0 & -Y_j^0 & \cdots \\
-Z_i^0 & 0 & X_i^0 & -Z_j^0 & 0 & X_j^0 & \cdots \\
Y_i^0 & -X_i^0 & 0 & Y_j^0 & -X_j^0 & 0 & \cdots \\
X_i^0 & Y_i^0 & Z_i^0 & X_j^0 & Y_j^0 & Z_j^0 & \cdots
\end{bmatrix}
\tag{B-37}
$$

in which each set of three columns corresponds to a point in the net.

If distance observations were present then the datum defect would be only six, and the last row in the matrix in Eq. B-37 could be eliminated. The X, Y, and Z terms in Eqs. B-36 and B-37 represent the current values of the approximations of the point coordinate parameters, as the solution proceeds to iterative convergence.

In order to demonstrate the plausibility of this solution, it will be shown that the rows of the matrix in Eq. B-36 are orthogonal to the coefficients of the angle condition equation. This condition equation represents the one most widely used in two dimensional, horizontal triangulation networks. For the clockwise angle at point i, in Fig. B-4, from point j to point k, the following row vector represents the coefficients of the linearized angle condition equation. For reference see Mikhail (1980).

$$
b = \begin{bmatrix}
\dfrac{\partial F_\theta}{\partial X_i} & \dfrac{\partial F_\theta}{\partial Y_i} & \dfrac{\partial F_\theta}{\partial X_j} & \dfrac{\partial F_\theta}{\partial Y_j} & \dfrac{\partial F_\theta}{\partial X_k} & \dfrac{\partial F_\theta}{\partial Y_k}
\end{bmatrix}
\tag{B-38}
$$

$b =$

$$
\begin{bmatrix}
\dfrac{Y_k^o - Y_i^o}{(S_{ik}^o)^2} - \dfrac{Y_j^o - Y_i^o}{(S_{ij}^o)^2} & -\dfrac{X_k^o - X_i^o}{(S_{ik}^o)^2} + \dfrac{X_j^o - X_i^o}{(S_{ij}^o)^2} & \dfrac{Y_j^o - Y_i^o}{(S_{ij}^o)^2} & -\dfrac{X_j^o - X_i^o}{(S_{ij}^o)^2} & -\dfrac{Y_k^o - Y_i^o}{(S_{ik}^o)^2} & \dfrac{X_k^o - X_i^o}{(S_{ik}^o)^2}
\end{bmatrix}
$$
$$
\tag{B-39}
$$

If one takes the inner product of b with the rows of C, the result is a vector of zeros. In other words, the rows of C are orthogonal to b.

$$
bC^T = \begin{bmatrix} 0 & 0 & 0 & 0 \end{bmatrix}
\tag{B-40}
$$

and

$$NC^T = \begin{bmatrix} 0 & 0 & 0 & 0 \\ 0 & 0 & 0 & 0 \\ 0 & 0 & 0 & 0 \\ 0 & 0 & 0 & 0 \\ 0 & 0 & 0 & 0 \\ 0 & 0 & 0 & 0 \end{bmatrix} \tag{B-41}$$

Thus we see that the rows of C are orthogonal to the rows of N, and therefore C can serve as a set of inner constraints for the matrix N. A similar demonstration can be made with the C of Eq. B-37 and the collinearity equations that are commonly used in 3-D networks in photogrammetry.

The following development, based upon Leick (1982), shows the geometrical meaning of using the inner constraints just described. The development will be summarized for the 2-D case. A simple extension can be made for the 3-D case. Consider a similarity transformation (four-parameter) between the adjusted coordinates, X_a, and the approximate coordinates, X_0,

$$X_a = T + (1 + k)R_\alpha X_0 \tag{B-42}$$

in which T is the translation vector, $(1 + k)$ is the scale factor, and R_α is the rotation matrix of a small angle. Written out,

$$\begin{bmatrix} x_a \\ y_a \end{bmatrix} = \begin{bmatrix} t_x \\ t_y \end{bmatrix} + (1 + k)\begin{bmatrix} \cos\alpha & \sin\alpha \\ -\sin\alpha & \cos\alpha \end{bmatrix}\begin{bmatrix} x^o \\ y^o \end{bmatrix} \tag{B-43}$$

Assuming a small angle and a scale factor near unity, and assuming that products of small quantities may be disregarded, we obtain

$$\begin{bmatrix} x_a \\ y_a \end{bmatrix} = \begin{bmatrix} t_x \\ t_y \end{bmatrix} + \begin{bmatrix} x^o \\ y^o \end{bmatrix} + \alpha\begin{bmatrix} y^o \\ -x^o \end{bmatrix} + k\begin{bmatrix} x^o \\ y^o \end{bmatrix} \tag{B-44}$$

Since this represents a step in the iterative solution,

$$\begin{bmatrix} x_a \\ y_a \end{bmatrix} = \begin{bmatrix} x^o + dx \\ y^o + dy \end{bmatrix} \tag{B-45}$$

we can combine the last two equations to obtain

$$\begin{bmatrix} dx \\ dy \end{bmatrix} = \begin{bmatrix} t_x \\ t_y \end{bmatrix} + \alpha\begin{bmatrix} y^o \\ -x^o \end{bmatrix} + k\begin{bmatrix} x^o \\ y^o \end{bmatrix} \tag{B-46}$$

Rearranging yields

$$\begin{bmatrix} dx \\ dy \end{bmatrix} = \begin{bmatrix} 1 & 0 & y^o & x^o \\ 0 & 1 & -x^o & y^o \end{bmatrix}\begin{bmatrix} t_x \\ t_y \\ \alpha \\ k \end{bmatrix} \tag{B-47}$$

We write these equations for every point in the network,

$$\begin{bmatrix} dx_1 \\ dy_1 \\ dx_2 \\ dy_2 \\ \vdots \\ dx_n \\ dy_n \end{bmatrix} = \begin{bmatrix} 1 & 0 & y_1^o & x_1^o \\ 0 & 1 & -x_1^o & y_1^o \\ 1 & 0 & y_2^o & x_2^o \\ 0 & 1 & -x_2^o & y_2^o \\ \vdots & \vdots & \vdots & \vdots \\ 1 & 0 & y_n^o & x_n^o \\ 0 & 1 & -x_n^o & y_n^o \end{bmatrix}\begin{bmatrix} t_x \\ t_y \\ \alpha \\ k \end{bmatrix} \tag{B-48}$$

If we consider this to be an overdetermined system of equations,

$$f \approx B\delta \qquad (B\text{-}49)$$

then it could be solved in the least squares sense by the usual normal equations,

$$\delta = (B^TB)^{-1}B^Tf \qquad (B\text{-}50)$$

Now, suppose that we would like to enforce the condition between the point coordinates before and after the iterative correction that there be no net shift, rotation, or scale change. In other words,

$$\delta = \begin{bmatrix} t_x \\ t_y \\ \alpha \\ k \end{bmatrix} = 0 \qquad (B\text{-}51)$$

This can be done by setting

$$B^Tf = 0 \qquad (B\text{-}52)$$

or, written out,

$$\begin{bmatrix} 1 & 0 & 1 & 0 & \cdots & 1 & 0 \\ 0 & 1 & 0 & 1 & \cdots & 0 & 1 \\ y_1^o & -x_1^o & y_2^o & -x_2^o & \cdots & y_n^o & -x_n^o \\ x_1^o & y_1^o & x_2^o & y_2^o & \cdots & x_n^o & y_n^o \end{bmatrix} \begin{bmatrix} dx_1 \\ dy_1 \\ dx_2 \\ dy_2 \\ \vdots \\ dx_n \\ dy_n \end{bmatrix} = \begin{bmatrix} 0 \\ 0 \\ 0 \\ 0 \end{bmatrix} \qquad (B\text{-}53)$$

The coefficient matrix here, it will be noticed, is identical in form to that of Eq. B-36. It was shown previously that this set of equations, considered as constraint equations, will form a basis for the null space of the datum deficient normal equations, and hence can be used as the inner constraint matrix. Thus the geometric interpretation of the inner constraint solution is that when advancing from one iteration to the next, there will be no net shift, rotation, or scale change between the approximate and the refined coordinate positions. Thus, rather than arbitrarily fixing two points (four coordinate components) out of many, one fixes four geometric relationships. All points then play equal roles in connecting the network to the coordinate system.

This can have dramatic effects upon the *a posteriori* confidence ellipses of the network points. If a point is fixed, then its confidence ellipse vanishes. For a three-point network, after fixing two points, all of the error is cast into the uncertainty of the third point. With the inner constraint or free net solution, however, each point has a finite confidence ellipse that reflects its strength of determination in the network. Furthermore, as was noted earlier, the inner constraint solution yields a minimum variance solution, and the trace of the variance-covariance matrix (the sum of the variances) will be a minimum among all constrained solutions. It should be emphasized that this property is most useful in network analysis and pre-analysis. In actual practice, the network must eventually be constrained at fixed points, in order to have the best consistency with the existing control points. The simple three-point network shown in Fig. B-4 is again depicted in Figs. B-6 and B-7 with confidence or error ellipses derived from two point constraints and from inner constraints, respectively. Of course, everything that has been described here relating to two-dimensional networks has an equivalent and obvious expression in three-dimensional networks, as found in

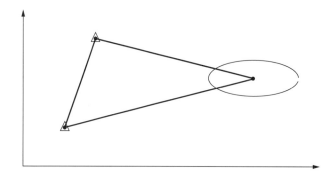

Figure B-6 Error ellipse with fixed point constraints.

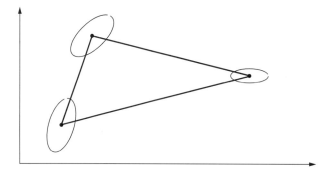

Figure B-7 Error ellipses with inner or free net constraints.

photogrammetry. The usual three-dimensional inner constraint matrix has been shown in Eq. B-37.

B.9   UNIFIED LEAST SQUARES ADJUSTMENT

In the techniques presented so far, a distinction was made between different groups of variables appearing in the mathematical formulation. Those variables for which covariance matrices are available *a priori* were designated *observables,* and the variables usually unknown in the adjustment were designated *parameters*. It was further specified that *condition* equations are those written as functions of both observations and parameters, and *constraint* equations are written only in terms of the parameters. Such classification of variables and equations follows what happens in reality and is therefore quite adequate for formulating many of the adjustment problems encountered in practice. However, recent rapid technical advances in photogrammetry have made it desirable to have an adjustment approach that can handle a mixture of input data in a more general and unified manner. As an example, computational orbital photogrammetry problems may involve regular photogrammetric data, orbital constraints, time measurements, auxiliary sensor information such as stellar photomeasurements, laser altitude and Doppler information, and gravity information. Although it is possible in principle to handle such complex adjustment problems by the techniques already described, an approach in which varying types of data can be handled with minimum effort and specialization would certainly be more appropriate.

The underlying premise for the unified least squares approach is the assumption at the outset that *all variables involved in the mathematical formulation are observations.*

With this assumption, no individual group classification becomes necessary as long as we introduce into the concept a mechanism by which a differentiation can be made whenever necessary. Such a mechanism is conveniently given by the *a priori* covariance, or weight, matrix of the *observations,* which in this case include all the variables in the model. To demonstrate the practicality of this approach, consider the following two extreme cases:

1. If an observation (in this case any variable in the model) is given an infinitely large variance, that is, its weight is $w = 0$, then it is allowed to vary freely in the adjustment and will therefore assume the role of an unknown parameter in the classical sense.

2. If an observation is given a zero variance, or a weight that approaches infinity, $w \longrightarrow \infty$, it is simply not allowed to change in the adjustment, with the consequence that its residual will be zero and it will assume the classical meaning of a constant.

Between these two bounding cases lies a large set of possibilities within which actual observations (in the classical sense) fit. Of course, one of the most important points to watch is that when the covariances of a heterogeneous set of variables are all collected together, care must be exercised with regard to units. This is particularly important if we choose to use cofactor matrices instead of covariance matrices. In such a case, only one common factor must be chosen to relate the different covariance matrices together.

Let the classically designated observational vector be l with *a priori* cofactor matrix Q and a vector of residuals v. The variables usually referred to as parameters will, in the present case, be treated as observations. For this purpose, we use the vector x with *a priori* cofactor matrix Q_{xx} and a vector of corrections Δ. There are n elements (observations) in l (and in v) and u elements (parameters) in x (and in Δ). In the classical approach the redundancy is $r = n - n_0$ and the total number of conditions is $c = r + u$, where n_0 is the minimum number of variables necessary to specify the model. In the unified concept there is a total of $(n + u)$ observed values. Therefore, the redundancy is given by

$$\text{redundancy} = \text{total number of observations} - n_0$$

or

$$(n + u - n_0) = r + u = c$$

If we carry u parameters, then the total number of equations is given by

$$\text{number of equations} = \text{redundancy} + \text{number of parameters}$$

$$= c + u$$

The first set of c conditions is that usually available from the geometric or physical conditions of the problem and formulated in Eq. B-13. Added to that we must have u more conditions to account for having *a priori* observations on the parameters. These may be formulated by recognizing that at the end of the adjustment, the value of the estimated parameters $\hat{x}$ must be identical to the estimated observations for the same variables, or

$$\hat{x} = x^o + \Delta = l_x + v_x$$

or

$$v_x - \Delta = f_x = x^o - l_x \tag{B-54}$$

where x^o is a vector of parameter approximations, and l_x is a vector of parameter observations. Combining Eqs. B-13 and B-54 gives

$$\begin{bmatrix} A & 0 \\ 0 & I \end{bmatrix} \begin{bmatrix} v \\ v_x \end{bmatrix} + \begin{bmatrix} B \\ -I \end{bmatrix} \Delta = \begin{bmatrix} f \\ f_x \end{bmatrix} \tag{B-55a}$$

or

$$\dot{A}\dot{v} + \dot{B}\Delta = \dot{f} \tag{B-55b}$$

for which the total cofactor matrix is $\dot{Q}$ given by:

$$\dot{Q} = \begin{bmatrix} Q & 0 \\ 0 & Q_{xx} \end{bmatrix}$$

Least squares may now be applied directly to Eq. B-55b. Thus

$$\dot{Q}_e = \dot{A}\dot{Q}\dot{A}^T = \begin{bmatrix} \dot{A}\dot{Q}\dot{A}^T & 0 \\ 0 & Q_{xx} \end{bmatrix} = \begin{bmatrix} Q_e & 0 \\ 0 & Q_{xx} \end{bmatrix} \tag{B-56a}$$

$$\dot{N} = \dot{B}^T\dot{W}_e\dot{B} = [B^T W_e B + W_{xx}] = (N + W_{xx}) \tag{B-56b}$$

$$\dot{t} = \dot{B}^T\dot{W}_e\dot{f} = B^T W_e f - W_{xx}f_x = (t - W_{xx}f_x) \tag{B-56c}$$

and thus

$$\Delta = \dot{N}^{-1}\dot{t} = (N + W_{xx})^{-1}(t - W_{xx}f_x) \tag{B-56d}$$

B.10 ASSESSMENT OF ADJUSTMENT RESULTS

After least squares adjustment, it is quite important in photogrammetry to analyze the results and provide a statement regarding the quality of the estimates. This operation is referred to as *post-adjustment analysis,* and applies various well-known statistical techniques.

Test on the Reference Variance

The first test is on the estimated reference variance, $\hat{\sigma}_0^2$. Let the *a priori* reference variance be σ_0^2, r the degrees of freedom (redundancy) in the adjustment, and assume that the residuals v_i are normally distributed. The statistic $r\hat{\sigma}_0^2/\sigma_0^2$ has a χ^2 distribution with r degrees of freedom. The two-tailed $100(1-\alpha)$ confidence interval for σ_0^2 is given by

$$(r\hat{\sigma}_0^2/\chi_{r,\,\alpha/2}^2) < \sigma_0^2 < (r\hat{\sigma}_0^2/\chi_{r,\,1-\alpha/2}^2) \tag{B-57}$$

If σ_0^2 is incorrect or the mathematical model used is improper or incomplete (does not adequately account for systematic errors), then σ_0^2 will fall outside this interval.

There are basically two broad categories to be investigated when the test on $\hat{\sigma}_0^2$ fails, one corresponding to the functional model and the other to the stochastic model. In the first category, one must ascertain that: (1) the computations are correctly performed; (2) the mathematical model is properly formulated, the equations are correctly written, and are correctly linearized (if nonlinear) etc...; (3) all possible systematic errors (model deficiencies) have been adequately corrected; and (4) any blunders or outliers have been identified, located, and eliminated. Once these matters have been accounted for, consideration is given next to the *a priori* value σ_0^2. Note that it is essentially a scale factor that indicates how realistic the chosen values of variances and covariances of the observations used in the adjustment were. If we are unable to ascertain the adequacy of σ_0^2 or when no *a priori* value σ_0^2 is available, the rest of the post-adjustment statistical evaluations will be performed using the *a posteriori* reference variance $\hat{\sigma}_0^2$ since it becomes the only available information to use. The value of $\hat{\sigma}_0^2$ is computed from the calculated observational residuals, v, which are estimates of the true errors in the observations, ϵ, and show only part of the true errors (Mikhail, 1979). Furthermore, $\hat{\sigma}_0^2$ combines all error sources, which cannot be separated, and therefore is a very limited statistic, so that its corresponding global statistical test is not very effective.

Test For Blunders or Outliers

If v_i is the ith residual and σ_{v_i} is its standard deviation, then

$$\overline{v}_i = v_i / \sigma_{v_i} \tag{B-58}$$

is called the *standardized residual*. Frequently, the effort involved in computing Σ_{vv} is quite extensive, and therefore an approximate estimate of σ_{v_i} may be obtained from

$$\hat{\sigma}_{v_i} = [(n-u)/n]^{1/2} \hat{\sigma}_0 \sigma_{l_i} / \sigma_0 = \left[\frac{n-u}{n} \right]^{1/2} \hat{\sigma}_0 q_{l_i} \tag{B-59}$$

in which n is the number of observations, u the number of parameters (thus $n-u=r$, the redundancy), σ_{l_i} is the *a priori* standard deviation of observation l_i, and q_{l_i} is the *a priori* cofactor, or relative standard deviation, of l_i. When σ_0^2 is known, $\overline{v}_i$ has a probability density function, or pdf, $N(0, \sigma_{v_i^2})$ and

$$\overline{v}_i = \left| v_i / \sigma_{v_i} \right| < N_{1-\alpha/2} \tag{B-60}$$

If σ_0^2 is not known, then

$$\overline{v}_i = \left| v_i / \hat{\sigma}_{v_i} \right| < \tau_{r, 1-\alpha/2} \tag{B-61}$$

in which $\hat{\sigma}_{v_i}$ is computed from Eq. B-59, and τ_r has a tau pdf with r degrees of freedom. If r is large as in photogrammetric or geodetic nets with extensive observations, τ_r may be replaced by a student t_r pdf or even a normal pdf.

Confidence Region For Estimated Parameters

The covariance matrix for the parameters as evaluated from the least squares adjustment is given by

$$\Sigma_{\hat{x}\hat{x}} = \sigma_0^2 N^{-1} \tag{B-62}$$

It can be shown that a region of constant probability is bounded by a u-dimensional hyper-ellipsoid centered at $\hat{x}$, if the parameters are assumed to have a multivariate normal pdf. The function

$$k^2 = (x - \hat{x})^T \Sigma_{\hat{x}\hat{x}}^{-1} (x - \hat{x}) \tag{B-63}$$

describes the hyper-ellipsoid. The quadratic k^2 has a χ_u^2 distribution with the probability for a point estimate being

$$P(\chi_u^2 < k^2) = 1 - \alpha$$

Table B-1 Values of k and Corresponding Probabilities for Error Ellipses

P	0.394	0.500	0.900	0.950	0.990
k	1.000	1.177	2.146	2.447	3.035

Table B-2 Values of k and Corresponding Probabilities for Three-Dimensional Error Ellipsoids

P	0.199	0.500	0.900	0.950	0.990
k	1.000	1.538	2.500	2.700	3.368

For the two-dimensional case (error ellipses), typical values for k and the corresponding probabilities are given in Table B-1. The values in Table B-1 are obtained from $k = \sqrt{\chi^2_{2,\,\alpha}}$, which is the limiting case of the more general $k = \sqrt{2F_{2,\,r,\,\alpha}}$. For the three-dimensional case (error ellipsoids), they are given in Table B-2. When $k = 1$, we usually call it the *standard* region, or *standard error ellipse* (for 2-D), and *standard error ellipsoid* (for 3-D).

Given Σ, for example, for a point in a plane, the semimajor axis, a, and semiminor axis, b, of the standard error ellipse are computed from the eigenvalues and eigenvectors (see Section A.3 in Appendix A). If one is interested in the 90% confidence region (i.e., significance level of $\alpha = 0.10$), the a,b are multiplied by 2.146. For the standard regions, there is a 0.683 probability that an adjusted point falls in a onedimensional interval, a 0.394 probability that it falls inside the standard error ellipse, and only a 0.199 probability that it falls within the standard error ellipsoid.

If the *a posteriori* reference variance, $\hat{\sigma}_0^2$, is used then

$$\hat{\Sigma}_{xx} = \hat{\sigma}_0^2 Q_{xx} = \hat{\sigma}_0^2 N_{xx}^{-1} \tag{B-64}$$

Whereas Σ_{xx} or $\hat{\Sigma}_{xx}$ provide the most complete description of the quality of the estimated values, such as target point coordinates, quite frequently there is a need for one representative assessment quantity. Such quantity is often taken as the mean of the variances, or

$$\overline{\sigma}_x^2 = \frac{1}{u}\mathrm{tr}(\Sigma_{xx})$$

$$\hat{\overline{\sigma}}_x^2 = \frac{1}{u}\mathrm{tr}(\hat{\Sigma}_{xx}) \tag{B-65}$$

in which $\overline{\sigma}_x^2, \hat{\overline{\sigma}}_x^2$ are estimates of the mean accuracy of the coordinates, and u is the dimension of the covariance matrix, or the number of the coordinates with which it is associated. Note that the trace of the covariance matrix is also equal to the sum of its eigenvalues, or the sum of the variances of the decorrelated covariance matrix or the transformed matrix such that all covariances are zero (i.e., all correlations have been eliminated).

Error Ellipses

The variance and standard deviation are measures of precision of the one-dimensional case of one random variable such as a distance, for example. In the case of two-dimensional problems, such as the position of an image point, error ellipses may be established around the point to designate precision regions of different probabilities. The orientation of the ellipse relative to the x,y axis system (Fig. B-8) depends on the correlation between x and y. If they are uncorrelated, the ellipse axes will be parallel to x and y. If the two coordinates are of equal precision, or $\sigma_x = \sigma_y$, the ellipse becomes a circle.

Considering the general case where the covariance matrix for the position of point P is given as

$$\Sigma = \begin{bmatrix} \sigma_x^2 & \sigma_{xy} \\ \sigma_{xy} & \sigma_y^2 \end{bmatrix} \tag{B-66}$$

the semimajor and semiminor axes of the corresponding ellipse are determined by the eigenvalues and eigenvectors as described in Section A.3 of Appendix A. They are computed in the following manner. First, a second-degree polynomial (called the characteristic polynomial, refer to Eqs. A-47 and A-48) is set up using the elements of Σ as

$$\lambda^2 - (\sigma_x^2 + \sigma_y^2)\lambda + (\sigma_x^2 \sigma_y^2 - \sigma_{xy}^2) = 0 \tag{B-67}$$

The two roots λ_1 and λ_2 of Eq. B-67 (which are called the eigenvalues of Σ) are computed and their square roots are the semimajor and semiminor axes of the *standard error ellipse*, as shown in Fig. B-8. The orientation of the ellipse is determined by computing the angle θ between the x-axis and the semimajor axis from

$$\tan 2\theta = \frac{2\sigma_{xy}}{\sigma_x^2 - \sigma_y^2} \tag{B-68}$$

Alternatively, the eigenvectors, corresponding to λ_1, λ_2, which are orthogonal, may be evaluated and used to determine the orientation of the error ellipse, see Section A.3. The quadrant of 2θ is determined from the fact that the sign of $\sin 2\theta$ is the same as the sign of σ_{xy}, and $\cos 2\theta$ has the same sign as $(\sigma_x^2 - \sigma_y^2)$. Whereas in the onedimensional case, the probability of falling within $+\sigma$ and $-\sigma$ is 0.683, the probability of falling on or inside the standard error ellipse in 2-D is 0.394. In a manner similar to constructing intervals with given probabilities for the one-dimensional case, different-size ellipses may be established, each with a given probability. It should be obvious that the larger the size of the error ellipse, the larger the probability of the actual value falling within the ellipse. Using the standard ellipse as a base, Table B-1 lists the scale multiplier k to enlarge the ellipse and the corresponding probability.

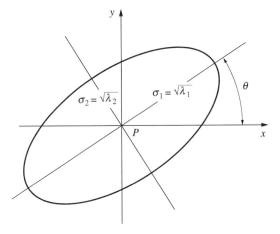

Figure B-8 Error ellipse.

As an example, for an ellipse with axes $a = 2.447a_s$ and $b = 2.477b_s$ where a_s and b_s are the semimajor and semiminor axes, respectively, of the standard ellipse, the probability that the point falls inside the ellipse is 0.95.

Often one is interested in the relative accuracy between two points in a horizontal network. Then the coordinate differences are

$$d_x = x_2 - x_1$$
$$d_y = y_2 - y_1$$

and the total covariance matrix for the coordinates is

$$\Sigma_{1,2} = \begin{bmatrix} \sigma_{x_1}^2 & \sigma_{x_1 y_1} & \sigma_{x_1 x_2} & \sigma_{x_1 y_2} \\ & \sigma_{y_1}^2 & \sigma_{y_1 x_2} & \sigma_{y_1 y_2} \\ & & \sigma_{x_2}^2 & \sigma_{x_2 y_2} \\ (Symmetric) & & & \sigma_{y_2}^2 \end{bmatrix}$$

Then,

$$d = \begin{bmatrix} d_x \\ d_y \end{bmatrix} = \begin{bmatrix} -1 & 0 & 1 & 0 \\ 0 & -1 & 0 & 1 \end{bmatrix} \begin{bmatrix} x_1 \\ y_1 \\ x_2 \\ y_2 \end{bmatrix} = JP_{1,2}$$

and

$$\Sigma_{dd} = J\Sigma_{1,2}J^T$$

from which

$$\sigma_{dx}^2 = \sigma_{x_1}^2 - 2\sigma_{x_1 x_2} + \sigma_{x_2}^2$$
$$\sigma_{dy}^2 = \sigma_{y_1}^2 - 2\sigma_{y_1 y_2} + \sigma_{y_2}^2 \qquad \text{(B-69)}$$
$$\sigma_{dxdy} = \sigma_{x_1 y_1} - \sigma_{x_1 y_2} - \sigma_{x_2 y_1} + \sigma_{x_2 y_2}$$

Introducing the variances, σ_{dx}^2 and σ_{dy}^2, and covariance, σ_{dxdy}, in Eqs. B-67 and B-68, the elements of a relative error ellipse can be readily computed.

In the three-dimensional case, where the elevation as well as the horizontal position of the point is involved, the precision region becomes an ellipsoid. Table B-2 gives the corresponding multipliers.

The concepts of error ellipse and error ellipsoid are quite useful in establishing confidence regions about points such as those determined by photogrammetric triangulation techniques. These regions are measures of the reliability of the positional determination of such points. They could also be specified in advance as a means of establishing specifications.

Although both absolute error ellipses (for points) and relative error ellipses (for lines) are used to evaluate adjustment quality, it is frequently more convenient to replace the two-dimensional representation by a single quantity (similar to $\bar{\sigma}$). In this case, a circular probability distribution is substituted for the elliptical probability distribution. Consequently, a single circular standard deviation, σ_c, is calculated from the two semi-axes of the error ellipse. The value of σ_c depends on the relative magnitudes of these axes.

Let

$$\Sigma_{xx} = \begin{bmatrix} \sigma_{\tilde{x}}^2 & \sigma_{xy} \\ \sigma_{xy} & \sigma_{\tilde{y}}^2 \end{bmatrix}$$

represent the covariance matrix for the x,y coordinates of a point. Then, $\lambda_1 = a^2$ and $\lambda_2 = b^2$ are the eigenvalues ($\lambda_1 > \lambda_2$) of Σ_{xx}, and a,b are the semimajor and semiminor axes of the error ellipse, respectively. Note that: $\mathrm{tr}(\Sigma_{xx}) = \sigma_x^2 + \sigma_y^2 = \lambda_1 + \lambda_2 = a^2 + b^2$. If $\sigma_{\min} = b = \sqrt{\lambda_2}$ and $\sigma_{\max} = a = \sqrt{\lambda_1}$, then the value of the ratio $\sigma_{\min}/\sigma_{\max}$ determines the relationship used to calculate σ_c.

When $\sigma_{\min}/\sigma_{\max}$ is between 1.0 and 0.6, then

$$\sigma_c \approx (0.5222\,\sigma_{\min} + 0.4778\,\sigma_{\max}) \tag{B-70}$$

A good approximation that yields a slightly larger σ_c (i.e., on the safe side) is given by

$$\sigma_c \approx 0.5(a + b) \tag{B-71}$$

which may be extended to the limit of $\sigma_{\min}/\sigma_{\max} \le 0.2$.

The mean accuracy measure, $\bar{\sigma}$ (see Eq. B-65), is

$$\bar{\sigma}^2 = \frac{1}{2}(\sigma_x^2 + \sigma_y^2) = \frac{1}{2}(a^2 + b^2) = \frac{1}{2}(\lambda_1 + \lambda_2)$$

and

$$\bar{\sigma}_c = \left[\frac{1}{2}(\sigma_x^2 + \sigma_y^2)\right]^{1/2} \tag{B-72}$$

is actually applicable only when $\sigma_{\min}/\sigma_{\max}$ is between 1.0 and 0.8, in which case it yields essentially the same value of σ_c as in Eq. B-71. As the ratio of $\sigma_{\min}/\sigma_{\max}$ decreases, $\bar{\sigma}_c$ from Eq. B-72 gets progressively larger than that from Eq. B-71 with a maximum increase of about 20% at $\sigma_{\min}/\sigma_{\max} = 0.2$.

Of course the probability associated with the standard error circle is the same as for the standard error ellipse, 0.394. The multipliers given in Table B-1 also still apply for circular errors of different probabilities. Figure B-9 shows several standard error ellipses and their corresponding circles for several $\sigma_{\min}/\sigma_{\max}$ ratios.

EXAMPLE B-4

The position of point A in Fig. B-10 is determined by the radial distance $r = 100$ m, with $\sigma_r = 0.5$ m, and the azimuth angle $\alpha = 60°$, with $\sigma_\alpha = 0°30'$. Compute the rectangular coordinates x and y and the associated covariance matrix for point A. (Assume r and α to be uncorrelated.) Then calculate the semimajor and semiminor axes of the standard error ellipse and its orientation.

SOLUTION

$$x = r\sin \alpha = 86.60\,\text{m}$$
$$y = r\cos \alpha = 50.00\,\text{m}$$

$$J = \begin{bmatrix} \dfrac{\partial x}{\partial \alpha} & \dfrac{\partial x}{\partial r} \\[2mm] \dfrac{\partial y}{\partial \alpha} & \dfrac{\partial y}{\partial r} \end{bmatrix} = \begin{bmatrix} r\cos \alpha & \sin \alpha \\ -r\sin \alpha & \cos \alpha \end{bmatrix} = \begin{bmatrix} 50 & 0.866 \\ -86.6 & 0.5 \end{bmatrix}$$

The covariance matrix of the known variables r and α is a diagonal matrix because these variables are uncorrelated. The covariance matrix is equal to

$$\Sigma = \begin{bmatrix} \sigma_\alpha^2 & 0 \\ 0 & \sigma_r^2 \end{bmatrix} = \begin{bmatrix} (0.0087)^2\,\text{rad}^2 & 0 \\ 0 & (0.5)^2\,\text{m}^2 \end{bmatrix}$$

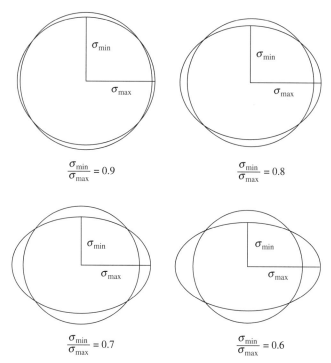

Figure B-9 Error ellipses and their corresponding error circles [From ACIC Technical Report No. 96, United States Air Force, February, 1962].

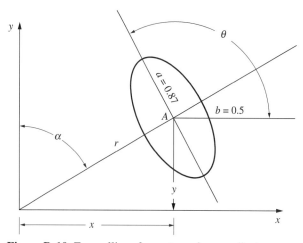

Figure B-10 Error ellipse for rectangular coordinates.

Applying covariance propagation, the covariance matrix of the Cartesian coordinates is

$$\Sigma_{\text{coord}} = \begin{bmatrix} 50 & 0.866 \\ -86.6 & 0.5 \end{bmatrix} \begin{bmatrix} 0.7569 \times 10^{-4} & 0 \\ 0 & 0.25 \end{bmatrix} \begin{bmatrix} 50 & -86.6 \\ 0.866 & 0.5 \end{bmatrix}$$

$$= \begin{bmatrix} 0.3767 & -0.2195 \\ -0.2195 & 0.6300 \end{bmatrix} \text{m}^2$$

This matrix expresses the reliability of the Cartesian coordinates of point A. According to Eq. B-67, the characteristic polynomial is

$$\lambda^2 - (0.3767 + 0.6300)\lambda + (0.3767)(0.6300) - (-0.2195)^2 = 0$$
$$\lambda^2 - 1.0067\lambda + 0.189141 = 0$$

from which the two roots (eigenvalues) are

$$\lambda_1 = 0.7567 \qquad \lambda_2 = 0.25$$

The semimajor axis is $a = \sqrt{\lambda_1} = 0.87$ m and the semiminor axis is $b = \sqrt{\lambda_2} = 0.50$ m. To get the orientation of the semimajor axis, use Eq. B-68.

$$\tan 2\theta = \frac{2\sigma_{xy}}{\sigma_x^2 - \sigma_y^2} = \frac{-2(0.2195)}{0.3767 - 0.6300} = \frac{-0.4390}{-0.2533} = 1.732$$

Because both $\sin 2\theta$ and $\cos 2\theta$ are negative, then 2θ is in the third quadrant, or $2\theta = 240°$, and $\theta = 120°$.

The results obtained here are rather important. They show, as demonstrated in Fig. B-10, that the semiminor axis is along the extension of r and is equal to $\sigma_r = 0.5$ m. Similarly, the semimajor axis is oriented normal to r and is equal to $r\sigma_\theta = (100)(0.0087) = 0.87$ m. Thus, the error ellipse axes are always oriented in the directions where the variables are uncorrelated, as in this case r and α. Along any other pair of directions the variables are correlated, as in the case of x and y, for which the correlation coefficient ρ_{xy} may be computed from

$$\rho_{xy} = \frac{-0.2195}{(0.6138)(0.7937)} = 0.45$$

For this example $\sigma_{min} = b = 0.5$ m, and $\sigma_{max} = 0.87$ m, and thus the ratio is $\sigma_{min}/\sigma_{max} = 0.57$. The radius of the standard error circle is, according to Eq. B-71,

$$\sigma_c \cong 0.5(a + b) \simeq (0.5(0.5 + 0.87)) = 0.685 \text{ m}$$

If Eq. B-72 is used then

$$\bar{\sigma}_c = [0.5(0.3767 + 0.63)]^{1/2} = 0.707 \text{ m}$$

which is about 3% larger than σ_c from Eq. B-71. If one can ascertain that the ratio $\sigma_{min}/\sigma_{max}$ is not below 0.2, then Eq. B-72 has the advantage of not having to compute the eigenvalues, and it produces a slightly larger error circle.

In a manner similar to circular errors for two-dimensional locations, a standard error sphere with radius σ_s may replace the standard error ellipsoid around a survey point in three-dimensional space.

$$\sigma_s = \frac{1}{3}(a + b + c) \tag{B-73}$$

If $a = \sqrt{\lambda_1}$, $b = \sqrt{\lambda_2}$, and $c = \sqrt{\lambda_3}$ where $\lambda_1 \geq \lambda_2 \geq \lambda_3$ are the eigenvalues of the covariance matrix Σ_{xx} of the coordinates X, Y, and Z, $a \geq b \geq c$ are the semi-axes of the error ellipsoid, then Eq. B-73 is used when $\sigma_{min}/\sigma_{max}$ is between 1.0 and 0.35, where σ_{min} and σ_{max} are the smallest and largest semi-axes of the ellipsoid, respectively. When the ratio $\sigma_{min}/\sigma_{max}$ is between 1.0 and 0.9, the square root of one third of the trace, or

$$\bar{\sigma}_s = \left[\frac{1}{3}(\sigma_x^2 + \sigma_y^2 + \sigma_z^2)\right]^{1/2} \tag{B-74}$$

will essentially give the same value as Eq. B-73. As this ratio gets smaller, $\bar{\sigma}_s$ will progressively get larger than σ_s, with an increase of about 16% as this ratio approaches 0.35 (depending of course on the value of the middle semi-axis, b). The spherical concept is not recommended when $\sigma_{min}/\sigma_{max}$ becomes less than 0.35.

REFERENCES

MIKHAIL, E. M. 1976. *Observations and Least Squares,* New York, NY: University Press.

MIKHAIL, E. M. 1980. *Analysis and Adjustment of Survey Measurements,* van Nostrand Reinhold, New York, NY.

LEICK, A. 1982. Journal of the Surveying and Mapping Division, Proceedings of the American Society of Civil Engineers, 108: no. SU2.

MIKHAIL, E. M. 1979. Review and Some Thoughts on the Assessment of Aerial Triangulation Results, Aerial Triangulation Symposium Proceedings, University of Queensland, Australia.

STRANG, G. 1988. *Linear Algebra and its Applications,* Fort Worth, TX: Saunders/Harcourt Brace Jovanovich.

Appendix C

Linearization of Photogrammetric Condition Equations

C.1   NONLINEAR CONDITIONS

All four condition equations used in photogrammetry—the projectivity, collinearity, coplanarity, and scale restraint equations—are nonlinear in the variables involved. Direct solution of such equations is quite impractical, particularly when they are used in conjunction with least squares estimation for redundant cases. Consequently, all equations are usually linearized by applying Taylor's series expansion as discussed in detail in Section A.6 of Appendix A. Linearization of each of the four photogrammetric conditions is developed in a separate section of this Appendix.

C.2   LINEARIZATION OF THE PROJECTIVITY EQUATIONS

The projectivity equations, developed in Section 4.1 and also given by Eq. A-71a can be written in the functional form as

$$F_1 = -y_1 + \frac{a_1 x_1 + b_1 x_2 + c_1}{a_0 x_1 + b_0 x_2 + 1} = -y_1 + \frac{r}{t} = 0$$

$$F_2 = -y_2 + \frac{a_2 x_1 + b_2 x_2 + c_2}{a_0 x_1 + b_0 x_2 + 1} = -y_1 + \frac{s}{t} = 0$$

(C-1)

Since the designation of which variables are observations (as opposed to parameters) depends on the specific application, Eq. C-1 will be linearized in the most general form considering all 12 variables involved. Consequently, applying Eq. A-81b gives

$$\underset{2 \times 12}{J} \quad \underset{12 \times 1}{\Delta} \quad = \quad \underset{2 \times 1}{f}$$

(C-2)

423

in which

$$
\underset{2 \times 12}{J} =
\begin{bmatrix}
\dfrac{\partial F_1}{\partial y_1} & \dfrac{\partial F_1}{\partial y_2} & \dfrac{\partial F_1}{\partial x_1} & \dfrac{\partial F_1}{\partial x_2} & \vdots & \dfrac{\partial F_1}{\partial a_0} & \cdots & \dfrac{\partial F_1}{\partial c_2} \\[3mm]
\dfrac{\partial F_2}{\partial y_1} & \dfrac{\partial F_2}{\partial y_2} & \dfrac{\partial F_2}{\partial x_1} & \dfrac{\partial F_2}{\partial x_2} & \vdots & \dfrac{\partial F_2}{\partial a_0} & \cdots & \dfrac{\partial F_2}{\partial c_2}
\end{bmatrix}
\tag{C-3}
$$

$$
\Delta = \begin{bmatrix} \Delta y_1 & \Delta y_2 & \Delta x_1 & \Delta x_2 & \vdots & \Delta a_0 & \Delta b_0 & \cdots & \Delta c_2 \end{bmatrix}^T
\tag{C-4}
$$

$$
f = \begin{bmatrix} -F_1^o \\[2mm] -F_2^o \end{bmatrix} = \begin{bmatrix} y_1^0 - \dfrac{r^o}{t^o} \\[3mm] y_2^0 - \dfrac{s^o}{t^o} \end{bmatrix}
\tag{C-5}
$$

The superscript o implies that the variables are evaluated at their approximate values.

In the case of a least squares adjustment made to satisfy the projectivity equations, the matrices J and Δ in Eq. C-2 would be partitioned to reflect the known measurements before the adjustment is applied. For example, if all four coordinates y_1, y_2, x_1, and x_2 are considered as measurements, partitioning would be as shown by the vertical dotted lines in Eqs. C-3 and C-4. In this case, the equations would take the form

$$
\underset{2 \times 4}{A} \; \underset{4 \times 1}{v} + \underset{2 \times 8}{B} \; \underset{8 \times 2}{\Delta} = \underset{2 \times 1}{f}
\tag{C-6}
$$

in which A and B are the submatrices of J, and v is the vector of four residuals. Least squares is then applied to Eq. C-6 as explained in Appendix B.

Evaluation of the elements of J is obviously an exercise in partial differentiation. The first four elements are very simple to obtain:

$$
j_{11} = -1, \; j_{12} = 0, \; j_{21} = 0, \; j_{22} = -1.
$$

For the remaining 20 elements, the following relationships apply for partial differentiation with respect to any one of the 10 variables, designated by p:

$$
\frac{\partial F_1}{\partial p} = \frac{1}{t}\left(\frac{\partial r}{\partial p} - \frac{r}{t}\frac{\partial t}{\partial p} \right) \quad \text{and} \quad \frac{\partial F_2}{\partial p} = \frac{1}{t}\left(\frac{\partial s}{\partial p} - \frac{s}{t}\frac{\partial t}{\partial p} \right)
\tag{C-7}
$$

Some of the elements are given here as examples, and the rest are left as exercises for the student.

$$
j_{13} = \frac{\partial F_1}{\partial x_1} = \frac{1}{t^o}\left(a_1 - \frac{r}{t}a_0 \right)^o
$$

$$
j_{25} = \frac{\partial F_2}{\partial a_0} = \frac{1}{t^o}\left(0 - \frac{s}{t}x_1 \right)^o = -\frac{s^o x_1^o}{t^{o^2}}
$$

$$
j_{17} = \frac{\partial F_1}{\partial a_1} = \frac{1}{t^o}(x_1 - 0)^o = \frac{x_1^o}{t^o}
$$

C.3   LINEARIZATION OF THE COLLINEARITY EQUATIONS

Although the collinearity condition takes different forms, that given by Eq. 4-24 is used most frequently, and will be considered here as an example. Equation 4-24 can be rewritten as

$$
F_1 = (x - x_0) + f\frac{U}{W} = 0
$$
$$
F_2 = (y - y_0) + f\frac{V}{W} = 0
$$
$$\tag{C-8}$$

The most common case is that in which the image coordinates x and y are considered the observations or measurements, the elements of interior orientation x_0, y_0, and f are considered known (without error) from calibration, and the remaining variables are considered unknown parameters. Consequently, the linearized form of Eq. C-8 is given by

$$\underset{2 \times 1}{v} + \underset{2 \times 9}{B} \underset{9 \times 1}{\Delta} = \underset{2 \times 1}{f} \tag{C-9}$$

in which

$v = \begin{bmatrix} v_x & v_y \end{bmatrix}^T$ = image coordinate residuals

B = the matrix of partial derivatives of the two functions in Eq. C-8 with respect to each of the six exterior orientation elements and the three coordinates of the object point

Δ = the vector of nine corrections to the approximations for the parameters

$$f = \begin{bmatrix} -F_1^o \\ -F_2^o \end{bmatrix} = \begin{bmatrix} -(x - x_0) - fU/W \\ -(y - y_0) - fV/W \end{bmatrix}^o$$

The elements of the B matrix will depend on how the orientation matrix M is constructed. Furthermore, the partial derivatives in B require the differentiation of M. Therefore, we will first select one set of three parameters for constructing M, then show how M is partially differentiated with respect to each of the parameters.

Consider as an example the sequence $\omega \longrightarrow \phi \longrightarrow \kappa$ (see Section A.4), where

$$M = M_\kappa M_\phi M_\omega$$

Therefore,

$$\frac{\partial M}{\partial \omega} = M'_\omega = M_\kappa M_\phi \frac{\partial M}{\partial \omega}$$

but

$$\frac{\partial M_\omega}{\partial \omega} = \begin{bmatrix} 0 & 0 & 0 \\ 0 & -\sin\omega & \cos\omega \\ 0 & -\cos\omega & -\sin\omega \end{bmatrix}$$

$$= \begin{bmatrix} 1 & 0 & 0 \\ 0 & \cos\omega & \sin\omega \\ 0 & -\sin\omega & \cos\omega \end{bmatrix} \begin{bmatrix} 0 & 0 & 0 \\ 0 & 0 & 1 \\ 0 & -1 & 0 \end{bmatrix}$$

or

$$\frac{\partial M_\omega}{\partial \omega} = M_\omega \begin{bmatrix} 0 & 0 & 0 \\ 0 & 0 & 1 \\ 0 & -1 & 0 \end{bmatrix}$$

Then,

$$M'_\omega = M \begin{bmatrix} 0 & 0 & 0 \\ 0 & 0 & 1 \\ 0 & -1 & 0 \end{bmatrix} \tag{C-10}$$

Next,

$$\frac{\partial M}{\partial \phi} = M'_\phi = M_\kappa \frac{\partial M_\phi}{\partial \phi} M_\omega$$

in which

$$\frac{\partial M_\phi}{\partial \phi} = \begin{bmatrix} -\sin\phi & 0 & -\cos\phi \\ 0 & 0 & 0 \\ \cos\phi & 0 & -\sin\phi \end{bmatrix}$$

$$= \begin{bmatrix} \cos\phi & 0 & -\sin\phi \\ 0 & 1 & 0 \\ \sin\phi & 0 & \cos\phi \end{bmatrix}\begin{bmatrix} 0 & 0 & -1 \\ 0 & 0 & 0 \\ 1 & 0 & 0 \end{bmatrix}$$

Hence

$$M'_\phi = M_\kappa M_\phi \begin{bmatrix} 0 & 0 & -1 \\ 0 & 0 & 0 \\ 1 & 0 & 0 \end{bmatrix} M_\omega$$

but

$$\begin{bmatrix} 0 & 0 & -1 \\ 0 & 0 & 0 \\ 1 & 0 & 0 \end{bmatrix} M_\omega = \begin{bmatrix} 0 & \sin\omega & -\cos\omega \\ 0 & 0 & 0 \\ 1 & 0 & 0 \end{bmatrix}$$

$$= \begin{bmatrix} 1 & 0 & 0 \\ 0 & \cos\omega & \sin\omega \\ 0 & -\sin\omega & \cos\omega \end{bmatrix}\begin{bmatrix} 0 & \sin\omega & -\cos\omega \\ -\sin\omega & 0 & 0 \\ \cos\omega & 0 & 0 \end{bmatrix}$$

Hence

$$M'_\phi = M \begin{bmatrix} 0 & \sin\omega & -\cos\omega \\ -\sin\omega & 0 & 0 \\ \cos\omega & 0 & 0 \end{bmatrix} \tag{C-11}$$

Alternatively, M'_ϕ can be derived as follows:

$$M'_\phi = M_\kappa \begin{bmatrix} -\sin\phi & 0 & -\cos\phi \\ 0 & 0 & 0 \\ \cos\phi & 0 & -\sin\phi \end{bmatrix} M_\omega$$

$$= M_\kappa \begin{bmatrix} 0 & 0 & -1 \\ 0 & 0 & 0 \\ 1 & 0 & 0 \end{bmatrix} M_\phi M_\omega$$

$$= \begin{bmatrix} 0 & 0 & -\cos\kappa \\ 0 & 0 & \sin\kappa \\ 1 & 0 & 0 \end{bmatrix} M_\phi M_\omega$$

or

$$M'_\phi = \begin{bmatrix} 0 & 0 & -\cos\kappa \\ 0 & 0 & \sin\kappa \\ \cos\kappa & -\sin\kappa & 0 \end{bmatrix} M_\kappa M_\phi M_\omega$$

$$= \begin{bmatrix} 0 & 0 & -\cos\kappa \\ 0 & 0 & \sin\kappa \\ \cos\kappa & -\sin\kappa & 0 \end{bmatrix} M \tag{C-12}$$

Finally,

$$\frac{\partial M}{\partial \kappa} = M'_\kappa = \frac{\partial M_\kappa}{\partial \kappa} M_\phi M_\omega$$

but

$$\frac{\partial M}{\partial \kappa} \kappa = \begin{bmatrix} -\sin\kappa & \cos\kappa & 0 \\ -\cos\kappa & -\sin\kappa & 0 \\ 0 & 0 & 0 \end{bmatrix}$$

$$= \begin{bmatrix} 0 & 1 & 0 \\ -1 & 0 & 0 \\ 0 & 0 & 0 \end{bmatrix} M_\kappa$$

Hence

$$M'_\kappa = \begin{bmatrix} 0 & 1 & 0 \\ -1 & 0 & 0 \\ 0 & 0 & 0 \end{bmatrix} M \tag{C-13}$$

From Eq. 4-21,

$$\begin{bmatrix} U \\ V \\ W \end{bmatrix} = M \begin{bmatrix} X - X_L \\ Y - Y_L \\ Z - Z_L \end{bmatrix} \tag{C-14}$$

This may be partially differentiated with respect to any of the nine variables ω, ϕ, κ (implicit in M), X_L, Y_L, Z_L, X, Y, and Z. As examples, one of each set of three is evaluated here, and the rest are left as exercises for the student.

$$\frac{\partial}{\partial\omega} \begin{bmatrix} U \\ V \\ W \end{bmatrix} = \frac{\partial M}{\partial\omega} \begin{bmatrix} X - X_L \\ Y - Y_L \\ Z - Z_L \end{bmatrix} = M^o \begin{bmatrix} 0 & 0 & 0 \\ 0 & 0 & 1 \\ 0 & -1 & 0 \end{bmatrix} \begin{bmatrix} X - X_L \\ Y - Y_L \\ Z - Z_L \end{bmatrix}^o$$

$$= M^o \begin{bmatrix} 0 \\ Z - Z_L \\ Y_L - Y \end{bmatrix} \tag{C-15}$$

$$\frac{\partial}{\partial Y} \begin{bmatrix} U \\ V \\ W \end{bmatrix} = M^o \frac{\partial}{\partial Y} \begin{bmatrix} X - X_L \\ Y - Y_L \\ Z - Z_L \end{bmatrix}^o = M^o \begin{bmatrix} 0 \\ 1 \\ 0 \end{bmatrix} = \begin{bmatrix} m_{12} \\ m_{22} \\ m_{32} \end{bmatrix}^o \tag{C-16}$$

$$\frac{\partial}{\partial Z_L} \begin{bmatrix} U \\ V \\ W \end{bmatrix} = M^o \frac{\partial}{\partial Z_L} \begin{bmatrix} X - X_L \\ Y - Y_L \\ Z - Z_L \end{bmatrix}^o = M^o \begin{bmatrix} 0 \\ 0 \\ -1 \end{bmatrix} = \begin{bmatrix} -m_{13} \\ -m_{23} \\ -m_{33} \end{bmatrix}^o \tag{C-17}$$

The elements of the B matrix may now be evaluated using relationships similar to those in Eq. C-7, or,

$$\frac{\partial F_1}{\partial p} = \frac{f}{W}\left(\frac{\partial U}{\partial p} - \frac{U}{W}\frac{\partial W}{\partial p}\right) \quad \text{and} \quad \frac{\partial F_2}{\partial p} = \frac{f}{W}\left(\frac{\partial V}{\partial p} - \frac{V}{W}\frac{\partial W}{\partial p}\right) \tag{C-18}$$

The partial derivatives of U, V, W in Eq. C-18 would be obtained from relations similar to those in Eqs. C-15 to C-17. Assuming that the sequence of variables is X_L, Y_L, Z_L, ω, ϕ, κ, X, Y, Z, the element b_{24} of B, for example, would be

$$b_{24} = \frac{\partial F_2}{\partial\omega} = \frac{f}{W^o}\left(\frac{\partial V}{\partial\omega} - \frac{V}{W}\frac{\partial W}{\partial\omega}\right)^o \tag{C-19}$$

The values of $\partial V/\partial\omega$ and $\partial W/\partial\omega$ are obtained from Eq. C-15 as

$$\frac{\partial V}{\partial\omega} = m_{22}^o(Z - Z_L)^o + m_{23}^o(Y_L - Y)^o$$

$$\frac{\partial W}{\partial\omega} = m_{32}^o(Z - Z_L)^o + m_{33}^o(Y_L - Y)^o$$

which may then be substituted into Eq. C-19 to yield the final form of the element b_{24}. All other elements of B can be obtained in a similar manner, which is left as an exercise for the student.

C.4   LINEARIZATION OF THE COPLANARITY EQUATION

The coplanarity condition is given in the form of a determinant by Eq. 4-48. Differentiation of the determinant was developed in Section A.6.5 of Appendix A for columns. Considering rows, the partial derivative of a determinant of order three with respect to a parameter p is equal to the sum of three determinants. If R_1, R_2, and R_3 are the three rows of a determinant D, then

$$\frac{\partial D}{\partial p} = \begin{vmatrix} \partial R_1/\partial p \\ R_2 \\ R_3 \end{vmatrix} + \begin{vmatrix} R_1 \\ \partial R_2/\partial p \\ R_3 \end{vmatrix} + \begin{vmatrix} R_1 \\ R_2 \\ \partial R_3/\partial p \end{vmatrix} \qquad (C-20)$$

Equation C-20 may be applied to Eq. 4-48 to evaluate all partial derivatives necessary for its linearization. Since in most cases the coplanarity condition is used for relative orientation, there are usually *four* observed image coordinates x_1, y_1, x_2, and y_2 and *five* parameters. Therefore, the linearized form of Eq. 4-48 is

$$\underset{1\times4}{A}\ \underset{4\times1}{v} + \underset{1\times5}{B}\ \underset{5\times1}{\Delta} = \underset{1\times1}{f} \qquad (C-21)$$

where

$A = \begin{bmatrix} \partial F/\partial x_1 & \partial F/\partial y_1 & \partial F/\partial x_2 & \partial F/\partial y_2 \end{bmatrix}$

v = the vector of four observational residuals

B = the matrix of the partial derivatives of F with respect to the five specific parameters selected, which depend on the type of relative orientation

Δ = corrections to the five specified parameters

$$f = -F^o = -\begin{vmatrix} b_X & b_Y & b_Z \\ u_1 & v_1 & w_1 \\ u_2 & v_2 & w_2 \end{vmatrix}$$

Each of the elements of A is composed of only one determinant, since each image coordinate appears in only one row. For example

$$a_{13} = \frac{\partial F}{\partial x_2} = \begin{vmatrix} b_X & b_Y & b_Z \\ u_1 & v_1 & w_1 \\ \dfrac{\partial u_2}{\partial x_2} & \dfrac{\partial v_2}{\partial x_2} & \dfrac{\partial w_2}{\partial x_2} \end{vmatrix}$$

in which the partial derivatives shown may be evaluated as

$$\frac{\partial}{\partial x_2}\begin{bmatrix} u_2 \\ v_2 \\ w_2 \end{bmatrix} = M_2^T\begin{bmatrix} 1 \\ 0 \\ 0 \end{bmatrix} = \begin{bmatrix} m_{11} \\ m_{12} \\ m_{13} \end{bmatrix}_2^o$$

and hence

$$a_{13} = \begin{vmatrix} b_X & b_Y & b_Z \\ u_1 & v_1 & w_1 \\ (m_{11})_2 & (m_{12})_2 & (m_{13})_2 \end{vmatrix}^o$$

The other elements of A can be similarly evaluated.

The elements of B depend on whether relative orientation is dependent or independent. Independent relative orientation solves for five of the six rotational elements of the two photographs, for example: ω_1, ϕ_1, κ_1, ϕ_2, and κ_2. For dependent relative orientation, the parameters of one photograph and one base component (usually b_X) are fixed, leaving five parameters to be solved for: b_Y, b_Z, ω_2, ϕ_2, κ_2. Taking the latter case as an example, the elements of B would be

$$b_{11} = \frac{\partial F}{\partial b_Y} = \begin{vmatrix} 0 & 1 & 0 \\ u_1 & v_1 & w_1 \\ u_2 & v_2 & w_2 \end{vmatrix}^o = - \begin{vmatrix} u_1 & w_1 \\ u_2 & w_2 \end{vmatrix}^o$$

$$b_{12} = \frac{\partial F}{\partial b_Z} = \begin{vmatrix} u_1 & v_1 \\ u_2 & v_2 \end{vmatrix}^o$$

$$b_{13} = \frac{\partial F}{\partial \omega_2} = \begin{vmatrix} b_X & b_Y & b_Z \\ u_1 & v_1 & w_1 \\ \dfrac{\partial u_2}{\partial \omega_2} & \dfrac{\partial v_2}{\partial \omega_2} & \dfrac{\partial w_2}{\partial \omega_2} \end{vmatrix}^o$$

$b_{14} = \partial F / \partial \phi_2$ and $b_{15} = \partial F / \partial \kappa_2$ are evaluated similar to b_{13}. The partial derivatives in b_{13} are given by

$$\frac{\partial}{\partial \omega_2} \begin{bmatrix} u_2 \\ v_2 \\ w_2 \end{bmatrix} = \frac{\partial M_2^T}{\partial \omega_2} \begin{bmatrix} x - x_0 \\ y - y_0 \\ -f \end{bmatrix}_2 = \left(M_2^o \begin{bmatrix} 0 & 0 & 0 \\ 0 & 0 & 1 \\ 0 & -1 & 0 \end{bmatrix} \right)^T \begin{bmatrix} x - x_0 \\ y - y_0 \\ -f \end{bmatrix}_2$$

$$= \begin{bmatrix} 0 & 0 & 0 \\ 0 & 0 & -1 \\ 0 & 1 & 0 \end{bmatrix} \left(M_2^{o\,T} \begin{bmatrix} x - x_0 \\ y - y_0 \\ -f \end{bmatrix}_2 \right) = \begin{bmatrix} 0 & 0 & 0 \\ 0 & 0 & -1 \\ 0 & 1 & 0 \end{bmatrix} \begin{bmatrix} u_2 \\ v_2 \\ w_2 \end{bmatrix}^o$$

or

$$\frac{\partial}{\partial \omega_2} \begin{bmatrix} u_2 \\ v_2 \\ w_2 \end{bmatrix} = \begin{bmatrix} 0 \\ -w_2 \\ v_2 \end{bmatrix}^o$$

Consequently, the final form of the b_{13} element would be

$$b_{13} = \begin{vmatrix} b_X & b_Y & b_Z \\ u_1 & v_1 & w_1 \\ 0 & -w_2 & v_2 \end{vmatrix}^o$$

In a very similar manner the student may derive the remaining two elements, b_{14} and b_{15}.

C.5   LINEARIZATION OF THE SCALE RESTRAINT EQUATION

The scale restraint equation was derived in Section 4.5.5 and is given by Eq. 4-72. The elements of the vectors a_1, a_2, and a_3 are given by Eqs. 4-60, 4-61, and 4-62. The two additional vectors, d_1 and d_2, in the condition equation are given by

$$d_1 = a_1 \times a_2 = \begin{bmatrix} a_{1y}a_{2z} - a_{2y}a_{1z} \\ a_{2x}a_{1z} - a_{1x}a_{2z} \\ a_{1x}a_{2y} - a_{2x}a_{1y} \end{bmatrix}$$

$$d_2 = a_2 \times a_3 = \begin{bmatrix} a_{2y}a_{3z} - a_{3y}a_{2z} \\ a_{3x}a_{2z} - a_{2x}a_{3z} \\ a_{2x}a_{3y} - a_{3x}a_{2y} \end{bmatrix}$$

The observations are $(x, y)_1$, $(x, y)_2$, and $(x, y)_3$, or the photo coordinates in the three photographs involved. The total possible parameters would be $(x_0, y_0, f; X_L, Y_L, Z_L; \omega, \phi, \kappa)_{1,2,3}$, where the subscripts 1, 2, and 3 indicate three sets of nine parameters, each set pertaining to one photograph. Therefore, the most general form of linearized scale restraint equation would be

$$\underset{1 \times 6}{A} \underset{6 \times 1}{v} + \underset{1 \times 27}{B} \underset{27 \times 1}{\Delta} = \underset{1 \times 1}{f} \qquad (C\text{-}22)$$

For convenience, we rewrite Eq. 4-72 in the short form:

$$F = \frac{R_1}{R_2} + \frac{R_3}{R_4} = 0 \qquad (C\text{-}23)$$

Designating by p any of the variables, the partial derivative of F with respect to p is

$$\frac{\partial F}{\partial p} = \frac{1}{R_2}\left(\frac{\partial R_1}{\partial p} - \frac{R_1}{R_2}\frac{\partial R_2}{\partial p}\right) + \frac{1}{R_4}\left(\frac{\partial R_3}{\partial p} - \frac{R_3}{R_4}\frac{\partial R_4}{\partial p}\right) \qquad (C\text{-}24)$$

where $\partial R_i/\partial p$, $i = 1, 2, 3, 4$, may be evaluated using differentiation of determinants, which was developed in Section A.6.5 in Appendix A.

The elements of the A matrix are

$$A = \begin{bmatrix} \dfrac{\partial F}{\partial x_1} & \dfrac{\partial F}{\partial y_1} & \dfrac{\partial F}{\partial x_2} & \dfrac{\partial F}{\partial y_2} & \dfrac{\partial F}{\partial x_3} & \dfrac{\partial F}{\partial y_3} \end{bmatrix} \qquad (C\text{-}25)$$

The B matrix is partitioned by photograph, or

$$\underset{1 \times 27}{B} = \begin{bmatrix} \underset{1 \times 9}{B_1} & \underset{1 \times 9}{B_2} & \underset{1 \times 9}{B_3} \end{bmatrix} \qquad (C\text{-}26)$$

For example, B_1 contains the partial derivatives of F with respect to the interior and exterior orientation elements of the first photograph, or

$$B_1 = \begin{bmatrix} \dfrac{\partial F}{\partial x_{0_1}} & \dfrac{\partial F}{\partial y_{0_1}} & \dfrac{\partial F}{\partial f_1} & \dfrac{\partial F}{\partial X_{L_1}} & \dfrac{\partial F}{\partial Y_{L_1}} & \dfrac{\partial F}{\partial Z_{L_1}} & \dfrac{\partial F}{\partial \omega_1} & \dfrac{\partial F}{\partial \phi_1} & \dfrac{\partial F}{\partial \kappa_1} \end{bmatrix} \qquad (C\text{-}27)$$

As an example, the first element of A is

$$a_{11} = \frac{\partial F}{\partial x_1} = \left[\frac{1}{R_2}\left(\frac{\partial R_1}{\partial x_1} - \frac{R_1}{R_2}\frac{\partial R_2}{\partial x_1}\right)\right]^o$$

with

$$\frac{\partial R_1}{\partial x_1} = \begin{vmatrix} (m_{11})_1 & d_{1x} & b_{1x} \\ (m_{12})_1 & d_{1y} & b_{1y} \\ (m_{13})_1 & d_{1z} & b_{1z} \end{vmatrix}^o + \begin{vmatrix} a_{1x} & [(m_{12})_1\,a_{2z} - (m_{13})_1\,a_{2y}] & b_{1x} \\ a_{1y} & [(m_{13})_1\,a_{2x} - (m_{11})_1\,a_{2z}] & b_{1y} \\ a_{1z} & [(m_{11})_1\,a_{2y} - (m_{12})_1\,a_{2x}] & b_{1z} \end{vmatrix}^o$$

$$\frac{\partial R_2}{\partial x_1} = \begin{vmatrix} (m_{11})_1 & d_{1x} & a_{2x} \\ (m_{12})_1 & d_{1y} & a_{2y} \\ (m_{13})_1 & d_{1z} & a_{2z} \end{vmatrix}^o + \begin{vmatrix} a_{1x} & [(m_{12})_1\, a_{2z} - (m_{13})_1\, a_{2y}\,] & a_{2x} \\ a_{1y} & [(m_{13})_1\, a_{2x} - (m_{11})_1\, a_{2z}\,] & a_{2y} \\ a_{1z} & [(m_{12})_1\, a_{2y} - (m_{12})_1\, a_{2x}\,] & a_{2z} \end{vmatrix}^o$$

noting that

$$\frac{\partial \boldsymbol{a}_1}{\partial x_1} = \begin{bmatrix} (m_{11})_1 \\ (m_{12})_1 \\ (m_{13})_1 \end{bmatrix} \text{ and } \frac{\partial \boldsymbol{d}_1}{\partial x_1} = \begin{bmatrix} (m_{12})_1\, a_{2z} - (m_{13})_1\, a_{2y} \\ (m_{13})_1\, a_{2x} - (m_{11})_1\, a_{2z} \\ (m_{11})_1\, a_{2y} - (m_{12})_1\, a_{2x} \end{bmatrix}$$

and that

$$\frac{\partial R_3}{\partial x_1} = \frac{\partial R_4}{\partial x_1} = 0$$

As another example, the fourth element of $\boldsymbol{B}_1$ is

$$b_{14} = \frac{\partial F}{\partial X_{L_1}} = \frac{1}{R_2}\left[\frac{\partial R_1}{\partial X_{L_1}}\right] = \frac{1}{R_2}\begin{vmatrix} a_{1x} & d_{1x} & -1 \\ a_{1y} & d_{1y} & 0 \\ a_{1z} & d_{1z} & 0 \end{vmatrix}$$

or

$$b_{14} = \frac{-1}{R_2}\begin{vmatrix} a_{1y} & d_{1y} \\ a_{1z} & d_{1z} \end{vmatrix}$$

The student can develop other elements using auxiliaries and applying the rules presented in this Appendix.

C.6   LINEARIZATION USING NUMERICAL DIFFERENTIATION

Another approach to linearization is to use numerical differentiation instead of analytically deriving the partial derivatives and then evaluating their numerical values by substituting the approximate values for the unknown variables. The difference between the two approaches is demonstrated by the following function of a single unknown variable:

$$F(x) = x^2$$

$$\frac{dF}{dx} = 2x \qquad \text{is the analytical derivative}$$

(C-28)

Thus, for an approximate value $x^o = 10$ for x, the numerical value of dF/dx is 20. The numerical value of the derivative can be evaluated directly using the following approximation

$$\frac{dF}{dx} \cong \frac{F(x^o + \delta x) - F(x^o)}{\delta x}$$

where x^o is as before the approximate value for x and δx is a small interval in the value of x. The smaller the value of δx, the closer is the approximation in the numerical derivative to that evaluated analytically. For our simple example, let $\delta x = 0.001$, then:

$$\frac{dF}{dx} = \frac{(10 + 0.001)^2 - (10)^2}{0.001} = 20.001$$

The same approach can be applied to any of the complex equations used in photogrammetry, by extending it to partial derivatives in the following manner. Let the function be of, say, three variables, or $F(x, y, z)$. Then

$$\frac{\partial F}{\partial x} = \frac{F(x^o + \delta x, y^o, z^o) - F(x^o, y^o, z^o)}{\delta x}$$

$$\frac{\partial F}{\partial y} = \frac{F(x^o, y^o + \delta y, z^o) - F(x^o, y^o, z^o)}{\delta y} \qquad \text{(C-29)}$$

$$\frac{\partial F}{\partial z} = \frac{F(x^o, y^o, z^o + \delta z) - F(x^o, y^o, z^o)}{\delta z}$$

in which x^o, y^o, and z^o are the approximations for the unknown variables and δx, δy, and δz are the corresponding small numerical increments. As an illustration, let us consider the collinearity equations, and the partial derivative with respect to X_L. First we write the pair of collinearity equations in the following functional form:

$$F_1(x, \text{IO}, X_L, \text{EO}_r) = 0$$
$$F_2(y, \text{IO}, X_L, \text{EO}_r) = 0$$

where x and y are the numerical values of the image measurements, IO are the elements of interior orientation, and EO_r are the five remaining elements of exterior orientation, $Y_L, Z_L, \omega, \phi, \kappa$. Then

$$\frac{\partial F_1}{\partial X_L} = \frac{F_1(x, \text{IO}^o, X_L + \delta X_L, \text{EO}_r^o) - F_1(x, \text{IO}^o, X_L^o, \text{EO}_r^o)}{\delta X_L}$$

$$\frac{\partial F_2}{\partial X_L} = \frac{F_2(y, \text{IO}^o, X_L + \delta X_L, \text{EO}_r^o) - F_2(y, \text{IO}^o, X_L^o, \text{EO}_r^o)}{\delta X_L}$$

in which the superscript o represents approximate values for the unknown parameters, and δX_L is a suitable numerical small increment in X_L, say 0.01 m. All other numerical partial derivatives of these equations, and for that matter any other condition equations, however complex, can be evaluated in a similar manner. The only critical matter is that of the choice of the small numerical increment in the value of the parameter under consideration. It should be chosen sufficiently small to minimize the effect of the approximation, and yet not so small as to cause numerical overflow, as explained at the beginning of this section.

Appendix D

Mathematical Description of Linear Features

D.1 INTRODUCTION

Image information represents different forms in the object space. In general, these forms can be classified as point features, linear features, and area features. Many photogrammetric reduction treatments have been developed based on object point features. Linear and area features were usually extracted from the photogrammetric model, but did not contribute to the solution.

At the present time, the conditions involved in photogrammetric activities are changing substantially. Digital imagery, either directly acquired or from digitized photography, is becoming much more available. Furthermore, more robust techniques for feature extraction from digital imagery are continually being developed. In particular, edges and other linear features are relatively easier to detect and extract from digital imagery than the traditionally used point features. Photogrammetric methodology, therefore, is expanding to accommodate features other than points, especially linear features. In particular, given the image description of an object linear feature on two or more images, the original linear feature may be rigorously derived, when image interior and exterior orientation is known. The derivation need not be dependent on the presence of corresponding image point features that lie on the linear feature. In fact, the feature description on various overlapping images may be of distinctly different segments of the linear object feature. In this case, conjugate image points are not possible. This also clearly implies that the description of the feature in the image is not the primary factor in the modeling; instead, it is the description of the feature in the object space. Once a general mathematical model for the linear feature is developed, it can be applied to various photogrammetric problems. For example, in photogrammetric resection of a single image, the linear feature is used as control in order to recover the sensor/platform exterior orientation parameters. For relative orientation of overlapping images, it acts as a *pass linear feature* in the same sense as a pass point. Space intersection of linear features involves the solution for the object space description of a linear feature given its image representation in two or more images and all the sensor/platform model parameters. Finally, photogrammetric triangulation is a simultaneous resection and intersection in which linear features may be used as control and/or pass features, thus allowing object completion in the 3-D object space.

In this Appendix, we will provide the mathematical representation of linear features in both two- and three-dimensional spaces. Although there are many types of linear features, consideration will be limited to the more practical forms: straight lines and circular lines. After presenting the mathematical description of these features in both two- and three-dimensional space, the *geometric constraints* that can exist among and between these two types of linear features will be discussed. Such constraints may provide purely *relative* information, such as parallel, perpendicular, coplanar, etc., or partial *absolute* information with respect to the object coordinate system, such as horizontal, vertical, etc.

D.2   STRAIGHT LINES IN TWO-DIMENSIONAL SPACE

A straight line in two-dimensional space is characterized by two independent parameters. There are many possible formulations, each of which has some weakness. For example, a straight line of the form

$$y_i = mx_i + d \tag{D-1}$$

where m is the slope of the line, d is the y-intercept (see Fig. D-1), and (x_i, y_i) are the coordinates of any point on the line, is undefined when the line is parallel to the y-axis ($m \longrightarrow \infty$). To avoid this degeneracy, the coordinate system must be rotated so that no straight lines of interest are parallel to the y-axis.

Another formulation uses the two parameters p and α (see Fig. D-1), with the equation

$$x_i \cos \alpha + y_i \sin \alpha = p \tag{D-2}$$

where p is the length of the perpendicular from the origin to the line and α is the angle measured counter-clockwise from the positive x-axis to that perpendicular.

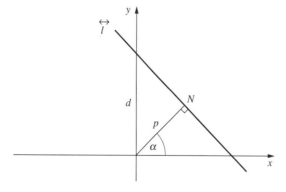

Figure D-1 A straight line in 2-D.

Yet another form of the straight line equation that is commonly used in image understanding is

$$a'x_i + b'y_i + c' = 0 \qquad (D\text{-}3)$$

When normalized by dividing the equation by the parameter c' (thus obtaining the required two independent parameters a and b), Eq. D-3 becomes

$$ax_i + by_i + 1 = 0 \qquad (D\text{-}4)$$

where

$$a = \frac{a'}{c'}$$

$$b = \frac{b'}{c'}$$

Equation D-3 is used to represent straight lines in homogenous coordinates. This is the reason behind its popularity in the image understanding community.

As can be seen in Eq. D-2 and Fig. D-1, α is undefined when a straight line passes through the origin ($p = 0$). Both Eq. D-3 and Eq. D-4 are also not valid for a straight line that passes through the origin. To avoid this degeneracy, the origin of the coordinate system can be shifted so that none of the lines of interest passes through the origin, ($p \neq 0$).

The representation of straight lines with descriptors implies that straight lines are of infinite length. This means that the position of a finite straight line segment is unknown. In those applications where it is important that the actual position be known, such as straight line matching, a straight line segment can be represented by its two end points, (x_1, y_1) and (x_2, y_2).

D.3   STRAIGHT LINES IN THREE-DIMENSIONAL SPACE

A common form of the equations describing a straight line in 3-D uses a set of six dependent parameters. Any point (X_i, Y_i, Z_i) on the line is determined by two vectors (see Fig. D-2). The first vector, $\vec{A} = \begin{bmatrix} A_x & A_y & A_z \end{bmatrix}^T$, is a vector along or parallel to the straight line. The second vector, $\vec{B} = \begin{bmatrix} X_n & Y_n & Z_n \end{bmatrix}^T$, is the vector from the origin to a point N with coordinates (X_n, Y_n, Z_n) on the line. Any point (X_i, Y_i, Z_i) on the straight line is obtained by adding a scalar multiple of $\vec{A}$ to $\vec{B}$, or

$$\begin{bmatrix} X_i \\ Y_i \\ Z_i \end{bmatrix} = \lambda_i \begin{bmatrix} A_x \\ A_y \\ A_z \end{bmatrix} + \begin{bmatrix} X_n \\ Y_n \\ Z_n \end{bmatrix} \qquad (D\text{-}5)$$

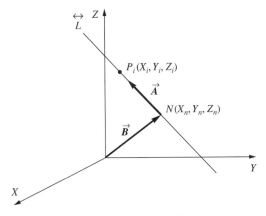

Figure D-2 A straight line in 3-D (dependent parameter form).

where λ_i is a scale factor that varies from point to point depending on the distance from N to the point (X_i, Y_i, Z_i), and the length of $\vec{A}$.

A straight line in 3-D has only four independent parameters. Consequently, two constraints must exist among the six parameters $A_x, A_y, A_z, X_n, Y_n,$ and Z_n. The first constraint is that $\vec{B}$ is perpendicular to the line, and the second is that A is a unit vector. These constraints remove dependency among the six dependent parameters. The constraint equations are

$$\vec{A} \cdot \vec{B} = 0 \qquad \text{or} \qquad A_x X_n + A_y Y_n + A_z Z_n = 0 \qquad (D\text{-}6)$$

and

$$\vec{A} \cdot \vec{A} = 1 \qquad \text{or} \qquad A_x^2 + A_y^2 + A_z^2 = 1 \qquad (D\text{-}7)$$

Another formulation of the straight line equation in 3-D uses coordinates of two points on the line as feature descriptors (see Fig. D-3).

$$\begin{bmatrix} X_i \\ Y_i \\ Z_i \end{bmatrix} = \lambda_i \begin{bmatrix} X_2 - X_1 \\ Y_2 - Y_1 \\ Z_2 - Z_1 \end{bmatrix} + \begin{bmatrix} X_1 \\ Y_1 \\ Z_1 \end{bmatrix} \qquad (D\text{-}8)$$

where (X_1, Y_1, Z_1) and (X_2, Y_2, Z_2) are the coordinates of two fixed points on the line.

The disadvantage of using dependent straight line parameters rather than independent straight line parameters is that it increases the complexity of the adjustment process. Therefore, it is advantageous to formulate the straight line equation using only four independent parameters. One way to accomplish this is to use Eqs. D-6 and D-7 to eliminate two parameters. For example, eliminating A_z and Z_n:

$$A_z = \sqrt{1 - A_x^2 - A_y^2} \qquad (D\text{-}9)$$

$$Z_n = \frac{-A_x X_n - A_y Y_n}{\sqrt{1 - A_x^2 - A_y^2}} \qquad (D\text{-}10)$$

Then Eq. D-5 will have only $A_x, A_y, X_n,$ and Y_n as the four independent parameters.

A more suitable alternative formulation of a straight line in 3-D is parameterized directly using four independent parameters $q, \beta_1, \beta_2,$ and β_3 (see Fig. D-4). The central idea of this formulation is that a straight line can be represented in a quite simple

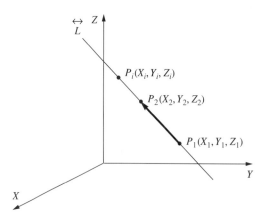

Figure D-3 A straight line in 3-D (two-points form).

form by only one parameter, q, in a different right-handed coordinate system, UVW, local to each straight line (see Fig. D-5).   The straight line equations in the XYZ coordinate system are obtained by transforming the straight line equations from the UVW coordinate system to the XYZ coordinate system.   This transformation involves only three rotation angles (β_1, β_2, and β_3).

The UVW coordinate system is constructed such that the straight line ($\overleftrightarrow{L}$) lies in the UW-plane and is parallel to the W-axis at a distance q from the origin (see Fig. D-5). Notice that the point N on the line at the shortest distance from the origin to the straight line is on the U-axis. The coordinates of any point (U_i, V_i, W_i) on the straight line in the UVW coordinate system can be written in the form of Eq. D-5 as

$$
\begin{bmatrix} U_i \\ V_i \\ W_i \end{bmatrix} = \lambda_i \begin{bmatrix} 0 \\ 0 \\ 1 \end{bmatrix} + q \begin{bmatrix} 1 \\ 0 \\ 0 \end{bmatrix}
\tag{D-11}
$$

In order to transform the UVW coordinate system into the XYZ coordinate system, three sequential rotations are needed. The first rotation brings the UVW coordinate

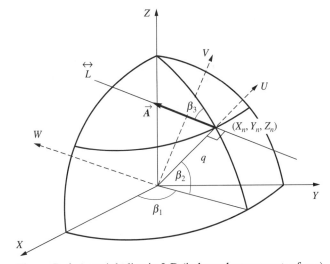

Figure D-4 A straight line in 3-D (independent parameter form).

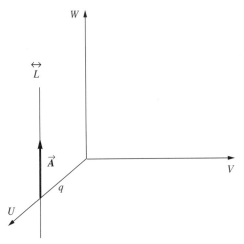

Figure D-5 A straight line in the UVW coordinate system.

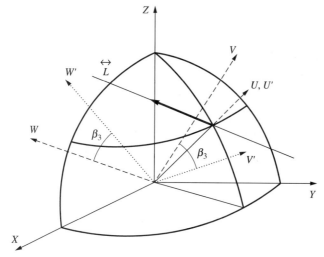

Figure D-6 The first rotation by β_3: $UVW \longrightarrow U'V'W'$.

system to the $U'V'W'$ coordinate system by rotating about the U-axis by the angle β_3, so that after the rotation the $U'W'$-plane is a vertical (or meridian) plane and the V'-axis is in the horizontal plane (see Fig. D-6). The angle β_3 is taken positive although it appears negative in Fig. D-6 for ease of illustration.

$$\begin{bmatrix} U_i' \\ V_i' \\ W_i' \end{bmatrix} = M_{\beta_3} \begin{bmatrix} U_i \\ V_i \\ W_i \end{bmatrix} \tag{D-12}$$

where

$$M_{\beta_3} = \begin{bmatrix} 1 & 0 & 0 \\ 0 & \cos\beta_3 & \sin\beta_3 \\ 0 & -\sin\beta_3 & \cos\beta_3 \end{bmatrix} \tag{D-13}$$

The second rotation brings the $U'V'W'$ coordinate system to the $U''V''W''$ coordinate system by rotating about the V'-axis by the angle β_2. This makes the W''-axis coincide with the Z-axis, with the U'' axis in the horizontal plane (see Fig. D-7).

$$\begin{bmatrix} U_i'' \\ V_i'' \\ W_i'' \end{bmatrix} = M_{\beta_2} \begin{bmatrix} U_i' \\ V_i' \\ W_i' \end{bmatrix} \tag{D-14}$$

where

$$M_{\beta_2} = \begin{bmatrix} \cos\beta_2 & 0 & -\sin\beta_2 \\ 0 & 1 & 0 \\ \sin\beta_2 & 0 & \cos\beta_2 \end{bmatrix} \tag{D-15}$$

The last rotation is about the W''-axis (or Z-axis) by the angle β_1, which brings the UVW coordinate system to coincide with the XYZ system (see Fig. D-8). Again, note that β_1 is positive in the equations, although it appears negative in Fig. D-8.

$$\begin{bmatrix} X_i \\ Y_i \\ Z_i \end{bmatrix} = M_{\beta_1} \begin{bmatrix} U_i'' \\ V_i'' \\ W_i'' \end{bmatrix} \tag{D-16}$$

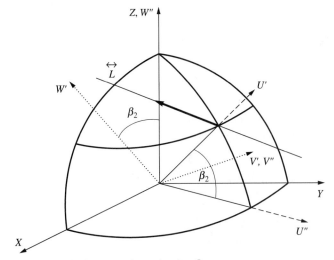

Figure D-7 The second rotation by β_2:
$U'\, V'\, W' \longrightarrow U''\, V''\, W''$.

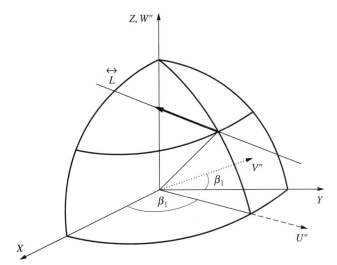

Figure D-8 The third rotation by β_1: $U''\, V''\, W'' \longrightarrow XYZ$.

where

$$M_{\beta_1} = \begin{bmatrix} \cos\beta_1 & \sin\beta_1 & 0 \\ -\sin\beta_1 & \cos\beta_1 & 0 \\ 0 & 0 & 1 \end{bmatrix} \tag{D-17}$$

The transformation can be summarized as

$$\begin{bmatrix} X_i \\ Y_i \\ Z_i \end{bmatrix} = M_\beta \begin{bmatrix} U_i \\ V_i \\ W_i \end{bmatrix} \tag{D-18}$$

where M_β is the total rotation matrix:

$$M_\beta = M_{\beta_1} M_{\beta_2} M_{\beta_3} \tag{D-19}$$

$$M_\beta =$$

$$
\begin{bmatrix}
\cos\beta_1\cos\beta_2 & \sin\beta_1\cos\beta_3 + \cos\beta_1\sin\beta_2\sin\beta_3 & \sin\beta_1\sin\beta_3 - \cos\beta_1\sin\beta_2\cos\beta_3 \\
-\sin\beta_1\cos\beta_2 & \cos\beta_1\cos\beta_3 - \sin\beta_1\sin\beta_2\sin\beta_3 & \cos\beta_1\sin\beta_3 + \sin\beta_1\sin\beta_2\cos\beta_3 \\
\sin\beta_2 & -\cos\beta_2\sin\beta_3 & \cos\beta_2\cos\beta_3
\end{bmatrix}
$$

(D-20)

To obtain the straight line equation in the *XYZ* coordinate system, Eq. D-11 must be transformed into the *XYZ* coordinate system:

$$
M_\beta \begin{bmatrix} U_i \\ V_i \\ W_i \end{bmatrix} = \lambda_1 M_\beta \begin{bmatrix} 0 \\ 0 \\ 1 \end{bmatrix} + q M_\beta \begin{bmatrix} 1 \\ 0 \\ 0 \end{bmatrix}
$$

(D-21)

$$
\begin{bmatrix} X_i \\ Y_i \\ Z_i \end{bmatrix} = \lambda_i \begin{bmatrix} \sin\beta_1\sin\beta_3 - \cos\beta_1\sin\beta_2\cos\beta_3 \\ \cos\beta_1\sin\beta_3 + \sin\beta_1\sin\beta_2\cos\beta_3 \\ \cos\beta_2\cos\beta_3 \end{bmatrix} + q \begin{bmatrix} \cos\beta_1\cos\beta_2 \\ -\sin\beta_1\cos\beta_2 \\ \sin\beta_2 \end{bmatrix}
$$

(D-22)

With this parameterization of straight lines, there exist two degenerate cases for which the angle β_1 is undefined. The first case is when a straight line passes through the origin ($q = 0$). The other case is when a horizontal line intersects the Z-axis. In the latter case, the X and Y coordinates of the point at the shortest distance from the origin (X_n, Y_n, Z_n) are both zero. These degenerate cases can be avoided by shifting the *XYZ* coordinate system. Similar formulation of a straight line in 3-D with independent descriptors can be found in Zielinski (1993).

D.4   CIRCULAR LINES IN TWO-DIMENSIONAL SPACE

A circle in 2-D can be described by three independent parameters, x_c, y_c, and r, where (x_c, y_c) are the coordinates of the circle center and r is its radius, as shown in Fig. D-9. A point (x_i, y_i) on the circle satisfies the following condition:

$$(x_i - x_c)^2 + (y_i - y_c)^2 = r^2$$

(D-23)

D.5   CIRCULAR LINES IN THREE-DIMENSIONAL SPACE

A circle in 3-D can be visualized as resulting from cutting a sphere with radius R centered at (X_c, Y_c, Z_c) by a plane having the normal vector $\vec{p} = \begin{bmatrix} p_x & p_y & p_z \end{bmatrix}^T$ through the sphere's center. This formulation attaches seven parameters to the circle:

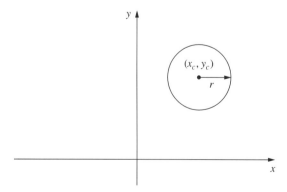

Figure D-9 A circle in 2-D.

X_c, Y_c, Z_c, R, p_x, p_y, and p_z (see Fig. D-10). Since there are only six independent parameters for a 3-D circle, the normal vector is constrained to be a unit vector, reducing the number of parameters for the normal vector to two. A convenient way to describe the normal vector is to use two angles, γ_1 and γ_2, as shown in Fig. D-11. The components of $\vec{p}$ can be written in terms of γ_1 and γ_2 as

$$p_x = \cos \gamma_1 \cos \gamma_2 \tag{D-24}$$

$$p_y = \sin \gamma_1 \cos \gamma_2 \tag{D-25}$$

$$p_z = \sin \gamma_2 \tag{D-26}$$

When the plane of the circle is horizontal, the value of the angle γ_2 is equal to $\pi/2$ and the value of γ_1 is equal to zero; thus $\vec{p} = \begin{bmatrix} 0 & 0 & 1 \end{bmatrix}^T$.

If we let

$$\vec{V} = \begin{bmatrix} X_i - X_c \\ Y_i - Y_c \\ Z_i - Z_c \end{bmatrix} \tag{D-27}$$

where (X_i, Y_i, Z_i) are the coordinates of any point on the circle and (X_c, Y_c, Z_c) are the coordinates of the circle center, then the following are the two equations for a circle in 3-D:

$$\vec{V} \cdot \vec{V} = R^2 \tag{D-28}$$

$$\vec{V} \cdot \vec{p} = 0 \tag{D-29}$$

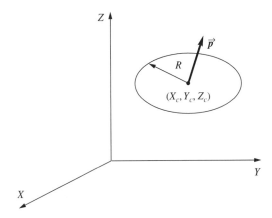

Figure D-10 A circle in 3-D.

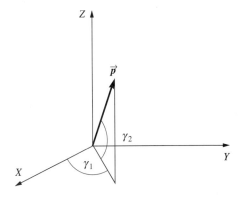

Figure D-11 Unit normal vector $\vec{p}$.

D.6 GEOMETRIC CONSTRAINTS AMONG STRAIGHT LINES

Horizontal Line

This constraint provides partial absolute information in one constraint equation.

A vector along the horizontal line is parallel to the XY plane and yields the constraint equation:

$$A_z = 0 \tag{D-30}$$

Line Parallel to the X-Axis

This constraint provides partial absolute information in two constraint equations.

The Y and Z components of a vector along the line are equal to zero:

$$A_y = 0 \tag{D-31}$$

$$A_z = 0 \tag{D-32}$$

Line Parallel to the Y-Axis

This constraint provides partial absolute information in two constraint equations.

The X and Z components of a vector along the line are equal to zero:

$$A_x = 0 \tag{D-33}$$

$$A_z = 0 \tag{D-34}$$

Line Parallel to the Z-Axis or Vertical Line

This constraint provides partial absolute information in two constraint equations.

The X and Y components of a vector along the line are equal to zero:

$$A_x = 0 \tag{D-35}$$

$$A_y = 0 \tag{D-36}$$

Parallel Lines

This constraint provides purely relative information in two constraint equations.

Two lines are parallel if the unit vectors along each line have the same component values.

$$A_{i_x} - A_{j_x} = 0 \tag{D-37}$$

$$A_{i_y} - A_{j_y} = 0 \tag{D-38}$$

$$A_{i_z} - A_{j_z} = 0 \tag{D-39}$$

Since a unit vector has only two independent components, only two equations result (any combination of two equations from the above three equations).

Perpendicular Lines

This constraint provides purely relative information in one constraint equation.

The dot product of two vectors along perpendicular lines is equal to zero:

$$A_{i_x}A_{j_x} + A_{i_y}A_{j_y} + A_{i_z}A_{j_z} = 0 \qquad \text{(D-40)}$$

Coplanar Lines

This constraint provides purely relative information in one constraint equation.

Let $N_i: (X_{n_i}, Y_{n_i}, Z_{n_i})$ be the point at the shortest distance from line i to the origin and $N_j: (X_{n_j}, Y_{n_j}, Z_{n_j})$ be the point at the shortest distance from line j to the origin. Then, for the two lines to be coplanar, three vectors must be on the same plane. These are: the vector $\overrightarrow{N_iN_j}$ and the two vectors $\overrightarrow{A_i}$ and $\overrightarrow{A_j}$ along the two lines. The coplanarity of these three vectors is expressed by

$$\overrightarrow{N_iN_j} \cdot \overrightarrow{A_i} \times \overrightarrow{A_j} = 0 \qquad \text{(D-41)}$$

or in the determinant form:

$$\begin{vmatrix} (X_{n_j} - X_{n_i}) & (Y_{n_j} - Y_{n_i}) & (Z_{n_j} - Z_{n_i}) \\ A_{i_x} & A_{i_y} & A_{i_z} \\ A_{j_x} & A_{j_y} & A_{j_z} \end{vmatrix} \qquad \text{(D-42)}$$

D.7 GEOMETRIC CONSTRAINTS AMONG CIRCLES

Horizontal Circle

This constraint provides partial absolute information in two constraint equations.

The normal vector is vertical, or

$$p_x = 0 \qquad \text{(D-43}a\text{)}$$

$$p_y = 0 \qquad \text{(D-43}b\text{)}$$

Circle in the XY-Plane

This constraint provides partial absolute information in three constraint equations.

In addition to Eqs. D-43a and D-43b, the center of the circle lies in the XY-plane, or

$$Z_c = 0 \qquad \text{(D-44)}$$

Circle in the XZ-Plane

This constraint provides partial absolute information in three constraint equations.

$$p_x = 0 \qquad \text{(D-45}a\text{)}$$

$$p_z = 0 \qquad \text{(D-45}b\text{)}$$

$$Y_c = 0 \qquad \text{(D-46)}$$

Circle in the YZ-Plane

This constraint provides partial absolute information in three constraint equations.

$$p_y = 0 \qquad \text{(D-47}a\text{)}$$

$$p_z = 0 \qquad \text{(D-47}b\text{)}$$

$$X_c = 0 \qquad \text{(D-48)}$$

Parallel Circles

This constraint provides purely relative information in two constraint equations.
  When two circles are parallel, their normal vectors are also parallel to each other.

$$p_{i_x} - p_{j_x} = 0 \qquad\qquad\qquad (\text{D-49})$$

$$p_{i_y} - p_{j_y} = 0 \qquad\qquad\qquad (\text{D-50})$$

$$p_{i_z} - p_{j_z} = 0 \qquad\qquad\qquad (\text{D-51})$$

Since a unit vector has only two independent components, only two equations result (any combination of two equations from the above three equations).

Two Circles in Perpendicular Planes

This constraint provides purely relative information in one constraint equation.
  The dot product of the normal vectors of two perpendicular circles is equal to zero.

$$p_{i_x} p_{j_x} + p_{i_y} p_{j_y} + p_{i_z} p_{j_z} = 0 \qquad\qquad (\text{D-52})$$

Coplanar Circles

This constraint provides purely relative information in three constraint equations.
  This constraint is a combination of two conditions. The first condition is equivalent to the two circles being parallel. This results in two constraint equations; any combination of two equations from Eqs. D-49–D-51. The second condition enforces the fact that the vector connecting the centers of both circles must be perpendicular to the normal vector of either circle. If we let C_i be the center of circle i, C_j be the center of circle j, and $\overrightarrow{C_i C_j}$ be the vector connecting them, then

$$\overrightarrow{C_i C_j} \cdot \overrightarrow{p}_i = 0 \qquad\qquad\qquad (\text{D-53})$$

or

$$(C_{i_x} - C_{j_x}) p_{i_x} + (C_{i_y} - C_{j_y}) p_{i_y} + (C_{i_z} - C_{j_z}) p_{i_z} = 0 \qquad (\text{D-54})$$

D.8 CONSTRAINTS BETWEEN STRAIGHT LINES AND CIRCLES

A Straight Line and a Circle are Coplanar

This constraint provides purely relative information in two constraint equations.
  This constraint consists of two conditions. The first condition is that the vector along the straight line, $\overrightarrow{A}$, is perpendicular to the normal vector of the circle, $\overrightarrow{p}$ (Eq. D-55). The second condition is that the vector $\overrightarrow{CN}$ connecting the circle center (X_c, Y_c, Z_c) and the point N on the straight line nearest the origin is perpendicular to the normal vector (Eq. D-56).

$$\overrightarrow{A} \cdot \overrightarrow{p} = 0 \qquad \text{or} \qquad A_x p_x + A_y p_y + A_z p_z = 0 \qquad (\text{D-55})$$

$$\overrightarrow{CN} \cdot \overrightarrow{p} = 0 \qquad \text{or} \qquad (X_n - X_c) p_x + (Y_n - Y_c) p_y + (Z_n - Z_c) p_z = 0 \quad (\text{D-56})$$

A Straight Line Perpendicular to the Plane of a Circle

This constraint provides purely relative information in two constraint equations.

The normal vector of the circle is parallel to the vector along the straight line.

$$A_x - p_x = 0 \tag{D-57}$$

$$A_y - p_y = 0 \tag{D-58}$$

$$A_z - p_z = 0 \tag{D-59}$$

Only two of the above three equations are independent (any combination of two equations from the above three equations).

A Straight Line Passing Through the Center of a Circle

This constraint provides purely relative information in two constraint equations.

In this case, we can write the equation of a straight line (Eq. D-5) to pass through the center of the circle as follows:

$$\begin{bmatrix} X_c \\ Y_c \\ Z_c \end{bmatrix} = \lambda_c \begin{bmatrix} A_x \\ A_y \\ A_z \end{bmatrix} + \begin{bmatrix} X_n \\ Y_n \\ Z_n \end{bmatrix} \tag{D-60}$$

After eliminating λ_c, Eq. D-60 reduces to

$$(X_c - X_n)A_z - (Z_c - Z_n)A_x = 0 \tag{D-61}$$

$$(Y_c - Y_n)A_z - (Z_c - Z_n)A_y = 0 \tag{D-62}$$

REFERENCES

BARAKAT, H. 1997. Photogrammetric Analysis of Image Invariance for Image Transfer and Object Reconstruction, Ph.D. Thesis, Purdue University, West Lafayette, IN.

MULAWA, D. 1989. Estimation and Photogrammetric Treatment of Linear Features, Ph.D. Thesis, Purdue University, West Lafayette, IN.

SAYED, A. 1990. Extraction and Photogrammetric Exploitation of Features in Digital Images, Ph.D. Thesis, Purdue University, West Lafayette, IN.

WEERAWONG, K. 1995. Exploitation of Linear Features for Object Reconstruction in Digital Photogrammetric Systems, Ph.D. Thesis, Purdue University, West Lafayette, IN.

ZIELINSKI 1993. Object Reconstruction with Digital Line Photogrammetry, Ph.D. Thesis, Royal Institute of Technology, Stockholm, Sweden.

Appendix E

Further Consideration of the Rotation Matrix

E.1 ALTERNATIVE METHOD FOR ROTATION PARAMETER ESTIMATION

There exist several geometric situations in practice, particularly in industrial applications, in which estimating ω, ϕ, κ (or similar sequential rotations) is quite difficult. The following method presents an alternative means to estimate the three independent parameters involved in the orientation matrix, M. The general technique should work with any set of three independent rotation parameters, but this description will focus on the algebraic parameters, because they have certain advantages with regard to nonsingularity and convergence. These developments were introduced by Pope (1970) and Hinsken (1988). Recall that the 3×3 rotation matrix is orthogonal; hence,

$$MM^T = I \tag{E-1}$$

Taking differentials of this and rearranging, we obtain

$$dMM^T + MdM^T = 0 \tag{E-2}$$

$$dMM^T = -MdM^T \tag{E-3}$$

$$dMM^T = -[dMM^T]^T \tag{E-4}$$

In other words, dMM^T is skew symmetric and can be written

$$dMM^T = S_w = \begin{bmatrix} 0 & w_3 & -w_2 \\ -w_3 & 0 & w_1 \\ w_2 & -w_1 & 0 \end{bmatrix} \tag{E-5}$$

or

$$dM = S_w M \tag{E-6}$$

This result states that a differential rotation matrix can be obtained from the original rotation matrix by premultiplying by a skew-symmetric matrix. For the conventional sequential rotation parameters, ω, ϕ, κ,

$$dM = \frac{\partial M}{\partial \omega}d\omega + \frac{\partial M}{\partial \phi}d\phi + \frac{\partial M}{\partial \kappa}d\kappa \tag{E-7}$$

For algebraic parameters, a, b, c, d (see Appendix A),

$$dM = \frac{\partial M}{\partial a}da + \frac{\partial M}{\partial b}db + \frac{\partial M}{\partial c}dc + \frac{\partial M}{\partial d}dd \tag{E-8}$$

In Eq. A-70, one representation of the rotation matrix was given in terms of three independent algebraic parameters. Another representation can be defined by four algebraic parameters,

$$M = \begin{bmatrix} d^2 + a^2 - b^2 - c^2 & 2(ab + cd) & 2(ac - bd) \\ 2(ab - cd) & d^2 - a^2 + b^2 - c^2 & 2(bc + ad) \\ 2(ac + bd) & 2(bc - ad) & d^2 - a^2 - b^2 + c^2 \end{bmatrix} \tag{E-9}$$

These parameters are dependent and therefore must satisfy the constraint

$$a^2 + b^2 + c^2 + d^2 = 1 \tag{E-10}$$

Combining Eqs. E-5 and E-8, we obtain

$$S_w = \frac{\partial M}{\partial a}M^T da + \frac{\partial M}{\partial b}M^T db + \frac{\partial M}{\partial c}M^T dc + \frac{\partial M}{\partial d}M^T dd \tag{E-11}$$

Evaluating each one of the matrices on the right side of Eq. E-11, selecting the terms that define w_1, w_2, w_3 on the left side, and simplifying the lengthy expressions yields

$$\begin{bmatrix} w_1 \\ w_2 \\ w_3 \end{bmatrix} = \frac{2}{d} \begin{bmatrix} d^2 + a^2 & ab + cd & ac - bd \\ ab - cd & d^2 + b^2 & bc + ad \\ ac + bd & bc - ad & d^2 + c^2 \end{bmatrix} \begin{bmatrix} da \\ db \\ dc \end{bmatrix} \tag{E-12}$$

Inverting this equation using the adjoint matrix and the determinant yields

$$\begin{bmatrix} da \\ db \\ dc \end{bmatrix} = \frac{1}{2} \begin{bmatrix} d & -c & b \\ c & d & -a \\ -b & a & d \end{bmatrix} \begin{bmatrix} w_1 \\ w_2 \\ w_3 \end{bmatrix} \tag{E-13}$$

The relevance of this result will be presented after a few more preliminaries. The technique of rotation parameter estimation will be illustrated using the collinearity equations. Linearization is an important step in the development of the technique; therefore, some material here may overlap with that presented in Appendix C. Recall the collinearity equations as shown in Eqs. 4-26 and 4-27,

$$\begin{bmatrix} F_x \\ F_y \end{bmatrix} = \begin{bmatrix} x \\ y \end{bmatrix} + f \begin{bmatrix} \dfrac{U}{W} \\ \dfrac{V}{W} \end{bmatrix} = \begin{bmatrix} 0 \\ 0 \end{bmatrix} \tag{E-14}$$

Linearizing this equation by the Taylor series yields

$$\begin{bmatrix} F_x^0 \\ F_y^0 \end{bmatrix} + \begin{bmatrix} dF_x \\ dF_y \end{bmatrix} = \begin{bmatrix} x \\ y \end{bmatrix} + f \begin{bmatrix} \dfrac{U^0}{W^0} \\ \dfrac{V^0}{W^0} \end{bmatrix} + \begin{bmatrix} dF_x \\ dF_y \end{bmatrix} \tag{E-15}$$

By evaluating the differential vector we obtain

$$
\begin{bmatrix} dF_x \\ dF_y \end{bmatrix} = \begin{bmatrix} dx \\ dy \end{bmatrix} + f \begin{bmatrix} \dfrac{1}{W} & 0 & -\dfrac{U}{W^2} \\ 0 & \dfrac{1}{W} & -\dfrac{V}{W^2} \end{bmatrix} \begin{bmatrix} dU \\ dV \\ dW \end{bmatrix} \tag{E-16}
$$

From Eq. 4-27 we obtain

$$
\begin{bmatrix} dU \\ dV \\ dW \end{bmatrix} = M \begin{bmatrix} dX - dX_L \\ dY - dY_L \\ dZ - dZ_L \end{bmatrix} + dM \begin{bmatrix} X - X_L \\ Y - Y_L \\ Z - Z_L \end{bmatrix} \tag{E-17}
$$

Substituting this into Eq. E-16 yields

$$
\begin{bmatrix} dF_x \\ dF_y \end{bmatrix} = \begin{bmatrix} dx \\ dy \end{bmatrix} + f \begin{bmatrix} \dfrac{1}{W} & 0 & -\dfrac{U}{W^2} \\ 0 & \dfrac{1}{W} & -\dfrac{V}{W^2} \end{bmatrix} \left[M \begin{bmatrix} dX - dX_L \\ dY - dY_L \\ dZ - dZ_L \end{bmatrix} + dM \begin{bmatrix} X - X_L \\ Y - Y_L \\ Z - Z_L \end{bmatrix} \right] \tag{E-18}
$$

Using Eq. E-6, we obtain

$$
\begin{bmatrix} dF_x \\ dF_y \end{bmatrix} = \begin{bmatrix} dx \\ dy \end{bmatrix} + f \begin{bmatrix} \dfrac{1}{W} & 0 & -\dfrac{U}{W^2} \\ 0 & \dfrac{1}{W} & -\dfrac{V}{W^2} \end{bmatrix} \left[M \begin{bmatrix} dX - dX_L \\ dY - dY_L \\ dZ - dZ_L \end{bmatrix} + S_w M \begin{bmatrix} X - X_L \\ Y - Y_L \\ Z - Z_L \end{bmatrix} \right] \tag{E-19}
$$

Using Eq. 4-27 again, we find that

$$
\begin{bmatrix} dF_x \\ dF_y \end{bmatrix} = \begin{bmatrix} dx \\ dy \end{bmatrix} + f \begin{bmatrix} \dfrac{1}{W} & 0 & -\dfrac{U}{W^2} \\ 0 & \dfrac{1}{W} & -\dfrac{V}{W^2} \end{bmatrix} \left[M \begin{bmatrix} dX - dX_L \\ dY - dY_L \\ dZ - dZ_L \end{bmatrix} + S_w \begin{bmatrix} U \\ V \\ W \end{bmatrix} \right] \tag{E-20}
$$

But

$$
S_w \begin{bmatrix} U \\ V \\ W \end{bmatrix} = \begin{bmatrix} 0 & w_3 & -w_2 \\ -w_3 & 0 & w_1 \\ w_2 & -w_1 & 0 \end{bmatrix} \begin{bmatrix} U \\ V \\ W \end{bmatrix} = \begin{bmatrix} 0 & -W & V \\ W & 0 & -U \\ -V & U & 0 \end{bmatrix} \begin{bmatrix} w_1 \\ w_2 \\ w_3 \end{bmatrix} \tag{E-21}
$$

Substituting Eq. E-21 in E-20 yields

$$
\begin{bmatrix} dF_x \\ dF_y \end{bmatrix} = \begin{bmatrix} dx \\ dy \end{bmatrix} + f \begin{bmatrix} \dfrac{1}{W} & 0 & -\dfrac{U}{W^2} \\ 0 & \dfrac{1}{W} & -\dfrac{V}{W^2} \end{bmatrix} \left[M \begin{bmatrix} dX - dX_L \\ dY - dY_L \\ dZ - dZ_L \end{bmatrix} + \begin{bmatrix} 0 & -W & V \\ W & 0 & -U \\ -V & U & 0 \end{bmatrix} \begin{bmatrix} w_1 \\ w_2 \\ w_3 \end{bmatrix} \right] \tag{E-22}
$$

Restoring the missing elements of the linearized condition equation as shown in Eq. E-15, and replacing the dx, dy with v_x, v_y we obtain

$$
\begin{bmatrix} v_x \\ v_y \end{bmatrix} + f \begin{bmatrix} \dfrac{1}{W} & 0 & -\dfrac{U}{W^2} \\ 0 & \dfrac{1}{W} & -\dfrac{V}{W^2} \end{bmatrix} \left[M \begin{bmatrix} dX - dX_L \\ dY - dY_L \\ dZ - dZ_L \end{bmatrix} + \begin{bmatrix} 0 & -W & V \\ W & 0 & -U \\ -V & U & 0 \end{bmatrix} \begin{bmatrix} w_1 \\ w_2 \\ w_3 \end{bmatrix} \right] = - \left[\begin{bmatrix} x \\ y \end{bmatrix} + f \begin{bmatrix} \dfrac{U^0}{W^0} \\ \dfrac{V^0}{W^0} \end{bmatrix} \right]
$$

$$
\tag{E-23}
$$

If we substitute G and H for two of the matrices in equation E-23, we obtain

$$v + fG\left[M\begin{bmatrix} dX \\ dY \\ dZ \end{bmatrix} - M\begin{bmatrix} dX_L \\ dY_L \\ dZ_L \end{bmatrix} + H\begin{bmatrix} w_1 \\ w_2 \\ w_3 \end{bmatrix}\right] = f \tag{E-24}$$

in which the bold face f is the right-hand-side vector, not the focal length. This result is put into the form of the least squares model of indirect observations (see Section B.4 in Appendix B), $v + B\Delta = f$.

$$v + fG[M|-M|H]\begin{bmatrix} dX \\ dY \\ dZ \\ dX_L \\ dY_L \\ dZ_L \\ w_1 \\ w_2 \\ w_3 \end{bmatrix} = f \tag{E-25}$$

Equation E-25 is solved for the parameter vector, including the "intermediate" rotation parameters, w_1, w_2, w_3. These intermediate parameters may then be transformed into the algebraic parameter corrections by Eq. E-13. During iteration, the updated parameters are used to generate a new matrix, M, and then one subsequently solves for the w's and repeats the procedure. In summary, this technique has particular advantages because of lack of singularity of the rotation parameters. Likewise, it is very robust and converges from a wide range of starting locations.

E.2   QUATERNION REPRESENTATION OF A ROTATION MATRIX

Euler's theorem states that any rotation matrix represents a rotation about a line in space, or about an axis by some angle (Kanatani, 1993). There are two equivalent ways of generating a rotation matrix from an axis and a rotation angle: a straightforward geometric representation and the quaternion representation.

Recall that the rotation matrix M, resulting from a rotation by an angle θ around an axis specified by a unit vector with direction cosines α, β, γ, was given in Appendix A by

$$M = \begin{bmatrix} \alpha^2(1 - \cos\theta) + \cos\theta & \alpha\beta(1 - \cos\theta) - \gamma\sin\theta & \alpha\gamma(1 - \cos\theta) + \beta\sin\theta \\ \alpha\beta(1 - \cos\theta) + \gamma\sin\theta & \beta^2(1 - \cos\theta) + \cos\theta & \beta\gamma(1 - \cos\theta) - \alpha\sin\theta \\ \alpha\gamma(1 - \cos\theta) - \beta\sin\theta & \beta\gamma(1 - \cos\theta) + \alpha\sin\theta & \gamma^2(1 - \cos\theta) + \cos\theta \end{bmatrix} \tag{A-68}$$

Conversely, given a rotation matrix, we can recover the axis parameters and rotation angle. The rotation angle is calculated from the trace of the rotation matrix (recalling that $\alpha^2 + \beta^2 + \gamma^2 = 1$):

$$\theta = \cos^{-1}\frac{\text{tr}M - 1}{2} \tag{E-26}$$

The direction cosines of the rotation axis are

$$\begin{bmatrix} \alpha \\ \beta \\ \gamma \end{bmatrix} = \frac{1}{2\sin\theta}\begin{bmatrix} m_{32} - m_{23} \\ m_{13} - m_{31} \\ m_{21} - m_{12} \end{bmatrix} \tag{E-27}$$

EXAMPLE E-1   *Rotation Matrix from Axis and Angle*

Given a rotation axis in the *XY* plane at a 45-degree angle counterclockwise from the *X*-axis and a rotation angle θ of 30 degrees, compute the corresponding rotation matrix.

SOLUTION   The direction cosines of the rotation axis are

$$\alpha = \frac{\sqrt{2}}{2}, \qquad \beta = \frac{\sqrt{2}}{2}, \qquad \gamma = 0$$

The rotation angle of 30 degrees yields

$$\cos\theta = \frac{\sqrt{3}}{2}, \quad \sin\theta = \frac{1}{2}$$

Substituting these values into the elements of *M* in Eq. A-68 results in

$$M = \begin{bmatrix} \dfrac{2+\sqrt{3}}{4} & \dfrac{2-\sqrt{3}}{4} & \dfrac{\sqrt{2}}{4} \\[2mm] \dfrac{2-\sqrt{3}}{4} & \dfrac{2+\sqrt{3}}{4} & -\dfrac{\sqrt{2}}{4} \\[2mm] -\dfrac{\sqrt{2}}{4} & \dfrac{\sqrt{2}}{4} & \dfrac{\sqrt{3}}{4} \end{bmatrix}$$

EXAMPLE E-2   *Axis and Rotation Angle from Rotation Matrix*

Determine the rotation axis and angle from the following matrix:

$$M = \begin{bmatrix} 0.8660254 & 0.35355339 & 0.35355339 \\ -0.5 & 0.61237244 & 0.61237244 \\ 0.0 & -0.70710678 & 0.70710678 \end{bmatrix}$$

SOLUTION   The trace of *M* is

$$\mathrm{tr}\,M = 2.1855046$$

Then from Eq. E-26 and E-27 we get

$$\theta = 0.936324 \text{ rad} = 53.6474 \text{ deg}$$

$$\alpha = \frac{1}{2\sin\theta}(m_{32} - m_{23}) = -0.81916073$$

$$\beta = 0.21949345$$

$$\gamma = -0.52990408$$

Quaternions are alternative representations for rotation around an axis that have several advantages over standard Euler angle (e.g., ω, ϕ, κ) parameterizations. A rotation around an axis is easier to visualize and control than a set of rotation angles, particularly for computer graphics applications (Shoemake, 1985; Barr et al., 1992). The quaternion representation is more efficient than standard matrix techniques in terms of operations required. Like the rotation around an axis just discussed, quaternions have the disadvantage that the axis is undefined when the rotation is 0. Also, each rotation is represented by two quaternions; a rotation around an axis in a given direction is equivalent to a negative rotation around the axis in the opposite direction.

E.2.1 The Definition and Manipulation of Quaternions

A quaternion consists of a scalar, q_s, and a three-element vector, (q_i, q_j, q_k) (Schut, 1959; Horn, 1987). The components of the vector are imaginary numbers:

$$q = \begin{bmatrix} q_s \\ q_i i \\ q_j j \\ q_k k \end{bmatrix} \tag{E-28}$$

The components follow the multiplication rules of imaginary numbers:

$$\begin{aligned} i^2 = j^2 = k^2 &= -1 \\ ij = k, \quad ji &= -k \\ jk = i, \quad kj &= -i \\ ki = j, \quad ik &= -j \end{aligned} \tag{E-29}$$

Multiplication of quaternions results in another quaternion and is done component-wise, taking into account the complex components. Due to the complex components, $pq \neq qp$ in general. Multiplication by a quaternion can be written in matrix form:

$$pq = \begin{bmatrix} p_s & -p_i & -p_j & -p_k \\ p_i & p_s & -p_k & p_j \\ p_j & p_k & p_s & -p_i \\ p_k & -p_j & p_i & p_s \end{bmatrix} q = Pq \tag{E-30}$$

For the product qp, the matrix P is written as

$$qp = \begin{bmatrix} p_s & -p_i & -p_j & -p_k \\ p_i & p_s & p_k & -p_j \\ p_j & -p_k & p_s & p_i \\ p_k & p_j & -p_i & p_s \end{bmatrix} q = \bar{P}q \tag{E-31}$$

Note that in this case the skew-symmetric 3×3 lower right submatrix is transposed and that in both cases the matrix is orthogonal.

The dot product of two quaternions is, as for regular vectors, the sum of the products of the individual components:

$$p \cdot q = p_s q_s + p_i q_i + p_j q_j + p_k q_k \tag{E-32}$$

The square of a quaternion's magnitude is similarly its dot product with itself:

$$\|q\|^2 = q \cdot q$$

The complex conjugate of a quaternion, q^*, is obtained by negating the three imaginary (vector) components:

$$q^* = \begin{bmatrix} q_s \\ -q_i i \\ -q_j j \\ -q_k k \end{bmatrix} \tag{E-33}$$

The product of a quaternion and its conjugate is a real number, equal to the dot product of the quaternion with itself:

$$qq^* = q_s^2 + q_i^2 + q_j^2 + q_k^2 \tag{E-34}$$

or the square of the magnitude of the quaternion. From this we can define the *inverse* of a quaternion as

$$q^{-1} = \frac{1}{q \cdot q} q^* = \frac{1}{\|q\|^2} q^* \tag{E-35}$$

showing that the inverse of a unit quaternion (i.e., a quaternion with unit length) is simply its conjugate.

E.2.2 Rotation Using Quaternions

If we express a coordinate vector in quaternion form,

$$r = \begin{bmatrix} x_p \\ y_p \\ z_p \end{bmatrix} = \begin{bmatrix} 0 \\ x_p i \\ y_p j \\ z_p k \end{bmatrix} \tag{E-36}$$

then rotation by a unit quaternion q is

$$r' = qrq^* \tag{E-37}$$

or, in matrix form,

$$qrq^* = (Qr)q^* = \bar{Q}^T(Qr) = (\bar{Q}^T Q)r \tag{E-38}$$

$$\bar{Q}^T Q = \begin{bmatrix} q \cdot q & 0 & 0 & 0 \\ 0 & q_s^2 + q_i^2 - q_j^2 - q_k^2 & 2(q_j q_i - q_s q_k) & 2(q_i q_k + q_s q_j) \\ 0 & 2(q_j q_i + q_s q_k) & q_s^2 - q_i^2 + q_j^2 - q_k^2 & 2(q_j q_k - q_s q_i) \\ 0 & 2(q_i q_k - q_s q_j) & 2(q_j q_k + q_s q_i) & q_s^2 - q_i^2 - q_j^2 + q_k^2 \end{bmatrix} \tag{E-39}$$

The 1,1 element of the matrix in Eq. E-39 is always 1 for unit quaternions. The lower right 3×3 submatrix is the rotation matrix that rotates r into r'.

The unit quaternions q and $-q$ model the same rotation; with the direction of the axis reversed, the direction of the rotation angle is also reversed.

To relate the quaternion representation to the axis/angle representation, it can be shown that

$$q_s = \cos\frac{\theta}{2}$$

$$\begin{bmatrix} q_i \\ q_j \\ q_k \end{bmatrix} = \sin\frac{\theta}{2} \begin{bmatrix} \alpha \\ \beta \\ \gamma \end{bmatrix} \tag{E-40}$$

From this, we can derive

$$\cos\theta = q_s^2 - (q_i^2 + q_j^2 + q_k^2)$$

$$\begin{bmatrix} \alpha \\ \beta \\ \gamma \end{bmatrix} = \frac{1}{\sqrt{q_i^2 + q_j^2 + q_k^2}} \begin{bmatrix} q_i \\ q_j \\ q_k \end{bmatrix} \tag{E-41}$$

EXAMPLE E-3 *Rotation Matrix Using Quaternions*

Write the quaternion corresponding to the rotation axis and angle in Example E-2.

SOLUTION   Applying Eqs. E-40 yields

$$q_s = \cos\frac{\theta}{2} = 0.892399$$

$$\sin\frac{\theta}{2} = 0.451247$$

$$q_i = 0.451247(-0.81916073) = -0.369644$$

$$q_j = 0.451247(0.21949345) = 0.0990458$$

$$q_k = 0.451247(-0.52990408) = -0.239118$$

$$q = \begin{bmatrix} 0.892399 \\ -0.369644i \\ 0.0990458j \\ -0.239118k \end{bmatrix}$$

E.2.3   Derivation of the Unit Quaternion Corresponding to a Rotation Matrix

By examining the diagonal elements of the quaternion rotation matrix in Eq. E-39 and making use of the relationship $q_s^2 + q_i^2 + q_j^2 + q_k^2 = 1$ for a unit quaternion, we can derive the following relations (Horn, 1987):

$$\begin{aligned} 1 + m_{11} + m_{22} + m_{33} &= 4q_s^2 \\ 1 + m_{11} - m_{22} - m_{33} &= 4q_i^2 \\ 1 - m_{11} + m_{22} - m_{33} &= 4q_j^2 \\ 1 - m_{11} - m_{22} + m_{33} &= 4q_k^2 \end{aligned} \qquad \text{(E-42)}$$

To maintain numerical precision, we calculate the quaternion component corresponding to the largest term (in absolute value). We then use the relationships between the symmetrically related diagonal elements (e.g., m_{32} and m_{23}) to derive equations for the remaining components:

$$\begin{aligned} m_{32} - m_{23} &= 2(q_j q_k + q_s q_i) - 2(q_j q_k - q_s q_i) = 4q_s q_i \\ m_{13} - m_{31} &= 4q_s q_j \\ m_{21} - m_{12} &= 4q_s q_k \\ m_{21} + m_{12} &= 4q_i q_j \\ m_{32} + m_{23} &= 4q_j q_k \\ m_{13} + m_{31} &= 4q_k q_i \end{aligned} \qquad \text{(E-43)}$$

EXAMPLE E-4   *Calculation of Quaternion from Rotation Matrix*

Determine the quaternion corresponding to the following rotation matrix:

$$M = \begin{bmatrix} 0.8660254 & 0.35355339 & 0.35355339 \\ -0.5 & 0.61237244 & 0.61237244 \\ 0.0 & -0.70710678 & 0.70710678 \end{bmatrix}$$

SOLUTION Applying Eq. E-42, we get

$$4q_s^2 = 3.1855038$$

$$4q_i^2 = 0.5465466$$

$$4q_j^2 = 0.03924013$$

$$4q_k^2 = 0.22870952$$

The largest term involves q_s, so we will use it in the remaining calculations, according to Eq. E-43.

$$q_s = 0.892399$$

$$q_i = \frac{(m_{32} - m_{23})}{4q_s} = -0.369644$$

$$q_j = \frac{(m_{13} - m_{31})}{4q_s} = 0.0990458$$

$$q_k = \frac{(m_{21} - m_{12})}{4q_s} = -0.239118$$

$$q = \begin{bmatrix} 0.892399 \\ -0.369644i \\ 0.0990458j \\ -0.239118k \end{bmatrix}$$

REFERENCES

BARR, A. H., CURRIN, B., GABRIEL, S., and HUGHES, J. F. 1992. Smooth interpolation of orientations with angular velocity constraints using quaternions. *Computer Graphics* 26(2):313–320.

HINSKEN, L. 1988. A Singularity Free Algorithm for Spatial Orientation of Bundles. *International Archives of Photogrammetry and Remote Sensing.* Commission V. Kyoto: ISPRS, p.262–272.

HORN, K. P. 1987. Closed-form solution of absolute orientation using unit quaternions. *Journal of the Optical Society of America A* 4(4):629–642.

KANATANI, K. 1993. *Geometric Computation for Machine Vision.* Oxford: Oxford University Press.

POPE, A. 1970. An Advantageous Alternative Parameterization of Rotations for Analytical Photogrammetry,

ESSA Technical Report C&GS 39. Rockville, MD: U.S. Dept. of Commerce.

SCHUT, G. H. 1959. Construction of orthogonal matrices and their application in analytical photogrammetry. *Photogrammetria* 15(4):149–162.

SHOEMAKE, K. 1985. Animating rotation with quaternion curves. *Computer Graphics* 19(3):245–254.

THOMPSON, E. H. 1958. A method for the construction of orthogonal matrices. *Photogrammetric Record* 3:55–59.

WHEELER, M. D., and IKEUCHI, K. 1995. *Iterative Estimation of Rotation and Translation Using the Quaternion.* Technical Report CMU-CS-95-215, School of Computer Science, Carnegie-Mellon University.

Appendix F

Orbital Photogrammetry

As more and more high-resolution, high-metric-quality images are obtained from satellites, it is useful to provide the mathematical concepts involved in imagery acquired by sensors in orbit. First, we begin with a review of basic mechanics. This review follows that of Beer and Johnston (1988).

F.1   MECHANICS REVIEW

A particle in motion along a curved path in a plane has tangential and normal unit vectors at positions p and p' as shown in Fig. F-1a. The change in the tangential vector $\boldsymbol{e}_t$ is

$$\Delta \boldsymbol{e}_t = \boldsymbol{e}'_t - \boldsymbol{e}_t \tag{F-1}$$

From the geometry of Fig. F-1b,

$$\left| \Delta \boldsymbol{e}_t \right| = 2 \sin\left(\frac{\Delta\theta}{2} \right) \tag{F-2}$$

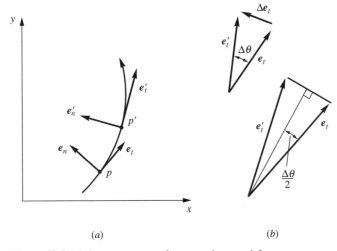

(a) (b)

Figure F-1 Trajectory vectors for a moving particle.

The magnitude of the vector $\Delta e_t/\Delta\theta$ approaches a limit of one as $\Delta\theta$ goes to zero, and the direction approaches the normal to the path,

$$\lim_{\Delta\theta\to0}\frac{2\sin\left(\frac{\Delta\theta}{2}\right)}{\Delta\theta}=\lim_{\Delta\theta\to0}\frac{\sin\left(\frac{\Delta\theta}{2}\right)}{\left(\frac{\Delta\theta}{2}\right)}=1 \tag{F-3}$$

Thus the normal unit vector is

$$e_n=\frac{de_t}{d\theta} \tag{F-4}$$

The velocity vector can be expressed as

$$v=ve_t \tag{F-5}$$

The acceleration vector is the time derivative of the velocity vector,

$$a=\frac{dv}{dt}=\frac{dv}{dt}e_t+v\frac{de_t}{dt} \tag{F-6}$$

Applying the chain rule yields

$$\frac{de_t}{dt}=\frac{de_t}{d\theta}\frac{d\theta}{ds}\frac{ds}{dt} \tag{F-7}$$

where s is the length of a differential curve element. From analytical geometry we have

$$\frac{d\theta}{ds}=\frac{1}{\rho} \tag{F-8}$$

where ρ is the instantaneous curve radius, and

$$\frac{ds}{dt}=v \tag{F-9}$$

Therefore the expression for the acceleration of the particle can be written in terms of its tangential and normal components:

$$a=\frac{dv}{dt}e_t+\frac{v^2}{\rho}e_n \tag{F-10}$$

In Fig. F-2a, position vector r corresponds to the instantaneous position of a point. The vector has radial and transverse components shown by unit vectors e_r and e_θ. From Fig. F-2b it can be seen that

$$\frac{de_r}{d\theta}=e_\theta$$

$$\frac{de_\theta}{d\theta}=-e_r \tag{F-11}$$

Now writing expressions for the time derivatives of e_r and e_θ,

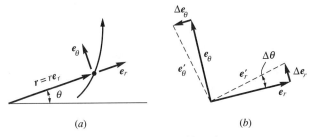

Figure F-2 Position and velocity with polar components.

$$\frac{de_r}{dt} = \frac{de_r}{d\theta}\frac{d\theta}{dt} = e_\theta\frac{d\theta}{dt} = \dot{\theta}e_\theta$$

$$\frac{de_\theta}{dt} = \frac{de_\theta}{d\theta}\frac{d\theta}{dt} = -e_r\frac{d\theta}{dt} = -\dot{\theta}e_r \tag{F-12}$$

The radial vector may be expressed as

$$\boldsymbol{r} = re_r \tag{F-13}$$

The velocity vector, with a substitution from Eq. F-12, is

$$\boldsymbol{v} = \frac{d\boldsymbol{r}}{dt} = \frac{d(re_r)}{dt} = \dot{r}e_r + r\dot{e}_r = \dot{r}e_r + r\dot{\theta}e_\theta \tag{F-14}$$

and the acceleration vector is

$$\boldsymbol{a} = \frac{d\boldsymbol{v}}{dt} = \ddot{r}e_r + \dot{r}\dot{e}_r + \dot{r}\dot{\theta}e_\theta + r\ddot{\theta}e_\theta + r\dot{\theta}\dot{e}_\theta \tag{F-15}$$

Making substitutions from Eq. F-12 yields

$$\boldsymbol{a} = \frac{d\boldsymbol{v}}{dt} = \ddot{r}e_r + \dot{r}\dot{\theta}e_\theta + \dot{r}\dot{\theta}e_\theta + r\ddot{\theta}e_\theta - r\dot{\theta}^2e_r \tag{F-16}$$

Decomposing Eq. F-16 into radial and transverse components yields

$$\boldsymbol{a} = (\ddot{r} - r\dot{\theta}^2)e_r + (2\dot{r}\dot{\theta} + r\ddot{\theta})e_\theta \tag{F-17}$$

The coefficient of the transverse component can be expressed as

$$2\dot{r}\dot{\theta} + r\ddot{\theta} = \frac{1}{r}\frac{d}{dt}(r^2\dot{\theta}) \tag{F-18}$$

This can be verified by evaluating the derivative on the right. Particle P with mass m moves along the curved path indicated in Fig. F-3. The *moment* of the momentum vector, $m\boldsymbol{v}$, is defined as

$$\boldsymbol{r} \times m\boldsymbol{v} = \boldsymbol{H}_0 \tag{F-19}$$

This is also called the *angular momentum*. Taking magnitudes, we obtain

$$H_0 = rmv\sin\phi \tag{F-20}$$

From Fig. F-4,

$$mv_\theta = mv\sin\phi \tag{F-21}$$

Substituting this into Eq. F-20 yields

$$H_0 = rmv_\theta \tag{F-22}$$

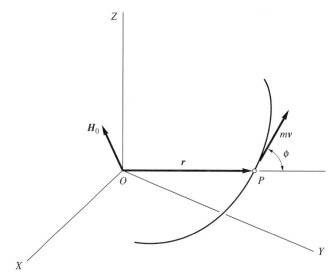

Figure F-3 Position, velocity, and momentum vectors.

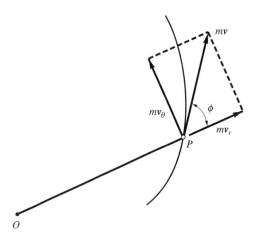

Figure F-4 Polar components of the momentum vector.

From Eq. F-14, $v_\theta = r\dot\theta$, so Eq. F-22 becomes

$$H_0 = mr^2\dot\theta \tag{F-23}$$

which is, dividing by m,

$$\frac{H_0}{m} = h = r^2\dot\theta \quad \text{or} \quad \dot\theta = \frac{h}{r^2} \tag{F-24}$$

where h is the magnitude of the angular momentum per unit mass. For a body moving under a central force, the angular momentum is constant in magnitude and direction. For a small-mass satellite in Earth orbit, the force, and therefore the acceleration, must be along the line between the centers of mass. With the co-ordinate origin at the center of mass of the Earth, and the position vector along the line joining the two bodies, the transverse component of the acceleration, and hence any force, must be zero. Finally, the radial component of the force must conform to Newton's law of gravitation, where the convention is adopted that a force directed toward the origin is negative.

$$F = -\frac{GMm}{r^2} \tag{F-25}$$

M denotes the mass of the Earth. From Eq. F-17,

$$F_r = -m(\ddot{r} - r\dot{\theta}^2) = \frac{GMm}{r^2} \tag{F-26}$$

From Eq. F-18,

$$F_t = m(2\dot{r}\dot{\theta} + r\ddot{\theta}) = m\left(\frac{1}{r}\frac{d}{dt}(r^2\dot{\theta})\right) = 0 \tag{F-27}$$

Equation F-27 can be integrated to obtain a result equivalent to Eq. F-24. In order to remove time from Eq. F-26, we use Eq. F-24 and the chain rule to make some substitutions:

$$\dot{r} = \frac{dr}{d\theta}\frac{d\theta}{dt} = \frac{h}{r^2}\frac{dr}{d\theta} = -h\frac{d}{d\theta}\left(\frac{1}{r}\right) \tag{F-28}$$

$$\ddot{r} = \frac{d\dot{r}}{dt} = \frac{d\dot{r}}{d\theta}\frac{d\theta}{dt} = \frac{h}{r^2}\frac{d\dot{r}}{d\theta} \tag{F-29}$$

$$\ddot{r} = \frac{h}{r^2}\frac{d}{d\theta}\left[-h\frac{d}{d\theta}\left(\frac{1}{r}\right)\right] \tag{F-30}$$

$$\ddot{r} = -\frac{h^2}{r^2}\frac{d^2}{d\theta^2}\left(\frac{1}{r}\right) \tag{F-31}$$

Substituting expressions for $\dot{\theta}$ and $\ddot{r}$ into Eq. F-26, and using $u = \frac{1}{r}$, yields

$$\frac{d^2u}{d\theta^2} + u = \frac{GM}{h^2} \tag{F-32}$$

The solution to this differential equation is

$$u = \frac{1}{r} = \frac{GM}{h^2} + C \, \cos(\theta - \theta_0) \tag{F-33}$$

This is the equation of a conic section in polar coordinates, with the origin at the center of mass of the Earth. If we set $\theta_0 = 0$, then this corresponds to measuring θ from the *perigee,* or point of closest approach of the satellite to the Earth. The *eccentricity* of the conic section is

$$e = \frac{Ch^2}{GM} \tag{F-34}$$

which determines the type of conic section,

$$e > 1 \text{ represents a hyperbola}$$

$$e = 1 \text{ represents a parabola}$$

$$e < 1 \text{ represents an ellipse}$$

Note that this use of the variable *e* should not be confused with the earlier use of *e* as a unit vector.

F.2 COORDINATE SYSTEMS AND ORBIT DESCRIPTORS

Figure F-5 shows the relationship between a quasi-inertial (i.e., nonrotating) geocentric Cartesian coordinate system, (X, Y, Z), and another system, (q_1, q_2, q_3), rotated into the plane of an elliptical orbit. The XYZ system has the Z-axis along the spin axis of the Earth, with the XY-plane in the equatorial plane, and the X-axis through the vernal equinox, or the first point of Aries. The origin coincides with one focus of the ellipse. Shown in the Fig. F-5 are some of the six Keplerian elements, $(a, e, T_0, \Omega, i, \omega)$, which can be used to describe a particular orbit: a is the semimajor axis of the ellipse, e is the eccentricity, T_0 is the time of perigee passage, Ω is the right ascension of the ascending node, i is the inclination, and ω is the argument of perigee. In Fig. F-6, f corresponds to

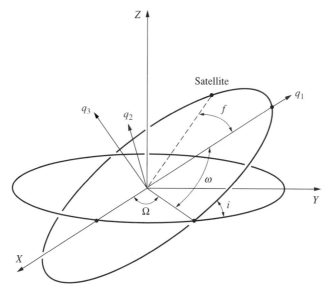

Figure F-5 Relationship of geocentric coordinate system, orbit plane coordinate system, and Keplerian elements.

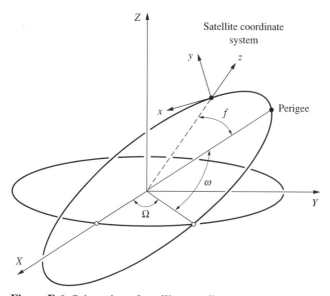

Figure F-6 Orientation of satellite coordinate system.

θ in the previous derivation, and is the angle in polar coordinates fixing a point on the ellipse. It is also called the *true anomaly*. The use of the word *anomaly* meaning angle is a vestige of Ptolemaic times, implying an angle which does not vary linearly with time. Rearranging Eq. F-33, letting $p = h^2/GM$, yields a more familiar form of the equation of a conic section,

$$r = \frac{p}{1 + e\cos f}. \tag{F-35}$$

where p is the semi-latus rectum, e the eccentricity, and f is the angle defining the satellite position on the ellipse, the true anomaly. The true anomaly and the eccentric anomaly, E, are related by

$$f = \tan^{-1}\left[\frac{\sqrt{1 - e^2}\,\sin E}{\cos E - e}\right] \tag{F-36}$$

and the mean anomaly is given by Kepler's equation,

$$M = E - e\sin E \tag{F-37}$$

The subsequent development follows that of Leick (1995). In order to obtain the Keplerian elements from the position and velocity of a satellite at a given time, we first obtain the components of the angular momentum vector from Eqs. F-19–F-24 assuming unit mass,

$$\boldsymbol{h} = \boldsymbol{r} \times \boldsymbol{v} = \begin{bmatrix} h_x \\ h_y \\ h_z \end{bmatrix} \tag{F-38}$$

With v representing the velocity magnitude, and $\mu = GM$, we obtain the following:

$$\Omega = \tan^{-1}\left(\frac{h_x}{-h_y}\right) \tag{F-39}$$

$$i = \tan^{-1}\left(\frac{\sqrt{h_x^2 + h_y^2}}{h_z}\right) \tag{F-40}$$

$$a = \frac{r}{2 - \left(\dfrac{rv^2}{\mu}\right)} \tag{F-41}$$

$$e = \sqrt{1 - \frac{h^2}{\mu a}} \tag{F-42}$$

$$\cos E = \frac{a - r}{ae} \tag{F-43}$$

$$\sin E = \frac{\boldsymbol{r} \cdot \boldsymbol{v}}{e\sqrt{\mu a}}$$

$$f = \tan^{-1}\left(\frac{\sqrt{1 - e^2}\,\sin E}{\cos E - e}\right) \tag{F-44}$$

Defining an intermediate coordinate system in the orbit plane with p_1 through the ascending node, and with $\boldsymbol{R}_i$ referring to the ith elementary rotation matrix, we can transform a position r in the geocentric XYZ system into the intermediate p-system by

$$p = \begin{bmatrix} p_1 \\ p_2 \\ p_3 \end{bmatrix} = R_1(i)R_3(\Omega)r \tag{F-45}$$

The argument of the perigee can be found from

$$\omega + f = \tan^{-1}\left(\frac{p_2}{p_1}\right) \tag{F-46}$$

and T_0 can be obtained from

$$M = \sqrt{\frac{\mu}{a^3}}(t - T_0) \tag{F-47}$$

To go in the other direction, that is, from Keplerian elements to position and velocity vectors, begin with Eq. F-47 to obtain the mean anomaly, then solve Eq. F-37 for E, and use Eq. F-36 to solve for f. The magnitude of the position vector is

$$r = a(1 - e\cos E) \tag{F-48}$$

The position vector in the orbit plane system is

$$q = \begin{bmatrix} q_1 \\ q_2 \\ q_3 \end{bmatrix} = \begin{pmatrix} r\cos f \\ r\sin f \\ 0 \end{pmatrix} \tag{F-49}$$

and the velocity vector is

$$\dot{q} = \frac{\sqrt{\frac{\mu}{a^3}}\, a}{\sqrt{1 - e^2}} \begin{bmatrix} -\sin f \\ e + \cos f \\ 0 \end{bmatrix} \tag{F-50}$$

A composite rotation matrix relates the orbit plane system and the spin axis/equinox system

$$r = Rq \tag{F-51}$$

where R is given by

$$R = R_3(-\Omega)R_1(-i)R_3(-\omega) \tag{F-52}$$

and $\dot{q}$ is transformed in the same way.

Before we can write the equations relating a terrestrial point and a satellite sensor, we need to be able to relate the quasi-inertial geocentric coordinate system XYZ with a rotating, Earth-fixed, geocentric system $X_eY_eZ_e$. The prime component of this transformation is the Earth orientation from its rotation, with smaller contributions from *polar motion*,

$$\begin{bmatrix} X_e \\ Y_e \\ Z_e \end{bmatrix} = R_2(-x_p)R_1(-y_p)R_3(\text{GAST}) \begin{bmatrix} X \\ Y \\ Z \end{bmatrix} = R_0 \begin{bmatrix} X \\ Y \\ Z \end{bmatrix}$$

$$\begin{bmatrix} X \\ Y \\ Z \end{bmatrix} = R_0^T \begin{bmatrix} X_e \\ Y_e \\ Z_e \end{bmatrix} \tag{F-53}$$

where GAST is the Greenwich apparent sidereal time, and x_p and y_p are the time-dependent parameters defining the polar motion. Equation 4-78 illustrates how these concepts are expressed in photogrammetric condition equations for a satellite pushbroom sensor.

REFERENCES

BATE, R., MUELLER, D., and WHITE, J. 1971. *Fundamentals of Astrodynamics.* New York: Dover.

BEER, F., and JOHNSTON, E. 1988. *Vector Mechanics for Engineers: Dynamics.* New York: McGraw-Hill.

BOULET, D. 1991. *Methods of Orbit Determination for the Micro Computer.* Richmond, VA: Willmann-Bell.

LEICK, A. 1995. *GPS Satellite Surveying.* New York: John Wiley & Sons.

MAKKI, S. 1991. Photogrammetric reduction and analysis of real and simulated SPOT imageries. Ph.D. thesis. West Lafayette, IN: Purdue University.

PADERES, F. 1986. Geometric modeling and rectifiction of satellite scanner imagery and investigation of related critical issues. Ph.D. Thesis. West Lafayette, IN: Purdue University.

Appendix G

Software for Photogrammetric Applications

The computer-related aspects of photogrammetry are becoming essential components of its practice. They include the use of the computer to perform numerical calculations, to manipulate and display digital imagery and other cartographic data, and to control equipment. But in today's world they also include the use of the computer as a communication and information appliance. As such, this appendix consists of two sections. Section G.1 describes a set of programs and data files that reside on the CD accompanying this book. They are useful to illustrate the translation of algorithms into explicit computer instructions. Being examples only, they make no pretense of being user-friendly or robust in the face of unexpected input. Users of the book are free to modify them and use them as they wish. Section G.2 gives a brief listing of some relevant and active (at the time of writing!) Web sites. This is done with the full realization that some of these may change or disappear in the near future. A browser-compatible version of this list of Web links is included on the CD in the file WEBSITES.HTM.

G.1 PROGRAMS AND DATA

The programs consist of C-language and MATLAB* code implementing some fundamental photogrammetric operations, as well as some higher-level operations that utilize the included imagery. None of the code is platform specific, and it has been run under both UNIX and Windows NT environments (one binary data file is platform specific, as noted in the documentation on the CD). Only source code is provided; no executables are given. The user needs an ANSI C compiler and a MATLAB license for version 5.3 or later. The Student Edition of MATLAB is sufficient to run everything that is provided. All image files that require program access are given in raw binary format.

_____

* MATLAB is a registered trademark of The MathWorks, Inc.

G.1.1   Installation

Make a working directory, load the CD into the drive, and copy the contents into that directory. There are about 275 MB of data, the bulk of that being imagery. See the file README.TXT for more information about the contents and procedures to compile and run the programs. See the file DOC.TXT for more details about the algorithms, chapter references, needed external functions, and data file formats.

G.1.2   Summary of C Programs

7_PAR.C

Calculate seven-parameter transformation between two 3-D coordinate systems (Appendix A).

DEP_RO.C

Two-photo dependent relative orientation using coplanarity equation (Chapters 4 and 5).

IMRECT.C

Rectify a frame photograph to a plane (Chapter 8).

INTERSECT.C

Calculate world coordinates by intersection of rays from two photos (Chapters 4 and 5).

ORECT.C

Orthorectify a frame photograph using USGS 3-arc-second DEM (Chapter 8).

ORIENT.C

Calculate various parameterizations of an orientation matrix (Appendix A).

PAIRRECT.C

Pairwise rectify (normalize) a pair of frame photographs and put into RGB file for anaglyph stereo viewing (Chapter 7).

RECTSPOT.C

Rectify a SPOT panchromatic image to a plane using rigorous sensor model (Chapters 4 and 8, and Appendix F).

USGS_TO_DEM.C

Convert USGS 3-arc-second DEM to float format (Chapter 8).

VLL2.C

Refine approximate DEM using vertical line locus method, an area-based correlation method, using two frame photos (Chapter 6).

WORLD_TO_IMAGE.C

Project world coordinates onto an image (Chapter 4).

G.1.3 Summary of MATLAB Programs

EX_ANG.M

Extract ω, ϕ, κ from 3×3 rotation matrix (Chapter 4 and Appendices A and E).

EX_ANG2.M

Extract sequential rotation angles in order z-x-z (Chapter 4 and Appendices A and E).

PAR4.M

Four-parameter transformation for inner orientation (Chapter 4 and Appendices A and B).

PAR6.M

Six-parameter transformation for inner orientation (Chapter 4 and Appendices A and B).

RESECT.M

Single frame image resection using collinearity equations (Chapters 4 and 5, and Appendix C).

SPOTRES3.M

Single SPOT image resection using rigorous 26-parameter sensor model (Chapter 4 and Appendix F).

DISP_DEM.M

Display a shaded rendering of the DEM created by VLL2.C (Chapters 6 and 8).

SPOT Satellite Imagery: copyright CNES 2000. Provided by SPOT Image Corporation. http://www.spot.com/spot-us.htm. This imagery is licensed for use in this educational product, and may not be copied or reproduced in any form without the specific permission of SPOT Image Corporation. Contact SPOT Image Corp. at 703-715-3100.

MATLAB is a registered trademark of The MathWorks, Inc.

For MATLAB product information, contact:

The MathWorks, Inc.
3 Apple Drive
Natick, MA, 01760-2098 USA
Tel: 508-647-7000
Fax: 508-647-7101
E-mail: info@mathworks.com

G.2   WORLD WIDE WEB SITES OF INTEREST FOR PHOTOGRAMMETRY

G.2.1   Sites Related to This Book

John Wiley & Sons

www.wiley.com/college/mikhail

Web site for **Introduction to Modern Photogrammetry**

http://www.cs.cmu.edu/~jcm/book

G.2.2   Photogrammetric Societies

American Society for Photogrammetry and Remote Sensing

www.asprs.org

International Society for Photogrammetry and Remote Sensing

www.isprs.org

G.2.3   Government Agencies

United States Geological Survey Spatial Data

http://mapping.usgs.gov/nsdi

Canadian Center for Remote Sensing

www.ccrs.nrcan.gc.ca

Danish Center for Remote Sensing

www.emi.dtu.dk/research/DCRS

NASA/JPL Imaging Radar

http://southport.jpl.nasa.gov

Alaska SAR Facility

www.asf.alaska.edu

U.S. Army Topographic Engineering Center

www.tec.army.mil

U.S. National Geodetic Survey

www.ngs.noaa.gov

German Remote Sensing Data Center—SRTM (Shuttle Radar Topography Mission) Site

www.dfd.dlr.de/srtm

JPL SRTM (Shuttle Radar Topography Mission) Site

www.jpl.nasa.gov/srtm

National Imagery and Mapping Agency

www.nima.mil

G.2.4 Academic Institutions

Purdue University Geomatics

www.ecn.purdue.edu/Geomatics
www.ecn.purdue.edu/muri

University of Southern California Institute of Robotics and Intelligent Systems

http://iris.usc.edu/USC-Computer-Vision.html

Ohio State University Dept. of Civil and Environmental Engineering and Geodetic Science

http://www-ceg.eng.ohio-state.edu

University of Maine Department of Spatial Information Science and Engineering

www.spatial.maine.edu

University of Wisconsin Civil and Environmental Engineering

www.engr.wisc.edu/cee

University of Florida—Geomatics

www.surv.ufl.edu

Ferris State University—Surveying Engineering

www.ferris.edu/htmls/colleges/technolo/surveyin.htm

University of Calgary Geomatics Engineering

www.ensu.ucalgary.ca

Carnegie Mellon University Digital Mapping Laboratory

www.maps.cs.cmu.edu

California State University, Fresno—Geomatics Engineering

www.csufresno.edu/geomatics

University of New Brunswick—Geodesy and Geomatics Engineering

www.unb.ca/GGE

California State Polytechnic University, Pomona—Surveying Engineering

www.csupomona.edu/%7Ece_surv

Carnegie Mellon University—Computer Vision Home Page

http://www.cs.cmu.edu/~cil/vision.html

French National Institute for Research in Computer Science and Control—INRIA

www.inria.fr/welcome-eng.html

Institute of Geodesy and Photogrammetry ETH, Zurich

www.igp.ethz.ch

University of Stuttgart Institute for Photogrammetry

www.ifp.uni-stuttgart.de

Technical University of Delft Geodesy

www.geo.tudelft.nl

Technical University of Stockholm Geodesy and Photogrammetry

www.geomatics.kth.se

Technical University of Munich, Chair for Photogrammetry and Remote Sensing

www.photo.verm.tu-muenchen.de/photo.html

ITC—International Institute for Aerospace Survey and Earth Sciences

www.itc.nl

University of Bonn Institute of Photogrammetry

www.ipb.uni-bonn.de

University of Hannover Institute for Photogrammetry and Engineering Survey

www.ipi.uni-hannover.de

University College London Geomatics Engineering

www.ps.ucl.ac.uk

University of Karlsruhe Geodetic Institute

www-gik.bau-verm.uni-karlsruhe.de

University of Nottingham Institute of Engineering Surveying and Space Geodesy

www.nottingham.ac.uk/~iszwww

University of Newcastle upon Tyne—Geomatics

http://geomatics.ncl.ac.uk

University of Glasgow—Geography and Topographic Science

www.geog.gla.ac.uk

University of New South Wales—Geomatic Engineering

http://www.gmat.unsw.edu.au

Curtin University of Technology Department of Spatial Sciences

www.cage.curtin.edu.au/spatial

University of Melbourne Geomatics Engineering

www.sli.unimelb.edu.au/intro.html

G.2.5   Resources

Terraserver

www.terraserver.com

MultiSpec Home Page

http://dynamo.ecn.purdue.edu/~biehl/MultiSpec

Web Server for Photogrammetry, Remote Sensing, and Land Surveying

`www.eng.upm.edu.my/home/shattri/html/website.html`

Curtin Geodesy Group—Web Links

`www.cage.curtin.edu.au/~geogrp/links.html`

Khoros: Large Image-Processing Environment

`http://www.khoral.com/`

Open Source Computer Vision Library

`http://www.intel.com/research/mrl/research/cvlib/`

Target Jr.: An Image-Processing and Computer Vision Development Environment

`http://www.targetjr.org`

REFERENCES

GLASSEY, R. 1993. Numerical Computation Using C. San Diego: Academic Press.

GYER, M., and SALIBA. 1981. *Automated Stereo Photogrammetric Terrain Elevation Extraction.* Technical Report GS-TR-81-1. Satellite Beach, FL: Gyer and Saliba, Inc.

MAKKI, S., 1991. *Photogrammetric Reduction and Analysis of Real and Simulated SPOT Imageries.* Ph.D. Thesis. West Lafayette, IN: Purdue University.

MIKHAIL, E. 1988. *Photogrammetric Modeling and Reduction of SPOT Stereo Images.* Civil Engineering Report CE-PH-88-4. West Lafayette, IN: Purdue University.

PADERES, F., and MIKHAIL, E. 1986. *Registration/Rectification of Remotely Sensed Data.* Civil Engineering Report CE-PH-86-4. West Lafayette, IN: Purdue University.

Index